Origin of Carbonate Sedimentary Rocks

Noel P. James

Department of Geological Sciences and Geological Engineering
Queen's University
Kingston, Ontario, K7L 3N6, Canada
james@geol.queensu.ca

Brian Jones

Department of Earth and Atmospheric Sciences
University of Alberta
Edmonton, Alberta T6G 2E3, Canada
Brian.Jones@ualberta.ca

Origin of Carbonate Sedimentary Rocks

WILEY

This edition first published 2016 © 2016 by Noel P. James and Brian Jones

Registered Office
John Wiley & Sons, Ltd, The Atrium, Southern Gate, Chichester, West Sussex, PO19 8SQ, UK

Editorial Offices
9600 Garsington Road, Oxford, OX4 2DQ, UK
The Atrium, Southern Gate, Chichester, West Sussex, PO19 8SQ, UK
111 River Street, Hoboken, NJ 07030-5774, USA

For details of our global editorial offices, for customer services and for information about how to apply for permission to reuse the copyright material in this book please see our website at www.wiley.com/wiley-blackwell.

Library of Congress Cataloging-in-Publication Data

James, Noel P.
  Origin of carbonate sedimentary rocks / Noel P. James, Brian Jones.
     pages   cm
  Includes bibliographical references and index.
  ISBN 978-1-118-65270-1 (cloth) – ISBN 978-1-118-65273-2 (pbk.)
1. Carbonate rocks.   2. Sedimentary rocks.   3. Diagenesis.   I. Jones, Brian (Geology professor)   II. Title.
  QE471.15.C3J37 2015
  552'.58–dc23
                                                        2014050127

A catalogue record for this book is available from the British Library.

Wiley also publishes its books in a variety of electronic formats. Some content that appears in print may not be available in electronic books.

Cover image: Underwater image of hard coral growth (mainly acroporids) on the atoll of Fakarara in the French Tuamotus in the South Pacific Ocean. The large fish in the center is a Moorish Idol (*Zanclus cornutus*). Photograph by D. Stokes. Reproduced with permission. Photomicrograph (plane polarized light) of Pleistocene limestone, Bermuda.

Printed in the UK

CONTENTS

PREFACE

This is a textbook. It is an overview highlighting the origin and preservation of carbonate sedimentary rocks. The approach is general and universal and draws heavily on fundamental discoveries, arresting interpretations of ancient limestones and dolostones, and keystone syntheses over the last five decades. The volume is designed as a teaching tool for upper-level undergraduate classes, a fundamental reference for graduate and research students, and a scholarly source of information for practicing professionals whose expertise lies outside this specialty. It is not a compendium of the current state of research, but is instead designed as a digestible synthesis of the science written for newcomers to the topic. We encourage those of you who have been drawn into the ancient world that is entombed in carbonate rocks to read the profound works upon which this book is based to see how this science evolved and to capture the fascinating details of the geology that our words here can only summarize. We have shamelessly borrowed from these fundamental works, but they are referenced at the end of this preface and at the end of each chapter. The genesis of carbonate rocks is far from completely understood – much remains to be discovered – and in this light we urge you to use this book as a stepping-stone towards new understanding.

The focus is on carbonate rocks and the sediments from which they are derived. To fully understand this science it is necessary to be as much of a natural scientist as possible and yet remain grounded in geology. We, your authors, have studied modern sediments and reefs where the deposits are born, using scuba in shallow tropical and temperate seas, submersibles to over 3 km in the deep ocean, slogged through rainy swamps, tiptoed across muddy tidal flats scorched by the desert sun, stumbled about in deep caves, and carefully sampled steaming and bubbling hot springs. Our study of ancient rocks spans those formed on the young Earth more than 3 billion years ago to those made just yesterday in the Holocene. This research has taken us from the high Arctic, to the Rocky Mountains of North America, to the steaming tropics of Australia, to the warm reefs of the Caribbean, to the shark-infested waters of the Southern Ocean, to hot and cold springs in New Zealand, Iceland, and China, to the barren hills of Arabia, and yes, to innumerable local rock quarries. Finally, we have also been lucky enough to study exquisite carbonate cores taken from the subsurface worldwide in the search for hydrocarbons. Having said that, it will quickly become apparent upon reading that the book is founded mostly on the modern and ancient environments that we have seen and the ancient carbonates that we have interpreted; this book is therefore a personal journey, and one that we hope you will enjoy.

Our approach is rigorous with every chapter following a similar format. Each one is designed as a separate lecture on a specific topic that is encased within a larger scheme. It is hoped that this approach allows a teacher to use the information as a base and to elaborate on aspects as they wish, while enabling the student to read and view the major concepts as a whole in an engaging fashion. With this in mind, there are no references in the narrative in order to keep the reader focused. The critical references are, however, present at the end of each chapter. There is repetition throughout by design in order to highlight and reinforce major concepts. We hope it works! Finally, we want to share our love of this challenging topic and our amazement that so much of the ancient marine world is preserved in these widespread rocks.

Key reference texts

This book contains a wealth of information that is useful background or extended reading. Each part of this volume contains further references for those readers who want to delve deeper into specific aspects of the science.

Bathurst, R.G.C. (1975) *Carbonate Sediments and their Diagenesis*. Amsterdam: Elsevier Science, Developments in Sedimentology.
A classic text with detailed descriptions of key modern environments.

Crevello, P.D., Wilson, J.S., Sarg, J.S., and Read, J.F. (1989) *Controls on Carbonate Platform and Basin Development*. Society for Sedimentary Geology, SEPM Special Publication no. 44.
Numerous papers on a wide spectrum of depositional systems several decades ago.

Demicco, R.V. and Hardie, L.A. (1994) *Sedimentary Structures and Early Diagenetic Features of Shallow Marine Carbonate Deposits*. Society for Sedimentary Geology, Atlas No. 1.
Beautiful illustrations of carbonate sedimentary rocks.

Einsele, G. (2000) *Sedimentary Basins: Evolution, Facies and Sediment Budget*. Second edition. Berlin, Heidelberg: Springer-Verlag.
The section on carbonate depositional systems is a required synthesis.

Flügel, E. and Munnecke, A. (2010) *Microfacies of Carbonate Rocks: Analysis, Interpretation and Application*. Second edition. Berlin, Heidelberg: Springer-Verlag.
The most up-to-date synthesis of carbonate rocks and a critical reference work.

Ginsburg, R.N. (ed.) (2001) *Subsurface Geology of a Prograding Carbonate Platform Margin, Great Bahama Bank; Results of the Bahamas Drilling Project*. Tulsa, OK; Society for Sedimentary Geology, SEPM Special Publication no. 271, vol. 70.
Numerous papers detailing this critical series of boreholes on the iconic Bahamian platforms.

James, N.P. and Dalrymple, R.W. (eds) (2010) *Facies Models 4*. St John's, Newfoundland: Geological Association of Canada, GEOtext 6.
The section on biochemical sedimentary rocks should be read in conjunction with this book.

Lukasik, J. and Simo, J.A. (2008) Controls on development of carbonate platforms and reefs. In: Lukasik J. and Simo J.A.T. (eds) *Controls on Carbonate Platforms and Reef Development*. Society for Sedimentary Geology, SEPM Special Publication no. 89, pp. 15–43.
A series of papers summarizing our understanding prior to the information in this book.

Milliman, J.D. (1974) *Marine Carbonates. Part 1 Recent Sedimentary Carbonates*. New York: Springer-Verlag.
A classic text that summarizes the attributes of modern carbonate systems and formed the basis for much subsequent research.

Moore, C.H. (2001) *Carbonate Reservoirs; Porosity Evolution and Diagenesis in a Sequence Stratigraphic Framework*. Netherlands: Elsevier, Developments in Sedimentology.
The text to consult when trying to understand the relationship between carbonate sedimentology-diagenesis and porosity.

Mutti, M., Piller, W.E., and Betzler, C. (eds) (2010) *Carbonate Systems during the Oligocene–Miocene Climatic Transition*. Wiley-Blackwell, International Association of Sedimentologists, Special Publication no. 42.
A relatively recent compendium of carbonate deposition systems during the middle Cenozoic, with applications to the older rock record.

Reading, H.G. (ed.) (1986) *Sedimentary Environments and Facies*. Second edition. Oxford: Blackwell Scientific Publications.
Several useful articles on carbonate facies.

Roehl, P.O. and Choquette, P.W. (eds) (1985) *Carbonate Petroleum Reservoirs*. New York: Springer-Verlag.
Numerous papers describing the details, in a uniform framework, of carbonate reservoir rocks.

Schlager, W. (2005) *Carbonate Sedimentology and Sequence Stratigraphy*. Society for Sedimentary Geology, Concepts in Sedimentology and Paleontology, vol. 8.
A book that successfully marries recent concepts in carbonate sedimentology with the latest advances in sequence stratigraphy.

Scholle, P.A. and Ulmer-Scholle, D.A. (eds) (2003) *A Color Guide to the Petrography of Carbonate Rocks: Grains, Textures, Porosity, Diagenesis*. Tulsa, Oklahoma: American Association of Petroleum Geologist, Memoir no. 77.
This is the book to read when you need to understand what is under the microscope and unravel paragenesis.

Scholle, P.A., Bebout, D.G., and Moore, C.H. (eds) (1983) *Carbonate Depositional Environments*. Tulsa, OK: American Association of Petroleum Geologists, Memoir no. 33.
An exquisite volume totally in color that beautifully illustrates a wide spectrum of depositional environments.

Simo, J.A.T., Scott, R.W., and Masse, J.-P. (eds) (1993) *Cretaceous Carbonate Platforms*. Tulsa, OK: American Association of Petroleum Geologists, Memoir no. 56.
Several key papers that illustrate carbonate deposition during this extraordinary period in geological history.

Swart, P.K., Eberli, G.P., and McKenzie, J.A. (eds) (2009) *Perspectives in Carbonate Geology. A Tribute to the Career of Robert Nathan Ginsburg*. Wiley-Blackwell, International Association of Sedimentologists, Special Publication no. 41.
A variety of introspective papers on carbonate sedimentology throughout geologic history.

Tucker, M.E. and Wright, V.P. (1990) *Carbonate Sedimentology*. Oxford: Blackwell Scientific Publications.
This scholarly text addresses all aspects of carbonate sedimentology and diagenesis.

Tucker, M.E., Wilson, J.L., Crevello, P.D., Sarg, J.R., and Read, J.F. (eds) (1990) *Carbonate Platforms: Facies, Sequences and Evolution*. Oxford: Blackwell Scientific Publications, International Association of Sedimentologists, Special Publication no. 9.
Numerous papers on a spectrum of carbonate platforms throughout geological history.

Wilson, J.L. (1975) *Carbonate Facies in Geologic History*. New York: Springer Verlag.
A text that established carbonate sedimentology by viewing the topic through the lens of geologic time.

ACKNOWLEDGEMENTS

This textbook would not have been possible without the wisdom of those who have gone before, the help of our dear friends and colleagues, and the accumulating knowledge of those working today in the field and laboratory. The book could never have been conceived without our mentors, namely Robin Bathurst (University of Liverpool), Phil Choquette (Marathon Oil Company), Bob Ginsburg (University of Miami), Eric Mountjoy (McGill University), and Jim Wilson (Rice University), who showed us so much about research, how to teach, and the complex world of science. Without them and their inspiration we would never have attempted this intimidating endeavor.

More specifically, the present organization of the manuscript and many of the illustrations are the work of Isabelle Malcolm to whom we owe an enormous debt of gratitude. She saw the project through to the end despite the many obstacles that we put in her way. Bill Martindale also deserves a huge vote of thanks for not only supplying many of the superb images but also for selflessly reading the book from beginning to end and offering his usual enormously wise and helpful suggestions. We are also indebted to Julia Jones-Bourque who took the time to proof-read the entire book and provide many insightful amendments.

Special thanks and appreciation are due to our colleagues who undertook the thankless task of critically reading large parts of the book to keep us honest, save us from scientific embarrassment, and make sure that our thoughts were clear and concise: Dan Bosence (RHBNC, University of London), Tracy Frank (University of Nebraska), John Grotzinger (California Institute of Technology), Kurt Kyser (Queen's University), Eric Hiatt (University of Wisconsin, Osh Kosh), Jeff Lukasik (Statoil, Brazil), Alex MacNeil (Osum Energy, Calgary), Ted Matheson (Queen's University), Cody Miller (Chevron Research, Houston), Peir Pufahl (Acadia University, Nova Scotia), Catherine Reid (University of Canterbury, New Zealand), Elizabeth Turner (Laurentian University, Sudbury), and Dana Ulmer-Scholle (New Mexico Tech, Socorro).

Our friends of many years, Peter Scholle and Dana Ulmer-Scholle (New Mexico Tech, Socorro), Gene Shinn and Pam Hallock (University of South Florida), Kurt Kyser, Guy Narbonne, and Bob Dalrymple (Queen's University), Yvonne Bone (University of Adelaide), Brian Pratt and Robin Renaut (University of Saskatchewan), Hal Wanless (University of Miami, Coral Gables, Florida), Pierre-André Bourque (Laval University, Québec), André Desrochers and Owen Dixon (University of Ottawa), Bob Stevens (Memorial University, Newfoundland), Catherine Reid (University of Canterbury, New Zealand), John Rivers (Exxon-Mobil, Houston), Mitch Harris (Chevron Research, Houston), Benoit Beauchamp (University of Calgary), Rachel Wood (Edinburgh University), Jon Clarke (Geoscience Australia), David Feary (University of Arizona), Paul Wright (British Gas), Lindsay Collins (Curtin University, Perth, Australia), Paul Taylor (British Museum, Natural History), Fiona Whitaker (University of Bristol), Clint Cowan (Carleton College, Minnesota), Wolfgang Schlager (Free University, Amsterdam), Jeff Lukasik (Statoil), Gene Rankey (University of Kansas), Jean-Yves Reynaud (Université de Lille, France), Cam Nelson (University of Waikato, New Zealand), and Eric

Hiatt (University of Wisconsin, Osh Kosh) have been partners in the research that led to many facets of this book. The numerous discussions and arguments with them have sharpened our reasoning and softened our biases. If we have omitted anyone, please forgive us.

This book is above all the result of extensive research and teaching. The Natural Sciences and Engineering Research Council of Canada, to whom we are most grateful, has financed our science over the years and around the globe. Shell Oil has for many years supported student field excursions in Bermuda, and many of the photographs here were taken during these field seminars. We would also like to acknowledge the support of the institutions where we have been lucky enough to carry out these endeavors, including the University of Miami, Memorial University of Newfoundland (St John's), Marathon Oil Company Research Laboratories (Littleton, Colorado), the University of Alberta (Edmonton), the University of Adelaide, (Australia), and Queen's University (Kingston, Ontario). It has, however, been the outstanding and demanding undergraduate students in these universities that, as we strove to deliver ever more current and understandable lectures, honed our thinking and increased our enthusiasm for this entrancing subject. Finally, we have been blessed with the best and brightest of research students; much of what is in this book is the result of their exceptional and dedicated research.

ABOUT THE COMPANION WEBSITE

This book is accompanied by a companion website:

www.wiley.com/go/james/carbonaterocks

The website includes:
* Powerpoints of all figures from the book for downloading
* Pdfs of all tables from the book for downloading

PART I
CARBONATE SEDIMENTOLOGY: AN OVERVIEW

Frontispiece Eocene limestone composed mainly of regular echinoid spines and bryozoans, Chatham Islands, New Zealand. Image width 7 cm.

Origin of Carbonate Sedimentary Rocks, First Edition. Noel P. James and Brian Jones.
© 2016 Noel P. James and Brian Jones. Published 2016 by John Wiley & Sons, Ltd.
Companion website: www.wiley.com/go/james/carbonaterocks

Introduction

Carbonate sedimentology is a biogeochemical system that has been in operation since the Archean. The system is a multitude of biological processes that take place within the constraints of global geological dynamics operating at unimaginably long timescales. What makes this system so interesting is that all these actions have ongoing feedback loops. What could be more intellectually demanding than trying to understand such a system? With this challenge in mind, the first part of the book outlines the fundamentals of carbonate sedimentology. This foundation provides the basis for learning the intricacies of carbonate depositional systems and then the diagenetic processes that transform sediments into hard carbonate rocks.

Carbonate minerals and their chemistry

The carbonate mineral system is relatively simple because it is dominated by just four minerals: calcite, aragonite, Mg-calcite, and dolomite. These minerals can precipitate in just about any marine or terrestrial environment. All of them, however, are relatively soluble and hence prone to dissolution, possibly even in the same environment in which they formed: therein lie the complexities and wonders of the carbonate system.

The carbonate factory

Using an industrial analogy, carbonates are continuously generated in a sediment factory that produces material across the grain-size spectrum from muds to sands to boulders and, uniquely, reefs. Most components are either biogenic or abiogenic (non-biological) or some combination of both, and precipitate from seawater or freshwater saturated or supersaturated with respect to $CaCO_3$. It now appears that many so-called "inorganic precipitates" have in fact precipitated with some degree of organic influence.

The most important attribute of carbonate sediments through geological time is that they are by and large produced in the environment of deposition. Thus, carbonate sediments are "born and not made" (Figure I.1)! That is to say, carbonate sediments, composed of a wide variety of biogenic and abiogenic components, are formed in the environment of deposition and not, like siliciclastic sediments, eroded from elsewhere and brought to the depositional environment. Although this axiom characterizes the sediment factory, once they begin to accumulate on

the seafloor the resultant particles do not remain inert. The grains can be quickly bored, macerated (disintegrated as the organic matrix decays), or dissolved. Particles can also be broken up and ingested by invertebrate animals or disturbed by burrowing organisms. Furthermore, because most depositional environments are shallow water, the sediments are subject to profound disturbance and redistribution by cyclonic storms. Finally, if accumulating on slopes, sediments can be episodically disturbed and redeposited into deeper water by sediment gravity flows.

The different carbonate factories

Carbonate sediment is produced in a variety of settings from tropical reefs to arid lakes; there are many different carbonate factories, each attuned to the ambient aqueous environment. The two ends of the environmental spectrum are marine and terrestrial. The marine system is either neritic or pelagic. The marine neritic (shelf) system has perhaps produced most of the carbonates in geologic history, and each depositional window has been framed by differences in water temperature, salinity, trophic resources, depth of the photic zone, siliciclastic, and sediment input. The first-order control seems to be water temperature where, in modern systems at least, 20°C is the critical interface. The pelagic (open-ocean) near-surface factory is subject to similar controls but has only been in operation as a carbonate factory since the Jurassic. Carbonates formed in terrestrial settings such as lakes, hot and cold springs, and caves appear to be most affected by water temperature and local climate.

Regardless of the specific factory, sediment particles are and were universally biogenic and composed of microbes, algae, and invertebrates or abiogenic and composed of coated grains, microcrystalline spheroids (pellets and peloids), and fragments of cemented sediment. Whereas abiogenic particles are relatively simple, mostly coated grains and muds, biogenic grains are produced by a spectrum of microbes, algae, and invertebrate animals that almost defies imagination.

Microbes and algae

Microbes can induce precipitation, can themselves be calcified (calcimicrobes), or can bind grains together to form mats and stromatolites. Algae, both green and red, have calcified skeletons and contribute mud as well as sand-sized sediment to the milieu; some even act as encrusters on actively growing reefs.

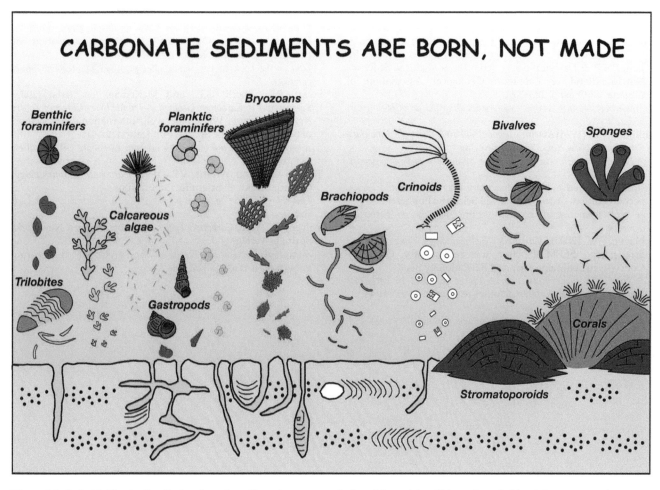

Figure I.1 A sketch illustrating that carbonate sediments, composed of a wide variety of biogenic and abiogenic components, are formed in the environment of deposition and not (like siliciclastic sediments) eroded from elsewhere and brought to the depositional environment. Source: James et al. (2010). Reproduced with permission of the Geological Association of Canada.

Invertebrate biofragments

Calcareous invertebrates are most easily thought of, in terms of sediment producers, as single cells, shells, or colonial skeletons. Single cells are mainly foraminifers or radiolarians; the latter are siliceous but are an important part of the depositional spectrum. Shells are derived from brachiopods, mollusks (gastropods, bivalves, and cephalopods), worms, or arthropods (barnacles, ostracods, and trilobites). By contrast, colonial skeletons are mostly corals, coralline sponges (stromatoporoids), and bryozoans. Highly segmented echinoderms are perhaps the most prolific sediment producers of all. Regardless of their affinity, biofragments are either a single piece (foraminifer test or gastropod shell) or two (brachiopod) or more (crinoid) pieces. Each component has its own distinctive attributes that are most easily identified in thin section.

Further reading

The following selected references are particularly useful for those readers who want further information on specific aspects of the basic carbonate systems. Additional references at the end of each chapter deal with specific topics discussed in that part of the text.

Bathurst, R.G.C. (1975) *Carbonate Sediments and Their Diagenesis.* Amsterdam: Elsevier Science, Developments in Sedimentology no. 12.
A classic text with detailed descriptions of key modern environments and the complexities of diagenesis.
Einsele, G. (2000) *Sedimentary Basins: Evolution, Facies and Sediment Budget.* Second edition. Berlin, Heidelberg: Springer-Verlag.
Contains a series of excellent chapters summarizing various aspects of carbonate sedimentology.
Flügel, E. and Munnecke, A. (2010) *Microfacies of Carbonate Rocks; Analysis, Interpretation and Application.* Second edition. Berlin, Heidelberg: Springer-Verlag.

The single-most authoritative compendium devoted to all aspects of carbonate sedimentology.

James, N.P., Kendall, A.C., and Pufahl, P.K. (2010) Introduction to biological and chemical sedimentary facies models. In: James N.P. & Dalrymple R.W. (eds) *Facies Models 4*. St John's Newfoundland & Labrador: Geological Association of Canada, GEOtext 6, 323–339.
The introduction to several chapters detailing the principles of carbonate deposition.

Milliman, J.D. (1974) *Marine Carbonates. Part 1 Recent Sedimentary Carbonates*. New York: Springer-Verlag.
A classic text with much information on modern carbonate depositional systems.

Reading, H.G. (ed.) (1986) *Sedimentary Environments and Facies*. Second edition. Oxford: Blackwell Scientific Publications.
A book with several excellent review chapters on carbonate deposition.

Schlager, W. (2005) *Carbonate Sedimentology and Sequence Stratigraphy*. SEPM (Society for Sedimentary Geology), Concepts in Sedimentology and Paleontology no. 8.
A book that successfully marries recent concepts in carbonate sedimentology with the latest advances in sequence stratigraphy.

Scholle, P.A. and Ulmer-Scholle, D.A. (eds) (2003) *A Color Guide to the Petrography of Carbonate Rocks: Grains, Textures, Porosity, Diagenesis*. Tulsa, Oklahoma: American Association of Petroleum Geologist, Memoir no. 77.
This is the book to use when attempting to identify carbonate particles.

Swart, P.K., Eberli, G.P., and McKenzie, J.A. (eds) (2009) *Perspectives in Carbonate Geology. A Tribute to the Career of Robert Nathan Ginsburg*. Wiley-Blackwell, International Association of Sedimentologists, Special Publication no. 41.
A recent compendium of articles on both carbonate sedimentation and diagenesis.

Tucker, M.E. and Wright, V.P. (1990) *Carbonate Sedimentology*. Oxford: Blackwell Scientific Publications.
This scholarly text addresses all aspects of carbonate sedimentology and diagenesis.

Wilson, J.L. (1975) *Carbonate Facies in Geologic History*. New York: Springer Verlag.
A text that established carbonate sedimentology by viewing the topic through the lens of geologic time.

CHAPTER 1
CARBONATE ROCKS AND PLATFORMS

Frontispiece Triassic carbonate rocks exposed in the Dolomites of northern Italy, cliff is ~300 m high.

Origin of Carbonate Sedimentary Rocks, First Edition. Noel P. James and Brian Jones.
© 2016 Noel P. James and Brian Jones. Published 2016 by John Wiley & Sons, Ltd.
Companion website: www.wiley.com/go/james/carbonaterocks

What are carbonate sedimentary rocks?

Carbonate sedimentary rocks (Frontispiece) are limestones and dolostones composed of the minerals calcite ($CaCO_3$) and dolomite ($CaMg (CO_3)_2$). They comprise roughly 20% of the surface sedimentary rocks on the planet with the oldest being ~3.5 Ga, almost as ancient as the Earth itself. Most of these rocks were once sediments that accumulated in aquatic environments that ranged from the warm, sunlit shallow seafloor to the cold, perpetually dark, deep ocean. Constituent particles precipitated directly from seawater, or were biologically formed from calcified microbes, calcareous algae, and the whole or broken skeletons of invertebrate animals. The spectrum of original sediments, ranging from muds, to deep-sea ooze, to reefs, to absurdly fossiliferous banks, is truly astonishing. Finally, carbonate rocks are not all marine. Carbonate deposits also form in association with terrestrial spring waters, in rivers and lakes, as calcareous soils, and in caves.

No one carbonate is like any other because the biosphere and the depositional environments have changed dramatically through time and space. The modern world is but a general guide as to how these amazing rocks originally formed.

Why should we care about studying these rocks?

Carbonate rocks are, because of their largely biological origin, unparalleled repositories of information about past life on our planet; they are our windows into deep time. As life has evolved, particularly in the ocean, so the composition of carbonate sediments and rocks has changed in tandem. No-one can fail to be amazed at the preservation of life that is hundreds of millions of years old and yet looks as though it was made yesterday. How can this be?

Many of the particles that comprise marine carbonate rocks are also proxies of ancient ocean composition because they were precipitated in equilibrium with ancient seawater. They contain within their skeletons mineralogic, petrographic, and geochemical vestiges of their birth. As the seas warmed and cooled, so the organisms changed in concert, thus preserving what these long-vanished oceans were like. The history of the marine realm is written in these rocks, but we must learn to read the text properly.

Carbonate rocks are profoundly important for human life as we know it. Many of our most historical and beautiful buildings are fashioned from limestone. The rocks are important freshwater aquifers worldwide. Limestones and especially dolostones also host a disproportionally large proportion of the world's metallic ores and hydrocarbons. To understand the nature of fluid flow through and the economic attributes of these rocks, we must know how they originated so that we can predict their attributes in regions that are yet to be explored.

What is the scientific approach?

Our objective is that after reading this book you will be able to interpret carbonate rocks, no matter their age and state of alteration. More specifically, we hope that you will be able to answer the fundamental questions as to how they were formed and how have they been altered after deposition before you are looking at them today. As stressed in the previous chapter, this is a profound intellectual challenge! What is the best way forward? The most useful approach seems to be one of looking carefully at the modern world of carbonate deposition (Figure 1.1) and diagenesis where we can discern the important *processes* that are operating and then apply these principles to the rock record. Ancient oceans, however, were not the same as they are today: biological evolution has been rampant through time so that the biosphere was different in the past; plate tectonics has altered the position and nature of depositional environments; and climate has induced profound changes in the atmosphere, in seawater temperature gradients, and in sea level itself. Is there any hope? The answer to that is *yes*, because knowledge of our oceans and environments is so sophisticated that we can apply our knowledge of modern analogs to interpret the products that we see in the rock record.

Figure 1.1 Students swimming over a Bermuda coral reef in ~2 m of water, the modern equivalent of many ancient limestones.

The carbonate continuum

Regardless of the age of the limestone you are trying to understand, there is always a recurring theme of deposition followed by diagenesis (Figure 1.2). Although this simple flow can be interrupted, reorganized, and recycled in many different ways, it provides an easy way in which to approach the topic.

Deposition

As emphasized above, the most obvious carbonate particles (Figure 1.3) are the broken and whole shells of invertebrates such as clams and corals that range in size from microns to meters. Biogenic components are mostly the skeletons left behind when an organism dies (e.g., a snail) or the crystallites embedded in the

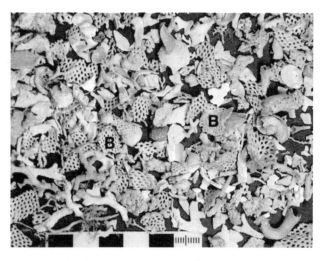

Figure 1.3 A carbonate sediment from the modern continental shelf off southern Australia composed mostly of bryozoans (B), centimeter scale.

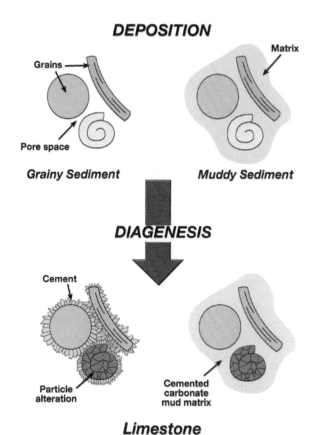

Figure 1.2 Sketch illustrating the carbonate continuum where sediment is transformed into limestone via diagenesis.

Figure 1.4 A cross-section of laminated hemispherical stromatolites in 1.9 Ga limestones from the Pethei Group, Great Slave Lake, Arctic Canada, 10 cm scale bar at base.

tissue are released when an alga expires (e.g., a green calcareous algae). Inorganic grains range from tiny crystals of mud that precipitate out of seawater or freshwater to sand-sized particles composed of numerous crystallites aggregated together (e.g., a coated grain). There are, however, many components that are formed by both biogenic and inorganic processes (e.g., a stromatolite; Figure 1.4).

The multifaceted, ongoing processes in the depositional environment not only generate particulate sediments such as carbonate sands and muds, but also construct large structures such as coral reefs. In the simplest terms, grainy sediments are composed of grains, matrix, and pore space (Figure 1.2). While grains are those particles described above, matrix is usually (for sand-sized sediment at least) composed of carbonate mud. Details of how such sediments and rocks are classified are outlined in Chapter 4. Reef rocks are classified in a different way because they are constructed of large skeletons that stack on top of one another with impressive internal cavities between, or are shed off as coarse debris (Chapter 15).

Diagenesis

Diagenesis is for the most part the processes of lithification and alteration. The term diagenesis is, for the purposes of this book, defined as the mineral, chemical, and fabric changes that carbonate sediments or rocks undergo at relatively low temperatures and pressures before they enter the realm of metamorphism.

Lithification is generally achieved by the precipitation of carbonate crystals in spaces between particles in clean sands or between the micropores of muddy sediments or matrix. Such carbonate is usually called *cement* (Figures 1.2, 1.5), not to be confused with the construction cement that is used to build houses and highways (which is quite different). Such crystals generally precipitate first on the particle surface and then grow towards the cavity centre where they can eventually fill the whole void. Cements can be precipitated in the environment of deposition, that is, at or just below the seafloor, shortly after deposition when fresh meteoric waters percolate through the rock, or much later when the whole succession is deeply buried beneath younger deposits. Most cementation and particle alteration can therefore be considered as taking place in different *diagenetic environments* that are defined by the nature of the waters involved.

Alteration is when the original composition of the sediment is changed. The range of potential modifications is wide. The simplest case is just mineralogical change where the texture and the fabric of the original deposit are largely unaltered. The more severe cases occur when the whole character of the sediment is altered such that most of the rock is composed of crystalline calcite or dolomite that obliterates the original fabric (fabric destructive).

Dolomitization of carbonate sediment or limestone is a common diagenetic process. The original calcareous material is partly or wholly transformed into dolomite (Figure 1.6). This occurs in a variety of ways and can happen at any time in the history of the deposit. The results range from complete preservation to total obliteration of

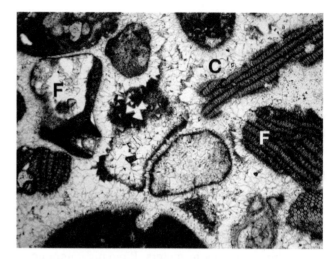

Figure 1.5 Thin-section image of a Bermuda Pleistocene limestone composed of biofragments (F) and interparticle cement (C), image width 1 cm.

Figure 1.6 Coarsely crystalline white (saddle) dolomite replacing Cambrian limestone, Main Ranges of the Rocky Mountains, Alberta, Canada; the inclined bedding is not cross-bedding but expansive hydrofracturing, centimeter scale. Photograph by W. Martindale. Reproduced with permission.

texture and fabric such that the rock becomes a mass of crystals that resemble brown sugar. Whereas the latter example causes most researchers distress, hydrologists and petroleum geologists rejoice because it is generally accompanied by a profound increase in rock porosity and permeability.

The bottom line is that whereas carbonates can precipitate easily, they are, because of this attribute, just as readily susceptible to profound change.

How do carbonate sediments form?

The sediments are born, not made

This deceptively simple phrase, as mentioned in the Introduction to Part I, encapsulates the main theme of carbonate deposition and highlights their differences from siliciclastic sedimentary rocks. Siliciclastic sediments are formed largely of detritus, derived from the disintegration of source rocks, which is transported to the depositional environment. Once there, the hydraulic regime dictates the nature of the facies because water movement is the primary control over transportation, deposition, and development of sedimentary structures. By contrast, carbonate sediments are *born* as precipitates or skeletons within the depositional environment. For carbonate deposition this means that:

- sediment composition is largely dictated by the skeletal composition of the resident biota;
- grain size is largely imposed by biological processes and variations in grain size and need not signal changes in the hydraulic regime;
- large structures such as platforms and reefs are self-generating and self-sustaining; and
- the temporal and spatial style of accumulation depends upon the nature of the sediments themselves.

The site of carbonate sediment production, especially in shallow illuminated aquatic environments, is commonly viewed as the *carbonate factory* (Chapter 3). Input to the factory is the Ca and CO_3 in seawater. Output is in the form of grains of all sizes and shapes that are generated by a spectrum of processes that commonly involve the direct or indirect intervention of the resident biota. Most sediments, which remain in or close to their place of formation, accumulate as widespread neritic or *subtidal* deposits or as reefs and mounds. Some of the abundant fine fraction (mud) is usually resuspended during storms and transported onshore where it forms muddy tidal flats that develop around highs on the platform and along the shoreline. As the storms wane, offshore currents will also move fine sediment seaward or basinward where it is deposited in deep water. Regardless of location, it is the factory that generates vast quantities of carbonate sediment.

The extent of marine carbonate factories is readily apparent, for example, on the satellite image that covers Cuba, the Great Bahama Bank, Florida Bay, and the Florida Shelf (Figure 1.7). On this image, areas of shallow-water carbonate sediment occur where there is no influx of detrital sediment from nearby landmasses (e.g., Florida Bay, Gulf of Batabano) or where the banks are isolated by surrounding deep oceanic waters (e.g., Great Bahama

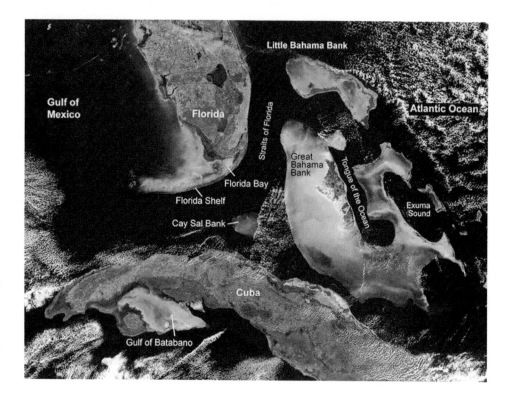

Figure 1.7 A satellite view of southern Florida, Cuba, and the Bahama Banks. The light blue areas, mostly <5 m deep, are regions of carbonate sediment formation. Source: Jones (2010). Reproduced with permission of the Geological Association of Canada.

Table 1.1 Sedimentary aspect of modern warm- and cool-water carbonate components and their ancient counterparts.

Component	Modern, warm-water	Modern, cool-water	Ancient counterpart
Large elements of reefs or biogenic mounds	Corals (with photosymbionts)	ABSENT	Corals, stromatoporoids, stromatolites, coralline algae, sponges, rudist bivalves, Archaeocyathans
Whole organism forms granule-size particles	Large benthic foraminifers (with photosymbionts)	ABSENT	Large benthic foraminifers (e.g., Fusilinids)
Remain whole or break apart into several pieces to form sand- and gravel-size particles	Bivalves, red algae	Bivalves, red algae, brachiopods, barnacles	Red algae, brachiopods, cephalopods, trilobites
Whole skeletons that form sand- and gravel-size particles	Gastropods, small benthic foraminifers	Gastropods, small benthic foraminifers	Gastropods, small benthic foraminifers
Spontaneously disintegrate upon death to form many sand-sized particles	Green (Codiacean) and red algae	Red algae, bryozoans, echinoderms	Phylloid algae, pelmatozoans and other echinoderms, bryozoans
Concentrically laminated or micritic sand-sized grains	Ooids, peloids	ABSENT	Ooids, peloids
Medium sand-sized and smaller particles in basinal deposits	Planktic foraminifers, coccoliths, pteropods	Planktic foraminifers, coccoliths, pteropods	Planktic foraminifers and coccoliths (post-Jurassic), stylolinids
Encrust on or inside hard substrates, to build up thick deposits or fall off upon death to form sand grains	Encrusting foraminifers, red algae, bryozoans, serpulid worms	Encrusting foraminifers, red algae, bryozoans	Red algae, calcimicrobes, encrusting foraminifers, bryozoans
Spontaneously disintegrate upon death to form lime mud	Green algae (Dasycladaceans)	Red algae, bryozoans, serpulid worms	Dasyclad green algae
Trap, bind, and facilitate precipitation of fine-grained sediment to form mats, stromatolites, and thrombolites	Bacteria and other microbes	Bacteria and other microbes	Calcimicrobes and microbes (especially pre-Ordovician)

Bank). Such factories can cover vast areas, for example the Great Bahama Bank has an area of ~96,000 km² and the Great Barrier Reef complex in Australia has an area of ~344,400 km². It should be noted, however, that some carbonate factories in past geological time covered even greater areas such as the vast shallow so-called epeiric seas that veneered most of continental North America during the Paleozoic.

Production in the carbonate factory is intimately linked to the resident biota. Thus, as animals and plants have evolved through geologic time, so too has the nature of the sediments changed. Some comparisons are obvious (Table 1.1); for example, the contrast between the Proterozoic microbial-dominated stromatolite factory and the Jurassic invertebrate-dominated coral factory is readily apparent. Other comparisons, especially those involving shorter time spans, are subtle with the differences being less obvious; for instance, the difference between Permian brachiopod-rich and Cretaceous bivalve-rich sediment.

Where are carbonates produced and where do they accumulate?

The marine realm

Marine carbonates have the potential to accumulate anywhere in the ocean except on the deepest abyssal plains and trenches where hydrostatic pressure is so high and seawater temperature is so low that they generally dissolve before reaching the seafloor (see Chapter 2). The main limiting factor in carbonate production in most places is abundant dirt, because too much siliciclastic sediment: (1) swamps the carbonates; (2) clogs the feeding apparatus of many calcareous organisms; and (3) decreases light penetration, a critical factor for photosynthetic biota. It used to be thought that carbonates were strictly warm-water tropical deposits but we now realize that whereas water temperature is important, carbonate production can extend from the equator well into polar waters. In summary, most carbonate sediment is produced in bright, relatively clear, near-surface ocean waters.

CARBONATE ACCUMULATION ZONES

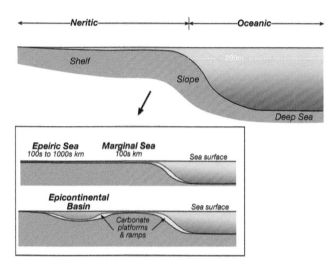

Figure 1.8 Sketch illustrating the different areas of carbonate sedimentation.

Carbonate production can take place on the shallow seafloor or in the shallow water column. The two major sediment-producing environments are the *neritic zone*, the part of the seafloor from the shoreline to the continental shelf edge at ~200 m water depth, or the *oceanic zone*, the shallow part of the water column to ~200 m depth (Figure 1.8). The major organisms in these two zones respectively are seafloor-dwelling (*benthic*) or water-column dwelling (*planktic*).

Carbonate deposition or accumulation in the neritic zone today is mostly benthic (both biogenic and physicochemical) with a minor pelagic component. The opposite is true in deep ocean basins where, even though some sediment can be washed into deep water via sediment gravity flows from neritic environments, most sediment is pelagic in the form of calcareous plankton that, upon death, sink to the deep seafloor; there is only a minor contribution from deep-water calcareous benthos.

Areas of neritic carbonate sedimentation (Figure 1.8) are either on the continents (intracontinental basins) or in the open ocean as spectacular shallow banks (e.g., Bahama Banks; Figure 1.7). Epicontinental carbonates accumulate in a variety of shallow basins, along continental margins, and in epeiric seas. An epeiric sea is defined as a vast shallow sea that was connected to the open ocean and stretched for hundreds to thousands of kilometers across a continent; sadly, there are no modern analogs to these environments. These are different from the more common, marginal seas that covered continental margins. Whereas epeiric seas were beyond the reach of oceanic tides,

marginal seas were well within the zone of such tides. The difficulty of interpreting carbonates deposited in epeiric seas is explored in Chapter 20.

The terrestrial realm

Not all carbonate sedimentary rocks are of marine origin; many originate on land in the terrestrial realm. The most extensive deposits accumulate in lakes. Settings range from marginal marine to truly terrestrial far from the ocean in large shallow structural basins, in rift valleys, or in mountain valleys. The most impressive lacustrine carbonates occur when the lakes are fed by internal drainage.

Calcareous soils are widespread in space and time. Such pedogenic horizons can develop in both siliciclastic and carbonate soils, generally under semi-arid climates. The interaction of soil and soil-related processes around the margins of lakes can also profoundly modify lacustrine deposits leading to palustrine carbonate deposits.

The ocean is a prolific and ongoing source of carbonate sediment and calcareous beach sediment can be blown into spectacular sand dunes called aeolianites. These dune complexes form along the shores of both warm-tropical and cool-temperate oceans.

The features that set terrestrial systems apart from marine carbonates are the dissolution and precipitation products associated with their interaction with fresh water, namely karst and springs. Karst features such as caves are produced by subsurface dissolution via water mixing and can contain impressive calcareous precipitates in the form of speleothems (e.g., stalagmites and stalactites). Carbonate springs, essentially carbonate saturated warm- or cold-subterranean waters that emerge onto land, precipitate carbonate in the form of tufa or travertine.

Tectonic settings and the nature of carbonate platforms

Carbonate accumulation through time in an area of crustal subsidence can lead to the formation of impressive rock bodies. Some of the more important are exposed in mountain belts such as the Rockies and the Alps, whereas others remain buried in the subsurface and are important hydrocarbon reservoirs. The term that is usually used for such structures is *carbonate platform* (Figure 1.9). This is a general term because there are three recurring end-members in the rock record: rimmed platforms; open platforms; and inclined platforms (usually referred to as *ramps*). The structures can be attached to land (e.g., Great Barrier Reef) or be detached and isolated in the open ocean or basin (e.g., Great Bahama

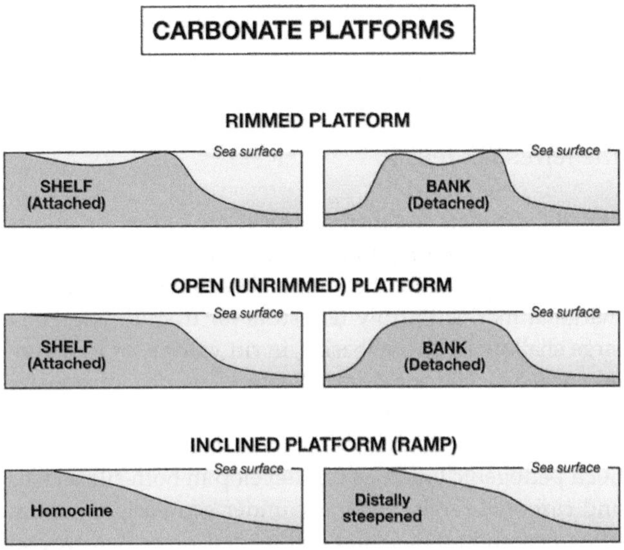

CARBONATE PLATFORMS

RIMMED PLATFORM

SHELF
(Attached)
— Sea surface —

BANK
(Detached)
— Sea surface —

OPEN (UNRIMMED) PLATFORM

SHELF
(Attached)
— Sea surface —

BANK
(Detached)
— Sea surface —

INCLINED PLATFORM (RAMP)

Homocline
— Sea surface —

Distally
steepened
— Sea surface —

Figure 1.9 Sketch illustrating the three main types of carbonate platforms.

Bank). A flooded craton many hundreds to thousands of kilometers across is called an *epeiric platform* when covered by shallow seawater (an epeiric sea) (Figure 1.8). A *bank* is an isolated platform surrounded by deep ocean water and cut off from terrigenous clastic sediments and terrestrial runoff. An *atoll* is a specific type of bank commonly developed on a subsiding volcano. Carbonate atolls and banks may be dominated by reefs such that their geological expressions are termed *reef complexes*.

The size, geometry, thickness, and stratigraphy of carbonate bodies are largely determined by tectonic setting, either extensional (Figure 1.10) or compressional (Figure 1.11). A few platform types can be present in each realm.

Extensional

Intracratonic. Such platforms are located in relatively shallow basins on the craton with low subsidence rates. They are relatively thin successions with subhorizontal boundaries that are largely determined by eustasy. Climate is important; interbedded siliciclastics occur in humid climates and coeval evaporites in arid climates. Platforms are rimmed around the basin margin with the basin center evaporites or shales giving a concentric or bull's eye facies pattern to the deposits. Platforms are generally progradational and <1 km thick. The Paleozoic Michigan Basin, the Permian Basin of North America, and the Cenozoic Murray Basin in Australia are excellent examples.

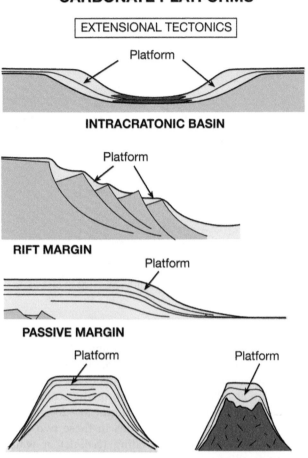

CARBONATE PLATFORMS

EXTENSIONAL TECTONICS

Platform

INTRACRATONIC BASIN

Platform

RIFT MARGIN

Platform

PASSIVE MARGIN

Platform

Platform

OFFSHORE BANK   **VOLCANIC PEDESTAL**

Figure 1.10 Sketches of carbonate platforms and atolls in different extensional tectonic regimes.

Rift basins. Carbonate platforms can develop on footwall highs or rotated fault blocks. Deposits are wedge-shaped and thicken down into hanging-wall dip slopes. They thin onto footwall highs. The footwall areas are commonly reef- or shoal-rimmed margins. A good example of these deposits is the Cenozoic along the Gulf of Suez.

Passive continental margins. These are some of the most extensive carbonate platforms in the rock record. Platforms are generally thick with their bathymetry inherited from underlying rift successions. Platforms can be immense: many thousands of kilometers in length and many hundreds of kilometers in width. Subsidence rates are generally low and so progradation is typical with facies belts parallel to the coastline. Stratigraphic geometries are generally subhorizontal because of domination by eustatic sea-level fluctuations (eustasy). The Paleozoic continental margin of eastern North America and the

CARBONATE PLATFORMS

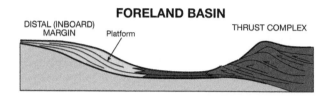

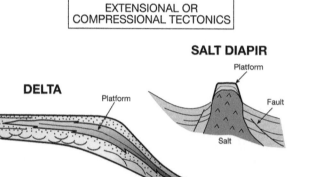

Figure 1.11 Sketches of carbonate platforms in compressional and extensional–compressional tectonic regimens.

Cenozoic northwest shelf of Australia are good examples of these systems.

Offshore banks. Large carbonate banks surrounded by deep basins are generally located in passive margin settings. As above, subsidence rates are usually low and progradational successions can dominate stratigraphy. These structures can have strongly differentiated margins with erosional or steep reefal windward margins and low-relief progradational leeward margins. The Bahama Banks is the quintessential example of a modern offshore carbonate bank.

Volcanic pedestals and island arcs. These platforms reflect their association with volcanic edifices, namely biogeographic isolation, variable subsidence, steep volcanic slopes, and tectonic instability. These relatively small platforms (Figure 1.10), 10–50 times smaller than continental platforms, range from fringing reefs around volcanoes to atolls composed of thick successions on top of thermally subsiding ocean volcanoes. Successions are typically aggradational and are surrounded by deep-water facies. The steep margins are characterized

by numerous slumps and redeposited carbonates. Thicknesses are highly variable from <100 m to >1000 m. Sequence boundaries are subhorizontal over the top and reflect eustasy. Biotas are highly endemic because of geographic isolation and typically of low diversity, with some components surviving longer than their cousins on continental margin platforms. They are common today in the tropical Indo-Pacific and in mountain belts where they have been obducted landward onto a craton (e.g., Mesozoic of Oman or Paleozoic of the Appalachians).

Compressional

Foreland. Platforms in foreland basin settings on the craton are either on the crest of advancing thrusts or along the passive inboard margin (Figure 1.11). Platforms on thrust tops are generally ribbon-like in plan view and relatively thin on rising thrust complexes with tectonics imparting an irregular unpredictable motif to carbonate sequences. Seismicity results in numerous redeposited slope deposits. Platforms on the inner distal margin are best developed when there is a semi-arid climate and minor input of siliciclastic sediment from the adjacent exposed craton. High subsidence rates result in locally thick successions (which can be >1 km) that onlap down into the axial trough with the platforms which generally, but not always, parallel the inboard shoreline. Platforms have a ribbon-like geometry and can eventually be buried by siliciclastics from the relentlessly advancing thrust sheets. The late Paleozoic Cordillera foreland basin in western North America and the Cenozoic foreland basin of the Alps are good examples.

Extensional and compressional

Salt diapirs. Platforms may be located on rising salt diapirs (Figure 1.11) and have ameboid to arcuate shapes. The internal geometry is generally complex and caused by a combination of the rise and fall of the diapir and subsurface salt dissolution. The sequences are generally shallowing upward, aggradational, and relatively thin (<500 m). The Persian Gulf off Abu Dhabi is an area with numerous such platforms.

Deltas. Platforms are typically irregular in this situation because of the complex interrelationships with siliciclastic deposition. They usually have an arcuate or lenticular shape and limited stratigraphic thickness. Platforms can coexist with coarse and intermittent siliciclastic sedimentation in arid settings. Only carbonate benthic communities that can tolerate fine sediment can thrive under humid climates that produce continuous fine-grained siliciclastic sediment. Good examples are the Mesozoic of Sicily and Pyrenees of Spain together with the Holocene of Borneo.

How do we study carbonate sediments and rocks?

To fully understand the sedimentology and diagenesis of carbonate rocks the researcher must be a natural scientist who is willing to use the main tools of stratigraphy, paleontology, petrography, and geochemistry to get at the answers.

It is probably a good idea to have several excellent books on paleontology at hand, particularly those that discuss not only the taxonomic attributes but also the ecological aspects of marine organisms. These books are very useful when working at the outcrop scale and searching for macrofossils.

There is no question, however, that the real understanding of the rocks comes from thin-section petrography. There are several excellent books (see the Further reading list) that illustrate microscopic attributes of the main rock-forming components and their alteration. Most workers have such books open and at hand near the microscope when describing and interpreting thin-sections. A key skill is being able to recognize organisms from random two-dimensional cuts through their skeletons.

Diagenetic aspects of the rocks can be greatly enhanced through the application of additional techniques. Stains are particularly useful for differentiating calcite and dolomite and for visualizing the iron contents of precipitates. Some trace elements, especially manganese, cause calcites and dolomites to luminesce when bombarded with electrons. Cathodoluminescence (CL) is particularly useful for determining the precipitation and alteration history of crystals and skeletons. Some features are just too small to be revealed by standard petrography and so scanning electron microscopy (SEM) of either broken or polished and etched surfaces can be extremely helpful.

Geochemistry is also a required research tool. The most useful trace elements are Sr, Ba, Fe, Mg, and Mn. Strontium is especially good for tracing the diagenetic evolution of young sediments and unraveling the genesis of dolomite. Sr isotopes have also been used for dating of carbonate rocks (especially in the Tertiary). Carbon and oxygen stable isotopes, because they are so abundant in carbonates and in the waters from which they precipitated, are indispensible in diagenetic studies. Most of these techniques and their utility are explained in Part III of the book.

Further reading

Bosence, D. (2005) A genetic classification of carbonate platforms based on their basinal and tectonic settings in the Cenozoic. *Sedimentary Geology*, 175, 49–72.

Crevello, P.D., Wilson, J.S., Sarg, J.S., and Read, J.F. (1989) *Controls on Carbonate Platform and Basin Development*. SEPM (Society for Sedimentary Geology), Special Publication no. 44.

Demicco, R.V. and Hardie, L.A. (1994) *Sedimentary Structures and Early Diagenetic Features of Shallow Marine Carbonate Deposits*. SEPM (Society of Sedimentary Geology), Atlas no. 1.

Einsele, G. (2000) *Sedimentary Basins: Evolution, Facies and Sediment Budget*. Second edition. Berlin, Heidelberg: Springer-Verlag.

Ginsburg, R.N. (ed.) (2001) *Subsurface Geology of a Prograding Carbonate Platform Margin, Great Bahama Bank; Results of the Bahamas Drilling Project*. Tulsa, OK: SEPM (Society for Sedimentary Geology), Special Publication no. 271.

James, N.P., Kendall, A.C., and Pufahl, P.K. (2010) Introduction to biological and chemical sedimentary facies models. In: James N.P. and Dalrymple R.W. (eds) *Facies Models 4*. St John's Newfoundland & Labrador: Geological Association of Canada, GEOtext 6, 323–339.

Lukasik, J. and Simo, J.A. (2008) Controls on development of carbonate platforms and reefs. In: Lukasik J. and Simo J.A.T. (eds.) *Controls on Carbonate Platforms and Reef Development*. SEPM (Society for Sedimentary Geology), Special Publication no. 89, pp. 15–43.

Mutti, M., Piller, W.E., and Betzler, C. (eds) (2010) *Carbonate Systems during the Oligocene–Miocene Climatic Transition*. Wiley-Blackwell, International Association of Sedimentologists, Special Publication no. 42.

Pomar, L. (2001) Types of carbonate platforms; a genetic approach. *Basin Research*, 13, 313–334.

Read, J.F. (1985) Carbonate platform facies models. *American Association of Petroleum Geologists Bulletin*, 69, 1–21.

Renaut, R.W. and Gierlowski-Kordesch, E.H. (2010) Lakes. In: James N.P. and Dalrymple R.W. (eds.) *Facies Models 4*. St John's, Newfoundland: Geological Association of Canada, GEOtext 6, pp. 541–576.

Simo, J.A.T., Scott, R.W., and Masse, J.-P. (eds) (1993) *Cretaceous Carbonate Platforms*. American Association of Petroleum Geologists, Memoir no. 56.

Swart, P.K., Eberli, G.P., and McKenzie, J.A. (eds) (2009) *Perspectives in Carbonate Geology. A Tribute to the Career of Robert Nathan Ginsburg*. Wiley-Blackwell, International Association of Sedimentologists, Special Publication no. 41.

Tucker, M.E., Wilson, J.L., Crevello, P.D., Sarg, J.R., and Read, J.F. eds (1990) *Carbonate Platforms: Facies, Sequences and Evolution*. Oxford: Blackwell Scientific Publications, International Association of Sedimentologists, Special Publication no. 9.

CHAPTER 2
CARBONATE CHEMISTRY AND MINERALOGY

Frontispiece Calcite crystals sitting on white sparry saddle dolomite, Devonian, Pine Point lead-zinc mine, Northwest Territories, Canada; image width 10 cm.

Origin of Carbonate Sedimentary Rocks, First Edition. Noel P. James and Brian Jones.
© 2016 Noel P. James and Brian Jones. Published 2016 by John Wiley & Sons, Ltd.
Companion website: www.wiley.com/go/james/carbonaterocks

Introduction

Carbonate sediments can accumulate wherever waters are supersaturated with respect to $CaCO_3$ and there is not much siliciclastic sediment input. Although forming largely in oceans, calcium carbonate also accumulates in lakes, rivers, around springs, in soils, and in caves. Once formed, carbonates can be extensively modified as recrystallization, replacement, dissolution, or cementation takes place as they are progressively emplaced into different diagenetic realms. In all situations, the principal control on limestone formation and diagenesis is chemistry, which appears simple but is extraordinarily complex in detail. The minerals formed by precipitation are more varied than expected from simple solution chemistry considerations. Similarly, the crystal morphologies involved are incredibly diverse.

Chemistry

The precipitation and dissolution of carbonate is governed by the following equation (Figure 2.1):

$$CaCO_3 + CO_2 + H_2O \leftrightharpoons Ca^{2+} + 2HCO_3^- \quad (2.1)$$

This general equation is, however, the result of four equilibria that relate the dissolution and precipitation of $CaCO_3$ in water to the input and extraction of CO_2.

$$CO_2 + H_2O \leftrightharpoons H_2CO_3 \quad (2.2)$$

This hydration reaction is relatively slow but is followed by the instantaneous reaction:

$$H_2CO_3 \leftrightharpoons H^+ + HCO3^- \quad (2.3)$$

Figure 2.1 The classic reversible carbonate reaction wherein solid carbonate, when reacting with water and carbon dioxide, can either dissolve or precipitate. The carbon dioxide and water typically combine to form carbonic acid (H_2CO_3), for example acid rain.

If the solution contains free CO_3^{2-} then the proton released in reaction 2 quickly reacts with the carbonate ion to give more bicarbonate:

$$H^+ + CO_3^{2-} \leftrightharpoons HCO_3^- \quad (2.4)$$

At the interface between water and solid $CaCO_3$, the equilibrium is:

$$CaCO_3 \leftrightharpoons Ca^{2+} + CO_3^{2-} \quad (2.5)$$

If these equilibria move to the right, both CO_2 and $CaCO_3$ are dissolved. If they move to the left, CO_2 is removed and $CaCO_3$ is precipitated (Figure 2.1).

Equation (2.1) is reversible in the sense that solid carbonate can be dissolved or precipitated; in reality however, because of crystal kinetics, dissolution is generally easier than precipitation. The situation is at equilibrium when both sides of equation (2.1) are satisfied, that is, the fluid cannot dissolve or precipitate carbonate. In most situations, CO_2 and H_2O combine to form weak carbonic acid (H_2CO_3). The real control on the reaction, however, is the addition or removal of CO_2. Adding CO_2 to the fluid will drive the reaction to the right causing dissolution until the solution reaches equilibrium; vice versa, if CO_2 is driven off, the reaction will be driven to the left and carbonate should precipitate. Temperature, pressure, and biology are all important in the movement of CO_2.

As temperature increases, CO_2 will bubble off (Boyle's Law) and precipitation will occur. For example, solar heating of a shallow saline pond makes precipitation easier. Conversely, cooling means the water can hold more CO_2 and thereby mediate carbonate dissolution by a process known as retrograde solubility. This seems to be against common expectation. For example, when you heat water you can usually dissolve a solute (e.g., sugar); in the carbonate system, the opposite is true! Pressure has a similar effect. Increasing the pressure means that the water can contain more CO_2 and thus promote dissolution, whereas decreasing the pressure results in escape of the gas. Opening a can of soda pop lowers the pressure and causes the contained CO_2 to bubble off.

This latter effect is most spectacular in carbonate springs and caves. When waters saturated with carbonate emerge as springs at the Earth's surface, the drop in pressure and temperature results in degassing and rapid precipitation of a variety of carbonate minerals. Likewise, when carbonate-charged waters high in CO_2 emerge into caves, they can degas and form a wide variety of carbonate precipitates (speleothems).

CO_2 levels are also modified by resident biota in many different ways. Shallow-marine light-dependent organisms (phototrophs, e.g., cyanobacteria, sea grass, green algae) can, for example, induce carbonate precipitation as they remove CO_2 from seawater during photosynthesis. In contrast, water percolating through soil will absorb CO_2 generated by soil microbes and become more acidic, thus capable of dissolving more of the underlying carbonate.

Mineralogy

From a chemical and material science perspective, there are six polymorphs of $CaCO_3$. The three anhydrous polymorphs are calcite (hexagonal), aragonite (orthorhombic), and vaterite (hexagonal), whereas the two hydrous forms are monoclinic ikaite ($CaCO_3.6H_2O$) and calcium carbonate monohydrate ($CaCO_3.H_2O$). Amorphous calcium carbonate (ACC – $CaCO_3.nH_2O$) can also occur. From a geological perspective, calcite and aragonite (Frontispiece, Figure 2.2) dominate the biogenic and abiogenic components of carbonate sediment. Vaterite, which is metastable with respect to calcite, is rare. Ikaite is associated with cold water (generally –3° to +3°C) whereas calcium carbonate monohydrate is very rare. ACC, which is formed of aggregates of small spheres (<1 μm diameter), has been found in some skeletons (e.g., echinoids,

corals), in some sediments, and in some precipitates associated with microbial mats in hot spring systems. ACC is highly unstable and rapidly transforms to either aragonite or calcite.

$CaCO_3$ polymorphs exist because of the size of the Ca^{2+} ion, which is 0.98 Å. Hexagonal crystals with six-fold metal-CO_3 co-ordination are the stable form if the ionic radius of the metal is <0.99 Å whereas orthorhombic crystals with 9-fold co-ordination are the norm if the metal ionic radius is >0.99 Å (Figure 2.3). Fe^{2+} and Mg^{2+} carbonates are all hexagonal because their ionic radius is much less than 0.99 Å, whereas Sr^{2+} carbonates are orthorhombic because it is larger. Ca^{2+} is so near the boundary that slight changes in precipitation kinetics can shift it one way or the other from hexagonal to orthorhombic.

According to thermodynamic theory, which is based largely on inorganic laboratory experiments, calcite is the only stable phase that should be precipitated from modern marine and fresh waters. That aragonite is commonly a major component of modern carbonate sediments can be attributed to the following factors:

- The Mg:Ca ratio of most seawater is ~3:1. Although Ca is less abundant than Mg, $CaCO_3$ is thermodynamically favored because Mg has a smaller ionic radius than Ca and is more tightly bonded to the O–H radical and therefore difficult to separate. In many settings the

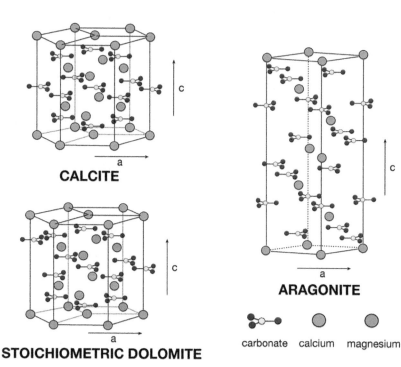

Figure 2.2 The chemical crystallography of calcite, aragonite, and dolomite.

CRYSTAL CHEMISTRY

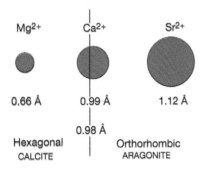

Figure 2.3 A simplified diagram illustrating the ionic radius of some important cations in carbonate minerals. The black line at 0.98 Å separates those cations that can fit into the hexagonal crystals structure (e.g., Mg^{2+}) as compared to those that can fit into the orthorhombic crystal structure (e.g., Sr^{2+}).

system locally defaults to the polymorph aragonite, a mineral that cannot incorporate much Mg^{2+} into the lattice because the ionic radius is too small to fit easily into the orthorhombic structure.

- Many plants and animals incorporate aragonite rather than calcite into their calcareous skeletons, apparently in defiance of thermodynamic principles! In these cases, precipitation takes place through metabolically mediated processes that operate within the plants and animals.

The other major mineral of marine sediments is magnesium calcite (Figure 2.4). Dehydration of some Mg ions allows incorporation of some of the Mg into the calcite, albeit in an irregular, random fashion that causes distortion of the crystal cell that, in turn, produces curved crystal faces. This magnesium calcite is an important component of many sediments. Although the teeth of some echinoids (Aristotle's Lantern) can contain up to 43 mole % MgCO$_3$, most Mg-calcite contains <18 mole % MgCO$_3$. The amount of Mg in non-skeletal Mg-calcite is a function of precipitation rate that, in turn, is dependent on other factors such as temperature. For example, the higher the water temperature, the higher the amount of Mg. These calcites are arbitrarily separated into low-magnesium calcite (LMC = 0–4 mole %), intermediate magnesium calcite (IMC = 4–12 mole %) and high-magnesium calcite (HMC = 12–18 mole %). Some authors refer to all calcite with <4 mole % as LMC and all above as HMC. LMC can occur as marine LMC (mLMC), especially in biogenic components and precipitates such as cements and ooids. In contrast, diagenetic LMC (dLMC) is generally precipitated in cold ocean waters, meteoric waters, or burial fluids.

CARBONATE MINERALOGY

1. **CALCITE** CaCO$_3$ *Hexagonal*
2. **MAGNESIUM CALCITE** CaCO$_3$ *Hexagonal* 4 - 18 mole % MgCO$_3$
3. **ARAGONITE** CaCO$_3$ *Orthorhombic*
4. **DOLOMITE** CaMg(CO$_3$)$_2$ *Hexagonal*

Figure 2.4 Crystallography and chemical composition of the four most common minerals found in carbonate sediments and carbonate rocks.

Dolomite (CaMg(CO$_3$)$_2$, Frontispiece, Figures 2.2, 2.4) has an hexagonal crystal structure and, when stochiometric, contains equal proportions of calcium and magnesium. In an ideal dolomite, the crystal structure is highly organized (ordered) with repeated stacked layers of Ca, CO$_3$, and Mg. In these ideal structures, the Ca and Mg are in discrete layers. With the high Mg:Ca ratio of seawater, it might be expected that dolomite should be precipitated instead of calcite or aragonite. This is, however, largely prevented by hydration. During crystal precipitation, both Ca–OH and Mg–OH radicals are attracted to the nucleation site but it is easier to dehydrate the Ca–OH radical and so CaCO$_3$ forms.

Calcite and aragonite occur in many different crystal forms. According to some studies, there are over 700 known varieties in calcite crystal forms. Although only about six forms are common, some of the more bizarre varieties of calcite include: (1) dendrite crystals that are characterized by numerous levels of branching (like a tree); and (2) skeletal crystals that are hollow (not to be confused with the crystals that form the skeletons of animals and plants). Variation in crystal morphology appears to be linked to the rate at which the crystals grow. Hence, precipitation of many of these bizarre crystal forms is commonly linked to waters that are highly supersaturated with CaCO$_3$. For example, rapid degassing of spring waters commonly elevates saturation levels to the point where rapid precipitation of crystals with bizarre crystal forms (e.g., dendrites) follows.

Relative solubility

Carbonate minerals exhibit different solubilities. Relative solubility is a function of several variables, namely crystal size, crystal heterogeneity, crystal defects, internal porosity and permeability, organic matrices, and mineralogy.

CARBONATE MINERAL RELATIVE SOLUBILITY

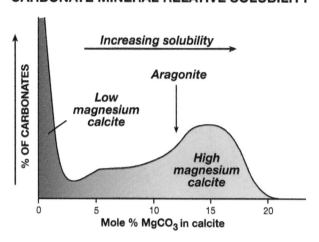

Figure 2.5 Sketch showing the relative solubility in fresh water of calcite with increasing Mg-content; aragonite has about the same solubility as 12 mole % $MgCO_3$ calcite.

Of these, mineralogy is probably the most important (Figure 2.5). Calcite solubility, for example, is a function of Mg content. Increasing amounts of Mg in the calcite lattice deform the crystal, making it progressively less structurally stable. Thus, the solubility of Mg-calcite increases with increasing Mg^{2+} content. Aragonite has about the same solubility in distilled water as HMC with roughly 12 mole % $MgCO_3$. Dolomite is the least soluble of all carbonates.

Carbonate precipitation and dissolution in the ocean

Seawater contains ~34.4 g of total dissolved solids (TDS) per kilogram of water (Figure 2.6). The dominant elements are, in order of abundance: chloride, sodium, sulfate, magnesium, bicarbonate, calcium, and potassium. Today, normal seawater has a salinity (total dissolved solids) of ~3.5% or, as it is more commonly expressed, ~35 parts per thousand (35‰).

Carbonate: evaporite precipitation

Carbonate precipitation from seawater will take place when the water becomes supersaturated with respect to either aragonite or calcite. This may be achieved through passive enrichment in dissolved solids whereby evaporation causes the loss of water without any loss of the dissolved solids. Although complex in detail, the precipitation of minerals from seawater via evaporation is generally as follows: (1) calcite precipitation after the loss of 50%

SEAWATER

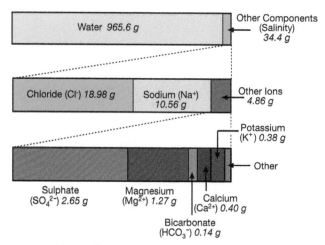

g = grams kilogram^{-1}

‰ = parts per thousand ‰

Figure 2.6 The most abundant components in 1 kg of seawater of 35‰ salinity; ions are represented in grams per kilogram, the same as parts per thousand (‰).

seawater; (2) gypsum precipitation after the loss of 66% of seawater; (3) halite precipitation after the loss of 90% of seawater; and (4) precipitation of bitter salts (K, Mg salts) after the loss of 95% of seawater (Figure 2.7).

Organic influences

Organisms have a profound effect on $CaCO_3$ precipitation. This is especially true in the marine realm where the resident microbes, algae, and animals can influence precipitation in two ways.

Biologically induced mineralization. This precipitation, mediated mostly by algae and microbes, takes place as their metabolic byproducts interact with the surrounding seawater and produce supersaturated conditions that lead to precipitation. It is difficult to precisely delineate the processes involved because this happens at a microscale. Although precipitated directly from seawater, the conditions leading to such precipitation are induced by the metabolic activity of the organism.

Biologically controlled mineralization. Vast amounts of $CaCO_3$ are locked into the skeletons of animals and plants that directly control precipitation. Organisms precipitate the minerals that best suit their needs at the time when they evolved. Some organisms precipitate pure aragonite skeletons (e.g., gastropods, modern

CONCENTRATION OF SEAWATER BY EVAPORATION

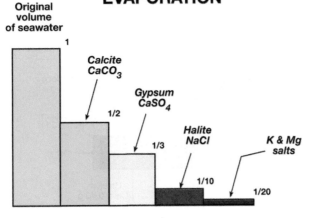

Figure 2.7 A histogram illustrating the various minerals that precipitate when seawater is gradually evaporated (e.g., gypsum precipitates when only 1/3 of the original seawater is left).

SEA WATER -

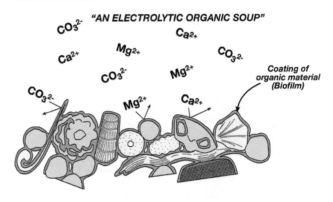

Figure 2.8 A sketch illustrating that all particles in seawater are coated with organic molecules that somewhat inhibit carbonate precipitation on their surfaces or, in some cases, promote such precipitation.

corals), others opt for HMC (e.g., benthic foraminifers, echinoderms), whereas a few precipitate LMC skeletons (e.g., brachiopods, planktic foraminifers). Some bivalves construct shells that are formed of alternating layers of aragonite and LMC. Irrespective of composition, the skeletal $CaCO_3$ crystals are typically permeated by organic material, and many crystallites are morphologically unlike inorganic precipitates.

Seawater is often described as "an electrolytic organic soup" (Figure 2.8) rather than just water because of the innumerable organic compounds (biofilms) found in seawater. Although these biofilms commonly influence $CaCO_3$ precipitation, they can also inhibit precipitation or dissolution by coating particles and thus sealing them off from the surrounding waters. Such is the dichotomy of processes in the carbonate realm.

Marine carbonate dissolution

For many years it was thought that carbonate dissolution in the ocean was a product of increasing cold and increasing pressure in the deep ocean. Recent studies, however, have confirmed that dissolution, promoted by microbial processes, also takes place in shallow-marine waters just below the sediment–water interface (Chapter 24). The extent of this dissolution process is only now being assessed.

What is not disputed, however, is that the combination of changing temperature and pressure has a profound effect on carbonates in the ocean. With increasing depth, the ever-decreasing temperature and increasing pressure

means that more carbonate can be dissolved and held in solution. From the perspective of carbonate sediments, two depths are important. The *lysocline* is the depth at which carbonate starts to dissolve and the *compensation depth* (Figure 2.9) is the depth where the water is so cold and the hydrostatic pressure is so high that all the carbonate has disappeared. If, however, the deep oceanic waters are later brought to the surface by processes such as upwelling, then as P decreases and T increases, particularly quickly in the shallow tropics, carbonate minerals can precipitate.

It is not only physical parameters that change CO_2. These levels in oceanic carbonate environments are also modified by the resident biota in many different ways.

Dissolution–precipitation zones

Each carbonate mineral, aragonite, LMC, and HMC has its own lysocline and compensation depth (Figure 2.9). Aragonite is more soluble than LMC and therefore starts to dissolve at higher T and lower P than LMC. Accordingly, the aragonite lysocline and aragonite compensation depth (ACD) are shallower than the calcite lysocline (LMC) and calcite compensation depth (CCD) (Figure 2.9). The factors that control the depths of the lysoclines and compensation depths are so complex that they vary from ocean to ocean in accordance with the physical and chemical processes that are operative in each region. Today, for example, the CCD is at a depth of ~5500 m in the Atlantic Ocean but at only 3500 m in the Pacific Ocean. It is hard to identify paleo-lysoclines and paleo-compensation depths in ancient

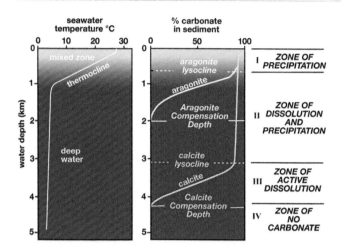

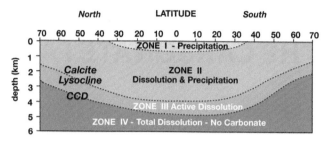

Figure 2.10 A generalized cross-section through the ocean from north to south illustrating the position of the various zones depicted in Figure 2.9. Source: Adapted from James and Choquette (1983). Reproduced with permission of the Geological Association of Canada.

Figure 2.9 Generalized profiles of (left) the change in seawater temperature with depth in the ocean and (right) the percentage of carbonate on the seafloor with increasing ocean depth. The zones of diagenesis are plotted on the right. Source: Adapted from James and Choquette (1983). Reproduced with permission of the Geological Association of Canada.

successions because diagenesis can mask the features that are used to define those depths.

In combination, the depths of the aragonite and calcite lysoclines and compensation depths effectively divide modern oceans into four depth-defined zones (Figure 2.10) as follows:

1. *Zone of Precipitation*: from ocean surface to aragonite lysocline, where both calcite and aragonite can be precipitated; this is divided on the basis of seawater temperature into the warm-water and cool-water depositional realms;
2. *Zone of Dissolution and Precipitation*: largely between the aragonite lysocline and ACD, where aragonite is dissolved but calcite can be precipitated;
3. *Zone of Active Dissolution*: between the calcite lysocline and CCD, where calcite is being actively dissolved; and
4. *Zone of No Carbonate*: which is beneath the CCD.

Only clay and siliceous ooze will be found on any seafloor below the compensation depth. Critically, these zones provide a broad-scale template for the distribution of aragonite and calcite in the oceans. It is important to realize that the decrease of temperature with depth in the tropical oceans parallels, to some degree, the latitudinal decrease in surface-water temperature from the equator to the polar regions of the Earth. The zone of precipitation is complex because neritic environments are areas of extensive precipitation in tropical latitudes, whereas in temperate latitudes neritic carbonate precipitation is much reduced in intensity and extent. Precipitation in polar latitudes is confined to neritic environments. Deeper-water carbonates are quickly dissolved because of the frigid water (Figure 2.10).

Further reading

Arvidson, R.S., Collier, M., Davis, K.J., Vinson, M.D., Amonette, J.E., and Lüttge, A. (2006) Magnesium inhibition of calcite dissolution kinetics. *Geochimica et Cosmochimica Acta*, 70, 583594.

Broecker, W.S. (2003) The oceanic CaCO$_3$ cycle. In: Elderfield H. (eds) *The Oceans and Marine Geochemistry*. Amsterdam: Elsevier, Treatise in Geochemistry Vol. 6, pp. 529–549.

Garrison, T. (2004) *Essentials of Oceanography*. California, USA: Brooks/Cole-Thompson Learning, Pacific Grove.

James, N.P. and Choquette, P.W. (1990) Limestone: The sea floor diagenetic environment. In: McIlreath I. and Morrow D. (eds.) *Diagenesis*. St John's, ND: Canada Geological Association of Canada Reprint Series 4, pp. 13–34.

Meldrum, F.C. (2003) Calcium carbonate in biomineralisation and biomimetic chemistry. *International Materials Review*, 48, 187–224.

Morse, J.W. (1985) Kinetic control of morphology, composition and mineralogy of abiotic sedimentary carbonates. *Journal of Sedimentary Petrology*, 55, 919–934.

Morse, J.W. and Arvidson, R.S. (2002) The dissolution kinetics of major sedimentary carbonate minerals. *Earth-Science Reviews*, 58, 51–84.

Reeder, R.J. (ed.) (1983) *Carbonates: Mineralogy and Chemistry*. Mineralogy Society of America, Reviews in Mineralogy Vol. 11.

CHAPTER 3
THE CARBONATE FACTORY

Frontispiece Reef dominated by acroporid corals, ~18 m water depth, Fakarava, an atoll in the French Tuamotus. Photograph by Dale Stokes. Reproduced with permission.

Origin of Carbonate Sedimentary Rocks, First Edition. Noel P. James and Brian Jones.
© 2016 Noel P. James and Brian Jones. Published 2016 by John Wiley & Sons, Ltd.
Companion website: www.wiley.com/go/james/carbonaterocks

Introduction

The carbonate sediment factory (Frontispiece) is a system that involves the production and early modification of carbonate sediments. It is truly amazing that the dissolved carbonate in freshwater and seawater can be transformed into such a plethora of different products. Production entails the biological and chemical precipitation of carbonate crystals, particles, and cements. Modification comprises a suite of synsedimentary biological, chemical, and physical processes that alter the particles before they are buried and become part of the stratigraphic record.

There are really two principal aqueous factories: the benthic (seafloor and lake floor) factory; and the pelagic (water column) factory (Figure 3.1). The major benthic factories are the lacustrine, intertidal, lagoon, subtidal, ooid sand shoal, and reef environments. Deposits produced by the benthic factory generally accumulate in place, although they can be retextured by waves or currents or periodically swept onshore or offshore into deeper water. The pelagic factory produces carbonate particles, mostly fine-grained, in the near-surface parts of the water column. These particles, in the form of calcareous phytoplankton and zooplankton, are most important in post-Jurassic seas. The skeletons and skeletal fragments rain down upon the seafloor and are most abundant in deep-sea or basinal

settings, where the benthic factory is not very productive. A third less-important factory is the subterranean and terrestrial realm where carbonates are produced as cave precipitates, within soils, and at the surface where carbonate-charged waters emerge as springs.

Sediment production

Sediments in the carbonate factory (Figure 3.2) are produced in a variety of ways by the precipitation of calcium carbonate out of water either directly, via organic compounds associated with microbes, bacteria, and algae (microbialites), as parts of invertebrate animals (Figure 3.3), or as byproducts of animal metabolic activities that have evolved through time.

Sediment particles

Sediment particles range in size from silt to boulder and are either sediment grains or large in-place skeletons such as modern reef corals. The dominant components (allochems) include microbialites, skeletons, ooids, coated grains, peloids, and intraclasts.

Microbialites. Microbes are active in almost every environment, forming all sorts of particles and structures as well as influencing carbonate precipitation (for details see Chapter 5). The most obvious microbial structures are stromatolites. These laminated, commonly domal or columnar structures, are particularly common in Archean and Proterozoic carbonates but are restricted to stressed environments in the Phanerozoic (Chapter 5). Photosynthetic microbes also form sediment-microbe mats in peritidal or lake margin environments and oncolites (microbial nodules) in tranquil subaqueous settings. Calcification of the microbe cells results in calcimicrobes that in turn form sediment grains and can coalesce to construct reefs. Microbes are also influential in the production of carbonate mud and synsedimentary cements.

Skeletons. Skeletal components are derived from calcareous and siliceous sheaths, shells, tests, and spicules of the myriad sessile, burrowing, drifting, or swimming invertebrates, and algae that live in shallow and deep water across the aqueous environmental spectrum (Figure 3.4). These components (Figure 3.5a) are evident as entire skeletons (Figure 3.2b) or as biofragments (Figure 3.5b–c). Entire skeletons, ranging from tiny foraminifers to large corals that can be up to 3 m in diameter, are commonly preserved intact. Biofragments

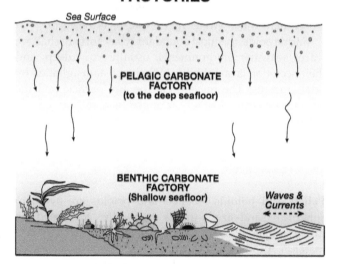

CARBONATE SEDIMENT FACTORIES

Sea Surface

• PELAGIC CARBONATE FACTORY (to the deep seafloor)

BENTHIC CARBONATE FACTORY (Shallow seafloor)

Waves & Currents

Figure 3.1 A sketch depicting the two major carbonate-sediment-producing factories: the pelagic factory dominated by calcareous plankton that rain down on the seafloor and the benthic factory composed of both biotic and abiotic carbonate components.

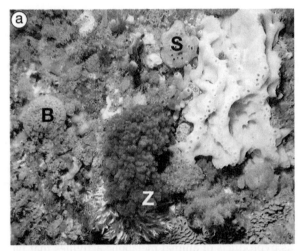

Figure 3.2 (a) A seafloor image of prolific marine growth at a depth of ~15 m off western Victoria, southern Australia. (S) sponge; (B) bryozoans; (Z) zoanthid. Image width 0.5 m. (b) A seafloor image of active reef growth at a depth of ~3 m off Bermuda. (C) coral; (H) hydrozoan (*Millepora* sp.); (G) gorgonian (soft coral with spicules); (O) soft oncolite (algal biscuit). Image width 1 m. Photograph by W. Martindale. Reproduced with permission. (c) Rippled seafloor of ooid sand in ~2 m of water on the Bahama Banks. The 20-cm-wide echinoderm in the foreground is an asteroid. Photograph by H. Wanless. Reproduced with permission.

are produced by: (1) the disintegration of skeletal components as the organic tissue that originally bound them together is lost to decay (e.g., calcareous algae, crinoids (see Figure 3.6), and bivalves); (2) biobreakage whereby various types of animals fragment the skeletons (e.g., crabs that break bivalve shells in order to eat the soft tissues inside); or (3) physical breakdown during high-energy events (e.g., storms). In general, skeletal disintegration and biobreakage produces most biofragments. Physical breakage is generally limited to beaches or devastation associated with cyclonic storms.

A fundamental difference between carbonate and siliciclastic particles is that the grain size of carbonate sediment is primarily a reflection of skeletal architecture rather than hydrodynamic conditions (as in siliciclastic sediments). A deposit of clean crinoid sand can, for example, be formed in deep water where currents are weak to non-existent, simply because of the *in situ* disintegration of their skeletons following death.

Ooids and pisoids.   Ooids and pisoids are coated grains characterized by a nucleus that is surrounded by a laminated cortex (Figure 3.7). Ooids are spherical to subspherical grains <2 mm in diameter, whereas pisoids are similar but generally >2 mm in diameter. This size division, equal to the upper limit of sand on the Wentworth Grain Scale, is somewhat artificial because many of these coated grain deposits can include grains with a size range that spans this boundary. Ooids and pisoids can form in marine environments, lakes, calcareous soil profiles, spring pools, discharge aprons, subterranean cave pools (cave pearls), water filtration plants, and even in a domestic kettle if it is used in an area where 'hard water' (water rich in $CaCO_3$) is the norm.

Marine ooids, the most abundant types, form in shallow, warm seawater environments, usually <6 m deep, and their accretion is largely due to their constant agitation by tidal currents. The nucleus can be any particle and the cortex is either tangentially arranged (generally aragonite; Figure 3.7) or radially disposed (generally Mg-calcite; Figure 3.8) crystallites. Cave ooids, lacustrine ooids, and pedogenic (soil) ooids are described in Chapters 8, 9 and 26 respectively.

Pellets and peloids.   Pellets and peloids are ovoid to spherical sand-sized grains generally composed of microcrystalline carbonate (Figure 3.7d). Pellets are fecal material produced by deposit and detritus feeding organisms that ingest mud and extract organic matter from that mud as they process the sediment through their guts, eventually excreting the non-digested mud in neatly packaged sand-sized particles. Sea cucumbers (Holothurians), for example, act like vacuum cleaners as they indiscriminately suck up sediment and process

BENTHIC CALCAREOUS ORGANISMS

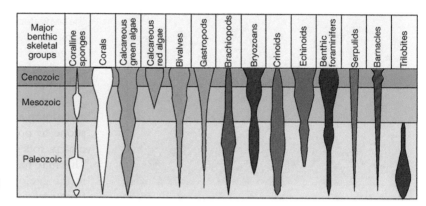

Figure 3.3. A plot of the relative importance of different calcareous organisms through geologic time; the width of the balloons is a general representation of their contribution to calcareous sediments. Source: Adapted from Wilkinson (1979). Reproduced with permission of the Geological Society of America.

SKELETAL GRAINS

• ALGAE & MICROBES   • ANIMALS

- *Microbes* (P & B)
- *Calcareous algae* (P & B)

- *Foraminifers* (P & B)
- *Molluscs* (P & B)
- *Coralline sponges* (B)
- *Sponges* (B)
- *Bryozoans* (B)
- *Echinoderms* (B)
- *Corals* (B)
- *Brachiopods* (B)
- *Trilobites* (B)

P = *pelagic*; B = *benthic*

Figure 3.4 A list of the major pelagic and benthic microbes, algae, and invertebrates that form most carbonate sediments throughout geologic history.

and extract organic materials in their intestinal tracts. Pellets can also be produced by rasping herbivores that ingest parts of the substrate that they are scouring and by filter feeders that eject the non-consumable carbonate parts of their prey. Although most commonly associated with benthic animals, some planktic organisms can also produce such pellets. Zooplankton that feed on calcareous phytoplankton, for example, produce pellets which are transported to the lake floor or seafloor that can be many kilometers below, to be preserved as pelagic carbonate.

Peloids have the same appearance as fecal pellets but are produced by: (1) the diagenetic alteration of other grains (biofragments, ooids; see Chapter 24); or (2) the spontaneous precipitation of Mg-calcite in synsedimentary cavities (in reefs) and in cave pools.

Aggregate grains and intraclasts.   Carbonate particles of all types, especially in tropical marine or saline lacustrine environments, can be cemented together by aragonite or Mg-calcite to form grain aggregates and intraclasts.

Aggregate grains are composed of biofragments, pellets, or ooids that are bound together by biofilms or encrusting organisms, or cemented together by aragonite or Mg-calcite. *Grapestone*, for example, refers to aggregates of ooids or pellets that are bound together to form grape-like clusters.

Intraclasts (intra = inside; clast = fragment), a term used to designate clasts formed in the environment of deposition (Figure 3.9), are formed as storm waves, strong tidal currents associated with channel avulsion, or seismic events break up sediments that have been partly or completely lithified on the seafloor. Other intraclasts are produced by erosion of lithified layers or muds that have been hardened and cemented during exposure on lakeshore, intertidal, and supratidal flats. These "rip-up clasts" are typically disc-shaped with their length and width being greater than their thickness. Fragments derived from limestone outside the depositional basin are called extraclasts.

Carbonate mud

Carbonate mud is not a product of erosion as in siliciclastic rocks, but instead is the result of a variety of organic and inorganic precipitation processes together with biological abrasion. The material is microcrystalline calcite or aragonite, generally shortened to *micrite* – a term that is roughly synonymous with *lime mud*. The official crystal size range is 1 and 4 μm, with material <1 μm called *minimicrite*. There are two types of micrite: (1) allomicrite, a sediment; and (2) automicrite, an in-place precipitate.

Allomicrite.   This mud is formed by: (1) the precipitation of carbonate crystals in the water column and subsequent fallout; (2) the disintegration of calcareous algae and invertebrate skeletons; (3) the byproducts of bioerosion; (4) secretions by fish; and (5) the accumulation of calcareous plankton.

Water column precipitation. When flying over many modern shallow-water tropical marine carbonate platforms, especially the Bahama Banks, it is common to see *whitings*

Figure 3.5 (a) Beach sand from southern Bermuda displaying a variety of abraded and somewhat rounded particles derived from the immediate offshore. Red grains are the encrusting benthic foraminifera *Homotrema rubrum*. This protist gives the sand a pink cast and thus the color "Bermuda pink". Image width 6 cm. Photograph by Catherine Reid. Reproduced with permission. (b) A monospecific accumulation of bivalves shells on a beach in the Chatham Islands, New Zealand. Image width 0.3 m. (c) Coarse sand from ~10 m water depth on the northwest shelf of Australia. The most numerous carbonate particles come from large benthic foraminifers (L), some of which are stained black, bryozoans (B), mollusks (M), and a variety of other smaller biofragments. Image width 6 cm.

(Figure 3.10a), clouds of milky water formed of innumerable micrometer-sized aragonite needles suspended in the seawater (Figure 3.10b). Although numerous interpretations have been proposed, the most current and compelling suggestion is that they are due to the induced precipitation of carbonate around picoplankton cyanobacteria cells. Similar precipitates are common in carbonate-rich lakes.

Disintegration and degradation. Calcareous green algae containing aragonite needles in their tissue are especially prone to post-mortem disintegration when the organic matrix rots and the micron-sized needles are released. This may have been a long-lasting source of marine mud in the geological record because green algae are known from rocks as old as the Ordovician. Delicate calcareous epibionts (both algae and invertebrates) that live on seagrass or algal blades can suffer the same fate once the grass or algae die.

The breakdown of delicate invertebrate skeletons is also promoted by the combination of organic matrix oxidation and microbial degradation (*maceration*). Many skeletons comprise carbonate crystallites surrounded by organic tissue (even such solid skeletons as bivalves). When this tissue rots the crystallites are freed to form part of the sediment.

Bioerosion. Carbonate components of all types ranging from large skeletons (e.g., corals) to sand-sized grains (e.g., ooids, shells) are commonly bored by microbes, algae, and various invertebrates (e.g., sponges, boring bivalves; Figure 3.11). This activity breaks down the substrates and usually results in the production of lime mud.

Teleost fish. Modern tropical marine fish produce and secrete various forms of fine precipitated (non-skeletal) carbonate in their guts. Such crystallites are <2 μm in size and, although calcite and aragonite are known, it is mostly HMC. It has been suggested that these crystals form a significant proportion of muds on the modern Bahama Banks. Given that fish first evolved in the Silurian, this may have been a significant source of mud throughout the Phanerozoic.

Calcareous plankton. There is also a profound contribution of carbonate mud from the marine pelagic realm. The main marine mud generators are pelagic green algae called coccoliths. Skeletons of these phytoplankton disintegrate into hundreds of micron-sized crystal platelets that produce copious amounts of mud which, when lithified, becomes chalk. Most of this fine material is seen in deep basinal settings because benthic mud production in shallower settings completely overwhelms the relatively minor contribution from such plankton.

Automicrite.   Automicrite is precipitated as aragonite or Mg-calcite cement on the seafloor or within sediment by inorganic and organically mediated processes

Figure 3.6 (a) A crinoid meadow at ~310 m water depth off northwestern Tasmania where each animal is ~1 m tall; each crinoid disintegrates into ~400 carbonate fragments. Reproduced by permission of CSIRO. (b) A coarse Silurian carbonate sand exposed on Anticosti Island, Canada, composed entirely of crinoid biofragments; such sands are a common component on many early and middle Paleozoic limestones. Image width 6 cm.

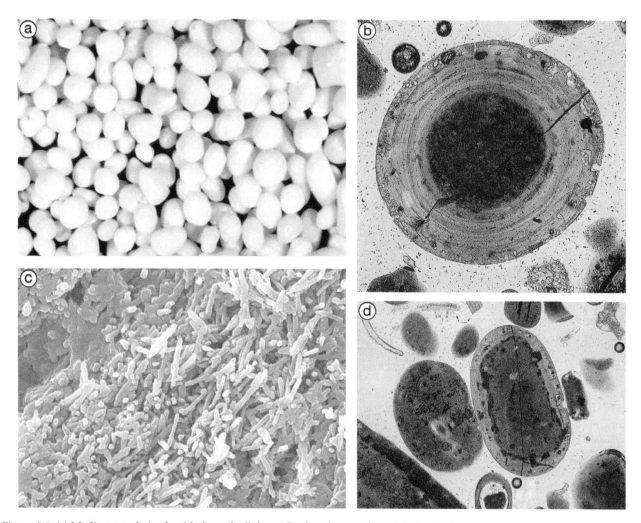

Figure 3.7 (a) Medium-sand-sized ooids from the Bahama Banks where each rounded grain is glistening because of recent carbonate precipitation. Image width 4 cm. (b) Thin-section photomicrograph (plane light) of an ooid from the northwest shelf of Australia. The nucleus is a carbonate peloid (likely a fecal pellet) that is surrounded by a cortex of tiny aragonite crystals, most of which are oriented tangentially to the nucleus. Holes at the margin are the result of boring microbes, likely cyanobacteria. Image width 0.6 cm. (c) Scanning electron microscope image of an ooid cortex such as that illustrated in (b), illustrating the aragonite needles. Image width 1.2 μm. (d) A thin-section photomicrograph (plane light) of an ooid with a thin cortex and peloid nucleus (right), often called a "superficial ooid", and a fecal pellet (left) composed of a variety of small carbonate fragments. Image width 0.8 cm.

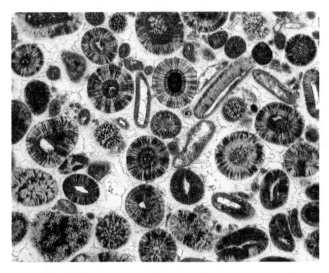

Figure 3.8 Middle Ordovician ooid grainstone in which most ooids have a radial cortex, Southern Ontario, Canada. Image width 8 mm.

Figure 3.9 Intraclasts of ooid sand ~5 cm in size, 2 m water depth Bahamas. Photograph by W. Martindale. Reproduced with permission.

(Figure 3.12). Much of this precipitation is associated with: (1) organic matrices (Ca-binding organic macromolecules); (2) metabolic processes of heterotrophic and chemotrophic bacteria and associated microbes; and (3) the metabolic processes of phototrophic cyanobacteria and algae. Although automicrite forms particulate sediment, it also occurs as peloidal, clotted, or laminated coatings around skeletons or lining the walls of cavities in biogenic mounds and reefs.

This microcrystalline carbonate is not obviously sediment, but neither is it obviously a synsedimentary cement.

The mud is nevertheless gravity-defying (i.e., it lines the roof of voids and cavities) and displays evidence of biological influence. It forms in the marine environment within growth cavities and within sediment. Automicrite is particularly common in reef mounds (especially mud mounds) and locally in stromatolites. It is related to biofilms (layers of complex organic molecules, also known as EPS or extracellular polysaccharides) and there is good evidence that it is the product of induced and supported organomineralization as well as the degradation of organism biopolymeric compounds (humification). Finally, automicrite is present in calcareous soils (calcrete) and lacustrine environments.

Ephemeral components

The carbonate factory also includes many important elements that have a low fossilization potential, namely microbes, marine plants, trees, and many different types of grass and algae (Figure 3.13). Grasses and algae shade the lake or ocean floor, reduce hydrodynamic energy by baffling the waves and swells, bind sediment in place, and serve as substrates for innumerable calcareous algae and invertebrates. Grasses (Figure 3.13a) are post-Early Cretaceous angiosperms and are restricted to sediment substrates in warm and warm-temperate marine environments. Their blades can be substrates for the growth of prolific calcareous epiphytes. Perhaps of more importance is the fact that the deep roots (rhizomes) of seagrasses (Figure 3.13b) limit significant erosion during cyclonic storms.

Marine macrophytes (Figure 3.13c) are red, brown, and green soft non-calcareous algae that thrive in cool- and cold-water environments. The most spectacular and arguably the most important are kelp, which are large brown macroalgae (phaeophytes) that commonly form marine forests in cool-temperate and cold, high-nutrient, high-energy, rocky seafloor environments. The dense, prolific growth of kelp reduces hydrodynamic energy. The holdfast (where the alga is attached to the rock) and the flexible blades can be encrusted by a spectrum of calcareous invertebrates and calcareous algae which contribute sediment when the alga dies. Although it is difficult to establish the age of the oldest significant kelp forests, it would seem to be middle Cenozoic.

Component modification

Once particles are formed and become part of the sediment, either as loose grains or as large in-place structures such as reefs, they can quickly be modified by a variety of biogenic, chemical, and hydrodynamic processes. Chemical alteration, generally manifested as cementation and dissolution, is detailed in Chapter 24.

Biogenic

Bioerosion. There are a host of organisms that bite off, drill into, and infest carbonate particles, thereby changing their composition and reducing them to even smaller grains (Figure 3.14). It appears that biobreakage is far more important than physical abrasion in marine carbonate systems. The two most important processes in this regard are feeding and boring.

Animals graze algae from hard substrates or bite off chunks of living coral. Herbivores, mainly gastropods and echinoids, scrape the algae but in the process also remove part of the substrate that is discarded as fragments. Fish, mostly marine parrotfish today (Figure 3.14a), ingest coral and excrete the undigested carbonate component as sediment particles.

Various organisms, including a wide variety of algae, fungi, bacteria, worms, sponges and bivalves (Figure 3.14b–c), bore into hard substrates as they seek

Figure 3.10 (a) Whitings or clouds ~40 m long of precipitated aragonite crystallites in the shallow waters of the Great Bahama Bank. (b) Crystallites of aragonite from a whiting in the Bahamas. Scale bar 1 µm.

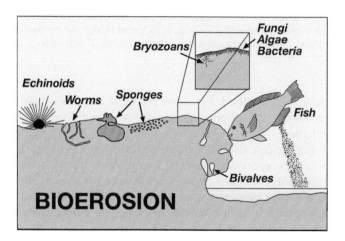

Figure 3.11 Sketch highlighting the different macro- and micro- (in box) invertebrates, microbes, and fish that break down carbonate substrates.

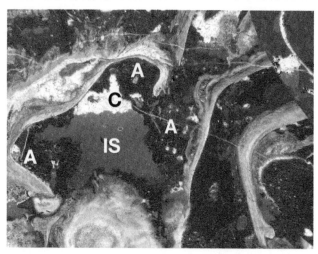

Figure 3.12 Oyster bioherm with numerous inter-skeletal cavities lined with automicrite (A), geopetal internal sediment (IS), and late burial cement (C), Jurassic, Morocco. Image width 9 mm. Photograph by F. Neuweiler. Reproduced with permission.

Figure 3.13 (a) A meadow of the seagrass *Posidonia* sp. in ~2 m of water off Perth, Western Australia where each blade is partially coated by calcareous epiphytes. Image width 20 cm. Photograph by P. Pufahl. Reproduced with permission. (b) Hurricane blowout of a *Thalassia* sp. meadow margin in ~1 m of water in Joulters Cays, Bahamas, exposing the complex rhizome mat that extends below the surface into the underlying sediment. Image width 1.5 m. Photograph by W. Martindale. Reproduced with permission. (c) A prolific macroalgal forest and diver at a depth of ~7 m off western Victoria, southern Australia. (d) The red mangrove *Avicennia* sp. at low tide in Florida Bay showing the numerous prop roots that extend down into the seawater-filled sediment. This is often called "the tree that walks" by tourists because of the leg-like roots. Tree ~5 m high.

nourishment, protection from predators, or protection from local environmental conditions. Penetration into these hard substrates, which includes lithified sediment (e.g., hardgrounds) or exposed limestones and hard skeletons (e.g., corals), is achieved through a combination of dissolution via organic acids, grinding by teeth, or rasping by shell margins as the animal rotates in its boring.

Microbes (Figure 3.14c) actively "drill" into skeletal and non-skeletal grains that lie on the seafloor (see Chapter 5). The holes, once vacated upon death of the formative microbe, may become filled with micrite that can be quickly lithified. Repeated cycles of this process can transform the original grain into a peloid formed of micrite.

Boring can also be a major sediment-producing process. The excavated material ranges from micron-sized fragments produced by worms and some bivalves to silt-sized chips generated by sponges. Amazingly, this process produces up to 30% of the carbonate mud in some Pacific atoll lagoons.

The preceding processes release carbonate back into seawater, break down the substrate into sediment grains, and weaken the host substrate, thereby making it more

Figure 3.15 The burrowing shrimp (*Callianassa*) has excavated a complex, open burrow system in the sediment below and ejected excavated material onto the seafloor where it forms conical mounds, 20 cm high, 2 m water depth, Bermuda.

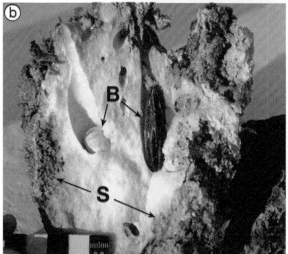

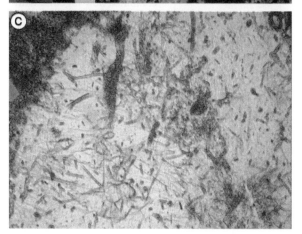

Figure 3.14 (a) A parrotfish whose strong teeth bite off both coral and algae. Organic matter is extracted in the gut and undigested carbonate particles are excreted as sand-size particles. Image width 8 cm. (b) A dead coral from the seafloor in Bermuda that has been broken open to illustrate boring bivalves (B) and boring sponges (S), centimeter scale. (c) Thin-section photomicrograph (plane light) of a bivalve shell that has been penetrated by innumerable microbe borings, ~2 µm in diameter.

amenable to physical breakdown. The prevalence of boring is commonly related to the nutrient levels in seawater because many of the boring organisms (e.g., sponges, bivalves) are filter feeders and therefore prefer relatively high-nutrient conditions.

Burrowing. Seafloor sediments can be actively burrowed by a variety of invertebrates that quickly and dramatically alter the nature of the deposit during the excavation that takes place as part of their feeding and dwelling activities (Figure 3.15). The most active burrowing animals include crabs, shrimps, worms, echinoids, and, in some areas, bivalves. Such animals burrow in search of food, to seek safety from predators, and to seek protection from environmental conditions. The consequences are homogenization of sediment with the loss of laminations and sedimentary structures, the loss of organic matter, and a decrease or increase in grain size (the latter by production of pellets).

Storms: above and below the ocean surface

Not all carbonate factories are tranquil shallow lagoons with seafloors colonized by a variety of sediment-producing communities that quietly operate undisturbed year-round. Shallow seafloors can be daily scoured by tidal currents, periodically ravaged by cyclonic storms, or disturbed by internal waves. The effects of tides in marine systems are outlined in Chapters 12 and 13 and

Figure 3.16 (a) Hurricane Fabian in Bermuda. (b) Coral rubble piled up near the reef crest as the result of hurricane transport.

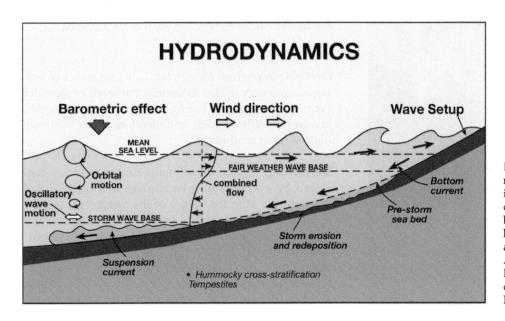

Figure 3.17 Sketch illustrating nearshore storm dynamics that involve wave setup and combined flow together with barometric effects to create hummocky cross-stratification and tempestites. Source: Adapted from Einsele (2000). Reproduced with permission of Springer Science+Business Media.

the following discussion is focused on storms and internal waves.

Cyclonic storms.   Hurricanes and other cyclonic tropical storms (Figure 3.16a) can wreak havoc on the shallow open seafloor where carbonates are forming and accumulating (Figure 3.16b). Seasonal storms can also profoundly affect lacustrine sedimentation. Catastrophic events can rapidly redistribute sediment that are seen in the subtidal carbonate stratigraphic record as units with distinctive depositional signatures referred to by the Shakespearian epithet as *tempestites*.

Waves and currents dominate hydrodynamics when strong storms sweep across a shallow lake or seafloor.

Strong winds drive water towards the shore or shallow marine shoals piling it up and pushing it into strandline and peritidal settings (Figure 3.17). The elevated water surface called *wave setup* or *storm surge* in turn generates a hydraulic head that forces a bottom current which flows seaward down the pressure gradient.

Both the wind-driven current toward shore and the return current into deeper water can be strongly deflected by Coriolis force. Thus, shoreward flow in the Northern Hemisphere can swing around to be at 90° to the right of the wind direction (to the left in the Southern Hemisphere). The return basinward flow will again move to the right in the Northern Hemisphere, such that this geostrophic current can eventually be subparallel to the shore. At the

CARBONATE TEMPESTITES

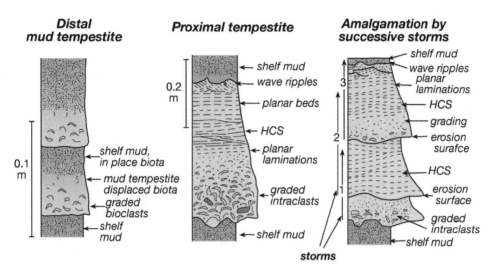

Figure 3.18 Sketch illustrating the change in tempestite stratigraphy with decreasing hydrodynamic energy. Source: Adapted from Einsele (2000). Reproduced with permission of Springer Science+Business Media.

same time, waves with their oscillatory motion impinge on the shallow seafloor. These coincident wave and current forces are called combined flow and result in the distinctive metre-scale unit or tempestite.

Tempestite.   A tempestite (Figure 3.18) is usually a few millimeters to several tens of meters thick, can be lenticular, and is typically interbedded with muddy carbonates. The base is conspicuously erosional with a variety of sole structures, channels, and gutter casts, most of which are unidirectional but can be locally bipolar. The deposit (Figure 3.18) itself has an upward-grading motif of: (1) a basal lag of coarse particles (clasts) and whole or broken skeletons; (2) parallel laminations; (3) hummocky cross-stratification (HCS); (4) flat laminations; (5) cross-laminations; and (6) mudstone that is typically burrowed.

The deposit reflects disruption of the seafloor by waves and currents, with the coarse lag deposit being, in part, transported organisms that were uprooted and locally redeposited during the storm and are now mixed with the organisms that lived there. It should be emphasized that the basal lag or graded portion of a tempestite is usually a mixture of infaunal and epifaunal organisms with bimodal sorting of storm-fragmented and complete shells, random shell orientation, and a mix of complete and disarticulated shells and biofragments (particularly bivalves, brachiopods, and crinoids). In some situations, the basal deposit can contain very few transported elements. The tops are, however, not always mudstone but can be a rippled

calcarenite or a hardground. Escape burrows are common, reflecting the efforts of the *in situ* infauna to move up through the accumulating sediment or transported shallow-water burrowers that have been carried into deeper water only to perish as they struggle upward. Laminated beds reflect rapid flow during the peak of the storm, whereas the HCS is the product of combined flow. The upper part of the unit is formed by waning currents as the storm diminishes.

The nature of these event beds changes in composition from shallow to deep water (Figure 3.18). Proximal units are thick-bedded, biofragment-dominated, and include many amalgamated and composite beds containing mixed and amalgamated biota. Distal tempestites are, by contrast, thinner, rare, and more burrowed in deeper environments. Many are mud-dominated with a planar base and contain only a few shell layers.

Hummocky cross-stratification and swaley cross-stratification.   Hummocky cross-stratification (HCS) and swaley cross-stratification (SCS) are diagnostic of tempestite deposits (Figure 3.19a). HCS is a suite of gently curved, low-angle cross-laminations of convex-up domal hummocks and concave-up bowl-like swales with pronounced curvatures developed in grainy rocks (Figure 3.19b). The dip is quaquaversal (in all directions) in contrast to current ripples and wave ripples. Cross-laminations dip at maximum angles of <10–15° with both erosional and non-erosional intersections. Hummocks are usually spaced 1–6 m apart and units are 0.1–2 m thick.

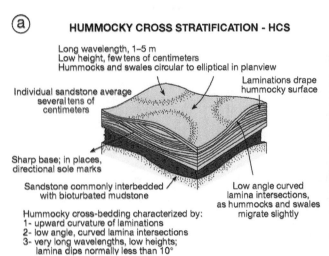

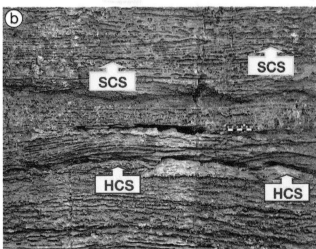

Figure 3.19 Hummocky cross-stratification. (a) Sketch illustrating the main attributes of HCS. Source: Walker (1982). (b) Outcrop of Middle Ordovician grainstones with both hummocky cross-stratification (HCS) and swaley cross-stratification (SCS), Crown Point, New York, scale 2 cm increments.

HCS is the result of combined flow with a strong oscillatory component but weak unidirectional current component, and where aggradational sediment rates are sufficient to preserve the hummocky bedforms. SCS develops when the tops of the hummocks are truncated and so only the swales remain (Figure 3.19b). The structures have the same controls as HCS but generally occur in shallower water where swales dominate because sediment aggradation is less and scouring is more frequent, thus removing the hummocks.

Variations and implications

The above is a common motif, but there are several important variations and implications.

- As storms sweep across shallow-water environments, mud is brought into suspension and grainy particles are redistributed across the seafloor. The suspended marine mud is the source of sediment for muddy tidal flats as it is swept onshore or can be carried out to sea through tidal passes in the marginal rim, forming peri-platform ooze (see Chapter 17).
- The shallow marine lagoon system quickly recovers from storms as the burrowing infauna returns and churns up the deposits, removing many physical sedimentary structures. One of the few records of such events is the filling of open burrows with coarse biofragmental sediment. Although such sediment can be later burrowed, there is usually a vestige that remains.

- Storms can also produce classic turbidites consisting of carbonate material, and so they should be expected in the suite of storm deposits.
- Flat pebble conglomerates (Figure 3.20a) composed of numerous tabular intraclasts, usually muddy, are a hallmark of many storm-influenced very early Paleozoic (Cambro-Ordovician) carbonates at a time before seafloor burrowing was rampant.
- The combined action of waves and currents can lead to the development of mud cracks in ooid-muddy facies, called *diastasis cracks* (Figure 3.20b), that strongly resemble synaeresis cracks in siliciclastic rocks. These are particularly common in early Paleozoic carbonates.
- Tempestites, because they are in part gravity driven, can transport sediment well beyond and below storm wave base.
- Storm deposits are more common where the environment faces into the storm track; leeward storm deposits are not as numerous.
- Given that grasses only evolved in the Cretaceous, they have only recently influenced the stabilization of shallow-water carbonates.

Subaqueous storms

Internal waves are waves that propagate below the ocean surface along the pycnocline (Chapter 4), the density boundary between deeper and shallower waters. Any perturbation of the pycnocline by surface waves, wind-stress fluctuations, tsunamis, or oceanic or tidal currents

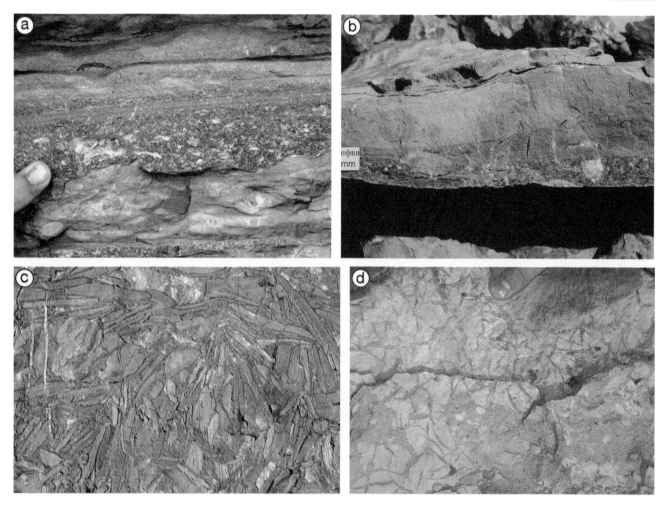

Figure 3.20 (a) Tempestite with a coarse basal unit, Upper Ordovician, Anticosti Island, Québec, finger scale ~1 cm wide. (b) Tempestite with scoured base into nodular mudstones; Upper Ordovician, Anticosti Island, Québec. (c) Bedding plane view of shallow subtidal flat clast conglomerate (rudstone) resulting from erosion and redeposition of partially cemented platform mudstones, Lower Ordovician, Wyoming. Image width ~30 cm. (d) Bedding plane of lime mudstone with diastasis cracks filled with calcite spar. These features are thought to form as a result of the interaction of storm waves with partially cemented or compacted muddy sediment at the seafloor; upper Cambrian, western Newfoundland. Finger scale ~1 cm wide.

will generate internal waves. The most common types are solitary internal waves or short trains of such waves called *solitons*. Most such waves are large and generated at the shelf edge between 100 and 200 m depth. They have amplitudes of 5–50 m, wavelengths of 0.5–15 km, periods of 5–50 minutes, and wave speeds that reach 0.5–2 m s^{-1}. The water motions associated with such waves are strongest where internal waves break on inclined subaqueous benthic surfaces producing episodic and repetitive high-turbulence events that could be capable of sediment erosion and transport (Figure 3.21).

Deposits associated with breaking internal waves are thought to include some or all of the following elements

(Figure 3.21): (1) a wave breaker zone (basal erosion surface); (2) a dumped 'heap' in the surf zone (mud-chip floatstone unit); (3) updip swash deposits (upslope-directed cross-laminated units); (4) backwash flow deposits (downslope-dipping cross-laminated units); (5) isolated starved ripples; and (6) thin sand laminae. The deposits of internal waves are not well understood, but are thought to be similar to tempestites and turbidites with an erosional base and subsequent depositional phase during the waning of turbulent conditions. Their deposits have been called *internalites*. It is suggested that they: (1) occur in distal mid-ramp settings detached from coeval shallow-water deposits; (2) do not have the

DEPOSITS OF INTERNAL WAVES

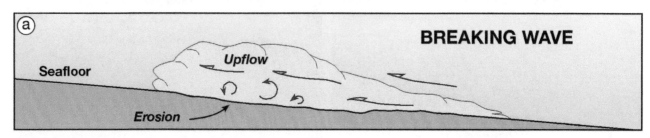

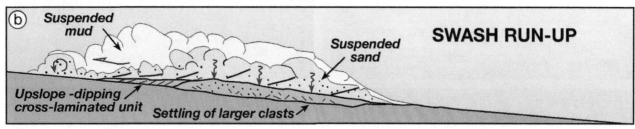

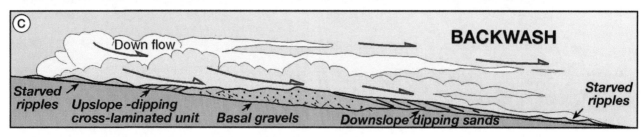

Figure 3.21 Sketch depicting sediment dynamics during breaking of internal waves. Source: Adapted from Pomar et al. (2012). Reproduced with permission of Elsevier.

coarsening-upward or thickening-upward trends of coastal storm deposits; (3) gradually thin out and disappear both updip and downdip where they are interbedded with mid-ramp mudstone; and (4) show little or no erosion towards shallower areas.

The origin of these deposits is controversial because much of the basis for their interpretation in the rock record is theoretical and laboratory based with few modern ocean measurements, even though internal waves have been well known of for over a century. Criteria for distinguishing internal-wave deposits from detached lowstand sediments have not yet been developed. It is also not yet clear if such deposits can form in large lakes.

Karst and carbonate spring precipitates

All of the above factories are freshwater or seawater subaqueous systems and produce most of the carbonate rocks in the geological record. There are, however, three other systems that also generate limestone, principally:

subterranean karst (see Chapter 8); carbonate springs (see Chapter 9); and soils (pedogenic carbonates, see Chapter 26).

As rainwater seeps through bedrock *en route* to a spring vent or a cave, it can be significantly modified as it interacts with bedrock or is heated by magma. Critically, these groundwaters can be heated to very high temperatures, become supercharged with CO_2, or become enriched with various elements (e.g., Ca, Mg, Si, Na) or both. Thus, precipitation in caves and around spring vents will be controlled largely by water chemistry and how the waters react with local climatic conditions. Rapid CO_2 degassing, which is common around many spring vents, can lead to supersaturation with respect to $CaCO_3$ and hence precipitation of calcite or aragonite or both. The resident microbial populations are important in these settings because they commonly dictate precipitation patterns. Microbes are also important in shallow soil horizons in semi-arid settings where they influence precipitation of carbonate crusts and "soil ooids," collectively forming a distinctive rock fabric termed calcrete or caliche.

Further reading

Aigner, T. (1985) *Storm Depositional Systems: Dynamic Stratigraphy in Modern and Ancient Shallow Marine Sequences.* New York: Springer-Verlag.

Bruggemann, J.H., van Kessel, A.M., van Rooij, J.M., and Breeman, A.M. (1996) Bioerosion and sediment ingestion by the Caribbean parrotfish *Scarusvetula* and *Sparisomaviride*: implications of fish size, feeding mode and habitat use. *Marine Ecology Progress Series*, 134, 59–71.

Einsele, G. (2000) *Sedimentary Basins: Evolution, Facies and Sediment Budget.* Second edition. Berlin, Heidelberg: Springer-Verlag.

James, N.P. and Clarke, J.A.D. (eds) (1997) *Cool-Water Carbonates.* Tulsa, OK: SEPM (Society for Sedimentary Geology), Special Publication no. 56.

Kennett, J.P. (1982) *Marine Geology.* Englewood Cliffs, USA: Prentice Hall.

Lees, A. and Buller, A.T. (1972) Modern temperate-water and warm-water shelf carbonate sediments contrasted. *Marine Geology*, 13, M67–M73.

Peryt, T.M. (ed.) (1983) *Coated Grains.* New York: Springer-Verlag.

Pomar, L., Morsilli, M., Hallock, P., and Bádenas, B. (2012) Internal waves, an under-explored source of turbulence events in the sedimentary record. *Earth-Science Reviews*, 111, 56–81.

Walker, R.G. (1982) Hummocky and swaley cross-stratification. In: Walker R.G. (ed.) *Clastic Units of the Front Ranges, Foothills and Plains in the area between Field, B.C. and Drumheller, Alberta.* International Association of Sedimentologists, Vol. 11, International Congress of Sedimentology, Field Excursion Guidebook 21A, pp. 22–30.

Wanless, H.R. (1981) Fining-upwards sedimentary sequences generated in seagrass beds. *Journal of Sedimentary Petrology*, 51, 445–454.

Wilkinson, B.H. (1979) Biomineralization, paleoceanography, and the evolution of calcareous marine organisms. *Geology*, 7, 524–527.

CHAPTER 4
MARINE CARBONATE FACTORIES AND ROCK CLASSIFICATIONS

Frontispiece Swimmers in oligotrophic, tropical crystal clear waters, North Lagoon, Bermuda.

Origin of Carbonate Sedimentary Rocks, First Edition. Noel P. James and Brian Jones.
© 2016 Noel P. James and Brian Jones. Published 2016 by John Wiley & Sons, Ltd.
Companion website: www.wiley.com/go/james/carbonaterocks

Introduction

The carbonate factory, because it is a delicately balanced biological-chemical system, is finely attuned to the surrounding environment, be it in oceans or in lakes. Sediment that accumulates in an ocean or lake is derived from both benthic and pelagic production. Given that the resident fauna and flora and the water chemistry dictate sedimentation, it is not surprising that factories produce an incredible range of deposits.

The factory is not one environment or one community but a spectrum of different carbonate-producing systems. These systems can be, for example, aggregations of benthic organisms on the open lake floor or seafloor, reefs rising above the lakebed or seabed, or rapidly shifting tidal dunes of ooid sand (Figure 3.2). All are sites of active biogenic and abiogenic carbonate precipitation that produce sediment components of all sizes. Organisms range from microbes and bacteria to diverse algae (Chapter 5) to a broad spectrum of invertebrates (Chapters 6 and 7).

Calcareous organisms, the most diverse elements of the sediment factory, have varied dramatically in terms of their makeup and relative importance through geologic time (Figure 3.3). Regardless, the calcareous benthos is composed of autotrophs (organisms that make their own food by photosynthesis or chemosynthesis), heterotrophs (organisms that derive their nourishment from consuming other organisms), or mixotrophs (heterotrophs containing photosymbionts that take over certain bodily functions that can allow increased calcification).

This biota adopts a variety of lifestyles, including *epilithic* (lives on the rock surface), *endolithic* (lives in the hard substrate), *epifaunal* (lives on the sediment), *infaunal* (lives in the sediment), *epiphytic* (lives on seagrasses and macroalgae), and *planktic* (floats in the water column). These organisms can be sessile and isolated individuals, prolific benthic aggregations that grow on top of one another, or mobile carnivores and herbivores that move across the seafloor or through the sediment.

Environmental controls

The principal controls that dictate carbonate sediment production in marine and lacustrine carbonate factories are: (1) light penetration; (2) salinity; (3) dissolved oxygen levels; (4) nutrient element supply; (5) water temperature; and (6) suspended sediment. For carbonate production and accumulation to be at a maximum, the environment must be just right and environmental parameters finely tuned, that is, not too deep, not too shallow, not too warm,

not too cold, not too fresh, not too saline, not too much terrigenous clastic sediment, and not too many or too few nutrients. This has been termed the "Goldilocks Window" after the familiar children's fairy tale.

The controls outlined here are *autogenic*, that is, they are local and intrinsic to each factory. These are different from *allogenic* controls that are imposed on the factory from outside, such as celestial mechanics and plate tectonics, which are described later (Chapter 10).

Light penetration

Given that so much carbonate production is associated with light-dependent (phototrophic) organisms, the intensity, wavelength, and depth of light penetration are critical. Illumination varies with water depth, latitude (which controls light refraction into the water), and water clarity. The photic zone is ~70 m deep in most oceanic settings (but can be 100 m in the gin-clear waters of the central Pacific Ocean). The influx of siliciclastic sediments into the oceans or lakes will, however, significantly reduce the depth of light penetration. In oceanic settings, light penetration can also be reduced by upwelling and high surface productivity, especially above the outer parts of platforms. The depth of light penetration in these situations can be reduced to a few tens of meters. On tropical marine platforms and banks, maximum carbonate production is usually greatest in water that is 10–20 m deep where seafloor illumination is at its highest.

The water column, at least for the Cenozoic, can be divided into: (1) the *euphotic zone*, shallow water that receives sufficient light for photosynthesis to occur and where organisms such as modern scleractinian corals and green algae thrive; (2) the *oligophotic zone*, the zone of less light where organisms such as the larger foraminifers and red algae are common; and (3) the *aphotic zone*, the zone of no light where invertebrates such as bryozoans, mollusks, crinoids, and sponges can grow in abundance (Figure 4.1). It must not be forgotten, however, that oligophotic and aphotic organisms can thrive in shallow water providing they live in the dim shadows or cavities where light cannot reach.

Water salinity

At first glance, it may seem that the separation between marine and lacustrine factories is simply a question of one being saltwater and the other freshwater. It must be remembered, however, that the water in lakes ranges from fresh to highly saline with salinity levels being controlled by net evaporation rates. This is a critical issue,

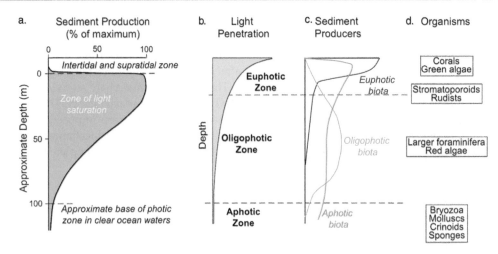

Figure 4.1 Plots of carbonate sediment production, light penetration, and different sediment-producing organisms against seafloor water depth. (a) Source: Jones (2010). Redrawn from Schlager (1998). (b–d) Source: Jones (2010). Redrawn from Pomar (2001).

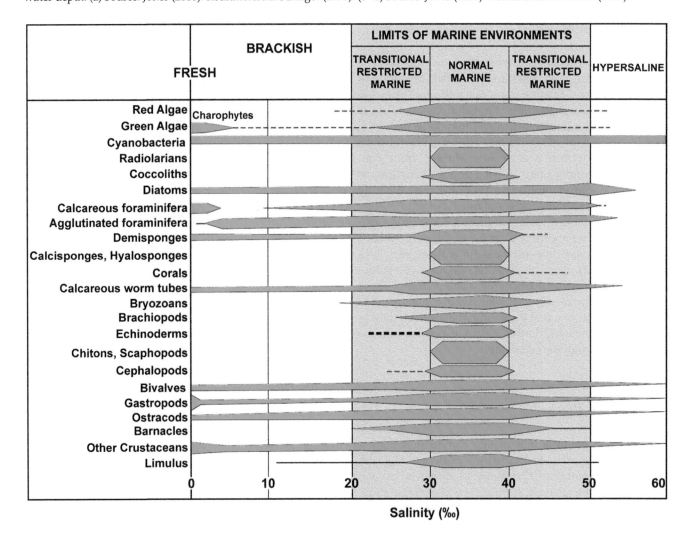

Figure 4.2 Salinity ranges of various animals and plants that contribute sediment production with the thickness of the line indicating relative abundance. Source: Heckel (1972). Reproduced with permission of Society for Sedimentary Geology.

because most plants and invertebrate animals (Figure 4.2) are highly sensitive to salinity levels.

Most marine biotas can only tolerate seawater of 30–40‰ salinity. In these settings, high-salinity levels are just as detrimental to their existence as are low-salinity levels. Some invertebrates such as agglutinated for-aminifers, demosponges, calcareous worms, mollusks, ostracods, bryozoans, and crustaceans can however live in brackish to fresh waters. Biotic diversity is nevertheless quickly reduced with increasing salinity, and most invertebrates disappear above 40‰. Calcareous algae can continue to be sediment producers where the salinity is >40‰, but they will also vanish as salinity rises much above that level. Occasionally, salt-tolerant skeletal invertebrates thrive in salinities too high for calcareous mud-producing algae. In Shark Bay (Western Australia), for example, enormous numbers of bivalves flourish despite the high-salinity levels but they are mostly all of one species. Crinoids are particularly sensitive to salinity fluctuations and are commonly used in the rock record as indicators of normal seawater; on the other hand, echinoids are not as sensitive. By contrast, the amount of carbonate generated by directly or microbially induced precipitation increases with increasing salinity and sediments can grade into fine-grained evaporitic carbonates.

The nutrient paradox

All living things need nutrient elements, particularly P and N, and in some settings they also need Fe and Si. Such elements enter the oceans and lakes from land via weathering and runoff, by means of diffusion from the atmosphere, via volcanism, or through emission at hydrothermal vents. Although it is perhaps natural to assume that those water bodies with abundant nutrients would be the most productive carbonate systems, this is not the case! This counterintuitive situation arises because increases in nutrient levels leads to an increase in water turbidity that, in turn, reduces the depth of light penetration and hence the growth of phototrophic sediment-producing organisms in the photic zone. Recall that all videos of tropical reefs show them growing in sparkling clear waters, whereas temperate algal forests and seas are characterized by green and turbid waters. Nutrients liberated in the depths of the oceans and lakes by the bacterial degradation of dead plankton are commonly recycled to surface waters via upwelling. The different trophic resource level systems are usually divided into four states (Figure 4.3).

Oligotrophic. Tropical marine environments far from land are typically nutrient-deficient because of their distance from nutrient sources (such as terrestrial run off). Near-surface waters are strongly temperature-stratified and upwelling of nutrient-rich waters is generally prevented. Such *oligotrophic* regions occur because organisms rapidly reuse all available bio-essential elements and few nutrients remain in the water column. Lacustrine environments are rarely oligotrophic because nutrients are supplied by surrounding rivers. Oligotrophic marine waters are typically blue and crystal clear (Frontispiece). It is paradoxical that these nutrient deserts are regions of extensive carbonate production. This situation is because $CaCO_3$ precipitates most easily in warm seawater and because many of the organisms (e.g., corals, sponges) have photosymbionts that promote calcification but are extremely sensitive to high nutrient levels. Whether or not similar organisms had symbionts in the past is a topic of active debate (Chapter 7).

Increased nutrients promote the active growth of symbionts (which is not good for the host coral or bivalve because they grow extremely rapidly and overtake the host) and plankton that in turn feeds and increases the population of heterotrophs biota. Increased nutrient levels also promote seagrass and macroalgal growth. All of these effects reduce the importance of the calcareous phototroph biomass. Such nutrients also promote the growth of phytoplankton to such an extent that their sheer numbers impede light penetration. Human introduction of nutrient elements into shallow tropical systems has sadly led to the demise of many modern reefs.

Mesotrophic. Those locations where nutrients are in good supply are dominated by filter-feeding invertebrates (heterotrophs) that, in marine settings, will include mollusks, bryozoans, sponges, brachiopods, and echinoderms. The waters are green and generally somewhat opaque because of suspended organic matter (Figure 4.4).

Eutrophic and hypertrophic. Environments where nutrients are overwhelming, especially in regions of intense coastal upwelling, at river mouths, in some lakes, and areas of human sewage outfall, are sites of prolific life. These environments generally lack a healthy benthic carbonate biota because oxygen is virtually absent, having been consumed by the prolific planktic heterotrophs. Instead, a suite of bacterial communities that thrive in organic-rich sediment colonizes the seafloor or lake floor.

There is also a connection to salinity here. Evaporation initially concentrates nutrients and major elements. Autotrophs thrive and are eaten by heterotrophs that consume oxygen, resulting in euxinic conditions. Increased salinities eliminate most scavengers, make bottom brines anoxic, and promote organic matter preservation.

NUTRIENT LEVELS

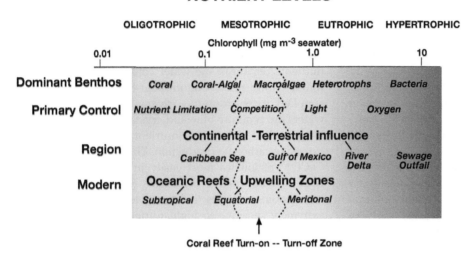

Figure 4.3 Nutrient gradients in low-latitude waters Source: Adapted from Mutti and Hallock (2003). Reproduced with permission of Springer Science+Business Media).

Figure 4.4 An underwater image of divers in green particle-rich temperate mesotrophic waters ~ 15m deep off southern Victoria, Australia.

Oxygen

Well-oxygenated waters are essential for the active growth of skeletal invertebrates, and any decrease in dissolved oxygen reduces the diversity and abundance of such organisms in a predictable way (Figure 4.5). Partial to complete anoxia can be induced in intermediate and bottom waters by: (1) stratification of the water column via pronounced temperature or salinity layering that reduces or arrests vertical mixing; (2) significantly increased salinity; or (3) dramatically increased nutrient supply to the ocean surface. The latter two cause intense primary productivity resulting in a rain of copious dead and

decaying phytoplankton to the seafloor (phytodetritus or pelagic snow). As this organic matter cascades through the water column and accumulates on the seafloor, its bacterial decay gradually depletes oxygen dissolved in the water.

Water temperature

Water temperature is a fundamental control on carbonate generation because it determines carbonate saturation levels and rates of chemical reactions. As outlined in Chapter 1, the warmer the water the more easily carbonate can precipitate; conversely, the colder the water the more easily it dissolves. In addition, the rates of chemical reactions increase exponentially with increasing temperature; biochemical precipitation is therefore much more rapid in warmer water than it is in colder water. The main factor that determines the attributes of any *benthic* carbonate factory is seawater temperature. This is not so with the *pelagic* factory, whose biogenic components extend from the equator almost to the poles (although production is highest in the tropics).

There is a prominent temperature stratification in the open ocean (Figure 4.6). This is especially pronounced in low-latitude tropical waters and becomes less important in middle and high latitudes. A second shallower seasonal thermocline is present in the tropics. The combination of rapidly decreasing water temperature with depth across the thermocline and rapidly decreasing salinity with water depth across the halocline (Figure 4.6a) results in a sharp increase in density. This zone of rapid increase in density is called the *pycnocline*.

SEAFLOOR ENVIRONMENT

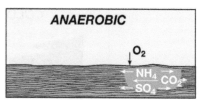

Figure 4.5 Sketch illustrating the difference in biota and sediments under aerobic oxygen-rich waters compared to those under anaerobic oxygen-depleted waters.

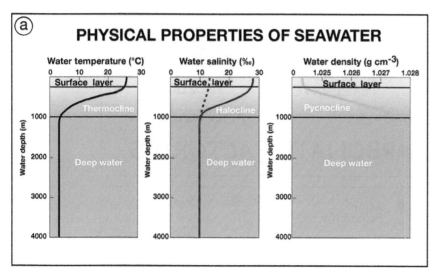

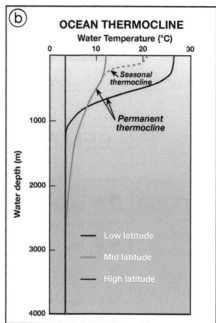

Figure 4.6 (a) Vertical profiles through the ocean water column illustrating general changes with temperature salinity and density with depth. (b) Vertical profile illustrating position of seasonal and permanent thermocline at different latitudes.

The shallow (neritic) marine biosphere is traditionally partitioned into the: (1) tropical (>20 °C); (2) temperate (10–20 °C); and (3) polar (<10 °C) zones with gradational boundaries separating one from the other (Figure 4.7). In terms of the carbonate factory, the big break is at ~20 °C. All carbonate producers thrive in the tropical zone, whereas the number of producers is much reduced in temperate and polar regions. Although there is a marked difference between tropical and temperate carbonates, the difference between temperate and polar deposits is not as profound. There are really two benthic marine neritic carbonate factories: one producing warm-water

carbonates and the other producing cool-water carbonates (see Chapters 10 and 11). The pelagic carbonate factory is not as affected by water temperature (see Chapter 13).

Seawater circulation

Water movement in the ocean is most affected by the combination of Coriolis force and wind blowing across the sea surface. *Coriolis force* is a difficult concept to grasp. The earth is a spinning sphere and so objects travel at different rates in different places. Objects at the equator

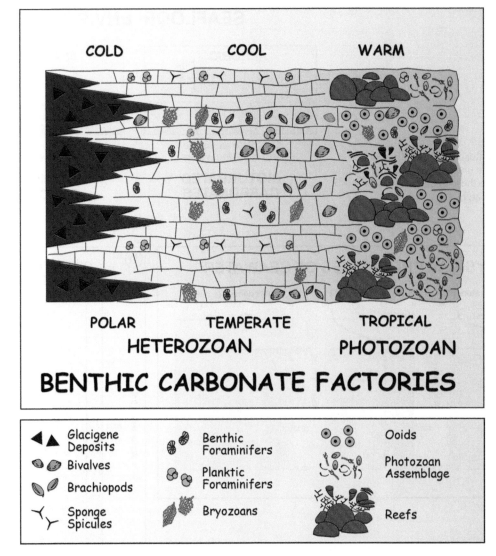

Figure 4.7 Sketch depicting the range of carbonate sediments produced in tropical to polar depositional environments and photozoan versus heterozoan factories. Source: Adapted from James & Lukasik (2010). Reproduced with permission of the Geological Association of Canada.

are, for example, moving eastward at ~1500 km hr⁻¹, whereas those at 45°N or 45°S are moving at ~760 km hr⁻¹. Objects not connected to land (e.g., air, water) that are pushed by the wind will tend to be deflected from their trajectory as they move north or south. In the Northern Hemisphere they will be deflected to the right compared to the direction of the force, whereas they will be deflected to the left in the Southern Hemisphere.

This results in strong westerly winds (winds from the west) in middle latitudes and strong easterly winds in the tropics. These winds interact with the ocean surface and result in the large-scale circular movement of surface water called *oceanic gyres*. These gyres circulate clockwise in the

Northern Hemisphere and counterclockwise in the Southern Hemisphere. They lead to *western boundary currents* that move warm water poleward along the western sides of oceans (e.g., the Gulf Stream). At the same time, they form *eastern boundary currents* along the opposite sides of the ocean that move cold water equatorward. Tropical carbonate systems therefore extend further poleward on the western sides of ocean basins (e.g., Bermuda) than they do on the eastern sides.

Much vertical water movement is driven by the Coriolis force. Wind blowing across the sea surface in the Northern Hemisphere results in movement of the water slightly to the right. This effect, called the *Ekman Spiral*, continues with depth with each deeper water layer deflected further to the right, but with decreased velocity until a depth of no movement is reached. The result is a net movement of water at 90° to the direction of the wind (Figure 4.8). The net result in the Southern Hemisphere is 90° to the left.

The Ekman Spiral effect results in upwelling and downwelling, especially along coastlines and at the equator

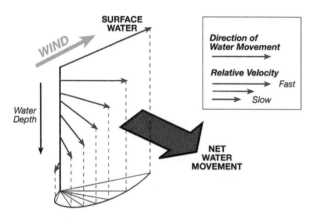

EKMAN SPIRAL (Northern Hemisphere)

Figure 4.8 A body of water can be thought of as a series of layers. The top layer is driven forward by the wind and each layer below is driven by friction. Each layer moves at a slower speed and at an angle to the layer immediately above it (to the right in the Northern Hemisphere and to the left in the Southern Hemisphere as a result of the Coriolis force), until the water movement at depth becomes negligible. The net flow of the water in the Northern Hemisphere is at 90° to the right of the prevailing wind direction.

(Figure 4.9a). If the wind, for example, blows from the north along the western side of a landmass in the Northern Hemisphere, resultant water movement will be to the right or offshore. This will cause sea level along the shore to be lowered and water will move up from depth to replace it. This is called *upwelling*. If the wind blows in the opposite direction then the water will pile up against the land causing *downwelling*. Since the wind blows from the east toward the equator in both hemispheres, there will be equatorial upwelling (Figure 4.9b).

These vertical movements of ocean water are important in terms of marine biology. Organisms living in the upper several hundred meters of the water column, mainly plankton, die and fall to the seafloor where they are broken down by bacteria into their original elements. These deep waters are enriched in chemical elements such as P, N, Si, and Fe that are critical for organism growth and reproduction. Upwelling brings these nutrient-rich waters back to the surface and thus promotes biological activity. Zones of high trophic resources, mesotrophic to locally eutrophic, are therefore mostly along the western sides of continents and along the equator. It also means that there are vast areas of the ocean that are oligotrophic.

Vertical near-surface water movement also takes place in large embayments and estuaries (Figure 4.10). If the local climate is humid, then freshwater from rivers flows out over seawater resulting in *estuarine circulation*, whereby seawater is drawn into the embayment and to the surface to mix with the fresh water. In arid regions, however, generally where there is no river, seawater flows into the embayment and evaporation results in increased salinity. Salinity and hence density are highest towards the head of the estuary. This

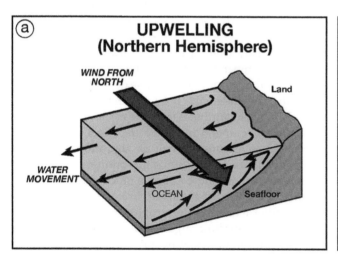
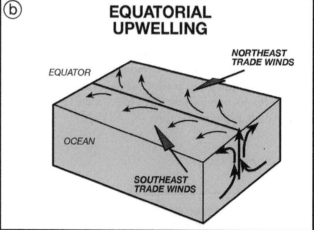

Figure 4.9 Upwelling. (a) Coastal upwelling in the Northern Hemisphere can be caused by winds blowing from the north along the west coast of a continent. Water moved offshore by Ekman transport is replaced by cold, deep, nutrient-laden water. Wind blowing from the south in the Southern Hemisphere will cause the same phenomenon. (b) Equatorial upwelling is the result of the trade winds blowing towards the equator at an angle that causes divergence at the equator, and thus upwelling.

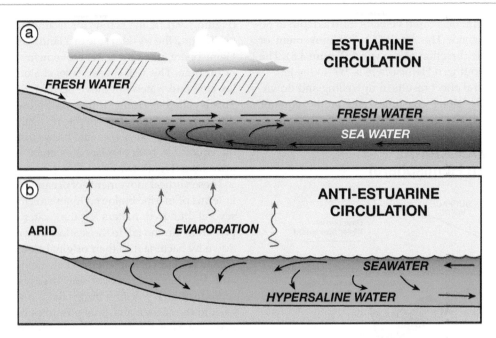

Figure 4.10 Estuarine circulation. (a) Estuarine circulation occurs in humid climates where freshwater flows from land and floats seawards on more dense seawater with gradual mixing taking place away from the shore. (b) Anti-estuarine circulation occurs in warm, arid climates where there is no fluvial input from land and evaporation results in dense, saline waters that sink and flow seaward beneath incoming, less dense, open ocean seawater.

water sinks and flows seaward along the embayment floor. This movement is called *anti-estuarine circulation*. Such water movement is not restricted to embayments, but is also present in lagoons and on open shelves.

Benthic marine factories

The benthic factory generates carbonate sediment on the ocean floor. This sediment-producing system has existed since the Archean, but its "modern" style only became operative during the late Neoproterozoic with the evolution of calcareous microbes and invertebrates. In the modern ocean, benthic marine factories can be divided into the Photozoan Factory, the Heterozoan Factory, and Intermediate Factories (Figures 4.7, 4.11).

The photozoan factory

This is the warm-water tropical factory, generally in oligotrophic environments, that comprises a complete spectrum of all the carbonate particles described in Chapters 5, 6, and 7. They are produced on shelves and in basins where water temperatures are generally >20°C year-round. The term photozoan signals the fact that the components are commonly dependent upon light (phototrophs or autotrophs) or are filter feeders containing photosymbionts

(heterotrophs with photosymbionts = mixotrophs), yet they also contain many non-light-dependent animals (heterotrophs) that produce carbonate sediment.

The heterozoan factory

This factory is bathed by waters that are mostly colder than 20°C. Neritic carbonate sediment of a different character is produced in these middle and polar latitudes. Such sediments, usually referred to as cool-water carbonates, are also called temperate carbonates, non-tropical carbonates, and extra-tropical carbonates. These deposits are produced in both temperate and polar environments, but with slightly different attributes (see Chapter 11).

Such low temperatures have the effect of excluding green algae, organisms such as zooxanthellate corals with symbionts, large benthic foraminifers with photosymbionts, ooids, and carbonate mud precipitates. The factory is dominated by invertebrate workers who are mostly filter feeders (heterotrophs), with the only autotroph being coralline algae. The organisms are slow growing and the rates of overall carbonate production are therefore lower than in photozoan factories.

Nutrients pose a complication here (Figure 4.11). As emphasized above, elevated nutrient levels favor increased phytoplankton and zooplankton growth. Benthic heterotrophs increase at the expense of benthic

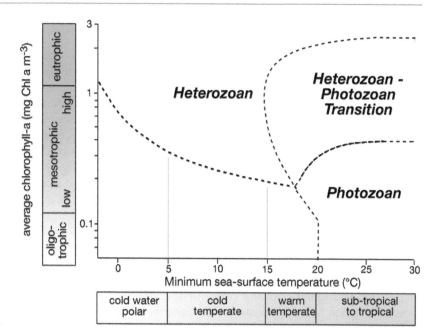

Figure 4.11 Compilation of shallow-water biotic associations from various carbonate environments put in context with minimum seawater surface temperatures and trophic resource levels. Source: Adapted from Halfar et al. (2006) and from James & Lukasik (2011).

autotrophs and mixotrophs. Benthic organisms in clear-water tropical environments are in fact so adapted to low nutrient levels that, when nutrients increase, the elements act as fertilizers promoting the prolific growth of phytoplankton, soft algae, and filter feeders. Corals and similar organisms adapted to low nutrient levels are simply overwhelmed and the photozoan assemblage is replaced by a heterozoan one. Whereas cool-water carbonates are always heterozoan, warm-water carbonates are not always photozoan.

Intermediate factories

The boundaries between the photozoan and heterozoan factories are commonly blurred. There is, for example, a common subtropical zone (18–22 °C) where most of the photozoan elements are reduced or minor, but large benthic foraminifers with photosymbionts prevail (see Chapter 12).

Pelagic marine factories

The pelagic marine factory (see Chapter 17), where sediment generation takes place in the water column largely by floating and swimming calcareous phytoplankton and zooplankton, is affected by the same controls as the benthic production system. The most important organisms are the green algae coccolithophorids (coccoliths), planktic foraminifers (with and without photosymbionts), and, to a lesser extent, pteropods (small swimming gastropods; Figure 4.12).

The marine pelagic carbonate factory has only been in existence since the middle Paleozoic, but is really a

Figure 4.12 Pelagic sediment composed of spherical planktic foraminifers and the cones of pteropods from 1600 m of water off Belize, Central America. Scale bar 0.5 mm.

post-Triassic phenomenon that developed following the appearance of calcareous plankton in the Jurassic. Paleozoic to middle Mesozoic marine carbonates are mostly shallow-water and produced by the benthic factory, whereas the pelagic factory presently accounts for two-thirds of all marine carbonate sedimentation worldwide.

Light is critical for coccoliths and planktic foraminifers with photosymbionts, so most of the production is limited to the photic zone. Salinity and oxygen constraints are similar to those for benthic carbonate producers. Moderately elevated nutrients (trophic resources) are beneficial to plankton growth. Temperature is again important because coccoliths are most prolific in tropical and warm temperate situations.

Limestone classification schemes

Given the spectrum of different sand- and mud-size carbonate sediments produced in the suite of factories described above, and the fact that they have changed through geologic time, it is imperative that a unifying nomenclature be used when discussing the various deposits. The scheme has to be simple and yet all–encompassing because we need to: (1) easily communicate with one another; (2) be concise as to what we mean; and (3) be forced into rigorous description. There are now several classification schemes in use, outlined in the following sections.

Before discussing specific classification schemes it is important to define the terms texture and fabric, words that are commonly muddled when discussing carbonate rocks. *Texture* generally refers to the size, shape, and arrangement of the constituent elements, especially grain size, whereas *fabric* describes the orientation in space of the particles, crystals, and cement.

Folk's classification

This classification scheme is based on the types of allochems (particles) and whether or not they are surrounded by micrite matrix or calcite cement. Folk (1962) based his classification on four basic types of allochems: bioclasts; pellets; ooids-pisolites; and intraclasts. This scheme has proven to be most useful when describing rocks in thin-section. The scheme has two tiers: (1) a graphic classification; and (2) a textural classification. The latter can be used as an adjective in the former (Figure 4.13).

The graphic classification has three parts: allochemical rocks; orthochemical rocks; and autochthonous reef rocks. Allochemical rocks are formed of different types of allochems that are surrounded by either micrite or cement and are used in the sample name (e.g., biosparite). Combinations of components can be used in the rock name (e.g., biopelmicrite). A dismicrite is a micrite with spar-filled blebs, generally burrows.

The textural scheme highlights sorting and rounding to emphasize a spectrum of deposits that accumulated under

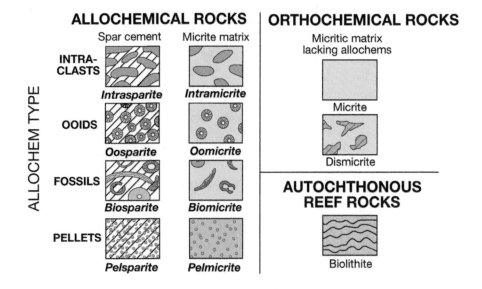

Figure 4.13 Classification of sand-size and smaller carbonate sedimentary rocks, proposed by Folk. Source: Folk (1962). Reproduced with permission of the American Association of Petroleum Geologists.

different energy conditions. The premise is that the gradual transition from muddy to grainy deposits represents a gradation from low-energy to progressively higher-energy accumulations. These terms can be combined with the graphic name to provide a more detailed description (i.e., sparse biomicrite or sorted biosparite).

This is an easy-to-use system that allows all levels of description and is utilized worldwide. Conversely, a microscope is often needed to accurately name the rock, it is a bit awkward to apply to modern sediments, it is not useful for describing reef rocks, and differentiating between micrite as a sediment versus a cement is commonly difficult.

Dunham's classification

Dunham (1962) proposed his classification at the same time as Folk, but with a different perspective that concentrated on the nature of the supporting framework between particles. As originally designed, the scheme (Figure 4.14) focused on the texture of the rock with the major division being between grain-supported (<50% mud) and mud-supported (>50% mud) fabrics. Mud-supported sediments with <10% grains are called *mudstones* whereas sediments with 10–50% grains are called *wackestones*. Sedimentary rocks formed entirely of grains and cement are called *grainstones*, whereas those with 10–49% interparticle mud are termed *packstones*. This elegant system is simple to apply because it is: (1) based solely on the components of the sediments and rocks; (2) independent of interpretations; and (3) largely devoid of semantic issues. The rock names derived from Dunham's classification scheme are commonly modified

by prefixes that recognize the dominant type of grain in the limestone; terms such as oolitic grainstone and bryozoan-crinoid wackestone are commonly used.

Images of the various rock types are illustrated in Figures 4.15–4.18. Reef rocks and associated deposits are described in Chapter 15.

A caveat

Classification systems in natural science fundamentally try to organize complex variables to make them understandable and thus interpretable. Given that such classifications

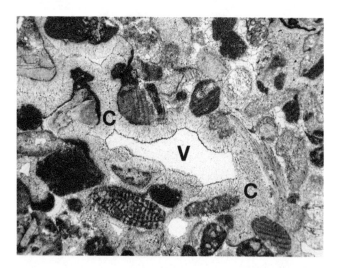

Figure 4.15 A partially cemented fossiliferous sparite (Folk, 1962) or fossiliferous grainstone (Dunham, 1962), Pleistocene Bermuda. C: cement; V: void. Image width 0.5 mm.

Grain Supported		Mud Supported	
No Mud	Has Mud	>10% Grains	<10% Grains
GRAINSTONE	**PACKSTONE**	**WACKESTONE**	**MUDSTONE**

CEMENT   MUD

sand   muddy sand   sandy mud   mud

Figure 4.14 Classification of sand-size and smaller carbonate sedimentary rocks proposed by Dunham. Source: Dunham (1962). Reproduced with permission of the American Association of Petroleum Geologists.

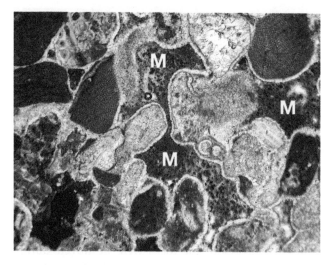

Figure 4.16 A fossiliferous micrite (Folk, 1962) or a fossiliferous packstone (Dunham, 1962), Pleistocene, Bermuda. M: peloidal mud matrix after early cement. Image width 0.3 mm.

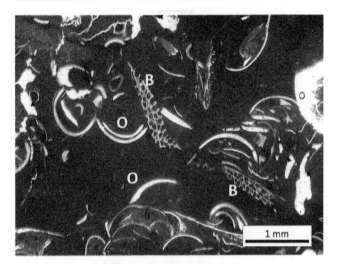

Figure 4.17 A packed biomicrite (Folk, 1962) or a fossiliferous wackestone (Dunham, 1962), Middle Ordovician, Kingston, Ontario, Canada. O: ostracod; B; bryozoan.

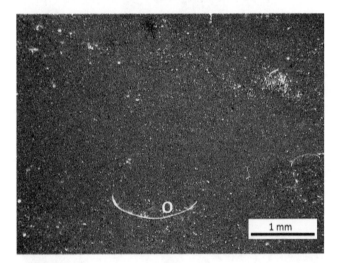

Figure 4.18 A sparse biomicrite (Folk, 1062), a mudstone (Dunham, 1962) or a calcistone; Middle Ordovician, Kingston, Ontario, Canada. O: ostracod.

are used to facilitate communication, it is important to know the limitations involved with such terms and fully understand the underlying premises involved in their application. The following carbonates are, for example, difficult to classify using these schemes:

- Limestones that have undergone extensive dolomitization can only be classified using the above classification scheme if the original fabrics were preserved. If the original fabrics of the rocks were destroyed during dolomitization, other classification schemes specific to dolostones must be used.
- Some organisms make classification difficult, for example, the rock can be a grainstone but due to the

three-dimensional architecture of the organism it can look like a wackestone or a mudstone.
- Carbonate rocks that formed in transitional areas between the zones of carbonate sediments and siliciclastic sediments are difficult to classify using the preceding schemes because they do not take into account the detrital components. In such cases, ternary diagrams that include the detrital components (e.g., quartz) must be used.
- Extensively brecciated carbonates that are commonly associated with karst terrains or the diagenetic dissolution of evaporites cannot be classified using these terms.

Further reading

Dunham, R.J. (1962) Classification of carbonate rocks according to their depositional texture. In: Ham W.E. (ed.) *Classification of Carbonate Rocks*. Tulsa, OK: American Association of Petroleum Geologists, Memoir no. 1, pp. 108–121.

Folk, R.L. (1962) Spectral subdivision of limestone types. In: Ham W.E. (ed.) *Classification of Carbonate Rocks*. Tulsa, OK: American Association of Petroleum Geologists, Memoir no. 1, pp. 62–84.

Garrison, T. (2012) *Essentials of Oceanography*. Pacific Grove, California USA: Brooks/Cole-Thompson Learning.

Halfar, J., Godinez Orta, L., Mutti, M., Valdez-Holguin, J.E., and Borges, J.M. (2006) Carbonates calibrated against oceanographic parameters along a latitudinal transect in the Gulf of California, Mexico. *Sedimentology*, 53, 297–320.

Hallock, P. and Schlager, W. (1986) Nutrient excess and the demise of coral reefs and carbonate platforms. *Palaios*, 1, 389–398.

Heckel, P.H. (1972) Recognition of ancient shallow marine environments. In: Rigby J.K. and Hamblin W.K. (eds) *Recognition of Ancient Sedimentary Environments*. Tulsa, OK: SEPM (Society of Sedimentary Geology), Special Publication no. 16, pp. 226–286.

James, N.P. and Clarke, J.A.D. (eds) (1997) *Cool-Water Carbonates*. Tulsa, OK: SEPM (Society for Sedimentary Geology), Special Publication no. 56.

James, N.P. and Lukasik, J. (2010) Cool- and cold-water neritic carbonates. In: James N.P. and Dalrymple R.W. (eds) *Facies Models 4*. St John's Newfoundland and Labrador: Geological Association of Canada, GEOtext 6, pp. 369–398.

James, N.P., Kendall, A.C., and Pufahl, P.K. (2010) Introduction to biological and chemical sedimentary facies models. In: James N.P. and Dalrymple R.W. (eds) *Facies Models 4*. St John's Newfoundland and Labrador: Geological Association of Canada, GEOtext 6, pp. 323–339.

Jones, B. (2010) Warm-water neritic carbonates. In: James N.P. and Dalrymple R.W. (eds) *Facies Models 4*. St John's Newfoundland and Labrador: Geological Association of Canada, GEOtext 6, pp. 341–369.

Mutti, M. and Hallock, P. (2003) Carbonate systems along nutrient and temperature gradients; some sedimentological and geochemical constraints. *International Journal of Earth Sciences*, 92, 465–475.

Pomar, L. and Hallock, P. (2008) Carbonate factories: A conundrum in sedimentary geology. *Earth Science Reviews*, 87, 134–169.

Schlager, W. (2003) Benthic carbonate factories of the Phanerozoic. *International Journal of Earth Sciences*, 92, 445–464.

CHAPTER 5
THE CARBONATE FACTORY: MICROBES AND ALGAE

Frontispiece Columnar, club-shaped living intertidal stromatolites, Shark Bay, Western Australia. Geological hammer 15 cm.

Origin of Carbonate Sedimentary Rocks, First Edition. Noel P. James and Brian Jones.
© 2016 Noel P. James and Brian Jones. Published 2016 by John Wiley & Sons, Ltd.
Companion website: www.wiley.com/go/james/carbonaterocks

Introduction

When viewing any modern carbonate environment, whether walking on a supratidal flat, wading in an open lagoon, or swimming around a reef, it is hard to appreciate the billions of microbes and algae that are actively involved in the generation and accumulation of calcareous sediment. Although largely invisible in modern settings and commonly difficult to detect in ancient carbonates, microbes are a critical element of these systems. Likewise, green and red algae merely look like soft marine plants but they are not; they are actually producing copious amounts of carbonate sediment.

Microbes and carbonates

Carbonate systems during the Archean and Proterozoic were dominated by microbes and associated organic molecules. Today, they are most obvious in stressed environments where they face little or no competition from grazing marine invertebrates. It is important to realize, however, that even in open marine habitats where marine invertebrates are so common, microbes are always in the background producing and altering calcareous sediment. Assessment of these simple life forms is critical to understanding the carbonate depositional system. Recent research has demonstrated that these organisms are far more important than we ever realized.

The diverse arrays of microbes found in carbonate systems, such as bacteria, fungi, and cyanobacteria, have a tremendous impact on all aspects of carbonate precipitation, sedimentation, and diagenesis. Prokaryotic bacteria, with spherical, rod, and helical cells up to $25\,\mu m$ long, are generally unaffected by salinity and some forms can thrive in temperatures up to $125\,°C$. Most prefer a pH range of 6–9 and usually avoid bright sunlight. These organisms can feed on preformed organic matter or synthesize organic material from inorganic CO_2. This latter process can involve inorganic chemical reactions (chemoautotrophy), chemolithotrophy (reactions with minerals and rocks), or photosynthesis via chlorophyll. Cyanobacteria, which were once called blue-green algae, are photosynthetic and have strongly pigmented cells that can be blue-green, olive-green, or red. The cells can be singular or arranged in colonies and protected by an organic sheath.

Microbes commonly (but not always) develop biofilms. In the simplest sense, a biofilm is a community of microbes that is held in a hydrogel formed largely of extracellular polysaccharides (EPS). The EPS can vary from a cohesive gel to a very loose slime, depending on the microbes involved and the local environmental conditions. Biofilms serve many purposes including: (1) binding the microbes to a substrate; (2) providing a stable and protective environment for the microbes; and (3) isolating the microbes from the external environment. Processes operative in biofilms, which will typically be on a submicron scale, can have a profound effect on the precipitation or dissolution of carbonates.

Specifically, these primitive organisms impact carbonate sedimentation by (1) acting as nucleation centers for carbonate precipitation; (2) inducing carbonate precipitation outside the organism by modifying their surrounding microenvironment; (3) binding sediment particles together with the potential of forming large organosedimentary structures such as stromatolites; and (4) mediating bioerosion that may reduce grain size and release Ca and CO_3 back into the surrounding waters.

Not all microbes live on exposed surfaces. Many microbes, for example, prefer to live in borings that they produce in hard substrates by secreting acidic fluids that are capable of dissolving $CaCO_3$ (Figure 3.14c). Although such individual borings are typically small (<1 μm diameter), there may be hundreds of drillholes per square centimeter. This boring activity, which can be in living (e.g., coral) or dead skeletons, weakens the substrate, destroys the original fabric of the substrate, releases elements into the surrounding water, and leads to early diagenesis (see Chapter 23).

Microbialites

Microbialites are defined as organosedimentary deposits that accreted as a result of a benthic microbial community trapping and binding detrital sediment or forming the locus of mineral precipitation. *Stromatolites* (Figure 5.1a) are laminated benthic microbial deposits. *Thrombolites* (Figure 5.1b) lack laminations but have a clotted fabric (the clots can be composed of calcimicrobes, especially in the Paleozoic). *Leiolites* are microbialites with a microcrystalline (aphanitic) microfabric. *Oncolites* (Figure 5.1c) are essentially spherical, concentrically laminated stromatolites that can contain numerous calcimicrobes.

Stromatolites have a wide spectrum of morphologies ranging from domal to columnar to branching and have been classified on the basis of geometric form (Figure 5.2). Underpinning such a classification is the notion that the different morphotypes reflect variable hydrodynamic conditions and different microbial communities.

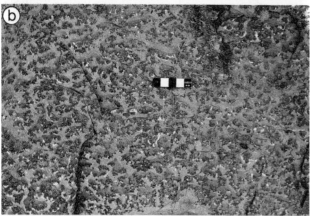

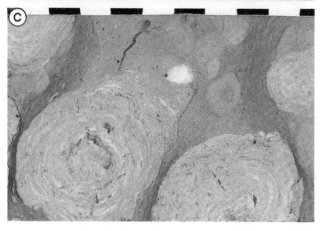

Figure 5.1 (a) Stromatolite complex from the Pethei Formation, Paleoproterozoic, Northwest Territories, Canada. Scale 15 cm. (b) Cross-section of a lower Cambrian thrombolite with a clotted cryptomicrobial fabric from the Amadeus Basin, central Australia, centimeter scale. (c) A subsurface core of Upper Devonian limestone composed of numerous oncolites with a distinctive concentric but irregular cortex, Western Canadian Sedimentary Basin, centimeter scale. Photograph by W. Martindale. Reproduced with permission.

Microbial mats

The basic element of a microbialite is the microbial mat (also known as an algal mat). This is a multilayered sheet of microorganisms, mainly bacteria and Archaea (light-dependent single cells that do not have a nucleus or many organelles). It is usually a surface film only a few centimeters thick at most, is attached to a substrate, and contains a wide range of chemical microenvironments.

The mat is really a community of bacteria, algae, and diatoms surrounded by a biofilm secreted by the microorganisms themselves. It is honeycombed with water channels that are important for fluid delivery, nutrient distribution, and waste removal.

The microorganisms that are embedded in the EPS are mostly phototrophic prokaryotes or Archaea. The cells can be arranged in colonies (coccoid forms) or threads (filamentous forms) that in turn can be branched. They need light to photosynthesize and, as such, their occurrence and distribution is controlled by light intensity (light quantity) and wavelength (light quality). If the light intensity becomes too high then they can: (1) move down to deeper water; (2) secrete pigments; or (3) add EPS. They photosynthesize during the day and respire at night.

Photosynthetic microbial mats display a macroscopic color banding in cross-section (Figures 5.3, 5.4) that can vary from micrometers to centimeters in thickness. The top layers exposed directly to sunlight are composed of diatoms, algae, and cyanobacteria and are highly pigmented by chlorophyll and carotenoids, resulting in a green or yellow color. Sulfur bacteria dominate the underlying red-purple and orange layers. A second, even lower green layer is populated by other sulfur bacteria and represents the lowest limit of light penetration. The lowermost layer is generally black and contains non-phototrophic sulfate-reducing bacteria.

Marine and marginal marine microbial mats (Figures 5.4, 5.5) today generally contain layers of sediment in the form of fine sand and mud, which are infested with non-calcified microbes and EPS. Benthic cyanobacteria dominate microbial mats in stressed environments today because they are highly tolerant of extremes in light intensity, water temperature, and nutrients that are fatal to most marine invertebrates.

The preservation potential of microbes is very low because they contain no hard parts and they are non-obligate calcifiers (they induce calcification but cannot calcify under low carbonate saturation conditions). In most settings, the microbes will indirectly induce carbonate precipitation as their metabolic processes or decaying organic remains modify the surrounding chemical microenvironment.

STROMATOLITES

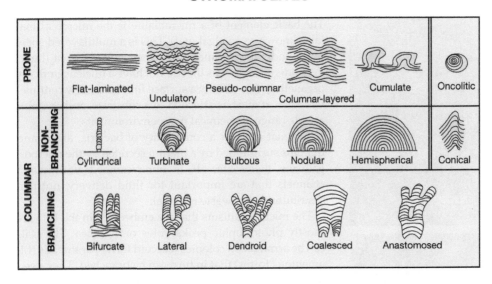

Figure 5.2 Different shapes of stromatolites. Source: Adapted from Grey (1989).

CYANOBACTERIAL MAT

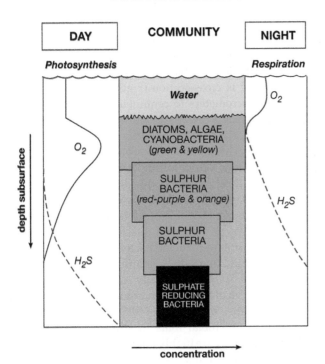

Figure 5.3 A generalized cross-section of the different microbial communities in a cyanobacterial mat. Source: Adapted from Konhauser (2006). Reproduced with permission of John Wiley & Sons.

Modern stromatolites

Today modern stromatolites (Frontispiece) are most abundant in stressed marine environments where high salinity (e.g., Shark Bay), fresh water (many lakes, e.g., Great Salt Lake; see Chapter 8), or hydrodynamics (e.g., Bahamas)

exclude most marine invertebrates. Numerous filamentous cyanobacteria (Figure 5.6a), commonly sheathed, are typically intertwined into a horizontal microbial mat that has a sticky surface akin to flypaper. When sediment grains are swept onto the mat by storms or migrating subaqueous sand dunes, the particles adhere to this sticky surface and cover the filaments (Figure 5.6b). The microbes, which prefer to live on the surface where they are bathed in sunlight, respond to burial by sediment grains by migrating upward and around the sediment particles (Figure 5.7). In doing so, they effectively bind the grains together and thereby hold them in place. Sheaths vacated by the upward-migrating filaments provide a fibrous reinforcement to the mat and the incorporated sediment. During periods when little sediment is being deposited, numerous taxa of filamentous cyanobacteria produce a new mat that is formed largely of prostrate interwoven filaments (Figure 5.6b). As time passes, non-motile slower-growing microbes colonize these mats.

Coccoid (subspherical-shaped) microbes can also become involved in the mats in hypersaline environments. The alternation of sediment-rich and microbial-rich layers produces characteristic undulating laminations, clotted micrite, and peloids. Although most of the sediment can be of detrital origin, delicate microcrystalline aragonite crystals can form crusts within these layers parallel to the mat surface. Critically, however, the preservation potential of the soft tissue is poor.

Shark Bay stromatolites

Since their discovery ~60 years ago, the spectacular stromatolites in Hamelin Pool, Shark Bay (Frontispiece) have been viewed as a window into the long-vanished

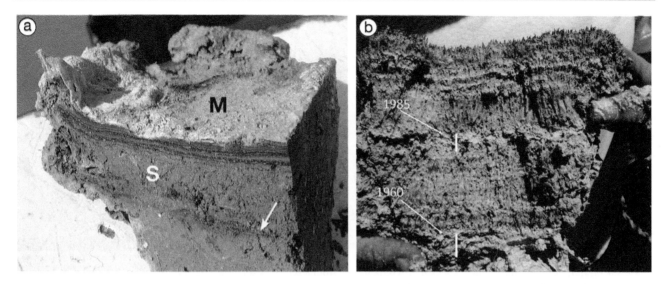

Figure 5.4 (a) Cross-section of sediment (S) below a living microbial mat (M) comprising several layers of different microbes, taken by shovel in the intertidal zone, near Ambergris Cay, northern Belize. Arrow points to older, buried and degraded mat. Image width 20 cm. Photograph by W. Martindale. Reproduced with permission. (b) A recently excavated intertidal mat from the Bahamas illustrating the different near-surface microbial communities and two hurricane layers of sediment deposited in 1960 and 1985 through which the microbes migrated. Photograph by H. Wanless. Reproduced with permission.

Figure 5.5 (a) Split and locally rolled flat microbial mats covering carbonate mud in the intertidal zone, Spencer Gulf, Australia, circled scale in 2 cm increments. (b) Tufted microbial mats and mangroves in the intertidal zone, Belize. Image width 2.5 m. Photograph by W. Martindale. Reproduced with permission.

Paleozoic and Precambrian worlds that were dominated by similar structures. They have, for this reason, been extensively studied and documented. These iconic organo-sedimentary features grow in a shallow water (<6 m deep) environment that is so saline (50–70‰) that all algal-grazing invertebrates and competitors, such as sea grasses and thallophytic algae, are excluded. Construction of the stromatolites is mediated by a variety of microbes (particularly cyanobacteria) that trap sedimentary grains and induce carbonate precipitation in a variety of complex ways. All of the stromatolites are nucleated on a hard substrate and most appear to have grown over the last 2000 years at rates of ~0.05–3 mm a^{-1}.

The stromatolites are strongly partitioned relative to the intertidal and shallow subtidal zones (Figure 5.8). Most workers recognize three major microbial mat types – pustular, smooth, and colloform or cerebral – as well as composite structures, with each having a characteristic microbial assemblage whose distribution is tidally controlled. Those above high tide are dead and stranded

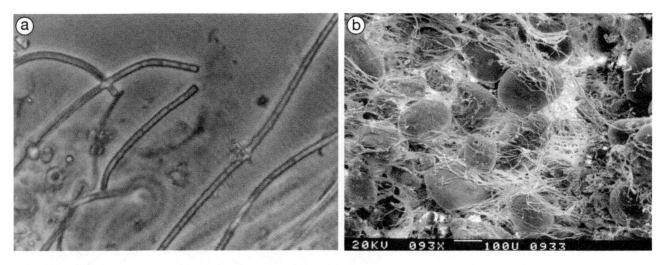

Figure 5.6 (a) Individual motile filaments of cyanobacteria teased from a sediment-rich subtidal stromatolite microbial mat dominated by this microbe. Plane light, image width 60 μm. (b) SEM image of sediment grains trapped in a mesh of intertwined filamentous microbes from a modern stromatolite in the Bahamas. Photographs by K. Brown. Reproduced with permission.

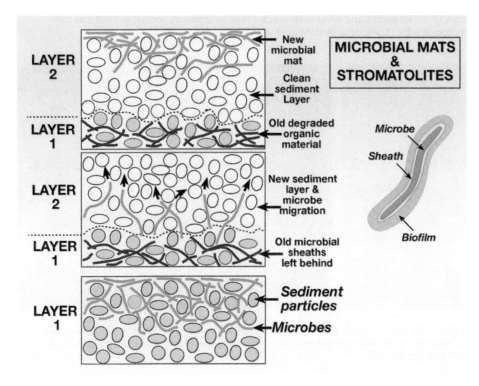

Figure 5.7 Sketch illustrating the structure of a filamentous microbe (right) and the progressive formation of microbially laminated sediment (left). With time, a layer of intertwined filamentous microbes on the surface is covered with sediment and the microbes move up through the sediment to colonize the new upper surface, leaving their sheaths behind in the process.

because sea level has fallen in this area since they grew several thousand years ago.

Pustular and smooth stromatolites (high-mid intertidal). These stromatolites are large columns and mounds up to 1 m wide and 40 cm high. The structures are highly influenced by hydrodynamics and range from club-shaped columnar structures to elongate wave-modified microbial ridges (Figure 5.9a). Most are unlaminated, composed of coarse fenestral networks, and have irregular outer surfaces. They are dominated by the coccoid cyanobacteria *Entophysalis* sp. and *Chroococcus* sp. that trap grains (peloids, ooids, quartz, bivalves, biofragments) by overgrowing and incorporating them into brown gelatinous pustules of mucilage. Rapid lithification takes place via microcrystalline aragonite precipitation in the polysaccharide gel that encases the cells.

Smooth stromatolites (low intertidal to shallowest subtidal). These relatively small smooth microbial mats form columns and mounds and can also colonize the tops of deeper-water colloform stromatolites when they grow into the intertidal zone. The structures are laminated (Figure 5.9b), finely fenestrate, and comprise subhorizontal millimeter-scale layers of alternating carbonate grains surrounded by exopolymers and layers of bacterial filaments and diatoms. The microbes are dominated by the filamentous cyanobacteria *Schizothrix* or *Microcoleus*. Grains are subject to severe microboring and generally suffer complete alteration to microcrystalline carbonate via the microbe *Solentia*. Laminations are preserved via intergranular microcrystalline aragonite precipitation that cements the microcrystalline grains together.

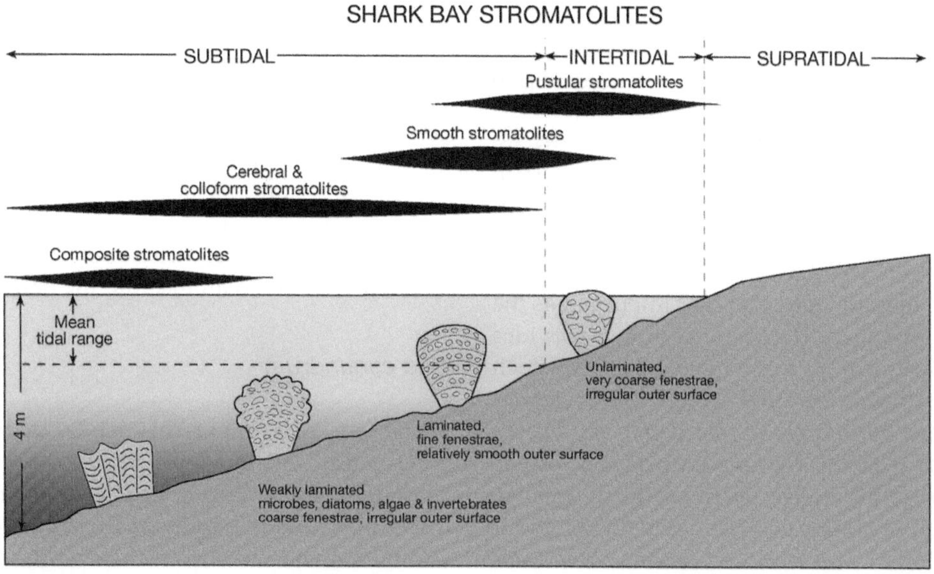

Figure 5.8 A sketch illustrating the different types of intertidal and subtidal stromatolites in Shark Bay. Source: Adapted from Playford and Cockbain (1976) and Playford et al. (2013).

Figure 5.9 (a) Image of elongate intertidal stromatolites composed of a lower pustular mat and upper film mat, Shark Bay, Western Australia. Circled hammer is 15 cm long. (b) A laminated columnar stromatolite nucleated on a fragment of bedrock that is impregnated with epoxy and slabbed to reveal the internal structure, Shark Bay, Western Australia.

Colloform and cerebral stromatolites (subtidal). These wholly subtidal structures (Figure 5.10a), up to 0.7 m high and in shallow–deep subtidal waters (4 m), are distinguished from shallower intertidal prokaryotic stromatolites by the fact that they are formed by a combination of both prokaryotes (microbes) and eukaryotes (algae). Colloform mats create individual elliptical to circular columns, domes, clubs, and compound masses up to 1 m high. An additional type of stromatolite is called "composite" and is composed of an aggregation of adjoining columnar heads. These complex structures can be up to 1 m high and 3 m across in subtidal waters up to 3.5 m deep.

The filamentous microbes in these subtidal forms are mostly *Microcoleus* and *Schizothrix* that, together with gelatinous colloform mats dominated by the coccoid cyanobacteria *Coccus* and *Chroococcus*, form the glutinous to leathery and largely ephemeral mats. The eukaryotic biota is a diverse diatom fauna that is as important as the microbes in mat formation. The irregular surface provides shelter for or attachment for many other eukaryotes such as bivalves, crustaceans, serpulid worms, micro-gastropods, foraminifers, green algae (*Acetabularia*), and brown algae (*Fucales* and *Gigartinales*). Small depressions on the top of the stromatolite receive much sediment and are bored by bivalves. Alteration of all fabrics to microcrystalline calcite forming a fused fabric is extensive. The resultant internal structure comprises layers of grains, many of which are surrounded by microcrystalline clots, alternating with layers of microcrystalline and needle aragonite, resulting in a coarse laminoid fabric.

It appears that the intertidal to shallow subtidal stromatolites are mostly formed by trapping and binding of grains, whereas wholly subtidal colloform structures comprise less trapping and binding but more carbonate precipitation. The deepest composite stromatolites are mostly composed of micrite precipitated by microbes.

Bahamian stromatolites

Large stromatolites, up to 2.5 m high, grow in tidal passes located between Pleistocene carbonate islands in the eastern Bahamas. There, strong tidal currents prevent settlement and growth of most invertebrates (Figure 5.10b). Algae and microbes such as cyanobacteria form the mats on the surfaces of these columnar structures. The stromatolites, like many in Shark Bay, are composed of ooid and peloid sand that is supplied by strong tidal currents that lifts grains into suspension and sweeps migrating subaqueous dunes across the columns. The microbes are mostly filamentous *Schizothrix* that, together with chlorophytes and diatoms, trap the particles.

The conundrum

There are two important points to remember when interpreting microbialites in the rock record. First, although superficially identical to their fossil counterparts, many modern (but not all) stromatolites are largely unlithified and would therefore not be preserved in the geological record. Recent studies have, however, confirmed that lithification does take place either simply as interparticle cement or more complexly via microbial alteration and induced precipitation that, by grain fusion, creates a thrombolite-like fabric (Figure 5.11a). The process is one of alteration by the microbe *Solentia* that alters not only the grain but all other particles that surround it

Figure 5.10 (a) Club-shaped colloform and cerebral subtidal stromatolites ~0.5 m high, Shark Bay, Western Australia. (b) Elongate subtidal stromatolites in the tidal race at Lee Stocking Island, Bahamas, diver scale.

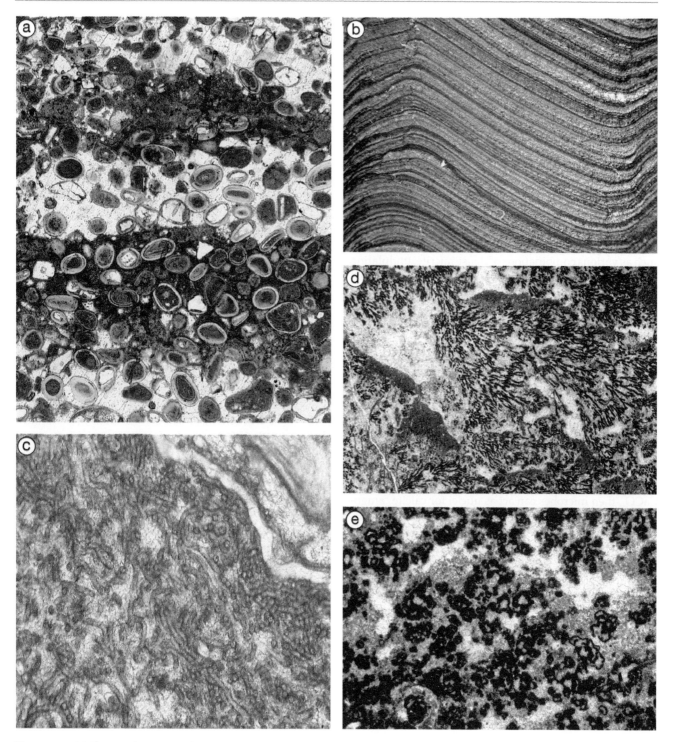

Figure 5.11 (a) Thin-section image (PPL) of layering in stromatolites from Shark Bay, Western Australia. The sediment is predominantly ooids with the dark layers cemented by microcrystalline aragonite. Image width 2 mm; (b) Thin-section image (PPL) of finely laminated Paleoproterozoic stromatolite from Pethei Formation, Great Slave Lake, Canada. The layers are alternating micrite and calcite cement. Image width 2 cm. (c) Thin-section photomicrograph (plane light) of microtubules of the calcimicrobe *Girvanella* sp., subsurface Upper Devonian, Western Canadian Sedimentary Basin. Image width 1.0 mm. (d) Thin-section photomicrograph (plane light) of the calcimicrobe *Epiphyton* sp., Lower Ordovician Cow Head Group, Newfoundland, Canada. Image width 6 mm. (e) Thin-section photomicrograph (plane light) of lunate clusters that comprise the calcimicrobe *Renalcis* sp., subsurface Upper Devonian, Western Canadian Sedimentary Basin. Image width 5 mm.

into a series of clots cemented together by microcrystalline aragonite.

The second point is that most modern marine stromatolites are composed of trapped particles that are bound by microbes and EPS. These structures are not known with confidence in rocks older than the Miocene. Most older stromatolites are therefore either direct precipitates from seawater, usually calcite spar (Figure 5.11b), or preserved biogenic remains via induced carbonate precipitation, or some combination of both. Many are alternating layers of microbial mats and seafloor cements. It can be very difficult to differentiate abiogenic precipitates and biogenic structures in these older stromatolites. In many situations, their biogenic character can only be inferred from the features and textures that are preserved such as various combinations of dense micrite, undulating micritic laminations, clotted micrite, calcimicrobes, and peloids. In short, most modern stromatolites are strikingly like older forms in terms of their macrostructure and mesostructure, but not their microstructure.

Calcimicrobes

Calcimicrobes form when microbes are replaced or coated by carbonate precipitates before the organic cells collapse and decompose (lyse). Given that organic material decays very quickly, such precipitation must take place while the organism is still alive or very soon after its demise. If the surrounding waters are supersaturated with respect to $CaCO_3$, then microbes can act as nuclei for calcite/aragonite precipitation and thereby become encased with carbonate. In other situations, calcification takes place beneath or within the microbial biofilm.

Typically, waters supersaturated with respect to $CaCO_3$ are a prerequisite for calcification. LMC will be precipitated in freshwater, whereas aragonite or HMC will be precipitated in marine settings. Precipitation in coccoid microbes usually takes place within the diffuse slime that fills spaces between cells. Calcification in filamentous microbes usually leads to the formation of a tube around the trichome (filament). In rare situations, the original sheath or cell wall can be replaced by calcite. Much of the calcite or aragonite precipitation can be mediated by an EPS matrix that encases the microbes in their host biofilm.

Identification of ancient calcimicrobes in terms of extant taxa is virtually impossible because: (1) calcification is typically incomplete and does not preserve all of the original parts of the organisms; (2) most calcimicrobes appear to have no living counterparts; and (3) most extant microbes are now classified on the basis of their DNA rather than their morphological attributes. As a result, ancient calcimicrobes are divided into groups based on their morphological attributes, which are usually only distinguishable in thin-section. In some cases, members of these groups have been further divided into various morphotypes that have been assigned generic or species names or both.

- *Girvanella* Group (Figure 5.11c), Paleozoic. Thin-walled tubes, tangled, prostrate growth in the form of nodules, oncolites, and sheets.
- *Epiphyton* Group (Figure 5.11d), Cambrian–Late Devonian. Dendritic filaments, chambered or tubular, erect or pendant.
- *Hedstroemia* Group, Cambrian–Cretaceous. Branching clusters to tubes with fan-like longitudinal sections.
- *Renalcis* Group (Figure 5.11e), Cambrian–Late Devonian. Clusters of hollow reniform bodies, micritic, fibrous or clotted walls which are erect or pendant, common in reefs.
- *Garwoodia/Mitcheldeania* Group, late Paleozoic–Cretaceous. Thin-walled radiating tube clusters.
- *Rothpletzella* Group, Silurian–Devonian. Flat, curved, and encrusting sheets of juxtaposed tubes that branch.

Calcareous algae

Algae are aquatic chlorophyll-containing photosynthetic benthic and planktonic eukaryotic organisms that range in size from minute unicellular forms to giant kelp several meters long. They are distinguished from plants by the absence of true roots, stems, and leaves. These organisms comprise the red (Rhodophytes), green (Chlorophytes), and brown (Phaeophytes) algae that are differentiated according to color and type of photosynthetic pigments. Not all of these algae are calcareous. Although most calcareous algae belong to the rhodophytes and chlorophytes, some of the phaeophytes (e.g., *Padina*) are lightly calcified. Their calcification, which is the subject of much research, can take place in the cell (e.g., coccoliths), in the cell walls (e.g., coralline algae), outside the cell (many green algae), or as a surface layer on the thallus. Calcified green and red algae have discrete skeletal ultrastructures.

Calcareous algae produce huge amounts of carbonate sediment. Despite their small size, the green algae *Penicillus*, *Halimeda*, and *Udotea* produce inordinate amounts of material. There is, for example, commonly much more sediment produced around coral reefs by calcareous algae than by the corals themselves. Various studies have also demonstrated that the volume of sediment found in lagoons and on platforms can be derived entirely from three common green algae (*Penicillus*, *Halimeda*, and *Udotea*) that thrive in those settings.

Mineralogy

Mineralogy is highly variable but generally taxonomically controlled. Green algae and some corallines (squamaraceans) are aragonitic but the coccoliths (planktic green algae) are LMC. All living red algae have skeletons formed of HMC (18–30 mole % $MgCO_3$) and some contain dolomite. Freshwater charophytes are formed of LMC and HMC.

Coralline and related algae

These red algae have contributed to the formation of limestones throughout the Phanerozoic. The taxonomic assignment of many fossil forms is problematic.

Coralline algae. Coralline algae (Figure 5.12) are the only known marine carbonate organisms that calcify their cellular membranes (Figure 5.12a, b). These algae include encrusting forms (Figure 5.12b) that coat hard substrates and branching geniculate (articulate branching) epiphytes (Figure 5.12a) that attach themselves to algae and sea grass. Nodular rhodoliths (Figure 5.12c, d) are coated grains that develop and grow as layers of encrusting corallines and associated organisms (e.g., foraminifera,

Figure 5.12 (a) The articulated (geniculate) coralline alga *Amphiroa* sp. (red is alive, white is recently dead), Yorke Peninsula, Southern Australia. Image width 8 cm. (b) Crustose coralline algal sheets over bedrock in ~10 m water depth, off Victoria, Southern Australia. Image width 40 cm. (c) Rhodoliths (coralline algal nodules) Miocene, southern France (Marseilles), pen for scale 10 cm long. (d) Broken modern rhodolith illustrating the dark limestone nucleus and white cortex of encrusting corallines, centimeter scale.

worms) that concentrically coat a nucleus. The size and shape of the rhodoliths is commonly dictated by the morphology of the nucleus, but they can be up to 10 cm or more in diameter.

Coralline algae grow in a wide spectrum of environments ranging from the tropics to the poles and from the intertidal zone to a maximum depth of ~80 m. Most prefer to grow in well-lit to slightly shaded, agitated, sediment-free waters with close to normal marine salinities.

Although coralline algae are known since the Late Jurassic, it has been suggested that they originated in the early Paleozoic. They are abundant carbonate producers with bioclasts derived from their skeletons being common in tropical, temperate, and cold-water settings. Throughout their vast environmental range, they form rhodoliths, rhodolith pavements, maerl (coarse particulate sediment composed of coralline algal fragments), algal ridges, algal cup reefs, and reefs. Many modern coral reefs contain significant numbers of encrusting coralline algae.

Heavy calcification via HMC makes these grains easily preservable and recognizable in the rock record (Figure 5.13a). Encrusting types have a basal prostrate series of cells and an upper erect series of cells containing reproductive organs (conceptacles) (Figure 5.13b), whereas branching forms are more symmetrical (Figure 5.13a).

Peyssonellid red algae.   This is a small group of encrusting reds (often with considerable interlayer pore space) that secrete an aragonitic skeleton (Figure 5.13c, d) which is mainly crustose but has extraskeletal dorsal aragonite cement botryoids (Figure 5.13e, f). They can also form rhodoliths, encrust soft sediment seafloors, and are intergrown within coral reefs. These algae live in temperate and tropical environments to ~100 m water depth but are most numerous in water <50 m deep. They first appeared in the early Cretaceous and are especially abundant in Paleocene and Eocene reefs. They might be the modern analog of some late Paleozoic phylloid algae such as *Archeolithoporella* (see 'Phylloid algae' below). These originally aragonitic algae are poorly preserved and can only be confirmed if HMC or some sediment has filled original pores (Figure 5.13d).

Solenopora.   This is an extinct encrusting algae (Cambrian–Miocene) with nodular and branching growth habits ranging from a few millimeters to several centimeters in size; the thallus consists of vertical calcified cells with divergent character. Their biological assignment is under debate; they are either red algae or possibly chaetetid sponges (demosponges).

Skeletons are characterized by radiating subparallel tubular or filamentous structures. Most are thought to have been calcitic but variable diagenesis suggests that some may have been aragonitic.

Green algae

This group (Figure 5.14) includes two living forms (Udoteaceans and Dasyclads) and three extinct forms (Gymnocodiaceans – Middle Devonian to late Cenozoic, especially late Paleozoic; Receptaculitids – Ordovician to Permian; and Cyclocrinitids –Ordovician to Silurian). The affinity of the Gymnocodiaceans is open to debate whereas the Receptaculitids and Cyclocrinitids appear to be related to the Dasyclads. All these algae are aragonite and so are not well preserved in the rock record.

Udoteaceans.   These algae, formerly called Codiaceans (Ordovician–Holocene) have a segmented thallus that is attached to the seafloor by a rhizome (root-like structure). They have an erect, filamentous (nodular) delicately branched, and phylloid (bladed) growth form. Calcification takes place by the induced precipitation of needle-shaped aragonite crystals within the thallus between tubules. Today, most of these algae live in shallow tropical seas. Extant forms, including *Penicillus*, *Halimeda*, and *Udotea*, collectively produce vast amounts of sand- and mud-size particles. They are preserved as molds with a general absence of specific features except for carbonate-encased filaments and tubules (Figure 5.15a). They are best preserved where the tubules were filled with HMC cement.

Dasyclads.   These unicellular millimeter to centimeter-size upright growing algae attach themselves to hard or soft substrates by rhizoids. The morphology is a long stem with whorls of lateral branches that break down into numerous fragments that are variably calcified by aragonite. They comprise a spherical hollow grain with radially oriented tubules. Most are present as fragments of an unbranched thallus. They are present from the Cambrian to the Holocene and are particularly prolific in the late Paleozoic, Late Jurassic–Early Cretaceous, and early Cenozoic; they were much more diverse and prolific in the rock record than they are today (e.g., there are 37 modern species but up to 830 fossil species). The algae are predominantly tropical but are also found in the Mediterranean Sea. They produce both sand- and mud-sizes particles. Some examples include *Cymopolia*, *Mizzia* (Figure 5.15b), and *Macroporella*.

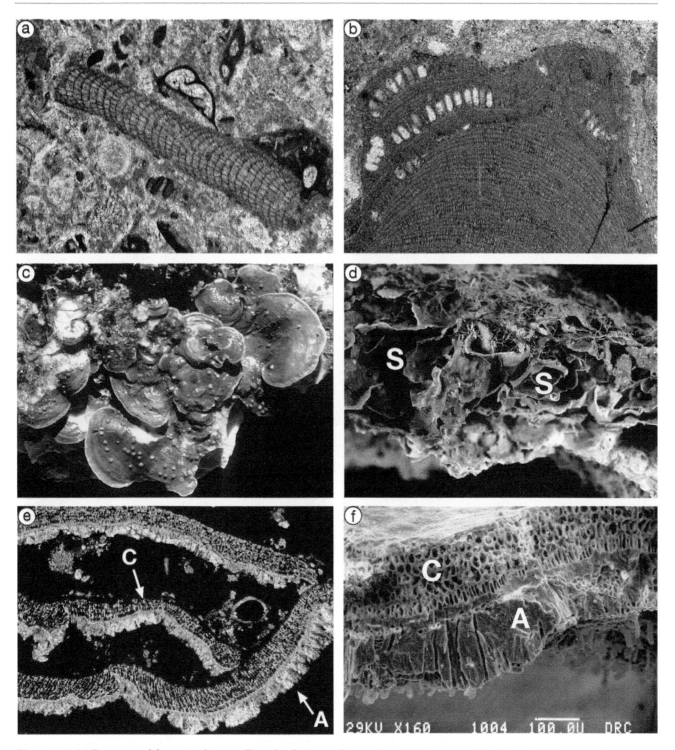

Figure 5.13 (a) Fragment of the geniculate coralline alga *Jania* sp., thin-section (PPL). Image width 0.8 mm. (b) The encrusting coralline alga *Sporolithon* sp. illustrating spar-filled conceptacles at top. Thin-section (PPL). Image width 1.8 mm. (c–f) Living Peyssonellid (aragonitic) red algae. (c) Plan view of overlapping plates, Bermuda. Image width 5 cm. (d) Cross-section illustrating the extensive pore space (S) between irregular plates. Image width 3 cm. (e) Thin-section (xpl) of plates with a calcified thallus (C) and basal layer of botryoidal aragonite (A), St Croix. Image width 1.6 mm. (f) SEM image of thallus with well-calcified cells (C) and basal botryoidal aragonite (A).

Figure 5.14 Green calcareous algae. A, B: different species of *Halimeda* sp., C: *Udotea* sp. and D: *Penicillus* sp.. All of centimetric scale. Photograph by W. Martindale. Reproduced with permission.

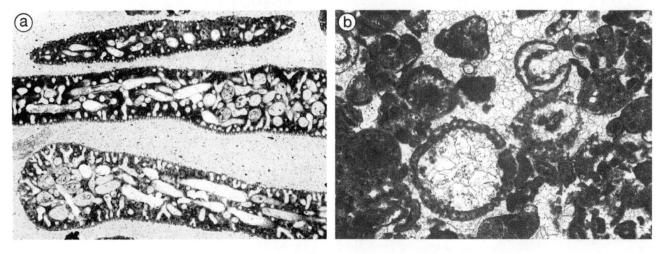

Figure 5.15 (a) Cross-sections of the green calcareous alga *Halimeda* sp., thin-section (PPL). Image width 3 mm. (b) Cross-section of the Permian green calcareous alga *Halimedia* sp. Thin-section (PPL). Image width 16 mm.

Phylloid algae

This is an informal name for a group of large Paleozoic calcareous algae with a leaf-like thallus resembling potato chips or crisps (Figure 5.16) and whose taxonomic affinities are poorly understood. The blades can be up to several centimeters long and as much as 1 mm thick. Most are placed in various groups of red (e.g., peyssonellid) or green algae. Some of the most numerous are *Eugoniophyllum*, *Ivanovia*, and *Archeolithoporella*. They were dominant rock constituents of the Pennsylvanian and Permian bedded carbonates and reef mounds, some of which are petroleum reservoirs. The blades are commonly open or cement-filled voids, indicating an original aragonitic mineralogy (Figure 5.16).

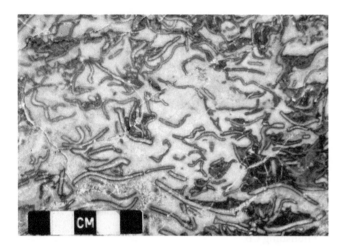

Figure 5.16 Limestone slab composed of phylloid (leaf-like) calcareous algae; the algae grew upright but when they died accumulated like a series of cornflakes, potato crisps, or tree leaves. Pennsylvanian, Paradox Basin, U.S.A. Photograph by P.W. Choquette. Reproduced with permission.

Coccoliths

Coccolithophorids are small planktic algae (Figure 5.17) composed of spherical cells 100 μm across enclosed by plates 5–25 μm in diameter. These plates are in turn composed of sub-micron-sized crystallites of LMC. The fossils are barely visible using a standard petrographic microscope and so are usually studied using the SEM. Both the plates and crystallites are constituents of carbonate mud, particularly chalk (see Chapter 17). Confined to the photic zone, these nannoplankton live in tropical to sub-arctic waters of mostly normal salinity, although their range is 18–40‰. They are most abundant in the tropics where, during blooms, their concentration can reach 100,000 cells per liter of seawater. Although evolving in the Triassic, they are most abundant from the Jurassic to present. They are the most numerous of the nannoplankton, which also includes extinct discoasters and calcitic cone-shaped nannoconids. It is estimated that coccoliths today form 30–45% of deep-sea carbonates in the humid tropics and 10–15% in the arid tropics.

Charophytes

These components are macrophytic green algae (Silurian–Holocene), several centimeters to a meter in size. They generally occur in oligotrophic freshwater environments and comprise an erect or branched thallus divided into a series of nodes with whorls of small branches and internodes. The preserved parts are the calcified (HMC or

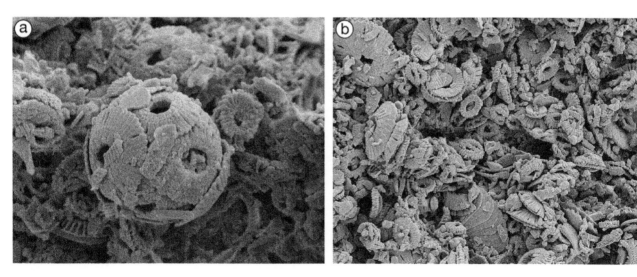

Figure 5.17 Coccolith carbonate sediment, seafloor ~2000 m water depth, Southwest Indian Ridge, Indian Ocean. (a) Coccosphere comprising several overlapping plates. Image width 10 μm. (b) Sediment from same locality composed entirely of coccolith plates and fragments. Image width 10 μm. Samples provided by Dr Xiaotong Peng.

LMC) female gametangia (oogonia). These are small (0.5–1 mm) spherical bodies with a circular central cavity surrounded by smaller ovoid features and a spiral-like external form.

Calcispheres

These small hollow calcareous spheres (<500 μm in diameter) occur in both Paleozoic and Mesozoic limestones. They are conspicuous in Devonian lagoonal and back-reef facies and in Carboniferous inner ramp deposits. Calcispheres are similar to modern dasyclad cysts and are thought to imply normal water salinity. They are also found in Jurassic–Cretaceous pelagic limestones where they are confirmed to be cysts of planktic algae.

Further reading

Burne, R.V. and Moore, L.S. (1987) Microbialites: organosedimentary deposits of benthic microbial communities. *Palaios*, 2, 241–254.

Dill, R.F., Shinn, E.A., Jones, A.T., Kelley, K., and Steinen, R.P. (1986) Giant subtidal stromatolites forming in normal salinity waters. *Nature*, 324, 55–58.

Golubic, S. (1991) Modern stromatolites: a review. In: Riding R. (ed.) *Calcareous Algae and Stromatolites*. Berlin: Springer-Verlag, pp. 541–561.

Grey, K. (1989) Handbook for the study of stromatolites and associated structures. In: Kennard J.M. and Burne R.V. (eds) *Stromatolites Newsletter*. Bureau of Mineral Resources, Geology and Geophysics, IGCP Project 261, Canberra, Australia, Vol. 14, pp. 82–171.

Jahnert, R.J. and Collins, L.B. (2013) Controls on microbial activity and tidal flat evolution in Shark Bay, Western Australia. *Sedimentology*, 60, 1071–1099.

Konhauser, K. (2006) *Introduction to Geomicrobiology*. Oxford, UK: Wiley-Blackwell Publishing Co.

Logan, B.W., Rezak, R., and Ginsburg, R.N. (1964) Classification and environmental significance of algal stromatolites. *Journal of Geology*, 72, 68–83.

Playford, P.E. and Cockbain, A.E. (1976) Modern algal stromatolites at Hamelin Pool, a hypersaline barred basin in Shark Bay, Western Australia. In: Walter M.R. (ed.) *Stromatolites*. New York: Elsevier Science Publishing, Developments in Sedimentology no. 20, pp. 389–413.

Playford, P.E., Cockbain, A.E., Berry, P.F., Roberts, A.P., Haines, P.W., and Brooke, B.P. (2013) *The Geology of Shark Bay*. Perth: Geological Survey of Western Australia, Bulletin no. 146.

Reid, P.R., James, N.P., Macintyre, I.G., Dupraz, C.P., and Burne, R.V. (2003) Shark Bay stromatolites: microfabrics and reinterpretation of origins. *Facies*, 49, 299–324.

Riding, R.E. and Awramik, S.M. (2000) *Microbial Sediments*. Berlin: Springer.

Walter, M.R. (ed.) (1976) *Stromatolites*. New York: Elsevier Publishing Co., Development in Sedimentology no. 20.

Wray, J.L. (1977) *Calcareous Algae*. New York: Elsevier Publishing Co.

CHAPTER 6
THE CARBONATE FACTORY: SINGLE CELLS AND SHELLS

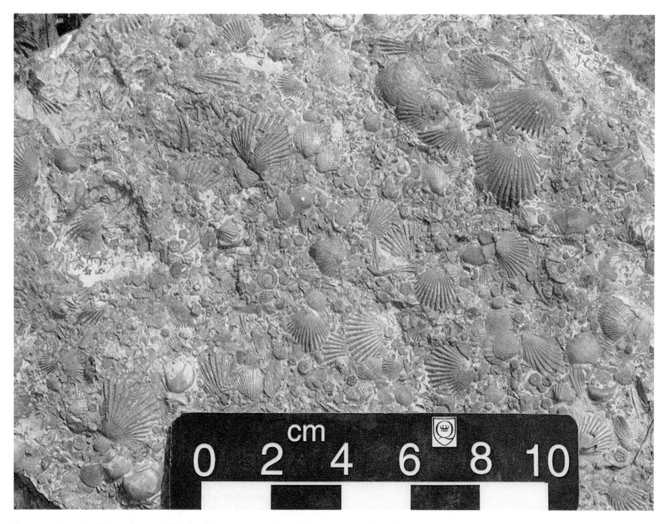

Frontispiece Bedding plane of Silurian limestone replete with numerous brachiopods, Anticosti Island, Gulf of St Lawrence, Eastern Québec, Canada.

Origin of Carbonate Sedimentary Rocks, First Edition. Noel P. James and Brian Jones.
© 2016 Noel P. James and Brian Jones. Published 2016 by John Wiley & Sons, Ltd.
Companion website: www.wiley.com/go/james/carbonaterocks

Introduction

Fossils (see Frontispiece), the preserved skeletal parts of past animals and plants, are of fundamental importance in carbonate rocks because: (1) they are often the most numerous and most obvious components of Phanerozoic limestones and dolostones; and (2) their recognition commonly carries important paleoecological information that provides critical insights into the depositional environments where they thrived.

Large foraminifers were among the first skeletons to be recognized as fossils by Aristotle when he looked at the Eocene nummulitic limestones used to build the Egyptian pyramids. On the other hand, Darwin first understood the enormity of accretionary tectonics when he saw relatively recent and well-known shallow-water bivalve shells in rocks near the top of the Andes in Peru. Skeletal particles have been a fundamental part of carbonate deposition throughout the Phanerozoic.

Whereas there are and have been a spectrum of calcareous invertebrates, the factories are dominated by relatively few phyla. This chapter and the following (Chapter 7) focus on these most numerous of carbonate producers. This chapter concentrates on a diverse group of animals that produce single tests, shells, and skeletal elements. Although they can be fragmented, many also remain whole. These comprise the foraminifers (benthic and pelagic) brachiopods, mollusks (gastropods, bivalves, and cephalopods), tentaculitids, calcareous worms, and crustacean arthropods. Chapter 7 focuses on quite different organisms such as echinoderms (crinoids and echinoids) and colonial invertebrates, namely sponges (spiculate and coralline), bryozoans, and corals (tabulate, rugose, and scleractinian).

A fundamental knowledge of paleontology is essential for identification of skeletal components, and the ability to recognize their taxonomic affinities based on random sections through skeletons is an essential skill.

Single-cell microfossils

Foraminifers

These small single-celled protists that construct chambered shells (tests) are either benthic (early Paleozoic–Holocene), living on the seafloor, or planktic (late Mesozoic–Holocene) residing in the upper 100 m of the water column. Benthic foraminifers (Figures 6.1, 6.2) form individual particles in grainy carbonates throughout geologic time, whereas planktic foraminifers, along

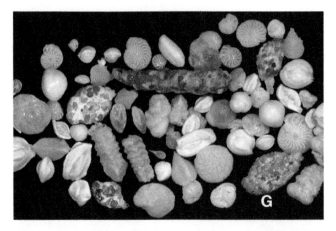

Figure 6.1 A suite of calcareous and agglutinated (G) small benthic foraminifers from the seafloor of Spencer Gulf, South Australia. Image width 1 cm. Photograph by L. O'Connell. Reproduced with permission.

with coccoliths, are the major constituents of chalks, a post-Jurassic rock type. A distinction is usually made between small foraminifers (foraminifers generally visible only with a hand lens) and large foraminifers (>0.5 mm) that are easily visible in the field. These small invertebrates are particularly useful in carbonate sedimentology because of their environmental sensitivity, rapid evolution (and thus biostratigraphic utility), and the fact that they are relatively small and therefore well preserved in subsurface cores and drill cuttings.

The eukaryotic cell is enclosed by a test that is either calcareous or is an agglutinated wall formed of all kinds of particles that are held together with organic matter or cement. These latter foraminifers, which are typically infaunal, are common in nearshore, brackish transitional environments or in the deep basin. Both agglutinated and calcareous foraminifers comprise one or several chambers with most being multi-chambered in the form of whorls. Modern foraminifers live from the intertidal zone to the deep sea. Although most grow in warm to cold neritic environments with normal salinity and normal oxygen levels, some have adapted to life in oxygen-depleted nutrient-rich waters.

Benthic foraminifers. Benthic foraminifers live free or attached to hard substrate, such as sediment grains, algae, and seagrasses. Small benthic foraminifers (Figure 6.1) range in size from 100 to 600 μm and live in all marine waters. Modern large foraminifers (Figure 6.2) are associated with tropical and subtropical shallow-water environments (generally <30 m, >15°C, oligotrophic), and have algal photosymbionts that restrict their habitats to the photic zone. Those with red

Figure 6.2 (a) A suite of large benthic foraminifers. Those that contain photosymbionts calcify and grow larger than most other protists. Image width 1.5 cm. Photograph by L. O'Connell. Reproduced with permission. (b) Outcrop image of large benthic foraminifers, mostly *Lepidocyclina* sp. in Eocene limestone, Mississippi. Centimeter scale.

and green endosymbionts are found in neritic habitats, whereas those with diatom symbionts are more common in the lower photic zone. Many tropical beaches today are formed entirely of foraminifer tests; take a close look next time you are on holiday in a warm climate. It is estimated that ~5% of the total world carbonate budget is benthic foraminifers, and it was much more so during the early Cenozoic.

Large benthic foraminifers (Figure 6.2) are especially common in late Paleozoic, Jurassic, and Cretaceous neritic limestones. Large benthic foraminifers were also particularly widespread in the early Cenozoic, forming spectacular sand shoals in neritic environments. Some of these foraminifera are huge with *Lepidocyclina* (Eocene) (Figure 6.2b) being 15–20 cm in diameter! It is hard to believe these large foraminifers were single-celled animals.

Benthic foraminifers in the early Paleozoic (Cambrian–Silurian) were mostly agglutinated types. Evolution and radiation of small calcareous benthic foraminifers in the Devonian resulted in these protists occupying numerous neritic to basinal environments, but never being abundant. Another radiation after the end-Devonian extinction resulted in protist rediversification such that benthic foraminifers (especially the large benthic fusilinids; Figure 6.3) were important limestone formers throughout the late Paleozoic. Many did not survive the end-Permian extinction, but those that endured diversified again such that benthic foraminifers were common from the Middle Triassic, through the Jurassic, and during the Cretaceous. During this time they spread into all of the environments that they inhabit today.

Benthic calcareous foraminifers are usually found as complete tests as opposed to fragments as in larger invertebrate skeletons; in thin-section they therefore appear as complete small animals (Figure 6.3a). The test is generally formed of HMC, LMC, or less commonly aragonite. The test wall itself is either: (1) microgranular (a dark-colored wall of tiny crystals); (2) porcelaneous (a translucent imperforate wall of randomly arranged rods, e.g., miliolids); or (3) hyaline (a transparent wall of interlocking crystals of calcite that are either oriented normal to the cell wall (radial) or microcrystalline (granular, e.g., rotaliinids *Orbitoides* and *Nummulites*) (Figure 6.3b). The radial hyaline forms display a pseudo-axial cross under plane-polarized light, whereas the granular crystals have a speckled appearance under similar illumination. Many of the large benthic and planktic foraminifers have hyaline walls, whereas late Paleozoic fusilinids (Figure 6.3c) have a microgranular wall.

Planktic foraminifers. Planktic foraminifers (Figure 4.12) evolved in the Middle Jurassic and became major constituents of limestones in the Cretaceous. Only five of the ~300 Cretaceous species survived the end-Cretaceous extinction, but after re-radiation became abundant again in the Cenozoic. Tests of planktic foraminifers are typically globular and arranged in expanding spirals. They are usually thin-walled with spines and a radial hyaline fabric (Figure 6.3d). All planktic foraminifers are LMC. Most live in the upper 50–100 m of the water column and many contain photosymbionts.

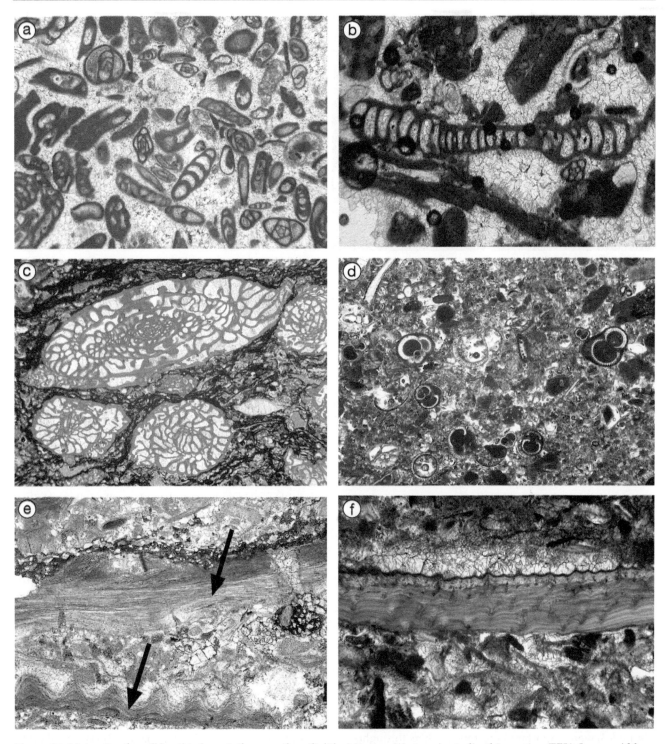

Figure 6.3 (a) A suite of small benthic foraminifers, mostly miliolids, Miocene, Western Australia, thin-section (PPL). Image width 3 mm. (b) The large benthic foraminifer *Sorites* sp., Pliocene, Western Australia, thin-section (PPL). Image width 4 mm. (c) Permian large benthic foraminifers (Fusilinids) from Arctic Canada, thin-section (PPL). Image width 7 mm. (d) Planktonic foraminifers in Holocene sediment, Tongue of the Ocean, Bahamas, thin-section (PPL). Image width 3.5 mm. (e) Cross-section of Middle Ordovician brachiopods with fibrous skeletal structure, southern Ontario, Canada, thin-section (PPL). Image width 1 cm. (f) Cross-section of Middle Ordovician pseudopunctate brachiopod, southern Ontario, Canada, thin-section (PPL). Image width 3 mm.

Radiolarians

These are small (typically <500 μm) zooplankton (Cambrian–Holocene), which secrete an amorphous (opal-A) silica skeleton. Although their preservation in limestones is generally poor, they are included here because they can be particularly numerous in fine-grained limestones and marls (mixtures of fine carbonate and siliciclastic clay). Their architecturally ornate tests are commonly in the form of a hollow sphere or vase with bars across the interior and spines projecting from their exteriors. They live from the surface to abyssal depths with warm-water species present since the Cambrian and cold-water species present since the Oligocene.

Macrofossils

Shells and exoskeletons that encase the soft body parts of the animals typify the macrofossils that contribute to carbonate sediments. Such components can be bivalved (brachiopods, mollusks, and ostracods), univalved (gastropods and serpulid worms), or formed of numerous plates (trilobites and barnacles).

Brachiopods

Brachiopods (Frontispiece) were the most diverse marine invertebrates of the Paleozoic. These exclusively marine organisms evolved in the early Cambrian and survived the Permian extinction but declined throughout the Mesozoic such that of the ~4500 fossil genera, only ~120 are known today. Their shell comprises two valves pulled together by an adductor muscle. Many are attached to the substrate by a muscular pedicle that extends from an opening in the pedicle valve. Fossil forms were either encrusting, pedically attached, semi-infaunal, or unattached.

The phylum includes both inarticulate and articulate forms. *Inarticulate brachiopods* (e.g., *Lingula*) are typically infaunal with most living in burrows in the intertidal zone. Modern types have a shell architecture that has remained virtually the same since the Cambrian. The shells of most articulated forms are markedly unequal in size and shape, but each shell is bilaterally symmetrical.

Articulate brachiopods (Frontispiece), unlike their inarticulate cousins, have the ability to separate their shells. Shells close as the adductor muscle contracts. These heterotrophic filter feeders derive their nutrients from ocean waters that enter their body cavity while the valves are open; as such, they are not tolerant of turbid waters.

Articulate brachiopods are architecturally diverse with shells of all sizes and shapes. One valve is always convex, but the other valve can be flat or concave (curves inwards). Although most brachiopods are 1–3 cm in size, some of the large Devonian forms (e.g., *Stringocephalus*) were up to 12 cm long. Some shells are smooth, others are ribbed, and still others are adorned with long spines. Large Permian forms looked like corals with spectacular spines attached to substrates or protruding into sediment, and lived in densely packed reef-like communities. Brachiopods were most abundant in the Devonian, declined in importance towards the end of the Permian, and were replaced by bivalves in many environments during the Mesozoic–Cenozoic. They are most numerous today in temperate and polar latitudes. In modern tropical oceans, small numbers of brachiopods are found in cryptic habitats such as protected cavities in coral reefs.

Their LMC skeletons are mineralogically stable and so are commonly well preserved and unaltered, even in Paleozoic rocks. They also secrete most of their shells in geochemical equilibrium with ambient seawater. With such excellent preservation, various geochemical characteristics (e.g., O isotopes) of their LMC shell material are commonly used as proxies for the interpretation of Paleozoic ocean composition (Chapter 23).

The shell structure is compound. Most forms have two and some have three layers. The thin outer or primary layer has crystals whose optic axes and long axes are oriented normal to the shell wall (commonly difficult to see). The thicker secondary layer is composed of long calcite fibers oriented at a low angle to the shell wall. Pentamerids and some spiriferids have a third, often thick, innermost layer composed of calcite prisms oriented like the primary layer.

There are four types of skeletal microstructure: (1) laminar; (2) impunctate (no perforations) fibrous (Figure 6.3e); (3) punctate shells with small holes that perforate the wall and are oriented perpendicular to the shell wall; and (4) pseudopunctate (Figure 6.3f) with stacked conical (parallel folds or ridges) or solid rods that end in the foliated structure and resemble punctae. Shells can be highly plicate (folded) giving them a wavy appearance. Detached spines are locally abundant in sediments (they are hollow with a two-layer fibrous structure).

Shell shapes are similar to and can be easily confused with bivalves with a foliate shell. A rule of thumb is that brachiopods grow a back and front (dorsal and ventral) valve with the line of symmetry cutting each valve in half, but bivalves grow a left and right valve with the line of symmetry along the margin of the valves.

Brachiopods were common in both photozoan and heterozoan Paleozoic carbonates. In the Cenozoic they were most abundant in temperate water accumulations. Their accumulation cannot be compared to bivalve beds because they are always LMC, they have a stronger tooth socket, and low vulnerability to borers and predators. Brachiopods have to use their internal muscles to open their shells. Upon death, the commonly stay together. By contrast, bivalves have a ligament to open and muscles to actively close their shells. Upon death, the shell opens and disarticulation into two shells follows ligament breakage.

Mollusks

Gastropods. Gastropods (Figure 6.4) (Cambrian–Holocene) are common contributors to limestones of all ages but were especially numerous during the Mesozoic and Cenozoic. Their unchambered univalved aragonitic shells are coiled about a central axis (Figure 6.5a) and come in all sizes, with some of the smallest (<1 mm long) being easily confused with foraminifera. Large gastropods such as the conch shell (*Strombus* sp.) are up to 30 cm long and have a shell of as much as 1 cm thickness around the lip.

These unchambered univalves are coiled about a central axis and some have an operculum of either conchiolin (organic material) or aragonite. They occur in sediment either as strongly curved fragments or whole shells. Skeletal microstructure is cross-lamellar (common) (Figure 6.5b), finely prismatic, homogeneous, or nacreous. With diagenesis, their original aragonitic shell is either replaced by LMC or dissolved away,

Figure 6.4 A suite of shallow-water modern marine gastropods. Specimens provided by Y. Bone.

leaving a mould that can subsequently be filled with cement (Chapter 32).

These ecologically diverse organisms, which are generally mobile, live in benthic marine, pelagic marine, freshwater, and terrestrial environments. They are herbivorous or carnivorous, and a few are filter feeders. They are particularly prolific in modern seagrass beds and kelp forests. One group, the vermetids, is important in the construction of cup reefs. The shells look more like a worm than a snail because they are not coiled, grow molded and cemented to a hard surface, and have irregular tubular elongate shells (see Chapter 14). Another group known as the pteropods (winged foot) are nektic, swimming in the upper part of the water column, and are significant contributors to Cretaceous and younger deep-water carbonates.

Bivalves

Bivalve shells (Figure 6.6) consist of two valves that can be either equal or unequal in shape and size. Their shells can be formed entirely of LMC, entirely of aragonite, or alternating layers of LMC and aragonite. Their complex mineralogy results in variable preservation, but a rule of thumb is that many epifaunal types are calcitic whereas many infaunal forms are aragonitic. Shells are typically symmetrical with the plane of symmetry passing between the two valves parallel to the hinge line. The valves open and close as the external ligament, located along the hinge line, contracts and relaxes. The pivot is provided by teeth and socket structures located in the hinge line and active contraction of the adductor muscles. Most bivalves will open upon death and disarticulate once the ligament is lost through decay.

Bivalves are common in limestones throughout the Phanerozoic. They inhabited a wide range of environments, adopting many different life styles with infaunal and epifaunal forms being most common. Others, including *Lithophaga*, employed an endolithic (boring) life style. Some, including the pectens (scallops), developed the ability to propel themselves through the water column, a form of crude "swimming". Additional forms are attached to solid substrates by a strong organic thread (byssus) that quickly decays after death, releasing the shells into surrounding sediment. They can also be cemented to the rock or other shells (e.g., oysters). Beds rich in bivalve shells (coquinas) are particularly common as event beds (storm deposits).

These animals are mostly filter feeders. They evolved in the early Cambrian and diversified in the Ordovician, with non-marine forms developing in the Devonian. Most Paleozoic forms were neither particularly diverse nor

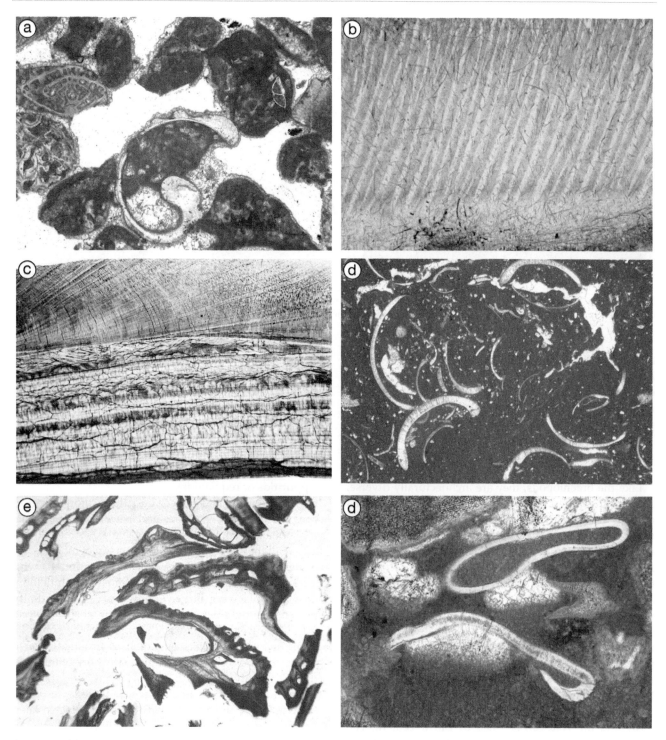

Figure 6.5 (a) Cross-section of a small Pleistocene gastropod, Bermuda, thin-section (PPL). Image width 2 mm. (b) Close view of cross-lamellar fabric in a modern gastropod, thin-section (XPL). Image width 0.5 mm. (c) Cross-section of a Pleistocene bivalve (*Glycymeris* sp.) illustrating the nacreous (top) and prismatic (bottom) skeleton layers, South Australia, thin-section (PPL). Image width 10 mm. (d) Cross-section of Middle Ordovician ostracod shells in carbonate mud, southern Ontario, Canada. Image width 6 mm. (e) Fragments of modern barnacle shells, Tasmania, thin-section (PPL). Image width 12 mm. (f) Pieces of Middle Ordovician trilobites, illustrating the typical 'shepherd's crook' cross-section, southern Ontario, Canada. Image width 2 mm.

Figure 6.6 (a) A suite of different modern bivalves. Specimens provided by Y. Bone. (b) A monospecific collection of pecten bivalves (*Sectipecten* sp.), Miocene, Chatham Islands, New Zealand. Hammer head 20 cm.

abundant, with most being bysally attached or burrowers in shallow neritic environments. They underwent a second radiation in the Mesozoic, with the evolution of a muscular foot and inhalant and exhalant siphons that gave them an ecological advantage over brachiopods. These features allowed them to more fully exploit the deep burrowing, intertidal, and endolithic (boring) life styles.

Like the brachiopods, they are architecturally diverse. Some of the largest shells are the tridacnids (giant clams of the tropical Pacific Ocean). These large bivalves can be up to 1 m wide and weigh as much as 180 kg, and most contain photosymbionts. Bivalve shells are smooth or ribbed, and some are adorned with spines. The basic bivalve shell structure comprises an inner lamellar layer and an outer prismatic layer. Aragonite elements are nacreous and cross-lamellar (Figure 6.5c). The nacreous layer comprises aragonite brick-like crystals in an organic matrix that gives the layer a lustrous appearance that is prized in jewelry making. The prismatic layer, by contrast, is generally calcite. The calcite shells or calcite portions of bimineralic shells are generally preserved during diagenesis, whereas the aragonite is typically dissolved away and the void can be filled by calcite cement. The shells are usually curved fragments but can show periodic growth lines (not found in ostracods), although some have perforations similar to brachiopod punctae.

Rudists. The bivalve group includes the rudists (Figure 6.7a), which were large, sessile, gregarious bivalves that thrived during the Late Jurassic through to the terminal Cretaceous. These were some of the most bizarre bivalves in the geological record, commonly with one huge valve and one small valve; they were, however, dominant sediment producers and limestone formers for almost 100 my. Rudists exhibit a variety of skeletal mineralogies. Some were almost entirely aragonitic (caprinids or polyconitids), some were mostly LMC (radiolarids), and still others were equal portions of each mineral (hippuritids or requienids).

They are usually divided into three morphological types: recumbents, clingers, and elevators. *Recumbents* were not attached, but the lower valve was stretched out flush with the substrate. They lived on top of mobile sand or shelly seafloors that were subject to continuous winnowing and deflation, but little sediment accumulation. Most animals were large, openly curved or radially splayed and had thick, mostly aragonitic, shells. *Clingers* had their lower valve entirely attached to the substrate (stable sediment, other shells, or hard grounds) over a broad zone and the shells were generally horizontally oriented. The valves could be spiral, prone, or expansive conical. *Elevators* were animals whose large lower valve was wholly involved in upward growth, trying to keep ahead of sediment accumulation. They lived implanted in muddy substrates that were periodically disturbed by storms. Individuals were locally stabilized by lateral attachments to their neighbors, both solitary and clustered, but commonly clustered. These organ pipe-like communities formed the most striking of all reef-like rudist aggregations. Although individual hippuritids and radiolitids are more than a meter in length, it is unlikely

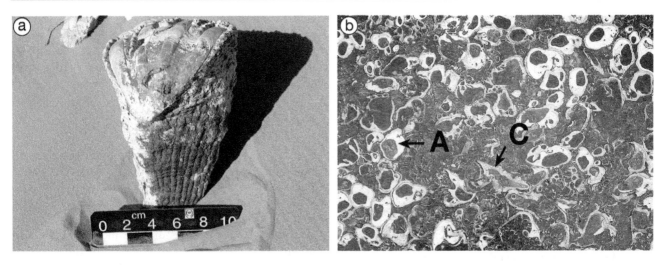

Figure 6.7 Rudist bivalves. (a) A single specimen of the Late Cretaceous (lower Campanian) hippuritid *Torreites* sp. from central Oman. There is convincing evidence that this particular form had photosymbionts in its tissue. (b) A polished slab of Urgonian Limestone (late Aptian–early Albian) from Spain with numerous *Horiopleura lamberti* (family Polyconitidae) illustrating the typical two-layered shell structure of rudists, with the calcitic outer shell (C) layer in pale grey, and the recrystallized aragonitic inner shell (A) preserved as white spar. Image width 1 m. Photograph by P. Skelton. Reproduced with permission.

Figure 6.8 (a) Cephalopod, Mingan Island, Québec, Canada. (b) Close-up of an ammonite illustrating the numerous sediment-filled chambers, Jurassic, Morocco. Image width 8 cm.

that during life they protruded more than a few centimeters above the seafloor sediment.

Rudists thrived in all neritic environments and produced large amounts of sediment at platform margins. It is doubtful that they ever formed a rigid reef framework, however. Finally, it is not confirmed if they harbored photosynthetic symbionts, like the giant clams in the modern Indo-Pacific tropics, but their large size compared to contemporary bivalves suggests that they did.

Cephalopods

Although most of these highly developed mollusks were active nektic predators, some were mobile benthic animals. They are most common in open-shelf and deep-water deposits. They comprise three different groups: the nautiloids, ammonoids, and coleiids. The nautiloids (Figure 6.8a) (late Cambrian–Holocene) have an external chambered shell with simple sutures between the chambers. The shell can be straight or coiled. They were active swimmers and lived close to the sea bed. Most are extinct, and the only coiled nautiloid found in the modern ocean is *Nautilus*, which lives in 150–300 m water depth in the south Pacific. Ammonoids (Figure 6.8b) (Late Devonian to Late Cretaceous) and goniatites (a type of ammonoid, Mississippian to Eocene) also have an external shell that is usually planispirally coiled with variable and more complicated sutures. Most skeletons are centimeters to decimeters in size, but some of the largest ammonoids were up to 2 m in diameter.

The aragonite shells of both nautiloids and ammonoids are chambered with an elongate siphuncle that connected the chambers. The shell wall is layered with an outer prismatic and an inner nacreous and prismatic layer. The animal lived in the outermost chamber where it was protected by an aptychi (cover) that could seal off the chamber. The aptychi can be the only preserved part of the skeleton because it was LMC.

Coleiids have an internal and reduced shell, and in some the shell is completely absent. These comprise all living cephalopods except *Nautilus* and include the modern squid, octopus, and cuttlefish. Belemnites (Late Mississippian–Late Cretaceous) had an internal LMC skeleton consisting of a solid chambered phragmacone and make up most fossil forms. Belemnites were common throughout the Paleozoic and particularly numerous in deep Silurian and Devonian environments with arrested sedimentation.

Tentaculitids

These small cone-shaped extinct pelagic fossils, typically <25 mm long and of uncertain affinity, are abundant in Ordovician–Devonian neritic and basinal carbonates. The slender conical calcitic shells have concentric thickenings on the outer wall. The shells occur in abundance, especially in argillaceous pelagic and hemipelagic limestones where they are often replaced by pyrite. Their calcitic shells can be confused with brachiopod spines.

Worm tubes

Worms with external protective tubes were present throughout the Phanerozoic. Tubes are calcareous (Figure 6.9) or agglutinated. The calcareous tubes are centimeter-sized, straight, curved, or spirally coiled and formed of calcite, aragonite, or both.

Serpulids and spirorbids are the most common and numerous calcareous tube-forming worms. The serpulids (feather worms) live in tropical to polar environments, where they have adapted to life in shallow and rocky coastal areas, on seagrasses, and on hard and soft seafloors. In some areas, including South Florida, they form small reefs. They are also common in Cenozoic lacustrine settings, where they construct small reefs in association with microbialites.

The serpulid skeleton, from <1 cm to 10 cm long and ~1 cm in diameter, is composed of HMC, aragonite, or a combination of both, with some being chitinous and phosphatic. They can be free-standing, can encrust other shells and can be straight, sinuous or spiral in style and can form intergrown clusters. Their distinctive tubular shape and encrusting morphology can be confused with that of vermetid gastropods, scaphopods (but these have a radial shell structure) and some tubular foraminifers. Their skeletal wall has a foliated microstructure and consists of a fine cone-in-cone structure. Aragonitic forms

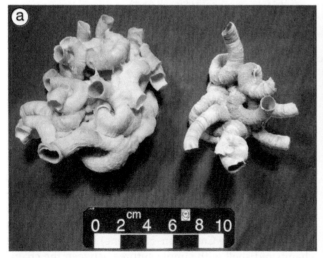 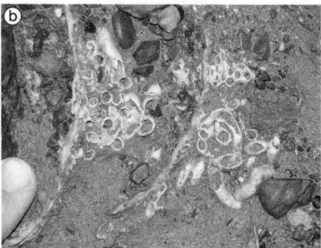

Figure 6.9 (a) A cluster of serpulid worm tubes (*Surpulorbis* sp.), intertidal, southern Australia. (b) Cross-section of serpulid worm tubes encrusting a bivalve shell in the Pleistocene marine glacial diamicts on Middleton Island, Alaska. Image width 6 cm.

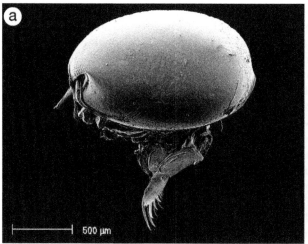
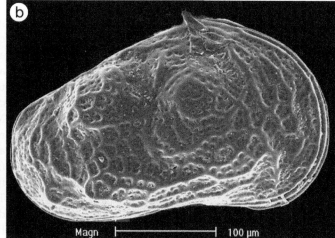

Figure 6.10 (a) A living ostracod with appendages extending out from the shell, southern Australia. (b) An ornamented ostracod shell, southern Australia.

have little preserved wall structure. The external skeletal structure of the serpulids can easily be confused with trilobite spines because it is characterized by a series of concentric rings.

Spirorbids are a group of very small (2–5 mm) post-Cretaceous polychaete worms that are a somewhat inconspicuous but important part of carbonate factories. They typically live attached to macroalgae and seagrasses, and their calcitic fragments accumulate with other carbonate particles when the host dies.

Crustacean arthropods

This group includes, among others, the ostracods, barnacles, and trilobites. Each animal produces five to seven shells throughout its life because the growth cycle includes numerous episodes of molting.

Ostracods. These late Cambrian–Holocene mobile arthropods, generally <1 cm long (but were up to 8 cm long in the Paleozoic), have a bean-shaped skeleton formed of two valves that are articulated by teeth along a hinge (Figure 6.10a). The skeleton is calcitic with an Mg-content that increases with rising water temperature. They live in almost all aquatic environments. Although most crawl or burrow into sediment, some modern forms are swimmers. They have the ability to thrive in stressed environments where few other animals live, and so are particularly useful for sedimentological analysis. They were exclusively marine throughout the Paleozoic but moved into lacustrine environments in the Mesozoic.

They molt several times in their lives. As a result, beds with thousands of specimens but low taxonomic diversity

are common. The carapace can be heavily calcified and ornamented with ribs, ridges, or tubercles (Figure 6.10b), or can be smooth and featureless. The shell is chitin and calcite (LMC to HMC) and is generally well preserved in the rock record. The shell wall has a prismatic to finely prismatic microstructure with crystal orientation perpendicular to the shell wall (Figure 6.5d). As a result, in contrast to most bivalves, they exhibit sweeping extinction. Like bivalves, however, they can have perforate walls. They do not, however, exhibit growth lines.

Barnacles. These sessile post-Cretaceous arthropods (Figure 6.11), formed of numerous plates, typically attach themselves to hard substrates such as rocks, shells, drifting

Figure 6.11 Two species of barnacles *Chthamalus* sp. (smaller) and *Balanus* sp. (larger) on boulders in the intertidal zone, Middleton Island, Alaska. Finger 1 cm wide.

Figure 6.12 A trilobite from the Devonian of Morocco. These arthropods typically disintegrate into many segments in carbonate environments. Image width 3 cm. Photograph by R. A. Fortey. Reproduced with permission.

numerous in early Paleozoic carbonates. Their segmented exoskeleton, formed of chitin, phosphate, and calcite (Mg-calcite), comprises a cephalon (head shield), a many-segmented thorax (abdomen 2–40 segments), and a pygidium (tail). Some were adorned with spines that can be confused with brachiopod spines (the latter consist of two calcite layers and are not homogeneous). Most were benthic and walked on the seafloor or burrowed in the sediment. A few were nektic and planktic and some were even blind. They are normally found as fragments, with the recurved nature of the shields resulting in a hook or shepherd's crook shape in thin-section. Their phosphatic nature allows them to stand out as dark-blue–grey particles against most other grey biofragments. Shells have a homogeneous prismatic structure with fine calcite prisms normal to the carapace surface. The wall appears homogeneous with no obvious crystals, but has sweeping extinction under cross-polarized light (Figure 6.5f). The shell structure is similar to that of ostracods and some bivalves, but they are generally larger than ostracods and more irregular in curvature than both.

wood, and even swimming organisms (e.g., whales). A few are swimmers. Their test is calcitic and they are especially abundant in modern temperate- and cold-water environments. A series of six to eight immovable compartmentalized wall plates are attached to a basal plate and house the organism, which is able to move its other opercular plates. They shed their triangular-shaped plates during molting to form coarse sands and gravels. The plates have a homogeneous granular or foliated microstructure that is distinctively hollow with numerous vesicles (Figure 6.5e) and, being HMC, they are typically well preserved.

Trilobites. These familiar but extinct (early Cambrian–late Permian) arthropods (Figure 6.12) are especially

Further reading

Armstrong, H. and Brasier, M. (2005) *Microfossils*, Second edition. Blackwell Publishing.

Boardman, R.S., Cheetham, A.H., and Rowell, A.J. (eds) (1987) *Fossil Invertebrates*. Palo Alto, CA: Blackwell Scientific Publications.

Flügel, E. (2004) *Microfacies of Carbonate Rocks: Analysis, Interpretation and Application*. Berlin: Springer.

Hallock, P. (1999) Chapter 8. Symbiont-bearing foraminifera. In: Sen Gupta B. (ed.) *Modern Foraminifera*. Amsterdam: Kluwer Press, pp. 123–139.

Hallock, P. (2011) Foraminifera. In: Hopley D. (es.) *Encyclopedia of Modern Coral Reefs*. Heidelberg: Springer, pp. 416–421.

Levin, H.L. (1999) *Ancient Invertebrates and Their Living Relatives*. Glenville, Illinois, USA: Prentice-Hall.

Milsom, C. and Rigby, S. (2010) *Fossils at a Glance*. Second edition. Oxford: Wiley-Blackwell.

Scholle, P.A. and Ulmer-Scholle, D.A. (eds) (2003) *A Color Guide to the Petrography of Carbonate Rocks: Grains, Textures, Porosity, Diagenesis*. Tulsa, Oklahoma: American Association of Petroleum Geologist, Memoir no. 77.

Stanley, S.M. (1970) *Relation of Shell Form to Life Habit of the Bivalvia (Mollusca)*. Geological Society of America, Geological Society of America, Memoir no. 125.

CHAPTER 7
THE CARBONATE FACTORY: ECHINODERMS AND COLONIAL INVERTEBRATES

Frontispiece A living fenestrate bryozoan (*Reteporella* sp.) from ~15 m water depth off Victoria, southern Australia. Image width 4 cm.

Origin of Carbonate Sedimentary Rocks, First Edition. Noel P. James and Brian Jones.
© 2016 Noel P. James and Brian Jones. Published 2016 by John Wiley & Sons, Ltd.
Companion website: www.wiley.com/go/james/carbonaterocks

Introduction

This chapter deals with how the skeletons of colonial animals such as sponges, bryozoans, and corals, together with echinoderms, are generated and their importance in the depositional spectrum. Most of these animals are sessile, being attached to the seafloor or other stable substrates. Such organisms, commonly large in size, frequently construct reefs and are important components of many limestones. Echinoderms, especially crinoids and echinoids, are prolific sediment producers.

Echinoderms

Echinoderms (Figure 7.1), known from Cambrian–Holocene environments, are exclusively marine invertebrates that mostly employ a benthic mode of life. The classification of echinoderms is continually changing, but in terms of sediment producers they can be broadly divided into: the *crinoids* (stalked forms, many of which are attached to the seafloor) and the *echinoids* (crawling and burrowing forms). The crinoid group includes true crinoids (sea lilies) and blastoids (the latter exclusively Paleozoic), whereas the echinoids embody the asteroids

Figure 7.1 (a) A suite of living echinoderms including asteroids (A), infaunal echinoids (I), and epifaunal echinoids (E). Specimens courtesy of Y. Bone. (b) Numerous echinoids on seafloor off southern California. Image width foreground 0.7 m. Photograph by Dale Stokes. Reproduced with permission. (c) The "decorator urchin" *Lytechinus* sp. ~15 cm in diameter with numerous spines holding onto a biofragment. Upon death, the urchin will form numerous carbonate particles. Found in water depths of ~1 m, Bermuda. (d) The holothurian *Stichopus* sp., ~20 cm long, whose only hard parts are small calcareous spicules but which ingests and excretes large amounts of muddy carbonate sediment (arrow).

(starfish), ophiuroids (brittle stars), echinoids (sea urchins), and holothurians (sea cucumbers) (Figure 7.1).

Many echinoderms, largely because of a unique hydraulic operating system that works best in normal seawater, are particularly sensitive to salinity changes. Modern echinoids live in waters that range from 22 to 38‰. Crinoids, by contrast, are restricted to waters close to normal salinity. This fact is particularly useful for interpretation of the rock record, because the presence of crinoid fragments confirm paleo-seawater of near-normal salinity.

The echinoderm skeleton, characterized by five-fold symmetry, is formed of plates and spines or sclerites. Each element is typically a single calcite crystal that is characterized by its crystallographic uniformity. This is unlike almost all other invertebrates that have skeletons composed of felt-like masses of tiny crystals. Plates also have numerous small (~10 μm) holes (stoma). These holes result in a particle of relatively low density. As a result, relatively large echinoderm particles have a lower hydraulic equivalence compared to dense grains of the same size. They can therefore be transported and sorted by much weaker currents than would be expected.

While the animal is alive, the skeletal elements are held together by organic material. Upon death, the organic material quickly decays (in a matter of days) and the skeleton rapidly disintegrates. It is estimated, for example, that one crinoid disintegrates into several hundred pieces (Figure 3.6b): the ultimate sediment producer! Echinoids are likewise significant sediment producers as their spines and the plates separate and are dispersed across the seafloor.

Crinoids and allied stalked echinoderms

The crinoidea comprise the crinoids (Ordovician–Holocene), the blastoids, and the cystoids, which did not survive the Permian extinction and are therefore found only in Paleozoic rocks. All of these filter feeders (Figure 3.6a) have five arms, a calyx, a stem, and a holdfast. Stem plates are circular to pentagonal with a hole in the center, whereas arm plates are characteristically U-shaped (Figure 7.2). The particles are well preserved because they were originally HMC to LMC.

They were particularly numerous during the Paleozoic and Mesozoic, with crinoids being the most abundant echinoderms in the Paleozoic. Today, stalked crinoids are largely restricted to deeper-water settings from ~200 to 4000 m but live across the temperature spectrum from the tropics to the Arctic. Some stalked crinoids have recently been seen "walking" across the seafloor. Stalkless (comatulid) crinoids (feather stars) evolved in the Mesozoic and today live in shallow tropical and temperate seas. Some are particularly adapted to living on shallow coral reefs.

Sediments produced by crinoids and associated organisms were particularly abundant between normal and storm wave bases on Paleozoic ramps. The particles are especially numerous on reef flanks, in storm deposits, and around mounds. Some Devonian–Pennsylvanian carbonates are formed entirely of crinoid ossicles; these are called encrinites (Figure 7.2). These sediments beautifully illustrate the principle that skeletal architecture and not necessarily hydrodynamics largely controls grain size; for example, coarse-grained deep-water crinoid

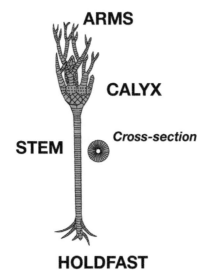
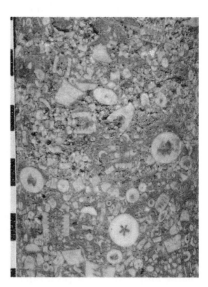

Figure 7.2 Sketch of a generic crinoid and subsurface core of Mississippian limestone, composed almost entirely of crinoid pieces formed by the spontaneous disintegration of the skeleton after death, Western Canadian Sedimentary Basin, centimeter scale.

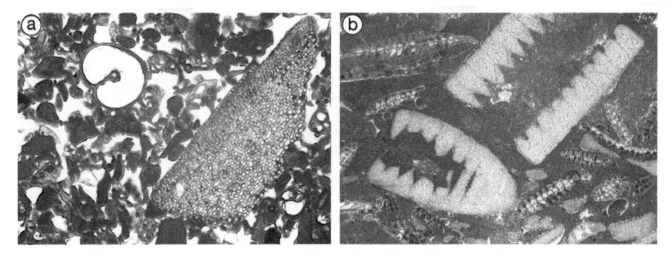

Figure 7.3 (a) Pliocene biofragmental grainstone containing a gastropod mold (upper left) and echinoid plate with empty stoma (holes), Western Australia, thin-section (PPL). Image width 2 mm. (b) Cross-section of two large crinoid pieces, each one being an optically single calcite crystal, Middle Ordovician, southern Ontario, Canada, thin-section (PPL). Image width 10 mm.

sands are found in waters that have few currents. Crinoids became profoundly less important as neritic carbonate sediment producers after the Cretaceous because of their shift in habitat to deep-water basinal environments.

These skeletons are by far the easiest of any biogenic grains to identify in thin-section. This is because the fragment extinguishes as a single calcite crystal under cross-polarized light and wall plates will have stomal texture present (Figure 7.3). Grains are typically surrounded by large epitaxial calcite cement crystals (Chapter 25).

Echinoids and allied stalkless echinoderms

Sea urchins, which evolved in the Ordovician, have been important sediment producers since the Jurassic. Modern echinoids (Figure 7.1a–c) occur in every conceivable marine environment from tropical to polar waters. Regular forms (almost perfect symmetry), which include the cidaroids (pencil urchins), live on hard bottoms. Irregular types with bilateral symmetry imposed on a pentameral (five-sided) structure, such as clypeasters (sand dollars) and spatangoids (heart urchins), are mobile and live within or on soft sediment. Even though they evolved in the Ordovician, echinoids were not significant carbonate producers until the Jurassic when they underwent rapid diversification. Less important in terms of sediment production are the stiff-armed starfish (asteroids) and delicate slender-armed ophiuroids (brittle stars).

Whole echinoids range from 1 to 10 cm in size. Their plates and spines are perforated with 15–25 μm pores.

Skeletons mostly break down into plate fragments and spines. A rule of thumb is that large spines come from regular urchins, whereas small spines come from the burrowing irregular echinoids. Like the crinoids and blastoids, they are easily identified in thin-section because their HMC to LMC skeletons are optically single crystals with common epitaxial cement. The particles are, however, more elongate than crinoid pieces. Only the teeth of regular urchins with up to 43 mol% $MgCO_3$ are polycrystalline.

Holothurians (sea cucumbers) (Figure 7.1d) are intensive and important sediment processors whereby they ingest soft mud-sized particles, remove the organic material, and excrete elongate pellets. In the process, their stomach acids can slightly dissolve some of the finer grains. Their only calcareous hard parts are silt-sized HMC stellate spicules.

Sponges

Modern sponges (Figure 7.4a) are largely marine and inhabit environments that range from intertidal to deep sea (5000 mean water depth or mwd). There are, however, some some sponges (mostly post-Cretaceous) that occur in freshwater settings including lakes. All sponges are filter feeders and thrive in nutrient-rich waters. Some of the shallow-water taxa have photosynthetic cyanobacteria and therefore have to live in the photic zone, whereas deep-water forms are heterotrophs. These most primitive multicellular animals with

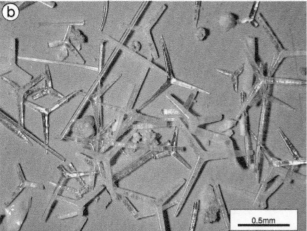

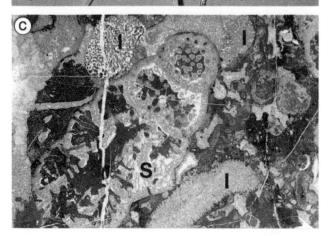

Figure 7.4 (a) A suite of shallow water spiculate sponges from temperate waters off Yorke Peninsula, southern Australia. (b) Opaline silica sponge spicules from modern temperate water sediments in the Great Australian Bight. (c) Thin-section image of Upper Triassic limestone from the Alps with inozoan (I) and sphinctozoan (S) calcareous sponges. Image width 1 cm. Photograph by R. Martindale. Reproduced with permission.

broadly cup-shaped bodies can be attached to hard or organic substrates, can grow in sand, and prefer quiet water environments.

Sponges can be classified as members of one of three groups: (1) calcarea with calcite spicules or cup walls that favor shallow-water environments; (2) demospongia with spongin or simple siliceous (opal) spicules embedded in a body wall that is generally unmineralized; and (3) hexactinellida that have silica (opal) spicules with a 6-fold symmetry and favor deep water (200–600 mwd) environments. Whole fossils of sponges are rare because they disintegrate into spicules upon death (Figure 7.4b). Sponges are attached to the seafloor by a holdfast that resembles roots or fine hairs that can attach to anything from hard substrates, to skeletons, to sand, and to mud. Apart from their spicules, the record of siliceous sponges is poor except for the Cretaceous where they are locally preserved in chert (flint) nodules.

Their impact on carbonate sediments comes largely from their skeletons and spicules (mainly siliceous). Some sponges, however, are borers and live in small submarine cavities that they excavate in hard substrates. In modern seas the sponge *Cliona* will attack any hard substrate, including coral heads, bivalve shells, and even concrete piers. Boring sponges (e.g., Figure 3.14b) are important because they: (1) produce large quantities of silt-sized grains (with characteristic scalloped faces) that are released into the areas around the substrate being attacked; (2) they bore by partly dissolving the substrate and thereby release Ca and CO_3 back into the ocean; and (3) they weaken substrates to the extent that subsequent storm activity can lead to substrate degradation.

The hexactinellids, demosponges, and calcareous sponges range from the Cambrian to the Holocene. From a sedimentological (but perhaps not taxonomic) perspective, they can be divided into spiculate and coralline sponges with the caveat that spicules can be fused into a skeleton (e.g., lithistid sponges). Some sponges had chambered segmented skeletons (sphinctozoans and inozoans) and were significant contributors to Permian–Late Triassic reefs (Figure 7.4c).

Spiculate sponges

Spicules (Figure 7.4b) occur as monaxons (single, commonly straight spicules), triaxons (three arms), tetraxons (four arms), and polyaxons (numerous arms). The symmetry and mineral composition of the spicules provides the basis for the classification of most sponges.

Demosponges, which comprise ~85% of all living sponges, have a spongin "skeleton" (a tough organic compound that readily decomposes after death), siliceous spicules, or a skeleton of fused opaline silica. Their spicules are monaxons, tetraxons, or polyaxons, but never triaxons. The hexactinellids, commonly called glass sponges, are characterized by siliceous tri-axon spicules. They form ~7% of modern sponges, and are commonly found in deep and cold ocean settings. The bodies of spiculate sponges have a relatively low preservation potential unless encrusted by microbes. Although such sponges were associated with reefs throughout the Paleozoic, they were particularly important in Jurassic reef facies. The original unstable opaline silica (opal-A $SiO_2.nH_2O$) usually converts to chert or chalcedony (SiO_2) or is replaced by calcite. Calcareous sponges have monaxon or tetraxon calcareous spicules that can be aragonite, LMC, or HMC. The spicules are generally optically single crystals.

Spicules are commonly hollow and therefore easily transported to basinal environments where they can accumulate in large numbers. Such spicules may be calcitized or diagenetically altered to chert (flint).

Some environments contain so many sponges that they produce spiculites, deposits composed almost entirely of siliceous spicules. These seafloor deposits are well developed together with microbial mats in very cold-water Arctic and Antarctic neritic environments. This is in contrast to most temperate or warm-water settings where spiculites are considered to be deep-water deposits. Recent discoveries of modern and Cenozoic spiculites in shallow inboard settings, however, suggest that this correlation is, not universal. Late Permian shallow-water neritic facies with numerous spiculites are interpreted to have been either the result of upwelling or local ocean acidification.

Coralline calcareous sponges

Coralline sponges are a polyphyletic group with affinities to both demosponges and calcareous sponges. Those that had a hard calcareous skeleton are today typified by sclerosponges. Unfortunately, their relationship to the sponges is still open to debate. Coralline sponges such as archaeocyathids are common in some lower Cambrian successions, whereas stromatoporoids were important biotic elements in many lower and middle Paleozoic as well as Mesozoic sequences.

The archaeocyaths (ancient cups; Figure 7.5) were a group of sessile organisms that lived in early Cambrian tropical to subtropical neritic environments, where they

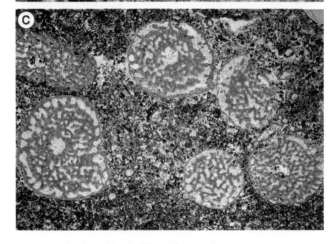

Figure 7.5 (a) Cross-section of a lower Cambrian archaeocyath cup, Siberia, thin-section (PPL). Image width 3 mm. (b) Cross-section of a Devonian laminar stromatoporoid, Western Canadian Sedimentary Basin, thin-section (PPL). Image width 8 mm. (c) Cross-section of several digitate Devonian stromatoporoids (*Amphipora* sp.), Western Canadian Sedimentary Basin, thin-section (PPL). Image width 16 mm.

commonly interacted with calcimicrobes to form the first true metazoan reefs. Their well-preserved calcitic cup-like cylindrical skeletons, which are up to 30 cm high and 20 cm in diameter, have a large central cavity that is surrounded by a double wall that is septate. A well-developed holdfast secured them to a solid substrate. The walls are very fine to microcrystalline calcite. They did not possess spicules.

The calcareous sphinctozoan (chambered and segmented) and inozoan (chambered but non-segmented) sponges (Figure 7.4c) with hard skeletons are particularly common in Carboniferous–Upper Triassic strata. They are important components of Permian and Triassic reefs. Most sphinctozoans had either an arrangement of irregularly clustered chambers or the chambers, were arranged in a subvertical series. Their walls are relatively thin and composed of microcrystalline calcite. The skeleton is not as well organized as that of corals. Superficially, they resemble some bryozoans in thin-section but they lack the fibrous wall structure of bryozoans.

The stromatoporoids (Figures 7.5, 7.6), known from Ordovician–Cretaceous rocks, were particularly abundant during the Silurian–Devonian. Once thought to be related to corals, they are now considered as sponges. They are morphologically diverse with growth forms that include delicate branching and dendroid (<1 cm diameter branches), laminar to wafer-like, large discs (up to 3 m diameter), and large bulbous growth forms (up to 1 m diameter). Some researchers have argued that their external morphology was controlled by environmental conditions with the delicate forms growing in quiet, lagoonal environments, whereas the more robust bulbous forms grew in higher-energy situations associated with reefs. Their large size and ubiquity in shallow-water environments has led to the suggestion that they may have harbored photosymbionts. This contention is, however, largely unresolved.

Their internal structure is distinctive and characterized by a cellular or lattice-like pattern of horizontal or concentric laminae and vertical or radial pillars (Figure 7.6d) composed of microcrystalline calcite. Good preservation suggests that they were LMC or HMC originally. The skeletal material generally has a brownish hue in thin-section due to the inclusion of organics. Aragonitic sclerosponges that today grow in water depths below the range of corals and in shallow marine caves are considered to be related to stromatoporoids.

Stromatoporoids dominated most Devonian neritic environments. Details of Devonian reefs are outlined in Chapter 15, but in brief the highlights are as follows. Lagoon environments were populated by innumerable stick-like growth forms of several genera (Figure 7.6a).

The reef crest was dominated by prone colonies that were commonly swept back into reef flat settings during storms (Figure 7.6b). The shallow fore-reef was a consortium of massive growth forms (Figure 7.6c), synsedimentary cements, and the calcimicrobe *Renalcis*. Only in the deeper fore-reef were stromatoporoids, mostly wafer-shaped colonies, associated with corals, brachiopods, and crinoids.

Middle Paleozoic stromatoporoids replaced most older coral–stromatoporoid associations and pushed the corals into relatively deep-water settings. This situation bears an eerie resemblance to the dominance of neritic environments by rudists in the Cretaceous, when corals were again confined to deep-water environments and mollusks ruled the shallow illuminated seafloor.

Bryozoans

Bryozoans (Frontispiece, Figure 7.7) range from the Ordovician to the Holocene but were particularly common from the Ordovician to the end-Permian. These predominantly marine, active suspension feeders never contain photosymbionts, form millimeter- to centimeter-sized colonies, and today live from intertidal to abyssal depths. They dominated many neritic paleoenvironments and today are most abundant between the intertidal zone and about 100 mwd. These organisms thrived in warm and temperate environments during the Paleozoic, but most lived in cool waters during post-Permian time. They have a distinct zonation of growth forms across temperate shelves where they can comprise >90% of the carbonate sediment.

The colony consists of connected individuals (zooids) that live in tube- or box-like chambers that have an opening at one end. All individuals are genetically identical. They feed via a lophophore, an array of tentacles that extract food from the water and have no circulation system or gills. Colonies range from millimeters to centimeters in size with individual zooecia usually less than a millimeter in diameter and length. The walls consist of laminar, foliated, or granular crystals of calcium carbonate.

There are two main groups: the stenolaemates and the gymnolaemates (Table 7.1). Stenolaemate skeletons are highly calcified and are characterized by elongate conical zooids. This group includes the trepostomates, cystoporates, cryptostomates, fenestrates, and tubuliporates (cyclostomes). These are the most common bryozoans found in Ordovician–Cretaceous limestones. Most stenolaemates were severely impacted by the end-Permian extinction, with few surviving through to the

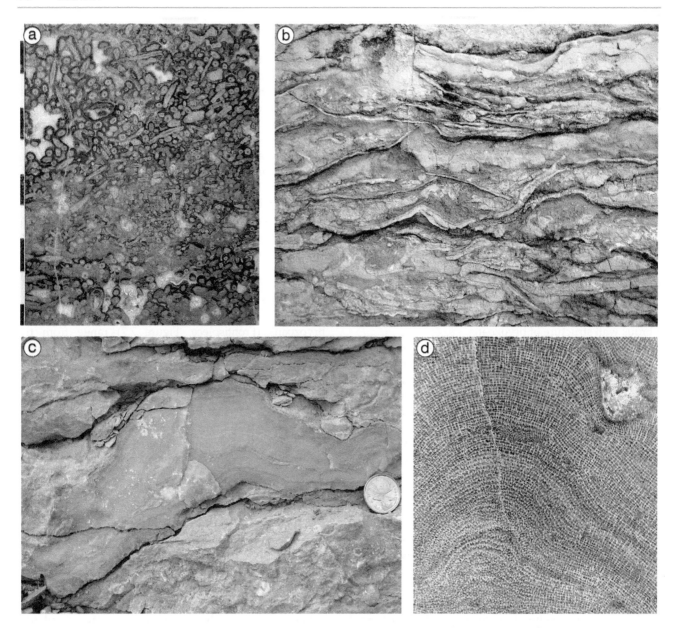

Figure 7.6 (a) The stromatoporoid *Amphipora* sp., a typical back-reef, lagoonal form, composed of many stick-like colonies in subsurface core, Upper Devonian, Western Canadian Sedimentary Basin. Centimeter scale. Photograph by W. Martindale. Reproduced with permission. (b) Laminar stromatoporoids from reef crest facies, Middle Devonian, Northwest Territories, Canada. Image width 30 cm. (c) A domal stromatoporoid from the Middle Devonian, Northwest Territories, Canada. Coin is 1 cm diameter. (d) Stromatoporoid in subsurface core illustrating the typical lamina and pillar structure of these coralline sponges, Upper Devonian, Western Canadian Sedimentary Basin. Image width 10 cm.

end of the Triassic (trepostomes and cystoporates). Only the cyclostomes, important in the Cretaceous, have persisted to the present. By contrast, the gymnolaemate (Silurian–Holocene) are generally less calcified, characterized by box-shaped zooids, and are morphologically more diverse. The group is dominated by the cheilostomes (Jurassic–Holocene), which are the main forms in modern oceans.

Most of the numerous colony forms developed by bryozoans have evolved repeatedly in different taxonomic groups and so vary in their sediment-producing potential. These forms are encrusting, dome-shaped, palmate, foliose, robust, branching, delicate branching, articulated, and free-living (Figure 7.8). The proportion of these morphotypes has changed significantly through time. Most Paleozoic bryozoan sediments come from:

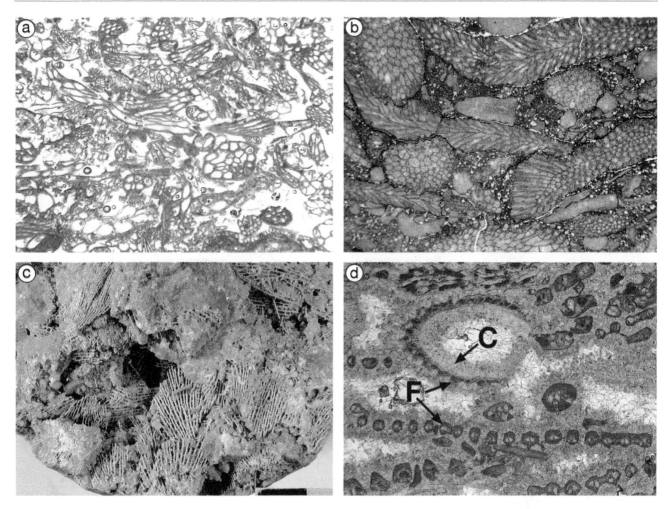

Figure 7.7 (a) Oligocene bryozoan grainstone, South Australia, thin-section (PPL). Image width 12 mm. (b) Middle Ordovician packstone, southern Canada, thin-section (PPL). Image width 16 mm. (c) A fenestrate bryozoan in subsurface core, Mississippian, Western Canadian Sedimentary Basin. Centimeter scale. Photograph by W. Martindale. Reproduced with permission. (d) Cross-sections of fenestrate bryozoans (F) surrounded by fibrous synsedimentary marine cement (C), Mississippian, Front Ranges, Rocky Mountains, Alberta. Image width 1 cm.

Table 7.1 Attributes of Phanerozoic bryozoans. Source: Taylor (2005). Reproduced with permission of Elsevier.

Class	Order	Morphology	Growth forms	Geological range
Gymnolaemata	Cheilostomata	Box-shaped zooids, often with complex frontal morphology and intricately shaped apertures	Diverse	Jurassic–Holocene
Stenolaemata	Cyclostomata	Cylindrical zooids with exterior frontal that are porous	Diverse	Ordovician–Holocene
	Trepostomata	Cylindrical zooids, no exterior frontal walls, non-porous interior walls	Encrusting, domal, erect branching and palmate	Ordovician–Triassic
	Cystoporata	Cylindrical zooids, no frontal walls, typically non-porous interior walls	Encrusting, domal, erect branching and palmate	Ordovician–Triassic
	Cryptostomata	Box-shaped to cylindrical zooids, no exterior frontal walls, non-porous interior walls	Encrusting, domal, erect branching, some articulated	Ordovician–Triassic
	Fenestrata	Box-shaped zooids, no exterior walls, non-porous interior walls	Narrow erect branched, fenestrate, zooids only open on one side of the branch	Ordovician–Permian

BRYOZOAN SEDIMENTOLOGY

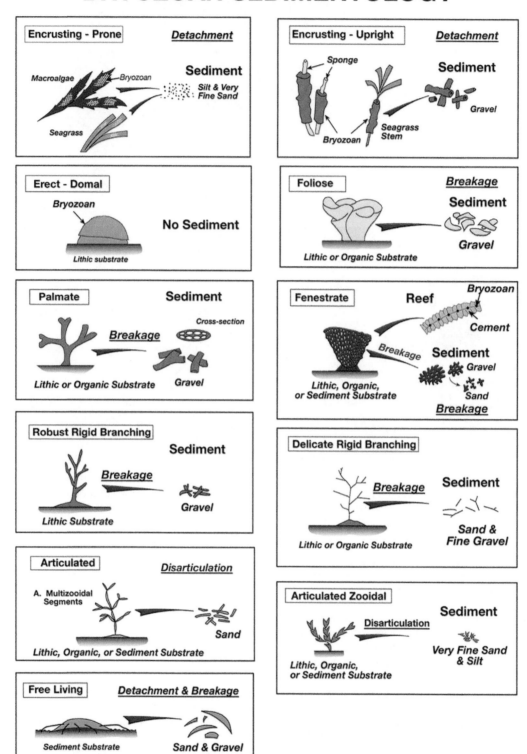

Figure 7.8 Typical post-mortem histories of bryozoans with different colony growth forms and their contributions to sediments. Source: Adapted from Taylor and James (2013). Reproduced with permission of John Wiley & Sons.

(1) domal, delicate-branching, robust-branching, and palmate forms; and (2) fenestrate forms (Figure 7.8). The former generate coarse sediment particles and were also components of stromatoporoid reefs in the early and middle Paleozoic. The delicate lattice-like fans of the fenestral bryozoa commonly formed the cores of late Paleozoic deep-water subphotic biogenic mounds (e.g., Waulsortian reefs). Nearly all post-Paleozoic bryozoan sediments are fragments of cyclostomes and cheilostomes, with many of the same growth forms but with the addition of free-living and colonies with articulated zooids. These latter types produced sand- and mud-size bryozoan sediment for the first time in geologic history.

The skeletal structure of stenolaemates, which is the most diverse group, is highly variable. Cyclostomes have very elongate zooecia and thin zooidal walls that are finely laminated or granular. Cystoporates are also thin-walled with short to long zooecia that can contain diaphragms. Pores characterize their laminated walls. Some (*Fistuliporina*) contain bubble-like pores (vesicles) between large zooecia and a granular to massive wall. Trepostomes are usually branching to encrusting colonies and have slender zooecia, with the central zooecial walls thin and fused together. Their walls are well developed and laminated, and thicken significantly towards the exterior (Figure 7.7b). The zooecial tubes contain numerous diaphragms that support the growing zooid in life, but act as internal cavities after death. Cryptostomes have short tubes that form delicate colonies and well-laminated walls that are thinner in the centre and become thicker outward. Fenestrates form mesh-like fronds that only contain a single layer of zooecia. They are quite distinctive in thin-section because they appear as a series of tiny circles or stitches (Figure 7.7d) as the section transects the wall.

In thin-section, bryozoans can be distinguished from other skeletons composed of calcareous tubes by the presence of polymorphic tubes of various widths and by a characteristic fan-like arrangement of the zooecia. The main differences in thin-section between bryozoans and corals are the smaller size of the bryozoan living chambers and the outward thickening of the zooidal walls.

Most evidence suggests that Paleozoic bryozoans were calcitic. Cyclostomes are LMC today, whereas cheilostomes can be either LMC, IMC, aragonite, or mixtures of calcite and aragonite.

Corals

Corals are common constituents of many limestones throughout the Phanerozoic (Table 7.2). Extinct Paleozoic tabulate and rugose corals were widely distributed in neritic paleoenvironments in both open subtidal and reefal settings. Mesozoic and Cenozoic scleractinian corals are important reef-builders but are also present in cool-water assemblages. Tabulates and rugose corals lacked a point of attachment, unlike the scleractinians that are commonly cemented to the substrate. The former two could not live in high-energy environments and were therefore most abundant in lagoonal and fore-reef environments. All of the corals have a similar basic structure: they grow as a calcareous cup (corallite) and within the corallite are vertical structures (septa) and horizontal plates (tabulae) to support the animal.

Table 7.2 Attributes of major coral taxa.

Feature	Rugose	Tabulate	Scleractinian
Age	Middle Ordovician–late Permian	Early Ordovician–late Permian	Middle Triassic–Holocene
Skeletal mineralogy	Calcite (most LMC)	Calcite (rare aragonite)	Aragonite
Morphological features	Solitary commonly horn-like; colonial variable architecture - branching massive	Colonial, with polygonal, oval, or circular corallites (up to 5 mm wide)	Colonial and solitary, highly variable growth forms; zooxanthellate or azooxanthellate
Distinctive features	Bilateral symmetrical arrangement of septa; well-developed outer wall with strong horizontal ridges or exterior	Corallites internally divided by horizontal tabulae; septa rare, short and spine-like	Skeletons with well-developed septa, generally arranged in six cycles giving hexameral symmetry
Size	Colonies up to 4 m diameter, solitary up to dm	Colonies up to 4 m high and wide	Highly variable sizes

Figure 7.9 (a) Large hemispherical favositid coral ~30 cm high, Middle Devonian, Twin Falls Formation, Northwest Territories, Canada. (b) Large rugose coral, Middle Devonian Alexandra Formation, Northwest Territories, Canada. Centimeter scale. (c) Large domical broken colony of a halysitid (chain coral) tabulate coral showing distinctive internal architecture (horizontal tabulae and vertical corallite walls), lower Silurian Anticosti Island, Canada. Note finger for scale.

Tabulates

The tabulate corals (Figures 7.9, 10a,b) occurred from the Early Ordovician to the late Permian and were particularly significant in the formation of many Silurian and Devonian reefs. Late Paleozoic tabulates are relatively rare. The skeleton of these colonial animals was calcitic and exclusively sheet-like, domal, erect branching, or chain-like in shape. Colonies were of variable size (4 mm to 4 m) and, within individual corallites, septa were commonly short or absent but numerous tabulae were present (Figure 7.9a). The corallites have slender polygonal, oval, or circular tubes with perforate walls. Corallites range from closely packed to loose aggregations. It is uncertain if these animals were mixotrophs and contained photosymbionts.

The wall structure is "fuzzy" in thin-section with fibers oriented normal to the plane of the tabulae. Whereas some have fibrous walls, others have walls of clear calcite crystals, and still others have a lamellar wall structure. Their individual living chambers are larger than those of bryozoans (Figure 7.9a, b).

Rugosans

These corals (Figures 7.9b, 7.10c), known from the Middle Ordovician to the late Permian, had their acme during the Silurian and Devonian and were both isolated and reef-associated organisms. Their numbers, however, declined significantly at the end of the Devonian. New families that evolved in the Mississippian persisted to the end of the

Permian and were more numerous than tabulates in the late Paleozoic.

The skeleton of these solitary or colonial, cup-shaped, and locally conical or cylindrical individuals was calcitic; colonies were horn-shaped to massively dendroid. Colonial types formed domes up to 1 m in diameter. The outside wall of the coral is generally strongly ribbed (rugae). There is no clear evidence as to whether or not these corals contained photosymbionts.

Their internal structure is characterized by abundant dissepiments and tabulae. Individuals have a distinctive bilateral symmetry of septal arrangement, and septa are better developed compared to the tabulates. The well-preserved wall structures are distinctly brownish and have a fuzzy fibrous fabric, whereas others have a clear calcite wall.

Scleractinians

These modern corals (Figures 7.10d, 7.11) have aragonitic skeletons and are solitary or colonial. They have been prominent reef-builders since their evolution in the Late Triassic. Photozoan tropical forms are mixotrophs and contain endosymbiotic zooxanthellae (photosymbiotic dinoflagellates, red and green algae, and cyanobacteria) that allow them to grow to a large size but restrict them to the photic zone (maximum water depth of ~100 mwd; Figure 7.11a, b). Corals are either zooxanthellate (with symbionts) or azooxanthellate (without symbionts) (Figure 7.11e). Zooxanthellate forms are often termed hermatypic (reef-building), whereas the non-azooxanthellate types are referred to

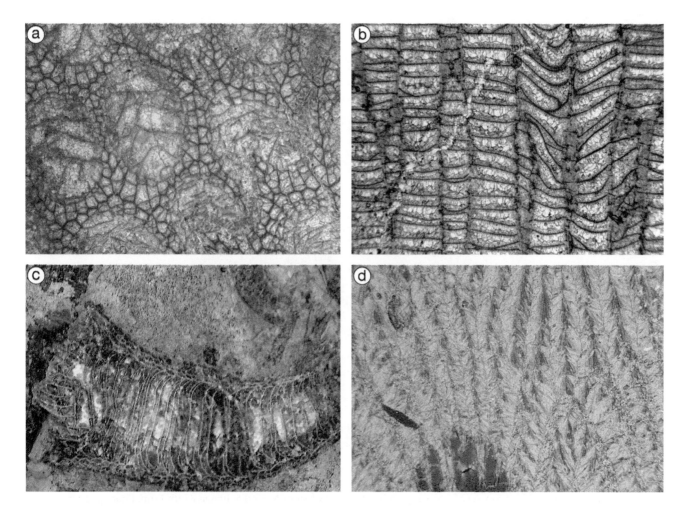

Figure 7.10 (a) Cross-section of Silurian tabulate coral, Anticosti Island, Québec, Canada, thin-section (PPL). Image width 20 mm. (b) Longitudinal section of a Silurian tabulate coral, Anticosti Island, Québec, Canada, thin-section (PPL). Image width 10 mm. (c) Cross section in outcrop of rugose coral (*Streptelasma rusticum*), Ordovician, Cincinnatian, Whitewater Mbr, Finneytown, Ohio. Photograph by D. Ulmer-Scholle. Reproduced with permission. (d) Cross-section of a modern scleractinian coral illustrating the spherulitic aragonite skeletal structure, Bermuda, thin-section (XPL). Image width 3 mm.

Figure 7.11 Living scleractinian corals. (a) *Diploria labyrinthiformis* (D) and *Montastraea* sp. (M) in aggressive contact, Bermuda. Image width 1 m. Photograph by W. Martindale. Reproduced with permission. (b) Branching *Acropora formosa*, Ningaloo Reef Western Australia. Image width 1.5 m. (c) *Montipora* sp., Ningaloo Reef Western Australia. Image width 1.0 m. Photograph by P. Pufahl. Reproduced with permission. (d) *Fungia* sp., a large single coral polyp that has the ability to move very slowly across the seafloor, Ningaloo Reef Western Australia. Image width 0.5 m. Photograph by P. Pufahl. Reproduced with permission. (e) Several specimens of the solitary azooxanthellate coral *Caryophyllia* sp. from ~110 m water depth on the Lacepede Shelf, Southern Australia. Image width 5 cm.

as ahermatypic (non-reef-building). The latter are generally small, relatively slow growing, and can exist in the deepest depths of the ocean where the waters are cold and perpetually dark.

Septa are arranged in six cycles (they are often called hexacorals) and dissepiments are shelf-like. The wall microstructure is distinctive when preserved. Fine spherulitic crystal arrays of fibrous aragonite crystallites radiate outward from a center of calcification. The spherulites are vertically stacked and appear as a fine dark line in longitudinal thin-section (Figure 7.10d).

Further reading

Fortey, R.A. (1982) *Fossils: A Key to the Past*. London: William Heinemann Ltd and British Museum (Natural History).

Levin, H.L. (1999) *Ancient Invertebrates and Their Living Relatives*. Glenville, Illinois, USA: Prentice-Hall.

Taylor, P.D. (2005) Bryozoans. In Selley R.C., Cocks L.R.M., and Plimer I.R. (eds) *Encyclopaedia of Geology*. Amsterdam: Elsevier, pp. 310–320.

Taylor, P.D. and James, N.P. (2013) Secular changes in colony forms and bryozoan carbonate sediments through geological history. *Sedimentology*, 60, 1184–1212.

PART II
CARBONATE DEPOSITIONAL SYSTEMS: AN OVERVIEW

Frontispiece Cliff of Upper Ordovician carbonate storm deposits, Anticosti Island, Québec, Canada, person for scale at left.

Origin of Carbonate Sedimentary Rocks, First Edition. Noel P. James and Brian Jones.
© 2016 Noel P. James and Brian Jones. Published 2016 by John Wiley & Sons, Ltd.
Companion website: www.wiley.com/go/james/carbonaterocks

Introduction

Carbonate depositional systems occur on land, across shallow marine settings, and onto the deep sea or basin floor (Figure II.1). Their attributes, as stressed in the previous chapters, are determined principally by ambient air or water temperature, topography or bathymetry, hydrodynamics, the saturation state of precipitating waters, and the evolving biosphere. These parameters control the specific suite of attributes and facies in each separate system.

Terrestrial systems

Carbonate rocks occur in a variety of terrestrial settings, with the most important being: (1) lakes, marshes, and soils; (2) caves; and (3) springs. Like their marine cousins, however, water is always involved, carbonate saturation states are important, and organisms are generally participating in one way or another. This part of the book concentrates on lake, lake margin, and spring depositional systems. The attributes of caves and calcareous soils, which involve much diagenesis, are covered in Chapter 26.

Lakes and marshes

Limestones and dolostones generated in these environments are by far the most extensive and economically important terrestrial accumulations. Deposits form in freshwater or saltwater lakes under a variety of climate regimens, ranging from environments that are highly vegetated in humid regions to evaporitic in arid areas. Lacustrine carbonates such as those in the Caspian Sea and Great Salt Lake rival many marine basins in size and stratigraphic thickness. Palustrine carbonates are mostly shallow freshwater deposits that have undergone extensive pedogenic modification, commonly in marshes and swamps; for example, those in the everglades of Florida.

Springs

Calcareous spring deposits develop wherever carbonate-charged subsurface waters emerge at the surface of the Earth. Having circulated at various depths within the crust, spring waters range from cold (<10 °C) to boiling (100 °C), from acidic (in some cases with pH<1) to alkaline (pH>8), and can contain many different dissolved solids at highly variable concentrations. With such variability, it is not surprising that many different types of precipitates can form from spring waters. Carbonate spring deposits are widespread, with the precipitation of calcite, aragonite, or both (and even dolomite in rare examples) being dictated by the manner in which the spring waters interact with local climate. As might be expected, microbes typically play a critical formative role in this style of carbonate precipitation.

Strandline systems

Muddy tidal flats

These low-energy peritidal depositional systems are widespread in time and space and, together with adjacent open marine sediments, can produce considerable thicknesses of meter-scale strandline cycles. They generally occur along somewhat protected shorelines and form as a result of the combined action of daily tides and cyclonic storms, the latter sweeping muddy waters onto the flats. Sequences comprise stacked subtidal, intertidal, and supratidal deposits. The character of such cycles is a function of local climate and the nature of the adjacent subtidal environments. Many such sequences

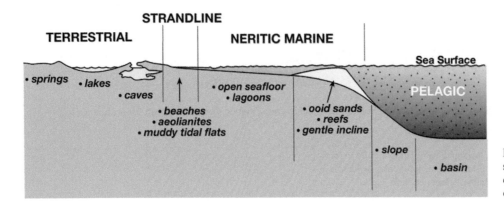

Figure II.1 A simplified cross-section of the suite of sedimentary environments encountered in carbonate depositional systems.

can be stacked on one another to form impressive successions of rocks.

Beaches and aeolianites

Beaches are generally adjacent to grainy high-energy shelf systems. They are best developed on open shelves and on inner ramps where waves and swells sweep unimpeded onshore. These deposits have the same general attributes as siliciclastic beaches. One important difference, especially in the tropics, is the presence of carbonate beachrock layers in the intertidal zone (see Chapter 24).

A visitor to many modern carbonate shorelines cannot help but be impressed by the windblown carbonate dunes (aeolianites), both modern and Pleistocene, that are present along and just behind the beach. These dunes form linear ridges with large cross-beds that can stand tens of meters above the beach. They are most active today under warm climates in areas with onshore winds that lie between 20° and 40° in both hemispheres. As such, their sediments can be formed by either photozoan or heterozoan components. Interbedded paleosols and calcareous soils (calcretes) are particularly common in Pleistocene aeolianite fields.

Marine systems

Open seafloor

This can be the widest and most areally extensive of all depositional systems. The deposits are often just referred to as *subtidal* and, in the past, have formed across unimaginably vast shallow illuminated ocean floors (epeiric seas). Under normal marine salinities, this environment is a prolific carbonate factory where a wide range of deposits is produced. If it is a lagoon that lies protected behind a barrier of some sort, the sediments are generally muddy. If the platform is unrimmed, then the sediments are grainy because of the large waves and rolling swells that sweep unimpeded across the shelf. The system is, however, profoundly different in warm tropical versus cool temperate and cold polar factories. Regardless, the open seafloor is where the widest diversity and abundance of benthic organisms thrive. They occur as epifaunal, infaunal and binding, boring and burrowing, constructing and destructing, autotrophs and heterotrophs. Finally, deposits can be profoundly altered when disturbed by cyclonic storms.

Shelf edge

The shelf edge or platform margin is critical because it effectively forms the boundary between the open ocean and the shallow-water shelf or lagoon. Reefs, sand shoals, or both characterize this zone in tropical systems. The shelf edge in cool-water systems is, by contrast, generally devoid of reefs because ocean conditions preclude the growth of corals and other reef-forming organisms. The shelf edge or margin on these temperate shelves is therefore generally deep with a gradually increasing gradient that passes imperceptibly into the slope proper. Communication between the shelf and the open ocean is good because there are no impediments to water movement between the two environments.

Slopes

Carbonate slopes pass from shallow to deep water and have inclines that range from <1° to >30° to subvertical. They are the middle- and outer-ramp environment and the generally steep slope between the shallow rimmed platform margin and basin. Deposition takes place either on the slope proper or sediment sweeps across the slope to accumulate at the base of slope along the basin margin. The deposits are a combination of pelagic carbonate fallout and sediment gravity flows.

Pelagic carbonates

These deep-water systems comprise fine-grained sediment and the skeletons of calcareous plankton. Deposition also takes place via the fallout of mud generated by storms on the platform or ramp and swept out to sea. Such deposits are rare in the Paleozoic because there were no calcareous plankton, but are prolific in post-Jurassic sediments because of the explosive evolution of these organisms.

FURTHER READING

In addition to those provided in Part I, the following selected references are useful for readers who would like further information on specific aspects of carbonate depositional systems. Additional references at the end of each chapter deal with more specific topics.

Ahr, W.M., Harris, P.M., Morgan, W.A., and Somerville, I.D. (2003) *Permo-Carboniferous Carbonate Platforms and Reefs*. Tulsa, OK, USA: SEPM (Society for Sedimentary Geology), Special Publication no. 78.
Numerous papers on carbonate sedimentation in this important period in geologic history.

Burchette, T.P. and Wright, V.P. (1992) Carbonate ramp depositional systems. *Sedimentary Geology*, 79(1–4), 3–57.
The first synthesis of ramps as part of carbonate sedimentology.

Ginsburg, R.N. (ed.) (2001) *Subsurface Geology of a Prograding Carbonate Platform Margin, Great Bahama Bank; Results of the Bahamas Drilling Project*. Tulsa, OK, USA: SEPM (Society for Sedimentary Geology), Special Publication no. 271, vol 70.
A series of papers discussing the three-dimensional aspects of a famous modern carbonate platform.

James, N.P. and Clarke, J.A.D. (eds) (1997) *Cool-Water Carbonates*. Tulsa, OK: SEPM (Society for Sedimentary Geology), Special Publication no. 56.

James, N.P. and Bone, Y. (2010) *Neritic Carbonate Sediments in a Temperate Realm, Southern Australia*. Dordrecht, Netherlands: Springer.

James, N.P. and Dalrymple, R.W. (eds) (2010) *Facies Models 4*. St John's, Newfoundland: Geological Association of Canada, GEOtext 6.
Chapters 13–19 and 21 are devoted to facies analysis of carbonate systems.

Lukasik, J. and Simo, J.A. (2008) Controls on development of carbonate platforms and reefs. In: Lukasik, J. and Simo, J.A.T. (eds) *Controls on Carbonate Platforms and Reef Development*. SEPM (Society for Sedimentary Geology), Special Publication no. 89.
A wide variety of articles on carbonate depositional dynamics.

Read, J.F. (1985) Carbonate platform facies models. *American Association of Petroleum Geologists Bulletin*, 69, 1–21.
A classic paper summarizing the main attributes of carbonate platforms and ramps.

Scholle, P.A., Bebout, D.G., and Moore, C.H. (eds) (1983) *Carbonate Depositional Environments*. Tulsa, OK: American Association of Petroleum Geologists, Memoir no. 33.
A compendium of most carbonate depositional systems with each chapter devoted to a specific paleoenvironment.

Tucker, M.E. and Wright, V.P. (1990) *Carbonate Sedimentology*. Oxford: Blackwell Scientific Publications.
Several chapters are devoted to detailed documentation of carbonate depositional environments.

Wright, V.P. and Burchette, T.P. (eds) (1998) *Carbonate Ramps*. London: Geological Society, Special Publication no. 149.
A follow-up to the classic Burchette and Wright (1992) synthesis with numerous papers.

CHAPTER 8
LACUSTRINE CARBONATES

Frontispiece Partially desiccated part of the Great Salt Lake, Utah. Shovel: 1 m high.

Origin of Carbonate Sedimentary Rocks, First Edition. Noel P. James and Brian Jones.
© 2016 Noel P. James and Brian Jones. Published 2016 by John Wiley & Sons, Ltd.
Companion website: www.wiley.com/go/james/carbonaterocks

Introduction

Carbonate sediments are present globally in perennial (permanently filled) or ephemeral (periodically dry) lakes. Lakes are the epitome of diversity because they can be shallow or deep, filled with saline or fresh water, cold or hot water, characterized by variable oxygen levels, and inhabited by many different types of organisms. With such a wide environmental spectrum it is not surprising that lacustrine carbonate sediments are extraordinarily diverse in terms of their mineralogy, facies, geochemistry, and stratigraphy. Some features of these sediments are akin to their marine brethren, whereas other aspects are unique.

Lacustrine and marine depositional systems differ in the following important ways:

- Lakes lack the tidal activity common to all marine basins.
- Lakes are closed systems that contain smaller volumes of water and sediment than marine basins; they are therefore prone to more rapid environmental change than marine systems.
- The relatively small size of lakes means that their water chemistry responds more rapidly to changes in local conditions and climate.
- The level of the lake outlet (spill point), which is a function of basin subsidence, inherited basin shape, and local tectonics, exerts a fundamental control over lake system accommodation, a situation that does not exist in ocean basins.

Modern lakes: Zonation and classification

The lake environment is divided into the *littoral*, *sublittoral*, *profundal*, and *pelagic* zones (Figure 8.1). The littoral zone encompasses the shallow water around the lake margin, whereas the profundal zone is the deep, aphotic part of the lake where little or no light penetrates. The littoral zone in many lakes is characterized by a shallow-water "bench" or "terrace" that extends out from the shoreline. The sublittoral zone refers to the transition region between the littoral zone and the deeper, profundal part of the lake. The water column above the profundal zone is called the pelagic zone.

Lake water is typically stratified with each layer having a distinctly different temperature or salinity, and hence different water chemistry (Figure 8.2). Such layering develops because of differential radiative heating, where surface waters are warmed by the sun while deeper waters remain isolated from solar radiation. In temperature-stratified freshwater lakes, the upper *epilimnion* layer is separated from the lower *hypolimnion* layer by the thermocline (also known as the *metalimnion* layer). The chemocline in saline or permanently stratified (meromictic) lakes separates the lower *monimolimnion* layer from the upper *mixolimnion* layer. Seasonal or permanent anoxia (lack of oxygen) can develop in both the hypolimnion or monimolimnion layers (Figure 8.2).

Mixing of the stratified water layers can be caused by winds, currents that are generated by water inflow from streams and rivers, or, most commonly, density contrasts. Water is densest at a temperature of 4°C. Once water in the epilimnion zone attains a temperature of 4°C, it can become denser than the water in the underlying hypolimnion and sinks to the lake bottom, forcing the warmer water to the surface (turnover). Limnologists use the frequency of mixing to classify modern lakes. Lakes that undergo only one phase of mixing per year are termed *monomictic*, whereas lakes that undergo two phases of overturning (typically in spring with warming and autumn with cooling) are termed *dimictic*. Polymictic lakes are shallow lakes with waters that mix freely throughout the year, but can become stratified for short periods of time. *Meromictic* lakes are lakes that remain permanently stratified.

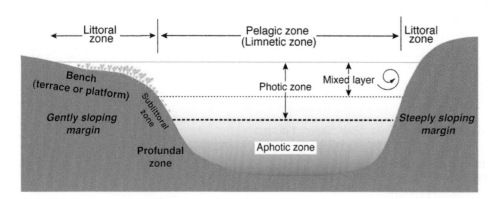

Figure 8.1 Zonation of lakes. Source: Adapted from Renaut and Gierlowski-Kordesch (2010). Reproduced with permission of the Geological Association of Canada.

a. Summer

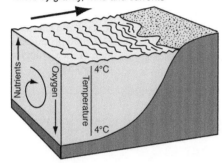

b. Autumnal cooling

Mixing of surface and bottom
waters by gravity, wind and currents

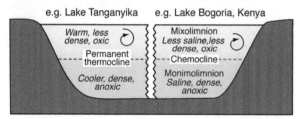

c. Meromixis and permanent stratification

Figure 8.2 Stratification in modern lakes. (a) Vertical profiles of temperature, pH, and oxygen in lake during summer months. The thermocline divides the warm water layer from the underlying cool water layer. (b) Water stratification is lost during winter months as different layers mix following cooling and action of wind-induced currents. (c) Meromixis develops when surface waters are consistently less dense than the deep waters. Source: Adapted from Renaut and Gierlowski-Kordesch (2010). Reproduced with permission of the Geological Association of Canada.

Controls on lake sedimentation

Carbonate sedimentation in lakes is determined by the interplay between different extrinsic and intrinsic parameters (Figure 8.3). The main extrinsic factors are climate and tectonics, whereas the dominant intrinsic factors are hydrology, sediment sources, and lake stratification and mixing.

Hydrology

Irrespective of location, lake hydrology is linked to the local climate and tectonics because these factors control the large-scale water input and hydrology of the region. Water enters a lake via precipitation (rain, snow), surface runoff, groundwater seepage, or sub-lacustrine springs. Critically, these waters are also responsible for bringing carbonates into the lake as bedload, as suspended load, or in solution. The potential for carbonate sedimentation is high when the bedrock around a lake is limestone or dolostone, but low if the bedrock is non-carbonate rock (e.g., siliciclastic sedimentary rocks and most igneous and metamorphic rocks).

Large amounts of carbonate can be introduced into lakes via subaqueous springs. In such circumstances, carbonate sedimentation can take place in a lake that is surrounded by non-carbonate rocks and therefore lacks carbonate input from rivers and surface runoff. The impact of subaqueous springs depends largely on the volume and chemical composition of the spring water that is introduced into a lake. Spectacular limestone tufa towers, such as those found in Mono Lake and Searles Lake (both California, USA), clearly attest to the potential importance of such springs.

As in any depositional setting, water movement can have a major impact on the manner in which the carbonate sediments form, accumulate, and are redistributed. Unlike marine settings, lakes are tideless and hence do not experience a regular short-term rise and fall in water level. This lack of tides could partly explain why the stratification of lake waters is not disturbed on a more frequent and regular basis. Waves and currents are instead largely wind-driven with wave amplitude and current strength being dependent on the size and orientation of the lake relative to the wind direction. Large lakes typically experience the highest frequency of water level fluctuations because the long fetches allow high-amplitude waves to build up. Currents in lakes can also be generated by inflow of rivers, density flows, and local variations in atmospheric pressure. The combination of wind shear and low atmospheric pressure can, for example, generate a *seiche*, which is a large-scale, long-period wave caused by the piling up of water at one end of a lake and its subsequent movement to the other end of the lake when the wind dies down.

Climate

Lakes with carbonate sediments are not restricted to specific climatic zones. Lacustrine carbonates are, for example, well known from Antarctica, Australia, Africa,

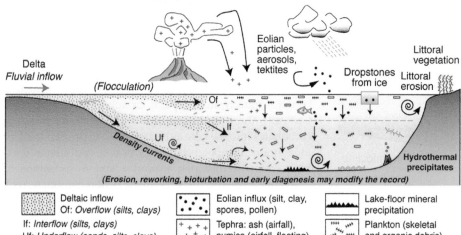

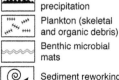

Figure 8.3 Potential sources of sediment in the central, open parts of lakes. Source: Adapted from Renaut and Gierlowski-Kordesch (2010). Reproduced with permission of the Geological Association of Canada.

and northern Canada. Climate is important because rainfall and temperature control the net water balance by determining water input via precipitation and water loss via evaporation. Climate can also influence carbonate sedimentation by dictating periods of biogenic activity (e.g., planktonic blooms), seasonal variance in calcite and aragonite precipitation, and varve development.

Sediment input

Lacustrine sediments are various mixtures of precipitated minerals, detrital material, and biological components. Abiogenic mineral precipitation is controlled by the composition of the water, whereas the biogenic components are intimately related to the resident biota.

Abiogenic precipitates include a diverse range of Ca-, Mg-, Mn-, Ba-, Sr-, Fe-, and Na-carbonates. There are at least 33 different carbonate minerals known from lacustrine deposits. The diversity of these precipitated minerals reflects the chemical diversity of lake waters. Although calcium carbonate (aragonite, calcite) and magnesium carbonates (magnesite, hydromagnesite) are the most common precipitates (Figure 8.4), dolomite is also found in many lake sediments. Precipitation of the carbonate minerals takes place when the waters become supersaturated due to evaporation, cooling, or CO_2 degassing. Degassing is generally restricted to the shallow parts of the lake and surface waters.

Figure 8.4 Succession of Miocene lacustrine carbonates formed largely of hydromagnesite and magnesite, marginal facies of the Kozani Basin, Greece. Outcrop: ~10m high. Photograph by R. Renaut. Reproduced with permission.

Lacustrine carbonate sediments also include detrital calcite, aragonite, and dolomite that are brought into the lake by local rivers or surface runoff. The presence of detrital carbonate grains depends on the composition of the bedrock around the lakes and whether or not any of the detritus from those rocks is fed into the lake. The detrital components can, however, be virtually impossible to identify once they have been mixed with the carbonates that are being actively produced in the lake itself.

Among the plethora of organisms that inhabit lakes, the most important sediment contributors are charophytes, ostracods, mollusks, insects, and diatoms. Charophytes, also known as stoneworts, are non-marine benthic green algae. Their distribution in the photic zone is controlled by environmental parameters such as temperature, salinity, water depth, energy levels, and substrate type. Although a soft-bodied organism that has no hard skeletal parts, biomineralization of their reproductive organs and stems means that they are common components of lacustrine carbonates. Indeed, their calcified gyrogonites (seeds) are used for taxonomic and stratigraphic purposes.

Ostracods are benthic and pelagic microcrustaceans characterized by a calcitic bivalved shell that is typically ~1 mm long with excellent preservation potential (Chapter 6). Providing the water is well oxygenated, epifaunal and infaunal bivalves and gastropods are found in many lake settings. The environmental ranges of the different taxa are determined primarily by temperature and seasonal changes. Insects, which are common inhabitants of lakes, can also contribute to carbonate sedimentation. Caddisfly larval or pupal cases, for example, are common components in many lacustrine successions. Chrominid and midge larva have also been found preserved in some microbialites.

Diatoms are photosynthesizing algae characterized by exoskeletons (known as frustules) of amorphous silica (opal-A – $SiO_2.H_2O$). Since their first appearance in the late Cretaceous, these microorganisms have become abundant biotic elements in many lakes. Such planktic, benthic, and epiphytic organisms can live in acidic (pH<3.5) to alkaline (up to about pH of 9) waters. These environmentally sensitive algae are important because they can be used to decipher the paleoenvironmental evolution of many lakes.

Ostracod shells, bivalve shells, gastropod shells, and diatom frustules are either scattered throughout the carbonate sediment or concentrated into individual laminae. Concentrations of diatom frustules, for example, produce thin siliceous layers that are intercalated with calcitic or aragonitic layers. Local reworking of the sediments can likewise produce interlayered coquinas (shelly layers) and muddy sediments.

Paleoenvironmental interpretations of lacustrine successions usually depend on their biogenic components. Such interpretations are based on: (1) comparisons with extant taxa that live in known environments; or (2) geochemical proxies (e.g., stable isotopes, trace elements) derived from the ostracod, gastropod, and bivalve shells.

Lake stratification and mixing

Lake stratification, which is controlled by the heating and cooling of surface waters, has a significant impact on the sediments that form and develop in the lake. Bottom waters in the aphotic zone of stratified lakes are isolated from the atmosphere and so rapidly become oxygen depleted through bacterial activity. This creates dysoxic or anoxic conditions that preclude habitation by most organisms, but promote preservation of organic matter. Other consequences of stratification and mixing include:

- recycling of nutrient elements (N, P) from the bottom waters to surface waters that typically stimulate planktic productivity;
- cooling or warming of bottom and surface waters, which can curtail or promote carbonate formation and biological activity; and
- renewed oxygenation of bottom waters, that can lead to mineral precipitation (e.g., Fe and Mn minerals) and promote sediment bioturbation by benthic organisms.

Lake sedimentation

Lake sediments are usually separated into a shoreline facies and a lake centre facies.

Shoreline facies

The development of carbonate deposits along the shoreline depends on how much carbonate is delivered by inflowing waters. Benches commonly form along steep lake margins (Figure 8.5), whereas ramps are associated with low-gradient shorelines. Irrespective of the lake floor configuration, sedimentation in this littoral zone is largely dictated by hydrodynamics.

Low-energy carbonate benches (Figure 8.5) are usually covered with marl (>50% carbonate mud) and carbonate sand with the constituents coming from: (1) the breakdown of ostracods, bivalves, and gastropods shells; (2) sand and mud precipitation associated with charophytes; or (3) bio-induced carbonate mud that precipitates in the water column. With ample sunlight, charophytes form dense carpets that cover the lower bench and slope. Oncoids (Figure 8.6a) and stromatolites (Figure 8.6b, c) can be abundant in shallow, nearshore waters. Strandline facies include beachrock, palustrine marshes, sand shoals and beaches, marl and micrite benches, marl and micrite ramps, or microbialites and microbial bioherms.

Figure 8.5 (a) Kelly Lake, British Columbia with a well-developed marginal bench (light-colored area) and deeper (dark blue) central basin. (b) Shallow bench, north shore of Kelly Lake, British Columbia, covered with carbonate sediments and microbialites, shallow bench is ~10 m wide. (c) Underwater image of modern calcareous microbialites on bench along north shore of Kelly Lake, British Columbia. Image width 2 m. Photographs by R. Renaut. Reproduced with permission.

Figure 8.6 (a) Oncoids from bench area, north shore of Kelly Lake, British Columbia. Coin 1 cm diameter. Photograph by R. Renaut. Reproduced with permission. (b) Stromatolites growing along shoreline area of Clifton Lake, Australia. Foreground width 10 m. Photograph by P. Pufahl. Reproduced with permission. (c) Underwater view of margin of a growing stromatolite, Clifton Lake, Australia. Image width 50 cm. Photograph by P. Pufahl. Reproduced with permission.

- *Beachrock*: This limestone forms as sheets of lithified sediment that accumulated as older deposits along the shoreline, cemented by calcite and aragonite (Figure 8.7). Beachrock typically develops with the lowering of lake levels and is usually associated with the input of groundwater along the lake margins.
- *Palustrine marsh*: These marshes, found around lake margins, are subject to highly variable environmental conditions (Figure 8.8). Deposits are muddy and include pedoturbation (disturbance of the soil by burrowing animals), various diagenetic fabrics that develop as a result of subaerial exposure (e.g., teepee structures - Figure 8.9) brecciation, rhizoliths, sparmicrite, and peats and coals.
- *Shoals and beaches*: Such deposits are generally coarse-grained ooids or biofragments and exhibit ripple cross-laminations or low-angle cross-stratification. Shell coquinas can be composed of ostracods, bivalves (Figure 8.10), and gastropods.
- *Marl-micrite benches*: Sediments are predominantly fine-grained marls and silts with grains derived from charophytes, mollusks, ostracods, and plant material (Figure 8.11). Oncoids, ooids, and microbialites can be numerous, whereas bioturbation is ubiquitous. Sediment gravity flow deposits including turbidites are common on slopes.
- *Marl-micrite ramps*: Typical sediments include structureless fine-grained carbonates, silts, and sands with associated biofragments, local coquinas, ooids, oncoids, charophytes, intraclasts, and plant material; microbialites are locally conspicuous.
- *Microbialites-bioherms*: Microbial buildups constructed by algae and bacteria can be up to decimeters in scale and sub-hemispherical, domal, bushy, columnar, or crust-like in morphology (Figure 8.12).

Figure 8.7 Fossil calcite-cemented beachrock (cemented sand and chert), south shore of Nasikie Engida, Magadi Basin, Kenya. Note the person for scale. Photograph by R. Renaut. Reproduced with permission.

Figure 8.8 (a, b) Modern palustrine area, Milk Lake, British Columbia. Mg carbonates are precipitating in the lake, whereas aragonite and dolomite are forming around the margin. Image width 20 m in foreground. Photographs by R. Renaut. Reproduced with permission.

Figure 8.9 Tepee structures forming in carbonate around the margins of a saline lake, Deep Lake, Yorke Peninsula, South Australia. Scale bar 10 cm.

Figure 8.11 Carbonate mud flat formed of aragonite, dolomite, and hydromagnesite, Milk Lake, British Columbia. Image width 20 m in foreground. Photograph by R. Renaut. Reproduced with permission.

Figure 8.10 Oyster biostrome, Galana Boi Formation (early Holocene), east Lake Turkana. Hammer handle: 30 cm long. Photograph by R. Renaut. Reproduced with permission.

Figure 8.12 Underwater view of thrombolites and carbonate sediment on bench area along north shore of Kelly Lake, British Columbia. Image width 2 m. Photograph by R. Renaut. Reproduced with permission.

Lake center and pelagic environments

The central parts of perennial lake floors, beyond the influence of shoreline processes, are usually covered with allochthonous or autochthonous fine-grained sediment. Water depth exerts a critical control over sediment generation and accumulation. All of the lake floor in shallow lacustrine systems can lie within the photic zone with plants covering the entire sediment surface. The sediment floor in deep lakes will lie below the photic zone where dark and even anoxic conditions can prevail. The combination of low water temperature and low oxygen levels in some lakes can lead to carbonate undersaturation and dissolution.

Deep lake sediments come from many different sources, including (1) skeletal grains derived from ostracods, bivalves,

gastropods, or charophyte debris; (2) detrital grains from the catchment area; and (3) bio-induced calcite and aragonite precipitation. The precipitation of aragonite as opposed to calcite is largely determined by: (1) ambient water temperature; (2) Mg: Ca ratio of the water; (3) dissolved organic matter; and (4) carbonate supersaturation levels.

Limestone facies of lake center and pelagic environments are structureless, laminated, and transported.

- *Massive (structureless) limestones*: These deposits include mudstones, wackestones, grainstones, and marlstones that may or may not contain scattered fossils. The structureless nature of the limestones is generally

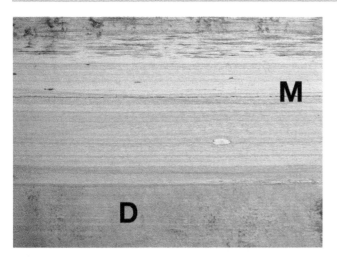

Figure 8.13 Lacustrine carbonate laminites, Eocene Green River Formation, grey organic-rich mudstones (M) and buff dolomitic carbonates (D). Image width 20 cm. Photograph by E. Gierlowski-Kordesch. Reproduced with permission.

attributed to bioturbation, where oxygenated waters allow an active infauna to burrow the sediments.

• *Laminated limestones*: These finely layered carbonates, with laminae of variable thickness and lateral extent, generally form from sediments that settle out of suspension (Figure 8.13). Laminations develop where there are temporal variations in the amount of sediment input and where the sediments, once deposited, remain largely undisturbed. Temporal variations in many lakes are seasonal with distinct differences between the summer and winter deposition. Terms used to describe these types of deposit include: (1) *varves*, which represent summer and winter couplets, that form on an annual basis in glacial lakes; (2) non-glacial varves that form in all other lakes; and (3) *rhythmites* or couplets of alternating composition that form over unknown time frames. Rhythmite laminae can be variable because they accumulate in response to both seasonal changes and other non-predictable variations in lake sedimentation. As such, care must be taken in the interpretation of such deposits because they can accumulate in response to: (1) seasonal changes in runoff patterns; (2) seasonal changes in water temperature and stratification that can control nutrient and saturation levels with respect to specific minerals; (3) seasonal changes in salinity; and (4) monsoons in tropical areas that affect rainfall and wind direction and intensity.

Lacustrine microbialites

Lacustrine microbialites, including stromatolites and thrombolites, are found throughout the world including the ice-covered lakes of Antarctica, the highly alkaline lakes in East Africa, the high-altitude lakes at elevations of 4000 m in the Andes, the hypersaline lakes in western Australia, and commonly in "normal" freshwater lakes. These microbialites are remarkable for their variance in size, their morphology, and their areal extent. It has been estimated, for example, that microbialite bioherms cover ~1000 km² in the Great Salt Lake. In other lakes, such as Lake Van in eastern Anatolia, Turkey, enormous tower-like microbialites up to 40 m high have developed in water more than 100 m deep. Elsewhere, much smaller but numerous lacustrine stromatolites are restricted to the shallow photic zone.

Microbialites are not found in every lake and the reasons for their presence or absence is controversial. Their development in lakes, as in the ocean, is usually attributed to the absence of grazing organisms, which is generally attributed to high salinity. This link is not, however, universal, because microbialites also grow in brackish water lakes (Lake Clifton; Figure 8.6) and freshwater lakes. In most cases, lakes with microbialites are located in limestone terrains, and the groundwater that enters the lakes is rich in calcium bicarbonate. Nutrient levels are also important because they dictate the nature of the lake microbiota. Low nutrient levels, for example, can be responsible for the lack of grazers that curtail the growth and spread of microbialites.

Lacustrine microbialites generally have poorly developed laminae; overall, thrombolites rather than stromatolites seem to be the most common types. Thrombolitic buildups can be impressive, for example, thrombolites in Lake Clifton in Western Australia have merged to form reefs that are up to 30 m wide and 5 km long. Thrombolites in Lake Clifton are formed largely of the aragonite that is precipitated in association with the filamentous cyanobacterium *Scytonema*.

Large mounds and towers can develop around sites where groundwater vents feed into the lake. This precipitation is usually triggered by fluids that become supersaturated with respect to calcite or carbonate as a result of CO_2 degassing, bacterial mediation, or mixing of fluids of spring and lake waters that have significantly different compositions. Such precipitation commonly leads to large, spectacular mounds. The tops of the 40 m high towers in Lake Van, for example, rise to within 8 m of the lake surface, where they are covered with dark green to black cyanobacterial mats and diatoms. Ca-rich groundwater in these towers is funneled upward through a porous axial column and aragonite precipitation takes place in the microbial mats as the groundwater starts to mix with the lake water.

Large towers can become visible when lake water level drops. Spectacular examples of these large structures are now exposed in Mono Lake, California; Searles Lake,

California; and Pyramid Lake, Nevada. Precipitates are characterized by complex crystal morphologies, many of which are poorly documented and understood.

Classification of ancient lake deposits

The classification scheme of modern lakes that is based on water stratification and the frequency of turnover cannot be applied to ancient lacustrine successions because of the difficulty of interpreting these attributes in the rock record. Instead, ancient lake successions are classified on the basis of their sequence-stratigraphy that in turn hinges on the balance between sediment supply, water supply, basin-sill height (spillpoint), and basin-floor depth. The tectonic setting is critical because it controls the temporal development of accommodation (basin shape and depth), sediment input from the surrounding area, and local drainage patterns. Evaporation rate, precipitation rate, and sediment supply are mostly influenced by regional climate. Using these factors, lake deposits are classified as: (1) *underfilled*; (2) *balanced-filled*; and (3) *overfilled*, with each type being characterized by well-known lithological successions that are characterized by differences in their facies architectures, fossils, and geochemical parameters. Although this classification system was developed largely on the basis of siliciclastic systems, carbonate sediments are found in each lake type.

Underfilled lake basins develop where water and sediment supplies are very low compared to the available accommodation. Drainage is closed, lakes are short-lived, and the position of the shoreline fluctuates frequently. Lowstand facies include evaporites, mudflat deposits with desiccation features, paleosols, and aeolianites. Highstand facies, which are common in many perennial saline lakes, can include laminites, carbonates, evaporites, sublittoral organic-rich mudstones, microbial bioherms, stromatolites, and beach deposits in the littoral zone. The biota in these lakes is generally limited by the high lake water salinity. Some of these lakes have a ring of peripheral mudflats around the lake center where evaporites are precipitating. The mudflat precipitates, which can include calcite, Mg-calcite, aragonite, and dolomite, precipitate when lake level is high and the brines are diluted, or during lowstand situations as a result of seepage and evaporation. Drying of the muds typically leads to the formation of intraclasts. Carbonate sediments composed of calcite, aragonite, hydromagnesite, or dolomite will accumulate in shallow lakes if the water never reaches saturation with respect to any of the evaporite minerals. Examples of this type of deposition are found in the Green River

Formation (Eocene) in Wyoming, Pleistocene–Holocene deposits associated with Lake Bogoria in the Kenyan Rift Valley, and many of the interior lakes in British Columbia, Canada.

Overfilled lake basins develop when the water and sediment supplies exceed the low rate of accommodation creation and the watershed remains open. Carbonates are generally rare in such lakes because the water remains undersaturated with respect to the carbonate minerals. Carbonates can, however, develop in lakes that lie in catchment basins formed of carbonate or volcanic rocks or both. The influx of siliciclastic sediment is low in these lakes and Ca saturation is attained relatively easily. Carbonate sediments on the marginal benches and ramps include charophytic sands, oncoids, microbialites, skeletal sands, and ooids. Carbonate precipitation in larger lakes is bio-induced by plankton and can lead to the production of varves (e.g., Lake Zürich). Climate is important because waters in cold regions will not achieve Ca saturation. Examples of this type of succession include those found in Kelly Lake (British Columbia), Green Lake (USA), and Lake Zürich (Switzerland).

Carbonate precipitation can be common in *balanced-fill lake basins*, which are intermediate in style between underfilled and overfilled basins. The many different carbonate deposits include grainstones, ooids, wackestones, mudstones, and microbialites. The production of carbonate sediment in these lakes is closely tied to lake levels. Carbonates precipitate during lowstand periods when the lake level is below the outlet sill and salinities increase during evaporation. Carbonate precipitation will cease when lake levels are high and waters overflow the lake sills and become undersaturated with respect to Ca. Examples of balanced-fill lake successions can be found in Green River, Wyoming.

Further reading

Alonso-Zarza, A.M. and Wright, V.P. (2010) Palustrine carbonates. In: Alonso-Zarza A.M. and Tanner L.H. (eds) *Carbonates in Continental Settings: Facies, Environments, and Processes.* Amsterdam: Elsevier, Developments in Sedimentology, vol. 61, pp. 103–132.

Gierlowski-Kordesch, E.H. (2010) Lacustrine carbonates. In: Alonso-Zarza A.M. and Tanner L.H. (eds) *Carbonates in Continental Settings: Facies, Environments, and Processes.* Amsterdam: Elsevier, Developments in Sedimentology, vol. 61, pp. 1–101.

Gierlowski-Kordesch, E.H. and Kelts, K.R. (eds) (2000) *Lake Basins through Space and Time.* American Association of Petroleum Geologists, Studies in Geology vol. 46.

Harris, P.M., Ellis, J., and Purkis, S.J. (2013) Assessing the extent of carbonate deposition in early rift settings. *AAPG Bulletin,* 97, 27–60.

Matter, A. and Tucker, M.E. (eds) (1978) *Modern and Ancient Lake Sediments*. Oxford: Blackwell Scientific Publications.

McNamara, K. (2009) Stromatolites: great survivors under threat. *Geoscientist*, 19, 16–22.

Perri, E., Tucker, M.E., and Spadafora, A. (2012) Carbonate organo-mineral micro- and ultrastructures in sub-fossil stromatolites: Marion lake, South Australia. *Geobiology*, 10, 105–117.

Renaut, R.W. and Last, W.M. (eds) (1994) *Sedimentology and Geochemistry of Modern and Ancient Saline Lakes*. SEPM (Society for Sedimentary Geology), Special Publication no. 50.

Renaut, R.W. and Gierlowski-Kordesch, E.H. (2010) Lakes. In: James N.P. and Dalrymple R.W. (eds.) *Facies Models 4*. St John's, Newfoundland: Geological Association of Canada, GEOtext 6, pp. 541–576.

Talbot, M.R. and Allen, P.A. (1996) Lakes. In: Reading H.G. (ed.) *Sedimentary Environments: Processes, Facies and Stratigraphy*, third edition. Oxford: Blackwell, pp. 83–124.

Tucker, M.E. and Wright, V.P. (1990) *Carbonate Sedimentology*. Oxford: Blackwell Scientific Publications.

CHAPTER 9
CARBONATE SPRINGS

Frontispiece Rimstone pools at Pamukkale, Turkey. Image foreground width about 12 m.

Origin of Carbonate Sedimentary Rocks, First Edition. Noel P. James and Brian Jones.
© 2016 Noel P. James and Brian Jones. Published 2016 by John Wiley & Sons, Ltd.
Companion website: www.wiley.com/go/james/carbonaterocks

Introduction

Spectacular carbonate deposits (Frontispiece, Figure 9.1) formed of calcite or aragonite or both characterize many spring systems (e.g., Yellowstone National Park, USA; Iceland; Kenya Rift Valley; Turkey). The scale and architecture of these deposits varies from those associated with spring-fed streams that radiate from vents to the more localized discharge aprons that develop around vents. Springs are multifaceted depositional systems with sedimentation and precipitation that are controlled by the interplay of both extrinsic and intrinsic factors. Although fundamentally related to the temperature and composition of the spring waters, deposition and precipitation is also strongly influenced by the resident biota that thrive in these settings.

Spring systems

Spring systems develop wherever subterranean water comes to the Earth's surface and is discharged subaerially or subaqueously. Spring water is typically meteoric, being derived from rainwater or snow that sinks into a subsurface aquifer and moves laterally along any available subsurface permeability system. Subsurface plumbing of spring systems is poorly understood because of the technical difficulties of detecting their pathways. Nevertheless, three aspects control the manner in which water eventually emerges onto the Earth's surface:

Figure 9.1 General view of calcite precipitates on discharge apron of *Pamukkale* (Turkish for "cotton candy"), formed from spring waters that are discharged at 32.5°C. Note people for scale.

1. If the subsurface waters pass by any heat source (e.g., a magma chamber), they will be warmed and the potential for hot springs is created. It is for this reason that hot springs are most commonly located in tectonically active areas where the geothermal gradient is high. Prime examples include hot springs in Yellowstone National Park, Iceland, the North Island of New Zealand, the Kenyan Rift Valley, and the Tengchong geothermal area of China. If the subterranean waters do not encounter any heat source during their journey, then cold-water springs will result. A prime example of a cold-water spring, with water temperature of 5–10°C, is found in Huanglong Valley in western China.

2. As subterranean waters flow through the aquifer, there are water–rock interactions that ultimately control water composition. These processes provide the Ca and other elements that ultimately precipitate on the spring discharge apron.

3. The subterranean plumbing system controls the manner in which the water flows to the surface of the Earth. If there are no restrictions in the plumbing system, water will flow with ease and eventually emerge from the spring vent. At the opposite end of the spectrum, geysers are characterized by periodic and spectacular eruptions that forcefully eject columns of water and steam out of their vents. Although thousands of geysers are known of worldwide, carbonate precipitation is only associated with a small proportion of them (opal-A is the most common precipitate). In the simplest sense, a geyser develops when there is a restriction in the subsurface plumbing system. As steam bubbles through the water, it causes a gradual buildup in pressure. Once confining pressure is exceeded, there is a sudden release of pressure and steam and water erupts from the geyser vent.

The size of a carbonate deposit that forms around a terrestrial spring is controlled largely by: (1) the volume of water that is discharged; (2) the physical and chemical characteristics of the discharged water; (3) the biota that lives on the discharge apron and along its margins; and (4) the flow paths of the water as it disperses from the spring vent. Some systems cover vast areas in highly scenic settings. The Antalya Travertine in Turkey for example, is up to 245 m thick and covers an area of ~630 km². The Huanglong system, located in northwest Sichuan Province of China, is a cold-water system that covers an area 7.5 km long and up to 2.5 km wide (Figure 9.2). Small-scale springs (Figure 9.3a, b), such as those in La Xin, Yunnan Province, China (Figure 9.3c), are at the opposite end of the spectrum. Subaqueous springs can also give rise to other extensive carbonate deposits, such as those at

Figure 9.2 Pools in upper part of the Huanglong spring system, NW Sichuan Province, China, with water temperature of 5–8°C. Image foreground width 10 m.

Mono Lake, California (Figure 9.3d). In some locations, spring waters that flow over steep cliffs produce spectacular deposits at the base of the waterfall (Figure 9.4).

Classification of springs

Springs have been classified according to many different criteria, including water temperature, comparison of the water temperature with ambient air temperature, and the source of the water. For active springs, the simplest approach is to base the definition on the highest temperature measured at the spring vent. This method divides springs into *cool springs* with water temperatures <20°C and *thermal springs* with water temperatures >20°C. Thermal springs can be divided into: (1) warm, with temperatures in the range 20–40°C; (2) mesothermal, with

Figure 9.3 (a) Cold-water spring system with biota dominated by bryophytes, Raven, Alberta, Canada. Image foreground width 10 m. (b) Perpetually spouting springs with 100°C water, forming calcite cones with microbial mats coating lower parts. Frog's Mouth Spring, Tengchong geothermal area, western Yunnan Province, China. Image width 3 m. (c) Boiling hot spring with calcite precipitates modified as a hot water source in the village of La Xin, western Yunnan Province, China. Pool is 7 m at widest part. (d) Springs of calcite formed around subaqueous springs when lake level was higher, Mono Lake, California, USA. Spires ~5 m.

temperatures of 40–75°C; and (3) hyperthermal, with temperatures >75°C (Figure 9.5). The use of water temperature in this manner is advantageous because it recognizes that water temperature: (1) can control the precipitation of the different $CaCO_3$ polymorphs; and (2) is one of the main controls on the distribution of biota (e.g., bacteria, green algae, and diatoms). In cool spring systems, there is little downslope variation in water temperature because

the water is <20°C at the vent and downslope cooling of the water is minimal. With hyperthermal springs, however, downslope cooling means that the discharge apron can be divided into the hyperthermal, mesothermal, warm, and cool zones. Although easy to apply in modern spring systems, this classification system is difficult to apply to dormant or ancient systems because water temperatures can only be inferred from the type of precipitate or the preserved biota.

Tufa, travertine, or sinter?

Different terms such as tufa, travertine, and sinter have been used for calcareous deposits that form in spring systems. The term *sinter* should only be applied to siliceous spring deposits and should never be used to describe calcareous deposits. Confusion exists with respect to the terms tufa and travertine because they are used without being precisely defined or are variously described according to water temperature, depositional setting, hardness, or age of the deposit. The term *tufa* should be applied to a spring deposit that is porous and displays clear evidence of forming in and around plants (but also see Chapter 18 on Precambrian carbonates). By contrast, *travertine* is dense, has low porosity, and is characterized by crystalline calcite with a variety of crystal morphologies. Some geologists avoid this issue by simply applying the term travertine to all calcareous spring deposits.

Figure 9.4 Waterfall fed by spring waters, flowing over bryophyte-covered slope, Güney Falls, Turkey. Image width 20 m.

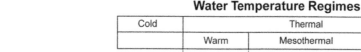
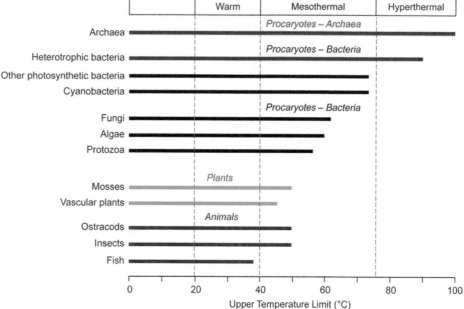

Figure 9.5 Temperature ranges for organisms that inhabit spring systems (from Brock, 1994) relative to spring water temperature regimes as defined by Renaut and Jones (2000). Source: Jones and Renaut (2010). Reproduced with permission of Elsevier.

Biota of spring systems

Diverse arrays of aqueous organisms inhabit spring systems, with their distribution being controlled by water temperature, pH, or water toxins. Of importance is the maximum water temperature that can be tolerated by each group of organisms (Figure 9.5). Heterotrophic bacteria and Archaea, for example, can thrive in water temperatures of >75°C, whereas diatoms will only be found where water temperatures are <45°C. The distribution of the biota on the discharge apron in hot spring systems that are fed by water that emerges with a temperature of ~100°C is related directly to the downslope cooling gradient of the spring water.

Bacteria, algae, fungi, and Archaea are common in many spring systems with mats formed by different microbial communities (Figure 9.6). Unfortunately, the preservation potential of microbes and other organisms in calcareous spring systems is low because most lack hard skeletal parts. Furthermore, if preserved, identification in terms of extant taxa is unlikely because: (1) only the general form of the microbe is fossilized; and (2) DNA signatures, by which most microbes are now defined, will not be maintained. Similarly, preservation of higher plants (trees, grasses, mosses) is generally rare despite the fact that they are encased in precipitated carbonate.

Carbonate precipitation in spring systems

Precipitation in spring systems requires nucleation centers and the delivery of solute to these nuclei. As spring water flows over a substrate, it constantly delivers the Ca and CO_3 to various nucleation sites. The driving force is the level of saturation in the water with respect to calcite, aragonite, and, in rare cases, dolomite. Most precipitation is related to the rate of CO_2 degassing because that process can quickly elevate saturation levels and trigger crystallization. Spring waters are commonly CO_2-rich and are therefore highly prone to CO_2 degassing as they flow down and across the discharge apron. Rates of CO_2 degassing from spring waters are influenced by: (1) the pCO_2 of the discharged spring water; (2) the area and depth of water on the discharge apron; (3) the degree of water agitation; or (4) aeration, jet-flow, and low-pressure effects.

The alternation between calcite and aragonite precipitation in many spring deposits has been attributed to temperature variations, humidity levels, sulfate poisoning, Mg poisoning, high Mg, Fe, and Mn concentrations, high Sr concentrations, saturation levels, and precipitation rates. Supported by laboratory experiments and detailed examinations of spring deposits throughout the world, water temperature is usually held responsible for the precipitation of aragonite as opposed to calcite in spring systems. According to this notion, aragonite

Figure 9.6 (a) Calcite spring deposit formed on riverbank below point source. Surface colors from microbial mats that coat surface of deposit. Deposit ~3 m high. (b) Cross-section through laminated calcite precipitates from deposit shown in (a), Sato-no yu (Spa), Ichi River, Japan. Image width 5 cm.

precipitation is associated with water >40°C and calcite with water temperatures <40°C. This simple division is not universally applicable because calcite is being precipitated from spring waters with a temperature of >90°C in New Zealand and intercalated calcite and aragonite precipitates are known from springs in the Kenyan Rift Valley where the water temperatures are 85–100°C.

Many spring deposits are characterized by a diverse array of precipitated crystals with unusual and bizarre morphologies (Figures 9.7, 9.8). Many of these crystals are precipitated rapidly in response to short-lived changes in the physiochemistry of spring waters. Deposits include dendrite crystals that have a tree-like morphology, skeletal crystals that are hollow because their walls grow before their cores, wheat sheaves that are bundles of needle-shaped crystals, and spherulites of needle-shaped crystals that radiate from a nucleus (Figure 9.9). Crystal morphology is largely related to the level of supersaturation with respect to $CaCO_3$, but precise limits or causes are difficult to determine. Scale is also an issue because many of these precipitates form in the micro-domains of microbial mats where physiochemical conditions can be vastly different from the conditions found in the spring water that flow over the mat. In all cases, however, these unusual crystals seem to form where some variant in the environment has triggered a rapid increase in the saturation levels with respect to $CaCO_3$. Crystal morphs are typically related to a "driving force", which may be a function of supersaturation, evaporation, degassing, or supercooling (Figure 9.9). Interpretation of spring deposits and laboratory experiments indicate that a progressive increase in the driving force will be matched by a change in crystal morphology from hollow (skeletal) crystals to dendrite crystals to fans, wheat sheaves, or spherulites (Figure 9.9).

Spring architecture

The architecture of spring deposits is controlled by volume of water, flow rate, water temperature, saturation level of the water with respect to aragonite/calcite, the pattern of water flow from the vent, the gradient of the land around the vent, and climate (Figure 9.10).

Rimstone pools that form behind rimstone dams are common features of many spring systems, irrespective of water temperature (Figures 9.11, 9.12). These pools, which typically have their widest axis perpendicular to water flow, are of all sizes, with the largest ones commonly being up to 20 m wide, 10 m long, and 3 m deep (Figure 9.11, Frontispiece). Only the water that flows into them on their upstream side disturbs the quiet water in the pools. Laminated, fine-grained sediment or precipitated calcite

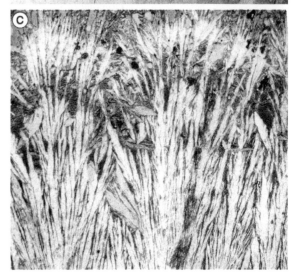

Figure 9.7 (a) Bushes formed of dendrite calcite crystals growing in shallow pool on discharge apron of modern spring at Lýsuhóll, Iceland. Image width 10 cm. (b) Arrays of calcite dendrite crystals forming roll-over structure, Fairmont Hot Springs, British Columbia, Canada. Image width 25 cm. (c) Thin-section photomicrograph of dendrite crystals from Shuzhishi Spring, Tengchong Geothermal Area, Yunnan Province, China. Image width 1 cm.

accumulates on the pool floors. Leaves and wood from the surrounding vegetation are commonly incorporated in pool deposits. Rimstone dams are built of calcite dendrite crystals that are precipitated from the turbulent waters that flow over them (Figures 9.7–9.10). The front wall of the rimstone dam progrades downstream as

Figure 9.8 Beds formed of dendrite crystals, Shuzhishi Spring, Tengchong geothermal area, western Yunnan Province, China. Image width 0.5 m.

calcite is either precipitated on the wall, or as stalactites on the front of the dam coalesce (Figures 9.12–9.14).

Low-gradient parts of discharge aprons are characterized by intricate arrays of micro-terraces or miniature rimstone dams and pools (Figure 9.15). The distribution and orientation of these structures typically reflect microscale variations in the directions of thin (commonly <1 cm) water flow across these surfaces.

Steep parts of the discharge apron are covered with calcite precipitates that form because of accelerated CO_2 degassing caused by water cascading down the slope. In some systems, such precipitation will be largely abiogenic and can involve the development of stalactites that locally merge to form carbonate curtains (Figure 9.13). In many cool spring systems, such as that at Huanglong China, bryophytes (mosses) actively grow on the surface, where they become encrusted with calcite (Figure 9.16).

The distal parts of spring systems with their cooler waters can have trees, bushes, grasses, and other plants growing along their margins or on isolated islands amid the flowing spring waters (Figure 9.10). In these settings, calcite is commonly precipitated around the plants even while they are alive (Figure 9.17). In most cases, the organic tissues are not calcified and will, in time, decay to leave pores and cavities that mimic the original plant morphology.

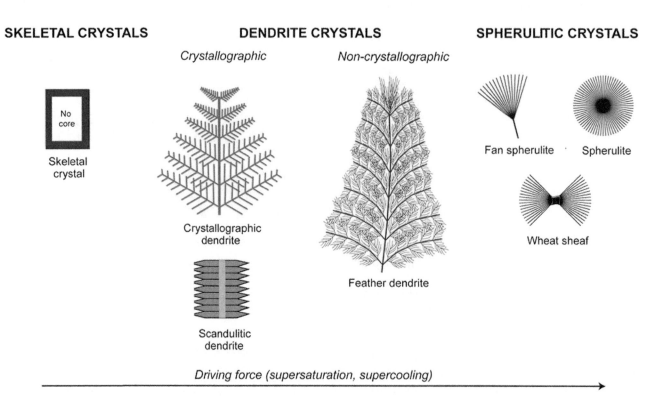

Figure 9.9 Diagram showing evolution of different crystal forms in calcite associated with increasing "driving force". Source: Jones and Renaut (2010). Reproduced with permission of Elsevier.

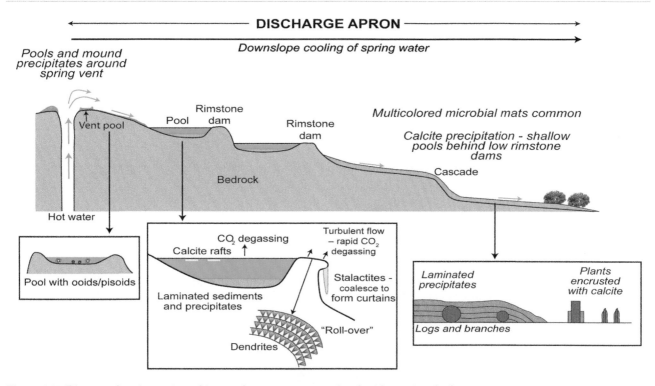

Figure 9.10 Diagram showing main architectural components associated with a spring discharge apron.

Figure 9.11 Series of rimstone pools in the Huanglong spring system, NW Sichuan Province, China. Pools in foreground 6 m wide.

Figure 9.12 Rimstone pools at Pamukkale, Turkey with stalactites forming along downstream sides of rimstone dams. Note the people for scale (circled).

Calcareous spring carbonate facies

Calcareous spring deposits (Table 9.1) are so diverse because they form in response to: (1) the composition of spring waters and their flow paths; (2) the local climate that controls cooling gradients as well as fluid dilution; and (3) the resident biota.

Plant and microbial tufa/travertine

Archaea, bacteria, cyanobacteria, green algae, diatoms, fungi, lichens, bryophytes, liverworts, and vascular plants (e.g., ferns, herbs, trees) influence the development

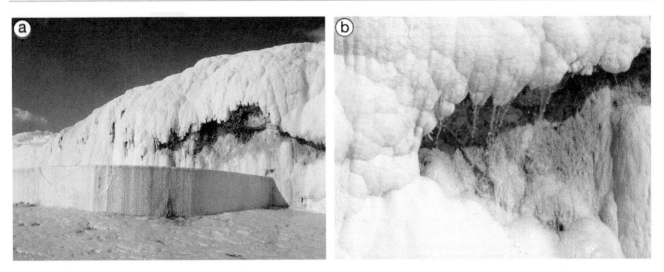

Figure 9.13 (a) Steep section in discharge apron at Pamukkale, Turkey covered with curtain formed by lateral merger of stalactites. Cliff face is ~5 m high. (b) Water flowing from tips of stalactites shown in (a). Image width 10 cm.

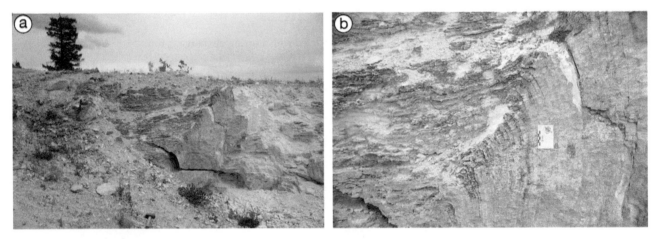

Figure 9.14 (a) Downstream cross-section through rimstone dam and pool. Image width 10 cm. Clinton travertine, British Columbia, Canada. (b) Enlarged view showing lateral merger of laminated pool deposits with vertical laminae of rimstone dam that are formed of calcite dendrites. Scale bar in centimeters.

of spring deposits. Highly porous tufas develop as $CaCO_3$ is precipitated around these components before the organic material is lost through decay. Preservation of the plants by calcification (permineralization) is rare. Stromatolites formed by various combinations of cyanobacteria, bacteria, fungi, green algae, and their biofilms develop as calcite or aragonite is precipitated on and around the microbes or by the trapping and binding of detrital grains.

Insect tufa/travertine

Insects can influence spring carbonate deposition. Larvae of chironomid flies (typically ~5 mm long, 0.3 mm wide) produce a tubular netting that becomes encrusted with calcite. Highly porous tufa develops where large numbers of these tubes are concentrated into small clusters. Similarly, caddisfly larvae form tubes that can be a major component of tufa, and blackfly commonly provide substrates for calcite precipitation on tufa barriers.

Shrub tufa/travertine

Calcitic shrubs, typically <3 cm high and characterized by an upward-expanding, irregular dendrite morphology, grow in terrace pools and depressions,

Figure 9.15 Surface of discharge apron showing intricate array of small-scale rimstone dams and pools, Pamukkale, Turkey. Image width 25 cm.

especially in sulfurous hot springs. These shrubs are formed of micrite and spar-rhombs or needle crystals. Although they have been attributed to both abiogenic and biogenic processes, bacteria are clearly involved in their formation.

Coated grains

Coated grains (ooids, pisoids) up to 5 cm in diameter develop on many spring discharge aprons (Figure 9.17c). Ooids are known from cool, warm, mesothermal, and hyperthermal springs. In spring pools at Tekke Ilica (Turkey), for example, aragonitic pisoids up to 5 cm in diameter grow in spring waters with temperatures of 60–80°C. In the Waikite Springs (New Zealand), calcitic pisoids up to 4 mm in diameter grow in spring waters with temperatures of >92°C.

Pisoids are ubiquitous and occur, for example, in the late Pleistocene travertines of Rapolano Terme (Italy) and Fairmont Hot Springs (British Columbia, Canada). The latter, up to 1 cm in diameter, are formed of radiating calcite dendrite crystals. Abiogenic and biogenic processes are both implicated in the growth of these coated particles.

Oncoids (oncolites) occur in many spring systems with accretion of the cortical laminae being attributed to the actions of various microbes. In extreme cases, microbialite-coated tree trunks, up to 60 cm wide and 4 m long, are entombed in the Oligocene tufas on Mallorca Island (Spain).

Figure 9.16 (a) Steep section in discharge system at Huanglong, NW Sichuan Province, China showing cold (5–8°C) water flowing down slope covered with bryophytes (mosses). Cliff face is ~6 m high. (b) Example of calcite-encrusted bryophytes, Güney Falls, Turkey. Image width 75 cm.

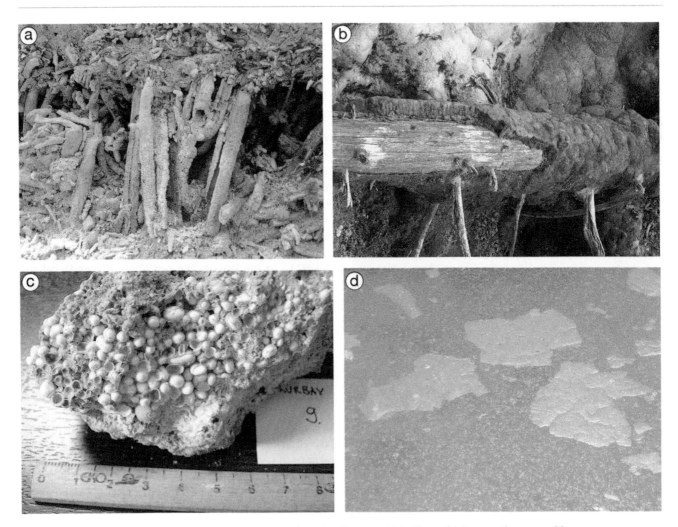

Figure 9.17 (a) Calcite encrusted reeds, Caerwys, North Wales. Image width 25 cm. (b) Log partly encased by calcite, Huanglong, NW Sichuan, China. Image width 2 m. (c) Pisoids from Çukurbağ spring, Pamukkale, Turkey. Photograph by M. Özkul. Reproduced with permission. (d) Calcite rafts in spring pool, Sato-no yu (Spa), Ichi River, Japan. Image width 25 cm.

Table 9.1 Classification of tufa, modified from Ford and Pedley (1996).

Autochthonous Phytoherm tufa	Allochthonous	
Boundstone sheets of micrite and peloids (stromatolite-like bacterioherms)	Micro detrital tufa	Macro detrital tufa
Microherm shrubby framework of bacterial colonies	Micrite tufa	Oncoid and cyanolith
Framestone: true "reef" framework of macrophytic coated with mixed micritic and sparry calcite fringes	Peloidal tufa	Intraclast tufa
	Sapropeltic tufa (organic rich)	Phytoclast tufa
	Lithoclast tufa (inorganic rich)	Lithoclast tufa

Stalactite tufa and travertine

Stalactites, identical in appearance to the stalactites found in caves, form on the downstream side of dams on terraced discharge aprons (Figure 9.13). Curtains develop where closely spaced stalactites coalesce during growth. Stalactite growth is favored by the turbulent flow over the lip of the dam, which increases the rate of CO_2 degassing. In some areas, microbial mats that thrive on the surfaces of the stalactites can also play an active role in their development.

Lithoclastic-bioclastic tufa and travertine

Phytoclasts (a product of the breakup of calcareous crusts formed on and around any plants), lithoclasts, and bioclasts are common in many spring systems. Erosion takes place when flow rates increase due to surges in spring flow rates or flooding after periods of heavy rain. Any such detritus that is washed downslope can accumulate in depressions or ponds, get caught on the upstream sides of dams, or become trapped on the surfaces of microbial mats.

Micritic tufa/travertine

Thick deposits of calcareous silt and mud, with numerous gastropods and ostracods, can cover the floors of pools and lakes behind tufa barrage dams. Examples of such deposits are found in the Wadi tufa at Kharga Oasis, Egypt and in pre-Quaternary travertines of the Itaboraí Basin of southeast Brazil. Disseminated pyrite, minor amounts of detrital quartz and feldspar, and scattered pebbles are present in some of these deposits. The micrite in these settings probably forms by biogenic processes, evaporation, or both.

Raft tufa/travertine

Rafts of calcite or aragonite commonly float on the surface of stagnant pools. Growth of these rafts is due to surface degassing of CO_2, which causes an increase in saturation levels. Filamentous cyanobacteria and diatoms can colonize these rafts and assist in $CaCO_3$ precipitation. Rafts are usually only a few millimeters thick and typically have a flat upper surface and a dentate lower surface that reflects the downward growth of crystals into the water. Disturbance of the water surface can cause the delicate rafts to break into pieces of various sizes and shapes and sink to the pool floor.

Coated bubble tufa/travertine

Gas bubbles are produced by: (1) microbial activity in the mats or in sediment on the floors of spring pools; or (2) by CO_2 or steam produced by pressure release or shallow boiling. Irrespective of their origin, these gas bubbles rise through the water column and can become coated with calcite, aragonite, or both (Figure 9.18). Although extremely delicate and easily broken, they can be preserved if they are trapped and protected in some way. Rocks formed from coated bubbles have been called "bubble limestone," "lithified bubbles," "honeycomb rock," or "foam rock."

Crystalline travertine

Layers of crystalline calcite or aragonite that form via rapid precipitation from fast-flowing waters occur in virtually every spring system. These layers conform to the topography of the discharge apron and vary from subhorizontal to almost vertical. Morphologically diverse and complex calcite crystals in these layers, which typically have their long axes normal to the growth surface, include ray crystals, feather crystals, and dendrites. Dendrite crystals, which are the product of rapid precipitation from waters that become supersaturated with respect to calcite because of rapid CO_2 degassing, are associated with spring waters of all temperatures. Newly formed crusts are typically soft, whereas older crusts are commonly hard.

Figure 9.18 Air bubbles partly covered with calcite (white), spring at Daimaru Ryokan, Japan. Image width 5 cm.

Further reading

Brock, T.D. (1994) *Life at High Temperatures*. Yellowstone National Park, Wyoming: Yellowstone Association for Natural Science, History and Education, Inc.

Capezzuoli, E., Gandin, A., and Pedley, M. (2014) Decoding tufa and travertine (fresh water carbonates) in the sedimentary record: The state of the art. *Sedimentology*, 61, 1–21.

Ford, T.D. and Pedley, H.M. (1996) A review of tufa and travertine deposits of the world. *Earth-Science Reviews*, 41, 117–175.

Jones, B. and Renaut, R.W. (2010) Chapter 4: Calcareous spring deposits in continental settings. In: Alonso-Zarza A.M. and Tanner L.H. (eds) *Carbonates in Continental Settings: Facies, Environments, and Processes*. Amsterdam: Elsevier, Developments in Sedimentology, Vol. 61, pp. 177–224.

Renaut, R.W. and Jones, B.J. (2000) Microbial precipitates around continental hot springs and geysers. In: Riding R.E. and Awramik S.M. (eds.) *Microbial Sediments*. Berlin, Heidelberg: Springer-Verlag, pp. 187–195.

CHAPTER 10
WARM-WATER NERITIC CARBONATE DEPOSITIONAL SYSTEMS

Frontispiece Aerial view of Belize Barrier Reef System. Photograph by W. Martindale. Reproduced with permission.

Origin of Carbonate Sedimentary Rocks, First Edition. Noel P. James and Brian Jones.
© 2016 Noel P. James and Brian Jones. Published 2016 by John Wiley & Sons, Ltd.
Companion website: www.wiley.com/go/james/carbonaterocks

Introduction

Carbonate production and accumulation can be spectacular in tropical environments where the sea is warm (>20°C) all year round, the ocean is supersaturated with carbonate, the clear waters are brightly illuminated, nutrients in the water column are low (oligotrophic), and carbonate production is both biotic and abiotic. These depositional settings are where ooids form, reefs thrive, carbonate muds are precipitated, synsedimentary cementation is rampant, and platform rims are formed and perpetuated. The best-known and most studied modern examples lie in the Caribbean (Florida, the Bahamas, numerous island complexes), Australia, the Persian Gulf, Pacific atolls, and Indonesia. These warm, clear tropical waters are what most sedimentologists envisage when they think of carbonate deposition in the geological record. They are right because this system produces the thickest and most widespread carbonate successions. They are also wrong because there is another major depositional realm: the cool and cold-water system (see Chapter 11).

The carbonate factory

Controls

The photozoan factory operates most efficiently under normal open marine conditions where the waters are oligotrophic, the illumination strong, salinity is close to 35‰, seawater temperature is above 20°C, and seasonal salinity fluctuations are low. Both decreased salinity due to fluvial input and increased salinity due to evaporation reduce biotic diversity and thus lower carbonate production rates. Long seawater residence times on particularly large platforms generally result in increasing salinity via evaporation. Upwelling of cold, nutrient-rich waters can profoundly affect an indigenous biota accustomed to warm, nutrient-poor waters. Upwelling waters, however, are also highly saturated with carbonate that can precipitate as cement. Finally, because of their location in the tropics, large but infrequent cyclonic depressions (hurricanes, typhoons, or cyclones) can be devastating to the shallow environment.

Organisms

Perhaps the most important groups of organisms are those that are photosynthetic or contain photosymbionts. Those that are restricted to shallow, warm modern tropical environments are green calcareous algae, large benthic foraminifers (with symbionts), scleractinian corals (with symbionts), tridacnid bivalves (giant clams with symbionts), and some siliceous sponges (with symbionts). This is also the only setting in which ooids form in abundance and carbonate mud precipitates from seawater. The same organisms are important in the rock record. There is still, controversy, however, if important invertebrates such as stromatoporoids, tabulate and rugose corals, and rudist bivalves had photosymbionts. The prolific productivity of such fossils suggests, but does not confirm, that they probably did.

Microbes are important, especially in facilitating and aiding in precipitation. It certainly appears that automicrite is most prevalent in warm-water carbonates, both today and in the past. They are also critical in altering many different particles by repeated grain infestation.

Bioerosion

Bioerosion by microbes, invertebrates, and fish is particularly prominent in warm-water carbonates. The most conspicuous macroborers are sponges and bivalves. Although bivalves erode the substrate and chemically release carbonate into the water column, sponges mine out silt-sized chips that add to the mud component of surrounding sediments. This process can be so important that it is estimated that as much as one-third of the mud in some Pacific atoll lagoons is generated in this way.

Carbonate precipitation

Carbonate precipitation in the form of ooids, water column microcrystals, pore-filling cement, and intraparticle cement is a hallmark of the warm-water neritic system (see Chapter 24). The precipitates are aragonite and HMC today, but could have been LMC in the past (see Chapter 20). At the smallest scale, microcrystalline cement fills microborings, allowing them to be preserved in the rock record. At a larger scale, the chambers of clonal skeletons (e.g., corals, stromatoporoids) and the internal cavities in reefs can be filled with similar precipitates. Intergranular precipitation and microbial binding is common between ooids forming granular clusters called *grapestones*. Such cement can also form between particles forming lithified lumps on the seafloor, termed *intraclasts*. Finally, intergranular precipitation of seafloor cement within sediment creates a *hardground* (see Chapter 24), a synsedimentary limestone that can form in only a few decades. Hardgrounds profoundly alter the nature of the seafloor, such that encrusting organisms, such as corals that prefer a hard substrate for attachment, quickly become established with some ultimately developing into reefs.

Carbonate dissolution

Whereas this setting is universally viewed as an environment of carbonate precipitation, recent work, as stressed above, has also demonstrated that dissolution is taking place in the shallow (<10 cm) subsurface just below the seafloor, especially in platform interior environments. Aragonite is particularly subject to such dissolution. It would appear that microbes, especially sulfate-reducing bacteria, are implicated in this process.

Depositional systems

Introduction

The major factories in the warm-water photozoan system are: (1) reefs; (2) ooid sand shoals; and (3) open seafloors or lagoons. Details of each of these systems are described in the chapters that follow. The carbonate structures that they construct are rimmed platforms, open platforms, and ramps (Figure 10.1).

The deposits produced on the platforms are autochthonous and remain largely in place after generation. By contrast, the tidal flat and slope depositional systems are composed of imported (allochthonous) sediment derived from the factories mentioned above.

When operating at full capacity, the carbonate factory continues to produce carbonate sediment through time. The ongoing accumulation of this material gradually raises the seafloor and brings it ever closer to sea level. This process results in the *shallowing-upward motif* of many limestone beds. More important perhaps is that this continued carbonate production, coupled with gradual tectonic subsidence, great thicknesses of rock are produced over millions of years. These are manifest as the impressive carbonates in major global mountain belts such as the Paleozoic of North America (Figure 10.2) and the Mesozoic of the European Alps. Perhaps the most imposing of such carbonate masses is the Bahama Banks, which has been producing carbonate since the Triassic and continues to do so, forming a structure ~6500 m in thickness and many thousands of square kilometers in areal extent.

Rimmed platforms and banks

These large carbonate structures are relatively flat-topped, possess a rim of reefs, islands, or sand shoals along their outer edge, and usually have steep outer slopes that descend into adjacent deep water (Figures 10.3, 10.4).

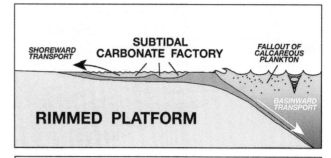
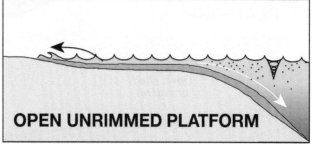
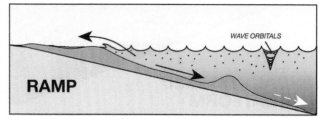

Figure 10.1 Sketch of the general attributes of a rimmed platform with highly differentiated facies, an open unrimmed platform where high-energy conditions prevail across the shallow seafloor creating a wide subtidal facies, and a ramp over which storms sweep into shallow waters. Most complex facies are located on the narrow inner part of the structure. Source: Adapted from James et al. (2010). Reproduced with permission of the Geological Association of Canada.

Figure 10.2 El Capitan and Guadalupe Peak, West Texas, USA, composed of ~400 m of massive Permian fore-reef dolomite and limestone overlying gentle slopes of basinal sandstone.

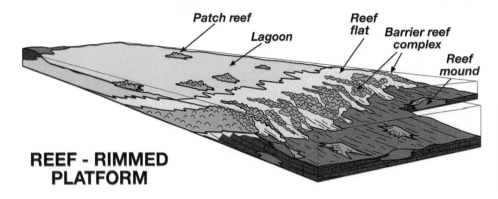

Figure 10.3 An isometric sketch of facies on a reef-rimmed platform. Source: James (1983). Reproduced with permission of the American Association of Petroleum Geologists.

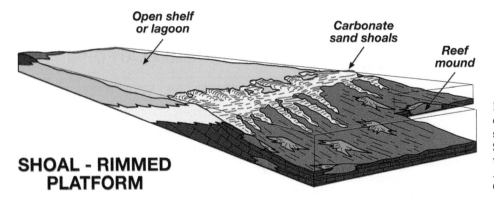

Figure 10.4 An isometric sketch of facies on an ooid or skeletal sand shoal-rimmed platform. Source: James (1983). Reproduced with permission of the American Association of Petroleum Geologists.

Platforms are connected to land but banks are isolated structures in a basin or the open ocean. They only form in tropical environments because the photozoan organisms that dominate these settings produce vast amounts of sediment, keep up with sea-level rise, and generate the rim facies. The elevated rim is critical because it absorbs most waves and swells that come from the open ocean, thereby protecting the relatively quiet-water leeward lagoon (Figure 10.5). Rimmed platforms and banks exhibit a recurring series of facies belts that are present regardless of the organisms that were growing at any given time.

Rim facies.   Reefs built by colonial animals are the most common facies of the rim, but large stromatolites occupied this niche in the Precambrian. Reefs are generally partitioned by channels that allow ocean waters to flow back and forth between the open sea and the shelf (Figure 10.5b). Reefal factories produce prolific amounts of sediment that can be swept together locally to form islands. Synsedimentary cementation, which is common especially in windward settings, increases the rigidity and robustness of the margin.

Ooid sand shoals, consisting of particles generated as tidal currents sweep back and forth across the shallow subtidal and intertidal sand bodies, accumulate rapidly to form segmented barriers along the edge of the platform. Synsedimentary islands are commonly a hallmark of these areas because the factory commonly overproduces sediment up to and above sea level.

Some of the islands along the edge of a platform are not only synsedimentary as above, but can be remnants of an old limestone landscape that formed during previous sea-level highstands. When sea level falls, the carbonates that accumulated along the platform margin are generally lithified as a result of meteoric diagenesis (see Chapters 25 and 26). When sea level rises again and submerges the landscape, some of the highest parts remain exposed as limestone islands.

Reefs on the leeward side of banks and atolls can be strikingly different compared to those on the windward edge, either because of reduced wave activity or because of "bad" water. Such foul water is generated on the platform by fine-grained sediment production, oxygen depletion, heating, evaporation or rainfall (Figure 10.6).

Figure 10.5 (a) A platform rimmed by continuous coral reef growth along the continental margin of Belize, looking south on an exceptionally calm day, with various facies named. Scale bar 50 m. (b) As for (a), rimmed by a segmented coral reef. Photographs by W. Martindale. Reproduced with permission. (c) The same barrier reef on a stormy day, where all of the wave energy is absorbed by the reef rim leaving the lagoon behind largely unaffected and tranquil.

It is usually driven off the bank downwind, across the leeward margin, inhibiting reef growth to a depth of several tens of meters. Such shallow leeward margins can therefore be bare rock floors covered by soft fleshy algae or hard coralline algae, with active reef growth taking place only in deeper water below this interface.

Lagoonal facies. The size and geography of lagoons on rimmed platforms and atolls have varied through time in accord with the changes in sea level. Although some of the modern lagoons appear large (e.g., Great Barrier Reef system), they are small when compared to the vast platforms that existed in past times (see Chapter 20). Water depth plays a critical role in these depositional systems because it controls the distribution of the photozoan assemblage. Tropical lagoons (Figure 10.7), with their seafloors in the photic zone are the epicenter of the carbonate factory, where vast amounts of sediment are generated from the algae and animals that thrive on the illuminated seafloors. Today, many of these areas can be divided into inner and outer shelf lagoons. Muddy sediments and scattered algae, but few corals, typically characterize the inner lagoon where *Thalassia* seagrass banks are extensive. Mangroves commonly grow along the coastal boundary of this zone. Patch reefs and skeletal sands formed from skeletons of the resident calcareous algae, and invertebrates typify the outer lagoon. Compared to the inner lagoon, sediments on the outer lagoon are generally coarser grained and contain a more diverse biota. Tropical storms can have a profound effect on the shallow-water platforms as vast amounts of sediments can be homogenized or moved around over very short time periods (see Chapter 3).

Lagoons covered by deep water are characterized by different sediments compared to their shallow-water counterparts, simply because the photozoan biota no longer dominates the factory. In the lagoons associated with atolls, for example, photozoans are the source of most sediment when the water is <30 m deep. With increased water depths, benthic foraminifera commonly become the dominant organism as the photozoan biota decreases because of ever-decreasing light levels. The floors of these deep-water lagoons, such as that associated with the Great Barrier Reef and many atolls, are typically covered by muddy sediments.

On some platforms, the inner lagoons can be characterized by terrigenous sand and mud that is brought into the area from the hinterland. As in parts of the Great Barrier Reef system and the Belize Shelf, there can be facies that are transitional between the siliciclastic and carbonate deposits.

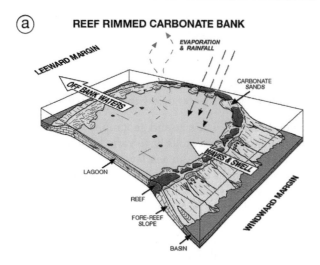

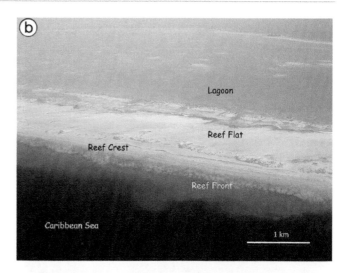

Figure 10.6 (a) An isometric diagram of a carbonate bank with well-developed reefs on the windward side and a poorly developed rim on the leeward side, due to saline and nutrient-rich waters formed in the lagoon by rainfall and evaporation. Source: Adapted from Hanford and Loucks (1993). Reproduced with permission of the American Association of Petroleum Geologists. (b) An aerial view of the leeward margin of Glovers Atoll, Belize with a poorly developed reef (compare to Figure 10.5a, b) that is leeward facing and swept by somewhat nutrient-rich warm and saline waters. Photograph by W. Martindale. Reproduced with permission.

Muddy peritidal facies. Active carbonate production in the lagoon is occasionally interrupted by cyclonic storms that stir up the shallow seafloor and transport the suspended mud landward, concentrating it on sand shoals or piling it up behind islands along the platform margin. It is common to see the lagoon waters looking like milk after the passage of such storms.

Muddy tidal flats (Figure 10.8) are a common product of fine-grained sediment production and landward transport on warm-water rimmed platforms, and are particularly common in the geological record. Sediment is spread onto the tidal flat by the cyclonic storms and then reworked by the day-to-day action of tidal currents. This is a harsh environment consisting of facies that are a product of both the adjacent marine factory and the terrestrial climate. Deposits are replete with microbial mats, a high abundance of low-diversity biota (particularly gastropods), storm deposits, desiccation features, dolomites, and evaporites.

Beach facies. Beaches are also common peritidal deposits (Figures 10.9, 10.10a). They are similar to their siliciclastic counterparts with the same hydrodynamic sub-environments of offshore, shoreface, foreshore, and backshore subfacies. The deposits are characterized by gentle seaward-dipping (5–15°) large-scale planar accretion beds. The *foreshore* portion is typified by alternating coarse- and fine-grained laminations, local inverse-graded laminations, and vertical burrows formed by mollusks,

worms, and crustaceans. Keystone vugs (Figure 10.10c), which are extra-large elongate to circular pores formed by escaping gas as the sediment is inundated by waves, are common. The *shoreface* is typified by small- to large-scale tabular to trough cross-beds. The *offshore* is a myriad of grainy to muddy sediments. The *backshore* can pass landward into dunes, washovers, or muddy supratidal complexes. One distinguishing feature of the carbonate beach environment is *beachrock* (Figure 10.10b), a series of seaward-dipping grainy limestone beds composed of beach sand that are cemented by aragonite or Mg-calcite (see Chapter 24).

Aeolianites. Holocene and Pleistocene marginal marine carbonate sand dunes (aeolianites) occur towards the northern and southern margins of the warm-water carbonate realm where the air is dry. Carbonate dunes are best developed where ocean waves sweep unimpeded onshore, mobilizing shallow sands and building beaches in the intertidal zone. Sand is then transported by trade winds to build dunes. They are not common in the tropics with a humid climate, nor are they well developed in reef-dominated environments. Some of the most extensive warm-water aeolianites are present in the Bahamas, where overproduction of ooids has led to sediment accretion above sea level and the formation of sand islands.

The deposits are shore-parallel bodies in the form of transverse ridges in which the dunes are predominantly oblique, parabolic, or barchanoid. Large-scale

Figure 10.7 (a) A shallow subtidal lagoon community of corals and green algae in ~1 m of water, Harrington Sound, Bermuda. Image width 30 cm. (b) A shallow subtidal shelf facies in ~2 m of water with sparse seagrass and echinoids, Belize. Image width 20 cm image. Photographs by W. Martindale. Reproduced with permission. (c) A shallow Paleozoic subtidal facies of brachiopods (Br) and stick-like bryozoans (By) in an Upper Ordovician limestone, Anticosti Island, Québec, Canada. Image width 15 cm.

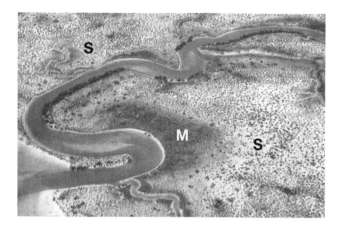

Figure 10.8 A tidal creek winding its way across the muddy tidal flats to the open ocean (left) in the Turks and Caicos Islands, bordered by dark intertidal microbial mats (M) and a wide supratidal zone (S). Image foreground width 80 m.

landward-dipping foresets (Figure 10.11a), grainflows, pinstripe laminations, slump scars, animal tracks, and numerous trough cross-beds related to blowouts along windward margins (Figure 10.11b, c) characterize these deposits. Pinstripe laminations are inversely graded laminations a few millimeters thick, formed by the lateral and vertical migration of wind ripples. Dunes in the Bahamas are generally between 20 and 30 m thick, with the highest being ~60 m. Pleistocene aeolianites also contain numerous interbedded paleosols (ancient soils) (Figure 10.11d, e) with land snails (Figure 10.11f) and calcretes, as well as many features of meteoric diagenesis (see Chapter 25).

Aeolian dunes are predominantly highstand deposits because they need a continuous supply of marine carbonate sand to form. Virtually all land in Bermuda, for example, is Pleistocene aeolianite. Likewise, almost all land above 7 m elevation in the Bahamas is formed by Pleistocene aeolianites and associated facies that formed during past eustatic highstands. This is important because these limestones control the nature and location of modern depositional environments throughout this extensive realm. Pleistocene islands determine the nature of inter-island tidal deltas (see Chapter 13) and act as barriers to waves and swells, permitting low-energy lagoons to develop in their lee.

It is somewhat puzzling that these sand bodies are not more prevalent in the geological record. This could be because they are very similar to marine sand bodies (see Chapter 13) and so have not been recognized as aeolianites; unlike siliciclastic aeolian dunes that are composed of mostly fine-grained sand, aeolian carbonate sands can contain a wide variety of grain sizes.

SIMPLE BEACH PROFILE

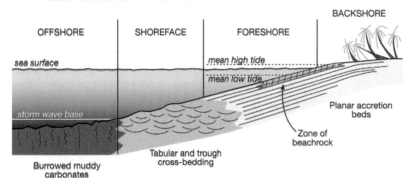

Figure 10.9 A sketch illustrating the different zones on a typical carbonate beach.

Figure 10.10 (a) A long carbonate beach with shallow foreshore (F) and offshore grassbed (G), southern Yorke Peninsula, Australia. Foreground image width 30 m. (b) Dark beachrock on a small island in the Bahamas (people for scale in background). (c) Keystone vugs (large holes) in a Holocene beach outcrop, Bahamas. Image width 8 cm. Photograph by W. Martindale. Reproduced with permission.

Slope facies. At the other end of the depositional spectrum, sediments on basin-facing slopes are largely derived from the platform; in post-Jurassic times, these deposits are also derived from the fallout of calcareous plankton from the overlying water column. Deposits form via a variety of sediment gravity-flow processes that result in a suite of calciturbidites and calcareous debrites interbedded with finely laminated to burrowed muds (Figure 10.12). The coarser particles come from reefs or ooid sand shoals along the margin, sands swept seaward through passes between reefs and shoals, and fragments from lithified upper slope carbonate. Muds are derived from fine lagoonal sediment swept seaward during cyclonic storms and leeward off banks during fair-weather processes. Once deposited, this fine-grained sediment can be lithified on the slope only to be ripped up by later events and redeposited downslope as flat-clast conglomerates and breccias. The only significant source of carbonate during the Paleozoic was the adjacent platform and those sediments are typically interbedded with shales, representing periods of non-carbonate transport and accumulation. This is not so during the middle Mesozoic and younger times, when the water column was a significant source of mud via the disintegration of calcareous plankton. As a result, these younger slope deposits tend to be more carbonate rich than their older counterparts.

Open platforms

These platforms lack a raised rim and so large ocean waves and swells can easily sweep across the whole shelf or bank. They are most prominent today in cool-water heterozoan settings where the waters are too cold to allow reefs or ooid sands to develop. They also occur in warm-water systems during times in geologic history when large reef-building organisms were absent. They typically face the open ocean along continental margins where underlying antecedent bathymetry has generated the

Figure 10.11 Pleistocene aeolianites, Bermuda. (a) Cliff section of aeolianite ~8 m high along south coast illustrating steep landward-dipping beds. (b) Eroded trough cross-bedded aeolianites, Castle Island. Cliff is ~8 m high. (c) Eroded trough cross-beds along the seaward margin of North Lagoon. (d) Road cut ~4 m high illustrating a lowstand paleosol between two aeolianites. (e) Accretionary soil (S) between beach deposits (B) and aeolianites (A) along the south coast. (f) Land snails in a paleosol, Northeast Coast. Centimeter scale.

margin morphology upon which the carbonates are now being deposited.

The facies spectrum is different from that on rimmed platforms. Not only is there no rim, but there is also no muddy open lagoon and the platform is characterized by grainy facies throughout, including beaches and aeolianites. These coarse sediments can be ooids or biofragments with microbialites in some tropical systems. Muddy tidal flat facies are not expected, except behind intraplatform islands that offer protection from the omnipresent waves and swells. The large amounts of sediment that are transported off the platform generate thick prograding clinoforms of predominantly neritic origin. Tropical banks

Figure 10.12 Cambro-Ordovician slope deposits in western Newfoundland, Canada, comprising ribbon limestones at left and a massive debrite with white platform margin blocks above. Note person (circled) for scale.

with no rims typically have a top characterized by very slow or arrested sedimentation and numerous lithified hardgrounds.

Ramps

A ramp is an unrimmed shelf that slopes gently basinwards at an angle of less than one degree. Nearshore wave-agitated facies grade outboard into deeper-water progressively lower-energy deposits and there is generally no discernable shallow-water break in slope (Figure 10.13), although some ramps can be distally steepened. Ramps are, by definition, tied to the hinterland and can therefore receive variable amounts of siliciclastic sediment from land.

Ramps are particularly well studied because they are often found in intracratonic basins (see below) and not fragmented in mountain belts in the same way as rimmed and unrimmed continental margin platforms. They are traditionally divided into hydrodynamically defined facies belts, quite unlike rimmed platforms (Figures 10.13, 10.14).

Inner ramp. The inner ramp (Figure 10.15a, b) is a complex zone, generally quite narrow (kilometer scale), comprising a series of sand shoals, narrow lagoons, and peritidal mud flats or beaches. This arrangement is like a narrow rimmed platform in many ways, except that it is up against the shoreline. Deposition is in the photic zone and above fair-weather wave base. Sand shoals, typically tidal, absorb much of the hydrodynamic energy. Depending on the situation they can be composed of ooids or, in the Paleozoic, crinoid sands. Other components can

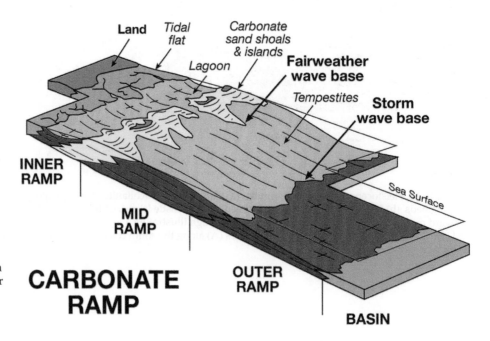

Figure 10.13 Isometric diagram of a carbonate ramp illustrating the major facies and location of the separate sectors, as defined by fair-weather wave base and storm wave base.

CARBONATE DEPOSITIONAL RAMP

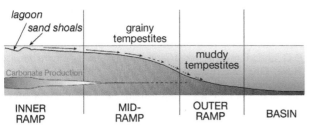

Figure 10.14 Cross-section of a carbonate ramp indicating the importance of the carbonate factory and location of different tempestites.

include bryozoans, green calcareous algae, and benthic foraminifers. These sands can also be concentrated into islands on the shoals, enhancing the hydrodynamic 'resistance' of these deposits, or instead be swept leeward into the narrow lagoon. Such cross-bedded sands pass shoreward into classic lagoonal facies, whose composition and biotic diversity depends largely on salinity levels. The peritidal facies are similar to those on rimmed platforms.

Mid-ramp. The mid-ramp is where the ramp system is dramatically different from rimmed and open platforms, because the facies are storm-dominated (see Chapter 13; Figure 10.15c). The seafloor here lies between fair-weather wave base and storm wave base and is generally more areally extensive than the inner ramp.

Figure 10.15 (a) Cross-bedded nearshore inner-ramp crinoidal limestones, Upper Ordovician, Anticosti Island, Québec, Canada. Note person (circled) for scale. (b) Finely interlaminated limestone and dolostone, lagoonal facies, Middle Ordovician Maryland, USA. Note 10cm long pen for scale. (c) Thinly bedded lenticular tempestites of the mid ramp, person scale, Upper Ordovician, Anticosti Island, Québec, Canada. Note person for scale. (d) Thinly interbedded muddy tempestites and shales, Upper Ordovician, Anticosti Island, Québec, Canada. Scale bar: 10cm increments.

Carbonate production decreases with increasing water depth. Day-to-day sedimentation is in the form of muddy sediment fallout, in addition to that generated from *in situ* organisms. The sediments are typically burrowed. Episodic storm deposition is in the form of tempestites with hummocky and swaley cross-stratification and episodic calciturbidites. Storm beds are typically lenticular and interbedded with muddy fair-weather deposits (Figure 10.15c).

Outer-ramp. The seafloor on the outer ramp is below storm wave base and dominated by suspension fallout of carbonate sediment that originated on the inner- and mid-ramp. In Paleozoic successions, these sediments are commonly interbedded with shale, and benthic muds in post-Triassic strata (Figure 10.15d). Although storm deposits can be present, they are generally rare and thin. Sediments can be extensively burrowed or laminated, depending upon the nature of the overlying water column. Anoxic shales are especially common in the Paleozoic.

Further reading

Abegg, F.E., Harris, P.M., and Loope, D.B. (eds) (2001) *Modern and Ancient Carbonate Eolianites: Sedimentology, Sequence Stratigraphy, and Diagenesis.* Tulsa, OK: SEPM (Society for Sedimentary Geology), Special Publication no. 71.

Brooke, B. (2001) The distribution of carbonate eolianite. *Earth-Science Reviews,* 55, 135–164.

Davies, P.J. (ed.) (1983) *Great Barrier Reef.* Bureau of Mineral Resources, Journal of Australian Geology and Geophysics, Vol. 8, p 291.

Enos, P. (1977) Holocene sediment accumulations of the South Florida shelf margin. In: Enos P. and Perkins R.B. (eds) *Quaternary Sedimentation in South Florida.* Boulder, Co: Geological Society of America, Memoir no. 147, pp. 1–130.

Frébourg, G., Hasler, C.-A., Le Guern, P., and Davaud, E. (2008) Facies characteristics and diversity in carbonate eolianites. *Facies,* 54, 175–191.

Handford, C.R. and Loucks, R.G. (1993) Carbonate depositional environments and systems tracts: responses of carbonate platforms to relative sea-level changes. In: Loucks R.G. and Sarg J.F. (eds) *Carbonate Sequence Stratigraphy: Recent Developments and Applications.* Tulsa, OK: American Association of Petroleum Geologists, Memoir no. 57, pp. 3–42.

James, N.P. (1983) Reef environment. In: Scholle P.A., Bebout D.G., and Moore C.H. (eds) *Carbonate Depositional Environments.* Tulsa, OK: American Association of Petroleum Geologists, Memoir no. 33, pp. 345–440.

James, N.P., Kendall, A.C., and Pufahl, P.K. (2010) Introduction to biological and chemical sedimentary facies models. In: James N.P. and Dalrymple R.W. (eds) *Facies Models 4.* St John's Newfoundland and Labrador: Geological Association of Canada, GEOtext 6, pp. 323–339.

Logan, B.W., Davies, G., Read, J.F., and Cebulski, D.E. (eds) (1970) *Carbonate Sedimentation and Environments, Shark Bay, Western Australia.* Tulsa, OK: American Association of Petroleum Geologists Memoir no. 13.

Purdy, E.G. and Bertram, G.T. (1993) *Carbonate Concepts from the Maldives, Indian Ocean.* Tulsa, OK: American Association of Petroleum Geologists, Studies in Geology no. 34.

Purdy, E.G. and Gischler, E. (2003) The Belize margin revisited. 1: Holocene marine facies. *International Journal of Earth Sciences,* 92, 532–551.

Purser, B.H. (ed.) (1973) *The Persian Gulf: Holocene Carbonate Sedimentation and Diagenesis in a Shallow Epicontinental Sea.* New York: Springer-Verlag, p. 471.

Rankey, E.C. and Reeder, S.L. (2010) Controls on platform-scale patterns of surface sediments, shallow Holocene platforms, Bahamas. *Sedimentology,* 57, 1545–1565.

Wanless, H.R. and Dravis, J.J. (1989) *Carbonate Environments and Sequences of Caicos Platform.* Washington, DC: American Geophysical Union, 28th International Geological Congress, Field Trip Guidebook T374, 75 pp.

Wilson, M.E.J. (2008) Global and regional influences on equatorial shallow-marine carbonates during the Cenozoic. *Palaeogeography, Palaeoclimatology, Palaeoecology,* 265, 262–274.

CHAPTER 11
THE COOL-WATER
NERITIC REALM

Frontispiece An underwater forest of the giant macroalga (kelp) *Macrocystis* sp. growing in the shallow (~15 m deep) cool ocean
waters off southern California. Photograph by Dale Stokes. Reproduced with permission.

Origin of Carbonate Sedimentary Rocks, First Edition. Noel P. James and Brian Jones.
© 2016 Noel P. James and Brian Jones. Published 2016 by John Wiley & Sons, Ltd.
Companion website: www.wiley.com/go/james/carbonaterocks

Introduction

Historically, most carbonates have been thought of as tropical in origin. Recent research has however shown that significant volumes of carbonate are or were deposited in cooler temperate and polar waters. A minimum average surface seawater temperature of ~20 °C provides a measurable boundary, albeit diffuse, between temperate (heterozoan) and tropical (mostly photozoan) carbonate depositional realms (Figure 11.1). The heterozoan biota (Figure 11.2) comprises benthic foraminifers, mollusks (bivalves and gastropods), echinoderms, barnacles, and bryozoans along with coralline algae, which are the only phototrophs. This biota extends from the mid-latitudes to the poles.

Cool- and cold-water calcareous sediments typically accumulate on open shelves or ramps (see Chapter 12). Sediments there are found from the strandline to upper slope. Many are sorted, swept into sand shoals, deposited in topographic lows, and transported in the same way as neritic siliciclastic sediments.

The *cool-water* realm is separated into *warm-temperate* and *cold-temperate* provinces that are divided, in the modern world, by a minimum seawater temperature of 15 °C (Table 11.1). Sediments in the warm-temperate province are largely heterozoan but can contain up to 20% photozoan elements such as poorly calcified green algae, zooxanthellate corals, and large benthic foraminifers with photosymbionts. Given that these sediments contain both heterotrophs and phototrophs, they can be called transitional heterozoan assemblages. Sediments in the cold-temperate province are formed entirely of heterotrophs.

The *cold-water* realm is entirely high-latitude polar with water temperatures <5 °C. The calcareous sediments, again entirely heterozoan, are always associated with marine glacigene deposits.

The Carbonate Factory

The Carbonate Factory is an active one, albeit operating at a lower output than the Photozoan Factory, and produces carbonates across the environmental spectrum.

Controls

Temperate carbonates are restricted to areas of low siliciclastic input or beyond the reach of terrigenous sediment transport on the outer parts of platforms. This is because rates of production and accumulation are

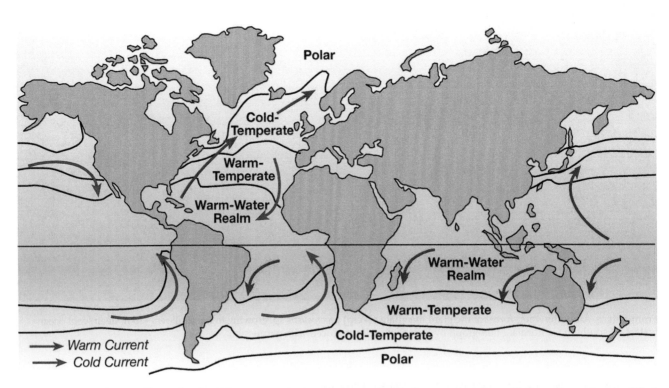

Figure 11.1 A world map illustrating the different seawater temperature realms in the ocean together with the major cool-water (blue) and warm-water (red) currents. Source: Adapted from James and Lukasik (2010). Reproduced with permission of the Geological Association of Canada.

relatively slow, and the sediments are easily swamped by input of siliciclastic material.

Temperate carbonates are generally located on the western sides of continents where arms of oceanic gyres advect cool waters equatorward. These are also areas of coastal upwelling where cool, nutrient-rich waters are brought up onto the shelf and promote prolific benthic bioproductivity. Cool-water carbonates can also form at high latitudes, for example, the Gulf Stream system injects relatively warm waters into the Norwegian polar realm (Figure 11.1).

Areas of elevated nutrient flux have high levels of sessile heterotrophic benthic productivity, resulting in enhanced biogenic sediment production. Places of such primary productivity are: (1) zones of upwelling;

(2) oceanographic fronts and shear zones; (3) areas of fluvial input; and (4) ice margins. Areas of downwelling and intervals of water column de-stratification in the open ocean are also important because they allow the transfer of nutrients from the surface plankton blooms to the deep seafloor.

With so few phototrophs living in the cool- and cold-water realm, light penetration would seem to be of little importance. Seafloor illumination is, however, critical because it determines the depth to which seagrasses, non-calcareous algae, and coralline algae can thrive. This is especially important at high latitudes where light vanishes for several months each year and ice covers the ocean surface.

The sediment-producing organisms live on and in open rippled sands, on hard rocky substrates, and on ephemeral grasses and seaweeds. The diversity and abundance of the calcareous epibenthos on open rippled sands is, however, almost an order of magnitude less than that on hard rocky substrates. The sediment factories can therefore be divided into: (1) a low production sediment plain; and (2) a highly productive rocky seafloor.

Sediment producers

Coralline algae. Coralline algae, which are the only benthic calcareous autotrophs, grow from tropical to polar latitudes in water from the surf zone to depths of ~80 m. They commonly form granule- to cobble-size rhodoliths that roll around on the seafloor with the currents. They can also encrust hard substrates, shells, and living seaweeds. Branches that break off shrub-like forms produce gravels of rod-shaped sticks called maerl. Articulated branching corallines are numerous on broad-bladed seagrasses as

Figure 11.2 A cool-water Oligocene limestone in southern Australia with numerous branching and fenestrate bryozoans. Image width 12 cm.

Table 11.1 The composition of heterozoan carbonates in different neritic marine provinces. Source: Adapted from James and Lukasik (2010). Reproduced with permission of the Geological Association of Canada.

Realm	Province	Attributes	Temperature (°C)
Cool water	Warm–temperate	Heterozoan and transitional heterozoan: open shelf and ramp minor photozoan elements (large benthic foraminifers, scattered zooxanthellate corals, poorly calcified green algae), minor carbonate mud, local hardgrounds, abundant corallines, seagrasses or macrophytes inboard (post-Early Cretaceous), palimpsest sediment, extensive bioerosion and maceration, bryozoan, mollusk, foraminifer factory.	15–20
	Cool–temperate	Heterozoan: open shelf and ramp, possible interbedded glacigene deposits, minor carbonate mud, local hardgrounds, palimpsest sediment, extensive bioerosion and maceration, corallines and macrophytes inboard (post-Cretaceous), conspicuous mollusks, and barnacles with bryozoans.	5–15
Cold water	Polar	Heterozoan: open shelf or oceanic bank, seasonal ice cover, interbedded glacigene deposits, IRD, glendonites, both biogenic carbonate (bryozoans, mollusks, barnacles), and biosiliceous (diatoms and sponges) sediment.	<5

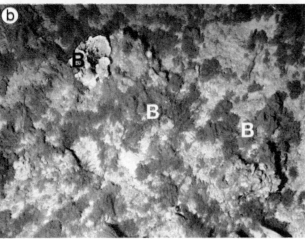

Figure 11.3 (a) A 15-m-deep rocky outcrop covered with a cool-temperate biota of fenestrate bryozoans off western Victoria, southern Australia. Image width 20 cm. (b) A seafloor image at 120 m depth off northwest Tasmania with numerous bryozoans (B) and a few sponges. Reproduced with permission of CSIRO. Image width 1.5 m.

well as on rocky substrates where they disintegrate to form innumerable rod-shaped, sand-sized grains.

Skeletal invertebrates. The smallest animals are benthic foraminifers that, together with ostracod shells and ascidian spicules, generate copious amounts of fine sand-sized sediment. Brachiopods are rare, except in some high-latitude settings. The most conspicuous components are bryozoans (Figure 11.3) that grow across the shelf and who, along with mollusks and foraminifers, are the dominant producers of the neritic temperate-water carbonate sediments. Regular and irregular echinoids generate

innumerable spines and plate fragments. Serpulid worms create a significant proportion of the sand and larger grains. Barnacles typically disintegrate into many plates and are especially numerous in high-energy and high-latitude environments. Carbonate mud produced by coccoliths forms only a minor part of the neritic sediments.

Although most of the sediments are calcareous, siliceous sponges are an important part of the benthic biota. Their spicules are easily transported because they are hollow and have a low specific gravity. As a result, they form an important part of the fine-grained fraction of the sediment.

Alteration

Sediments that accumulate in cool- and cold-water realms undergo modification by bioerosion and maceration, cementation, and dissolution. Collectively, these processes can significantly modify the original sediments.

Bioerosion and maceration. With the high biomass, predation is rampant in cool-water environments; bryozoans and mollusks are eaten by crabs and durophagus fishes with jaws and teeth strong enough to crush bones and shells, bivalves are drilled and ingested by carnivorous gastropods, starfish prey upon epifaunal bivalves, and phaeophytes are grazed by echinoids. Agents of bioerosion (bivalves, gastropods, worms, sponges, and microbes) are also profuse. This results in production of mud by the boring process (especially by sponges) and disintegration of grains into fine sand and mud fragments by repeated sponge infestation. Since synsedimentary cement is not widespread and fine-grained mud is scarce, the holes left by boring microbes remain largely unfilled and grain outlines are therefore not preserved in the rock record, there are, no micrite envelopes. Particles held together by organic tissue also fall apart *in situ* after the decomposition of their organic matrix (maceration). This process is accelerated by the action of microbes that consume intercrystalline organics.

Cementation. The absence of non-skeletal particles such as ooids, the lack of obvious precipitated lime mud, and the relative scarcity of marine cement leads to the notion that cool-water settings are not environments of early, abiotic Mg-calcite or aragonite precipitation. Although largely true, there is also contrary evidence for episodic hardground formation via intergranular precipitation of synsedimentary carbonate cement.

Dissolution. The role of carbonate dissolution on cool-water shelves is complex and poorly understood. Aragonite skeletons that become 'chalky' in many cool-water settings

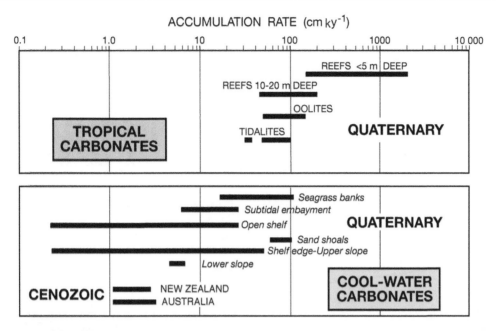

Figure 11.4 A comparison of carbonate sediment accumulation rates. Source: Adapted from James (1997). Reproduced with permission of the SEPM Society for Sedimentary Geology.

are signals of partial dissolution. There is also evidence that Mg is lost from Mg-calcite. Significant amounts of aragonite in sediment on the modern seafloor vanish, probably due to dissolution that is associated with oxidation of the organic material. As a result, part of the aragonitic biota disappears soon after deposition, leaving the sedimentary record strongly biased towards calcite components.

Production and accumulation

In general, net rates of carbonate production in temperate waters are about one order of magnitude less than those in warmer-water settings (Figure 11.4). There are, however, some exceptions. The shallow seagrass, kelp, and coralline algal facies of the cool- and cold-water realm are locally prodigious sediment producers whose production rates rival those from the warm-water realm. In addition, because of the overall relatively low sedimentation rates, temperate water deposits can contain older particles, particularly *relict grains* and *stranded grains*. Relict grains are those particles that formed during relatively recent sea-level highstands, were exposed and locally altered during subsequent sea-level lowstands and then reincorporated into new deposits during the next sea-level rise. Stranded grains were deposited in shallow environments during the last eustatic sea-level lowstand, only to be marooned in somewhat deeper water as sea level rose to its present high position. Deposits can therefore be mixtures of modern, relict, and stranded grains.

Depositional settings

The most common depositional settings are: *open ocean shelves*, as represented by continental margins or offshore banks; *interior basins*, as represented by seas such as the Mediterranean, together with a spectrum of epicontinental, foreland, and intracratonic depocenters and *seaways*; as well as narrow marine passages that link larger bodies of water such as oceanic straits, elongate intra-platformal channels, or inter-island pathways (Figure 11.5). Each of these systems is somewhat different in warm-temperate, cool-temperate, and polar settings.

Open ocean shelf

Open ocean environments, located along continental margins or on isolated banks, tend to be wave-dominated depositional systems. Fair-weather wave base is deep, commonly 50–60 m, such that sediment to these depths is moved almost constantly (sometimes called the zone of wave abrasion) and transported both offshore and onshore. The water column is well-mixed in the winter but somewhat stratified in summer with a shallow thermocline. Water temperatures below the thermocline (the zone in the water column of rapid temperature decrease) are usually <15 °C. Nutrients are mostly supplied via seasonal upwelling onto the outer shelf or even to the shoreline. The middle to outer shelf depositional regions

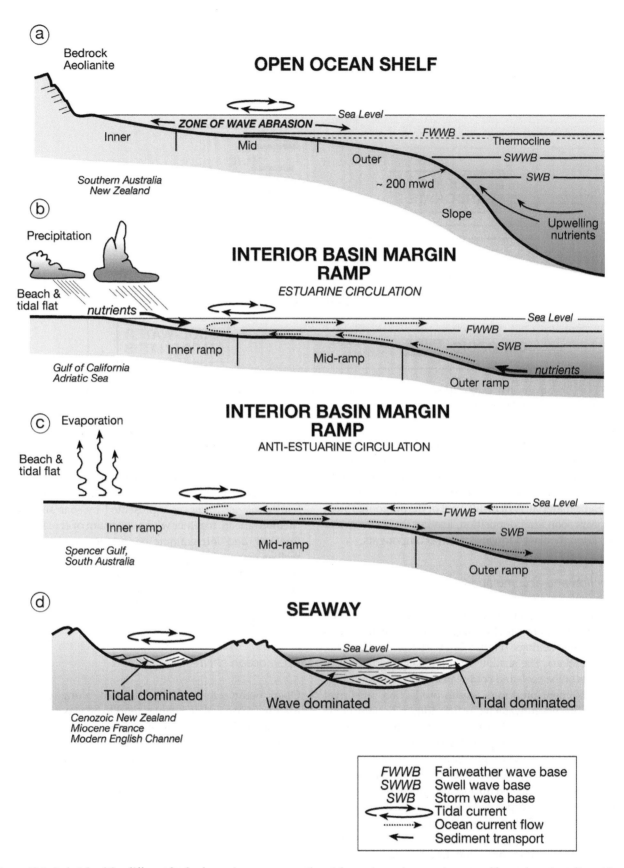

Figure 11.5 A sketch of the different hydrodynamic zones on continental margin environments covered by carbonate sediment in cool-water open ocean, interior basin, and seaway environments. Source: Adapted from James and Lukasik (2010). Reproduced with permission of the Geological Association of Canada.

are largely similar because they all lie below the shallow thermocline. By contrast, shallow inboard facies are more complex, partly because the thermocline that moves up and down annually changes seafloor temperatures.

Interior basin

Interior basins include settings such as enclosed seas (e.g., Mediterranean), large semi-isolated embayments (e.g., South Australian Gulfs, Gulf of California, Adriatic Sea), epicontinental to epeiric basins (e.g., Murray Basin, Australia), and foreland basins (e.g., New Zealand). All are relatively low-energy environments with ramps or distally steepened ramps. Waves are subdued across the ramps and swells are not important, although environments are perturbed by seasonal storms. Water circulation patterns can be estuarine or anti-estuarine. Tides can be important because of basin geometry. There are seasonally wide fluctuations in both temperature and salinity such that elevated salinity is important, especially in nearshore regions. Although nutrients can be ocean-derived, most come from the adjacent land and the basins are highly sensitive to changes of trophic resource levels. In short, these are very different environments compared to open water settings. Interior basins can therefore have a wider spectrum of deposits over smaller distances compared to open ocean settings.

Seaway

Where conditions are optimum for seawater flow acceleration, facies are strongly current bedded. Such currents are bipolar tidal or unidirectional ocean flow. These conditions can develop in oceanic straits, in passages between islands, or in elongate seaways that range from tens to hundreds of kilometers in scale. Cross-sectional geometry of seaways, together with the prevailing oceanic tides and wave regime appear to determine whether the system will be wave- or current-dominated (see Chapter 13).

Warm-temperate carbonates

High-energy open ocean systems

Sediment is produced throughout these systems. Although some sediment accumulates on the shelf (Figure 11.6a), high-energy conditions mean that large amounts of sediment are transported onshore to form *aeolianites* (calcareous aeolian dunes) or moved offshore to form prograding slope sediment wedges.

Strandline carbonates in temperate locations or in semi-arid climates are usually backed by extensive aeolianites.

There, beaches are pounded by large waves and swells that sweep onshore across unrimmed platforms. Beach sands are then blown onshore by strong seasonal winds (Figure 11.7a). This situation is especially so in warm-temperate marine environments with luxuriant seagrass growth and attendant rapid carbonate production in shallow waters offshore. The most extensive systems are along the coasts of southern and western Australia and South Africa, as well as areas in the Mediterranean.

The vast scale of aeolianites occurring over a distance of 3000 km in southern Australia has amazed geologists ever since Darwin visited the area in the late 1800s. He was so impressed with their scale that he thought that the amount of carbonate tied up in these systems rivaled that of the Pacific reefs he had seen. The Pleistocene structures in the cool-water region occur both as stacked and prograding complexes, the latter as old as the Pliocene. Modern shoreline exposures of these Pleistocene limestones with interbedded paleosols are typically truncated by erosion and sculpted into high cliffs (Figure 11.7b, c).

The particles generated in the offshore heterozoan factory are variable in size and heterogeneity, making them particularly difficult to differentiate from marine carbonate sands.

The *inner shelf* typically has a steep shoreface that descends to 10 mwd or more where the seafloor is covered with rippled sand, bedrock with kelp, or areas of thin-bladed to local broad-bladed seagrasses. Bryozoans, mollusks (mostly infaunal bivalves), echinoids, and small benthic foraminifers produce the biofragmental sand and gravel, whereas coralline red algae encrust most bedrock surfaces. The *middle shelf* is wave-rippled sand (Figure 11.8) that can develop locally into large subaqueous dunes. This is a zone of sediment production but little *in situ* accumulation due to wave sweeping. The *outer shelf* is a zone of carbonate production as well as accumulation. Bryozoans, sponges, and other heterotrophs thrive where the shelf rolls over onto the upper slope. The *shelf edge* is usually a gentle roll-over of the open shelf margin into progressively deeper and muddier sediments with a corresponding decrease in macrobiota. The *upper slope* is covered with muddy sediment that is formed of silt-grade skeletal material, including carbonate skeletal fragments and siliceous spicules swept from the shelf and mixed with pelagic fallout, especially coccoliths.

Low-energy interior basin systems

These depositional environments (Figure 11.6b) are generally more tranquil than open ocean settings, where the depth of fair-weather wave base (FWWB) is negligible

TEMPERATE CARBONATE FACIES SPECTRUM

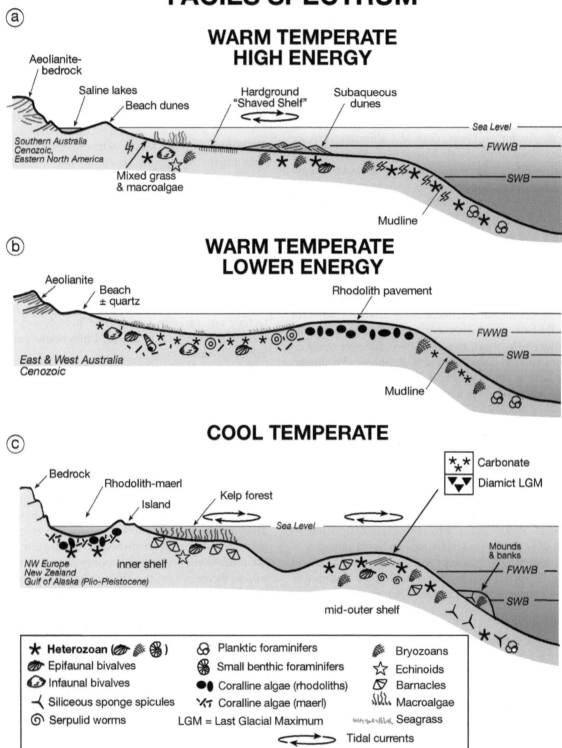

Figure 11.6 A sketch of the different hydrodynamic zones and main calcareous biota on open-ocean continental margin environments in warm-temperate and cool-temperate marine settings. Source: Adapted from James and Lukasik (2010). Reproduced with permission of the Geological Association of Canada.

Figure 11.7 Aeolianites, South Australia. (a) Modern beach dune complex on a quiet day, Robe. Dunes are ~18 m high. (b) A 20-m-high cliff of eroded Pleistocene aeolianite illustrating discrete internal dune morphologies, Ethel Beach. (c) A 35-m-high cliff composed of stacked Pleistocene aeolianites and interbedded red paleosols, Cape Spencer.

Figure 11.8 A seafloor image of coarse rippled biofragmental sand at 130 m depth off northwest Tasmania. Reproduced with permission of CSIRO. Image width 1.6 m.

compared to the open ocean and winter storms, although violent, are generally local and of short duration. Such lower-energy settings contain a facies spectrum similar to their higher-energy counterparts, but are generally finer grained. Interior basin platforms are in the form of ramps or open shelves.

The *inner platform* is a diverse array of environments, including rocky shorelines and beaches, barrier islands, and leeward muddy lagoons with numerous bivalves, ostracodes, and gastropods. There are also muddy tidal flats in more protected settings with all of the attributes of typical peritidal deposits (see Chapter 12). Seaward of these environments are a suite of cross-bedded sand shoals (packstone and grainstone) composed of abraded biofragmental grains and extensively burrowed, muddy seagrass beds (Figure 11.9a). The *mid-platform* is characterized by sub-wave base facies that are typified by red algae in the form of encrusted pavements or fields of rhodoliths and articulated coralline algae associated with bryozoans, mollusks, and foraminifers. Deposits there range from burrowed and thick-bedded to conspicuously cross-laminated grainstone and packstone beds. The *outer platform* is typically characterized by muddy facies with a biota of filter-feeding and infaunal calcareous organisms, together with conspicuous planktonic foraminifers.

Seaways

Most seaways are dominated by warm-temperate facies where current-dominated, large-scale simple and compound subaqueous cross-bedded dunes (up to 10 m high) can form in water 40 to 60 m deep. The sediments are generally heterozoan grainstone and rudstone with

Figure 11.9 (a) A seafloor image at 6 m of seagrass and biofragmental sand in a warm-temperate environment, Spencer Gulf, South Australia. Image width foreground 1 m. (b) A seafloor image at 3 m of brown macroalgae and purple coralline algal encrusted bedrock in a cool-temperate environment, western Victoria, Australia. Image width 25 cm.

coralline algae and variable amounts of terrigenous clastics. These facies can alternate with deposits of progressively more finer-grained, commonly bioturbated, flat-bedded carbonates as the influence and velocity of currents decreases and waves become the dominant depositional process. The ends of seaways can also be sites of flood tidal delta deposition in the form of shallowing-upward cycles (Chapter13).

Cool-temperate carbonates

All cool-temperate systems are open ocean (Figure 11.6c). Unlike the tropics, large parts of the shallow seafloor in cool-temperate and polar latitudes have been glaciated during the Neogene such that many shelves are covered with glaciomarine sediments (diamictites). With retreat

of continental glaciers after the Last Glacial Maximum (LGM), many of the Northern Hemisphere shelves have become sites of heterozoan carbonate deposition. Minimum surface seawater temperatures on these open shelves range from ~5 °C to 15 °C. These carbonates were deposited between periods of glaciomarine sedimentation, leading to interbedded carbonates and glaciomarine deposits. Today, such carbonates form a thin layer of biogenic material on top of LGM diamicts and are most numerous in the Northern Hemisphere. Sediments are almost completely heterozoan, and nearshore kelp forests are typical.

The *inner shelf* to depths of 50 mwd or less is typically dominated by: (1) kelp forests (Figure 11.9b); and (2) coralline algal (maerl) banks. There are no broad-leafed seagrasses. Kelp forests are located in ocean-facing, high-energy environments to depths of 30 mwd. The hard substrates between phaeophyte holdfasts are colonized by numerous barnacles that, with epilithic bivalves and echinoderm debris, generate large amounts of barnacle-rich sediment that covers the adjacent seafloor in the form of rippled sands. The maerl facies is found in somewhat protected environments behind islands. Rigid branching corallines grow on hard surfaces. They can also be broken and fragmented and subsequently swept into shoals of cross-bedded maerl to form banks (reefs) of interlocking branches and algal crusts. The *mid- to outer shelf* is subphotic in the form of bedrock or winnowed glacigene boulder lags that are densely colonized by bivalves, bryozoans, azooxanthellate solitary corals, and serpulids, together with echinoids, sponges, and benthic foraminifers. Their remains are swept off to form rippled and burrowed sands that are habitats for infaunal and epifaunal bivalves. Gravels composed of serpulid worm tubes are widespread where currents are subdued along the shelf edge. These facies are also present on isolated offshore banks that can contain patches and mounds of the azooxanthellate coral *Lophelia* in deeper water.

Cold-water, polar carbonate systems

Shelves in these high-latitude environments have water temperatures that rarely rise above 5 °C, and are dark and ice-covered during winter months. Icebergs frequently scour their sediment-mantled shelves (Figures 11.10, 11.11a). The heterozoan carbonate deposits are characterized by the presence of ice-rafted debris (IRD) (Figure 11.11b, c) and a large biosiliceous component composed of diatom frustules and sponge spicules (Figure 11.11d).

Antarctica has vast floating ice shelves that cover much of the deep (average 500 mwd) continental margins.

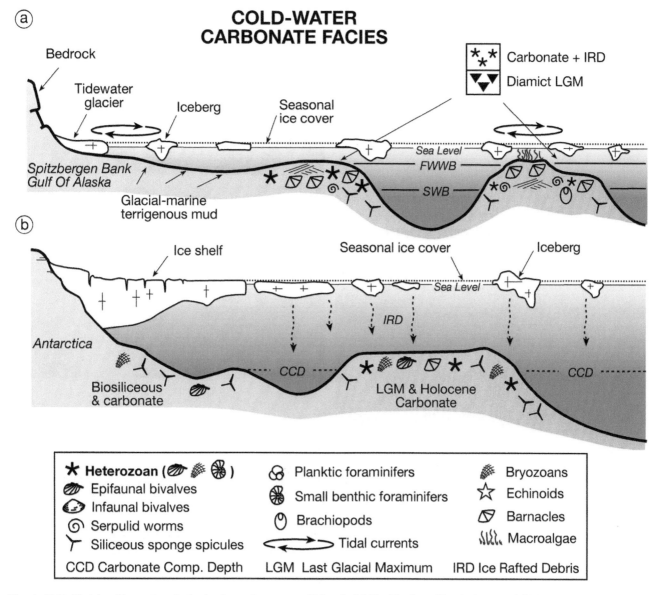

COLD-WATER CARBONATE FACIES

Figure 11.10 Sketches illustrating the hydrodynamic zones and biota in (a) The Northern Hemisphere; and (b) Antarctic cold-carbonate marine environments. Source: Adapted from James and Lukasik (2010). Reproduced with permission of the Geological Association of Canada.

Icebergs are numerous and IRD plus windblown sediment are the major sources of terrigenous sediment. These shelves are sites of extensive upwelling and high seasonal primary productivity. The shallowest parts of the shelf at ~60 mwd are current-swept with epibenthic suspension feeders growing on hard substrates. The carbonates in these areas are typically formed of bivalve, bryozoan, echinoid, gastropod, and benthic foraminifer remains.

There are several distinct features of this environment that arise because of the low sedimentation rates and the absence of durophagus fishes: (1) epibenthic suspensions feeders grow on soft and not hard substrates; (2) sponges

can cover more than 50% of the seafloor with their spicules forming dense mats 1–2 m thick; (3) there is virtually no infauna of deposit-feeding or burrowing animals; and (4) mollusks are conspicuously rare.

The rock record

Climate has fluctuated dramatically through geologic time (Figure 11.12) and the world ocean, which is inexorably linked to the atmosphere, has also changed profoundly over short and long time periods. The first-order

Figure 11.11 (a) Dirty icebergs laden with sediment in Otto Fiord, Ellesmere Island, Arctic Canada. Image width foreground 100 m. (b) Core from the Ross Sea, Antarctica of Pleistocene ice-rafted debris (I) and carbonate biofragments, mainly hydrocorals and bryozoans. (c) Numerous Pecten bivalve shells and other biofragments in Pleistocene marine glacigene sediments on Middleton Island, Alaska. Hammer head 15 cm. (d) Siliceous sponge spicules in a core from the Ross Sea, Antarctica. Image width 6 cm.

Figure 11.12 Chart illustrating global sea-level curves, general icehouse and greenhouse epochs, and cool- versus warm-marine periods through Phanerozoic time. Source: Adapted from Frakes et al. (1992). Reproduced with permission of Cambridge University Press.

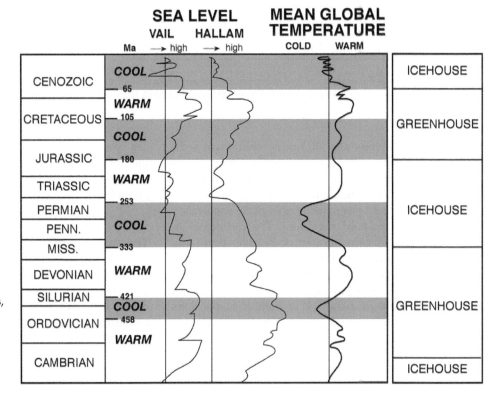

greenhouse and icehouse epochs are variably partitioned into cool and warm periods. Thus, cool-water carbonates can be expected at mid-latitudes at specific times throughout the Phanerozoic.

The preceding discussion has focused on modern depositional systems, which are applicable to the Cenozoic and much of the Mesozoic, but recognition of Paleozoic cool-water carbonates is not, however, straightforward. This is because the heterozoan benthic organisms were fundamentally different during the Paleozoic. Overall, it was an epibenthic-dominated community with relatively little predation. Many post-Paleozoic cool-water heterozoan components are missing because of evolution or were not significant as sediment producers. Ephemeral components such as seagrass and large phaeophytes were absent, and the role of other macroalgae is unknown. Although coralline algae, echinoids, bivalves, and serpulids were present, they were not significant contributors to the carbonate sediment. Conversely, the crinoids and brachiopods were fundamental components. Bryozoans remain abundant through time, although in the Paleozoic they were dominated by erect-rigid, non-articulated forms. Acorn barnacles, the common types in modern seas, had not yet evolved. The ecological requirements such as light, seawater temperature, and nutrient requirements for rugose corals, tabulate corals, and stromatoporoids are highly controversial (see Chapter 7). Whereas small benthic foraminifers were present throughout the Phanerozoic, large benthic photosymbiont-bearing foraminifers (fusilinids) only appeared in the late Paleozoic. Finally, ocean chemistry was different with higher silica contents because diatoms had not evolved and higher carbonate saturation because

calcareous plankton had not evolved. On balance, the core of the Paleozoic heterozoan assemblage was brachiopods, bryozoans, crinoids, and small benthic foraminifers.

Most Paleozoic cold-water carbonates identified to date are Mississippian–Permian in age (Figure 11.13), a global icehouse period related to world cooling and the late Paleozoic Ice Age (LPIA). The rarity of non-tropical carbonates in the early–middle Paleozoic is likely due to greenhouse conditions that dominated that period, much like the Cretaceous. Paleozoic cool- and cold-water carbonates documented to date can be related to lower seawater temperature, stratification, or large-scale climatic change. At present, there are two major documented systems: the thermocline system and the *climate change system*. The *thermocline system* is one where shallow-water photozoan facies graded downslope into deeper, cooler, heterozoan facies (Figure 11.14). Climate change is recorded at the end of the LPIA by either regional cooling or regional warming. In the Northern Hemisphere (Sverdrup Basin), regional cooling led to the transition through time from warm-temperate to cool-temperate and eventually cold-carbonate deposition, with spiculites common towards the end of the Permian.

By contrast, in the Southern Hemisphere (Australia) the cold- and cool-temperate paleoenvironments at the end of the LPIA give way to warm-temperate deposition. This seemingly contradictory situation is thought to be related to plate readjustment and changing global ocean current systems.

The Precambrian is the most difficult period in Earth's history in which to assess carbonate facies in terms of seawater temperature because there are no calcareous invertebrates to act as proxies. It is clear, however, especially

Figure 11.13 (a) An ice-rafted igneous boulder in Permian cold-water carbonate limestone containing numerous bryozoans, Maria Island, Tasmania. Scale increments 2 cm. (b) Outcrop surface of cold-water Permian limestone containing numerous branching and fenestrate bryozoans, Ellesmere Island, Arctic Canada. Image width 10 cm.

THERMOCLINE-STRATIFIED RAMP

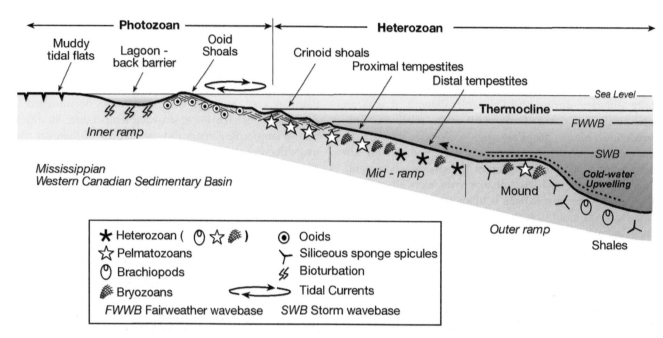

Figure 11.14 Sketch of marine paleoenvironments on an interpreted temperature-stratified ramp as represented by Mississippian carbonates in the Western Canadian Sedimentary Basin. Source: Adapted from Martindale and Boreen (1997). Reproduced with permission of the SEPM Society for Sedimentary Geology.

in the Neoproterozoic (see Chapter 18), that there were wide fluctuations in climate to the point that there may have been global glaciation ("Snowball Earth"). Glacigene sediments are typically overlain by carbonates (cap carbonates) that contain no clear evidence of cold- versus warm-water deposition. The presence of glendonites (pseudomorph of ikaite, a mineral that precipitates from seawater generally <7 °C) in oolitic carbonates between glacigene deposits, however, implies that very cold seawater temperatures were present during what would otherwise be interpreted as a time of warm-water sedimentation. It is likely that this unusual situation occurred because Neoproterozoic seawater was so highly supersaturated that ooids could form even in cool-water settings. There is much to learn about carbonate deposition in these occult environments.

Further reading

Beauchamp, B. (1994) Permian climatic cooling in the Canadian Arctic. *Geological Society of America Special Papers*, 288, 229–246.

Betzler, C., Brachert, T.C., and Nebelsick, J. (1997) The warm temperate carbonate province: a review of facies, zonations and delineations. *Courier Forschungsinstitut Senckenberg*, 201, 83–99.

Carannante, G., Esteban, M., Milliman, J.D., and Simone, L. (1988) Carbonate lithofacies as paleolatitude indicators: problems and limitations. *Sedimentary Geology*, 60, 333–346.

Frakes, L.A., Francis, J.E., and Syktus, J.I. (1992) *Climatic Modes of the Phanerozoic*. Cambridge: Cambridge University Press.

James, N.P. (1997) The cool-water carbonate depositional realm. In: James N.P. and Clarke M.J. (eds) *Cool-Water Carbonates*. SEPM Special Publication no. 56, pp. 1–20.

James, N.P. and Bone, Y. (2010) *Neritic Carbonate Sediments in a Temperate Realm, Southern Australia*. Dordrecht, Netherlands: Springer.

James, N.P. and Lukasik, J. (2010) Cool- and cold-water neritic carbonates. In: James N.P. and Dalrymple R.W. (eds) *Facies Models 4*. Geological Association of Canada, GEOtext 6, pp. 369–398.

Lees, A. and Buller, A.T. (1972) Modern temperate-water and warm-water shelf carbonate sediments contrasted. *Marine Geology*, 13, M67–M73.

Martindale, W. and Boreen, T.D. (1997) Temperature-stratified Mississippian carbonates as hydrocarbon reservoirs: examples from the Foothills of the Canadian Rockies. In: James N.P.

and Clarke M.J. (eds) *Cool-Water Carbonates*. SEPM (Society for Sedimentary Geology), Special Publication no. 56, pp. 391–409.

Nelson, C.S. (ed.) (1988) *Non-Tropical Shelf Carbonates: Modern and Ancient*. Amsterdam: Elsevier, Sedimentary Geology, no. 60.

Pedley, H.M. and Carannante, G. (eds) (2006) *Cool-Water Carbonates; Depositional Systems and Palaeoenvironmental Controls*. London: Geological Society, Special Publication no. 255.

Rogala, B., James, N.P., and Reid, C.M. (2007) Deposition of polar carbonates during interglacial highstands on an early Permian shelf, Tasmania. *Journal of Sedimentary Research*, 77, 587–606.

CHAPTER 12
MUDDY PERITIDAL CARBONATES

Frontispiece Desiccation cracks in peritidal dolomite, Lower Ordovician, Newfoundland, Canada, hammer scale 15 cm long.

Origin of Carbonate Sedimentary Rocks, First Edition. Noel P. James and Brian Jones.
© 2016 Noel P. James and Brian Jones. Published 2016 by John Wiley & Sons, Ltd.
Companion website: www.wiley.com/go/james/carbonaterocks

Introduction

Muddy tidal flat carbonates are a hallmark of carbonate deposition throughout geologic history. They are distinguished in the field and in core by the rapid alternation of facies (Figure 12.1), many of which contain features indicative of subaerial exposure. The muds from which they are built come largely from the adjacent shelf or ramp. Facies develop in response to both marine and terrestrial processes so that ancient peritidal deposits contain critical information about the paleoclimate as well as the shallow marine environment.

The flats either flank an adjacent hinterland or lie offshore on the leeward side of islands and shoals (Figure 12.2). There are three major depositional environments in the system (Figure 12.3). The *intertidal zone* is the region influenced by the daily movement of tides that can be diurnal (roughly one tidal cycle per day) or semi-diurnal (roughly two cycles per day). The

Figure 12.1 A series of many shallowing-upward peritidal cycles spanning a total stratigraphic thickness of ~200 m, each of which comprises an argillaceous limestone base and resistant limestone cap, Pennsylvanian, Ellesmere Island, Arctic Canada.

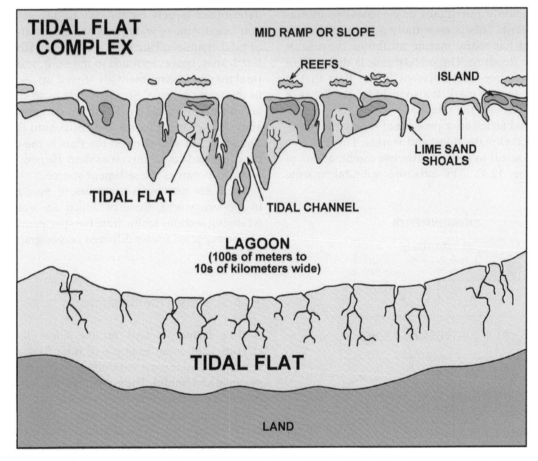

Figure 12.2 Plan view of the main elements of a muddy tidal flat. Source: Adapted from James (1979). Reproduced with permission of the Geological Association of Canada.

THE PERITIDAL ENVIRONMENT

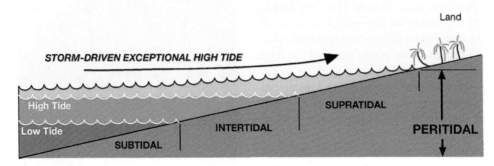

Figure 12.3 Sketch illustrating the main zones of a muddy carbonate tidal flat.

tidal range, which is determined by a combination of celestial mechanics (mainly the Sun and Moon) and geometry of the surrounding marine basin, is generally classified as microtidal (0–2 m range), mesotidal (2–4 m range), or macrotidal (>4 m range). The *supratidal zone* is that part of land that is inundated when strong storms (commonly hurricanes or cyclones) push seawater landward. This is essentially a terrestrial environment that has many marine attributes because of this episodic flooding. The *subtidal zone* is that part of the marine environment adjacent to the flats that is below the low water mark. It also includes the floors of channels that dissect the tidal flat.

Depositional facies are a product of the marine environment, local climate, tides, and storms. The first two controls can result in several different combinations of facies (Figure 12.4). The offshore subtidal marine

environment can range from normal marine to saline with an attendant suite of different sediments. The supratidal can range from humid with good seasonal rainfall and commonly an algal marsh, to semi-arid that is devoid of much vegetation, to arid and desert-like (a sabkha). The morphology of peritidal flats is determined largely by the day-to-day action of tides that usually move water on and off the flats via a series of tidal channels. During large cyclonic storms such as hurricanes, however, muds in the shallow offshore subtidal marine environment are stirred up and suspended in the water column. Strong winds push the sediment-laden water up tidal creeks (storm surge) where it rises above the channel banks and spills onto the flat itself. The source of sediment on the flats is the ocean; it has been likened to a "fluvial system turned inside out," with the ocean as the sediment source.

There are numerous examples of muddy tidal flats in modern world, three of which are detailed in the following sections to illustrate the spectrum of facies that can be expected under different oceanographic–climatic regimens.

Andros Island: The Bahamas

Andros Island, located on the Great Bahama Bank (Figure 12.5), is an example of tidal flats adjacent to a nearly normal marine environment where the climate is seasonal and humid. These well-studied and often visited flats lie along the western coast of Andros Island. Viewed from the air, the mosaic of channels and ponds that dissect the flats into a series of complex sub-environments is readily apparent (Figures 12.6, 12.7). The peritidal flats are characterized by a complex lateral and vertical facies architecture that reflects the ever-changing microscale variations in depositional environments. Subtidal facies develop in the tidal channels and in the

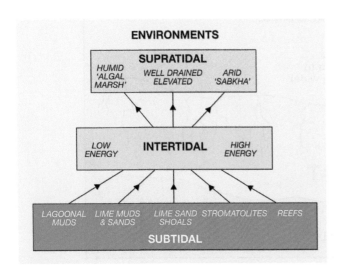

Figure 12.4 A diagram showing the different segments of a muddy tidal flat, emphasizing the spectrum of subtidal and supratidal facies. Source: Adapted from James (1979). Reproduced with permission of the Geological Association of Canada.

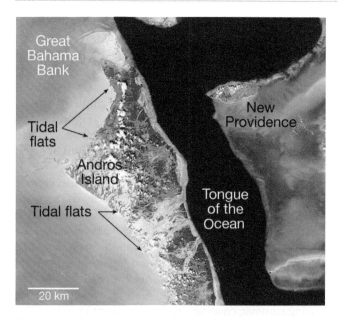

Figure 12.6 A low-level oblique aerial image of the three-creeks area located in northern part of Figure 12.5 with different sub-environments highlighted. Image width foreground 1 km.

Figure 12.5 Satellite image of the central Bahamas with the location of the tidal flats illustrated in Figure 12.6 arrowed. Copyright © 2014 Esri, i–cube. Reproduced with permission.

ponds. Intertidal facies exist along the margins of the channels and around the pond margins. Supratidal facies form along the channel levees and in the algal marsh. Facies of this type in the rock record are thin and of limited areal extent.

Offshore subtidal zone

Lying on the inner part of the Bahama Bank, the seafloor is shallow (typically <2 m deep), slightly more saline that the open ocean that experiences poor circulation during the summer months, and muddy with a restricted macrobiota of green algae, larger foraminifers, gastropods, and bivalves. Whitings are common across the bank in this region. Coarser particles are washed onshore during storms to form local beach ridges.

BAHAMIAN TIDAL FLAT DEPOSITIONAL ENVIRONMENTS
(Three Creeks Area)

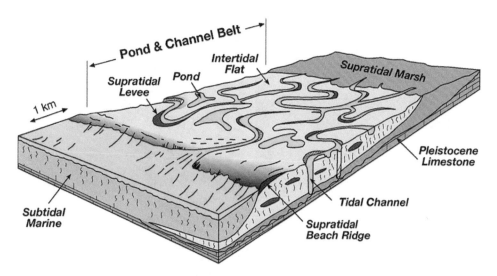

Figure 12.7 An isometric sketch of the three-creeks area. Source: Adapted from Shinn (1983). Reproduced with permission of the AAPG.

Pond and channel belt

This complex region comprises the tidal channels, their elevated banks, ephemeral ponds, widespread microbial mats, mangrove trees, and hard-lithified crusts. The day-to-day flow of the tides is up the creeks, into the ponds, and then back out again. During storms, the sediment-laden waters are forced up the creeks where they commonly spill over the banks, lose velocity and therefore deposit sediment to form levees that stand a few centimeters above the overall surface. In effect, they are a supratidal deposit that forms on levees along the channel margins (Figure 12.8a). These muddy sediments on the levees are typically covered with microbial mats and, if buried by a storm deposit, the microbes simply migrate upward through the sediment and form a new surface mat. This results in a distinctly laminated sediment where the organic material that remains behind decays, creating gas that opens up

sheet-to-ovoid holes in the sediment that are called fenestral pores (Figure 12.8b). The sediments in this oxidized zone are typically buff colored. Sediment layers, exposed for months or even years before being covered again, can be dry and hard or even semi-lithified such that subsequent storms may rip them up and thereby form flat-clast conglomerates (Figure 12.8c). Levees are locally called sidewalks because you can walk along them without sinking into the mud or stand and angle for bonefish in the channels.

Expansion and contraction of the ponds with every tide allows the ponds to fill and partially empty during each cycle. The intertidal zone around the margin of each pond is covered with microbial mats that thrive because they are wetted and dried each day. Desiccation cracks found in the upper part of the zone develop because the tides are variable (Figure 12.8d). Extreme heating of the pond water (the water can burn your feet) leads to elevated salinity during summer, creating a very stressed environment. It is

Figure 12.8 (a) A tidal creek supratidal levee with numerous mangrove pneumatophores (aerial roots), Bahamas. Creek is 30 m wide. Photograph by W. Martindale. Reproduced with permission. (b) Fenestral pores in Permian muddy supratidal dolostones, West Texas, USA. Image width 7 cm. (c) Laminated desiccated supratidal sediment broken into flat clasts by a hurricane, Bahamas. Hammer scale 50 cm long. (d) A recently desiccated hurricane mud layer in the high intertidal, Bahamas. Hand lens 3 cm in diameter.

devoid of mats but occupied by a thriving low-diversity high-abundance biota of herbivorous gastropods. These gastropods graze on the mats but cannot live in the shallower intertidal zone because it is too harsh. Locally, the innumerable shells, upon death, are swept into the channels where they accumulate in small elongate shelly sand bars.

Supratidal algal marsh

This environment is mostly a freshwater system that is only inundated by saline waters and marine-derived mud during storms. These episodic events result in centimeter-thick microbial mats (algal marsh) and alternating centimeter-thick storm-deposited mud layers. In the rock record, the mats would be evident as thin organic layers and the muds would form most of the record. The microbial mats are locally lithified and can contain microdolomite. They are typically broken by plant and tree growth and can easily be remobilized during storms.

Shark Bay: Western Australia

Shark Bay is a large embayment on the west coast of Australia that faces the Indian Ocean (Figure 12.9a). The sharks are mostly dolphins, but you have to keep an eye out for very poisonous sea snakes that lurk in this area. Hamelin Pool (Figure 12.9b), which is one arm of the embayment, is partially cut-off from the open ocean by an extensive seagrass bank (Faure Sill). Given that the climate in the area is semi-arid, salinities in Hamelin Pool commonly exceed 50‰ and salinities on the intertidal flats can be as high as 70‰. Small kilometer-scale

embayments around the margins of Hamelin Pool are characterized by suites of peritidal deposits that vary according to hydrodynamics and have been the subject of numerous studies. Peritidal environments range from: (1) rocky headlands fronted by a narrow (100–200 m wide), relatively high-energy, intertidal zone; to (2) wider (200–800 m) beaches and sand flats in arcuate bights that are fronted by a shallow intertidal platform where sand sheets swept by waves and tidal currents are actively prograding; to (3) broad bights, usually in the centre of an embayment, with a well-defined low-energy muddy tidal flat zonation.

Lagoonal subtidal environment

The elevated salinity here excludes most normal marine organisms and the seafloor is mostly covered with oolitic or skeletal sand (low diversity of benthic foraminifers and bivalves) and spectacular stromatolites (see Chapter 5). The stromatolites are world famous and have commonly been used as a general analog for Precambrian microbial structures.

Intertidal environment

The intertidal zone is covered with variably desiccated microbial mats, similar to those in the Bahamas. The surface is dissected with small tidal creeks. Stratiform flat-lying sheets are the dominant microbial structures in this protected environment. The mats comprise a series of recurring types, each composed of a distinct microbial community.

Locally, the surface is littered with numerous storm-generated intraclasts cemented by aragonite marine

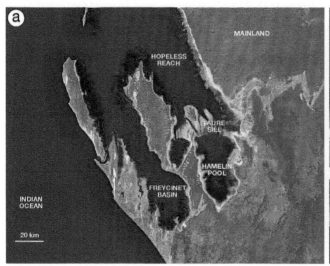
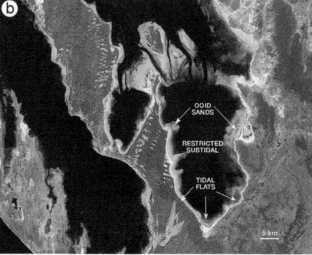

Figure 12.9 Satellite images of (a) Shark Bay, Western Australia showing the location of Hamelin Pool; and (b) Hamelin Pool with sedimentary facies highlighted. Copyright © 2014 Esri, i–cube. Reproduced with permission.

Figure 12.10 (a) A teepee developed in the supratidal zone of Hamelin Pool. Image width in foreground 3 m. (b) Leached gypsum laths in Middle Ordovician peritidal limestone, Kingston, Ontario Canada.

Figure 12.11 Satellite image of the muddy tidal flats and associated facies along the Trucial Coast, United Arab Emirates. Copyright © 2014 Esri, i–cube. Reproduced with permission.

cements precipitated from the warm and saline waters. There are also distinctive teepees, chevron-like structures of lithified carbonate crusts (decimeter-scale) forced upward because of excessive intergranular carbonate precipitation.

Supratidal environment

This wide flat is covered with desiccated microbial mats as well as numerous salt-tolerant terrestrial plants. Tidal channels floored with coarse debris are found in the lower supratidal with intervening areas being characterized by numerous cemented crusts and pavements. The supratidal muds are distinguished by aragonite precipitation, intraclasts, teepees (Figure 12.10a), and the growth of gypsum

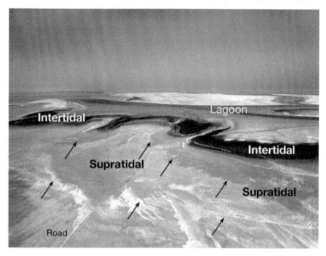

Figure 12.12 A low-level oblique aerial image of the United Arab Emirates muddy tidal flats with the intertidal zone covered with microbial mats in contrast to the barren supratidal zone with different storm tide lines (arrows), road scale. Photograph by P.A. Scholle. Reproduced with permission.

laths (Figure 12.10b) in the sediment from rising saline groundwater during the long, hot, evaporative summers.

The United Arab Emirates: Persian Gulf

The southern margin of the Persian (Arabian) Gulf is adjacent to the vast deserts of Saudi Arabia. The mesotidal (1.2–2.1 m range) coast of the United Arab Emirates is kept free of siliciclastic sand by winds (the Shamal) that blow from the north. The system is a classic ramp along the cratonward side of a foreland basin (Figures 12.11, 12.12). The muddy tidal flats face a narrow lagoon (the Khor al

Bazam) that is bordered on the oceanward side by a series of carbonate ooid sand shoals that pass gulfward into mid- and outer-ramp facies.

Offshore subtidal environment

Although the Gulf waters have slightly elevated salinity (35 to 40‰), they are inhabited by a relatively diverse biota. Sediments on the outer- and mid-ramp are largely bivalve and foraminiferal-rich muds with seagrasses in shallow waters. Whitings are common. The inner ramp is dominated by a series of ooid-skeletal sand shoals with

tidal passes between them, and coral reefs growing in front and around some of them. The lagoon, with salinities up to 50‰, is floored by gastropod–foraminifera peloidal sands and muds.

Intertidal environment

Muddy flats composed of peloidal sands and muds are dissected by tidal creeks in the lower parts and covered with microbial mats in the upper parts where the environment is too stressed for gastropods. Sediments (Figure 12.13a) are microbially laminated (Figure 12.13b)

Figure 12.13 (a) Intensively desiccated microbial mats with upturned edges filled with salt in the high intertidal zone (the salt will be washed away at the next very high tide). Scale is a 10 cm diameter core tube. Photograph by P.W. Choquette. Reproduced with permission. (b) A section through the modern intertidal sediments showing stacked microbial mats separated by innumerable thin and several thick storm events in the form of grey mud. Centimeter scale. Photograph by P.A. Scholle. Reproduced with permission. (c) A pit in the supratidal zone along the United Arab Emirates coastal tidal flats illustrating nodular anhydrite (A) that has grown porphyroblastically in the sediment. Photograph by P.W. Choquette. Reproduced with permission. (d) Nodular anhydrite in dolomitic supratidal mudstone with a slabbed core of similar nodular anhydrite from the Permian of West Texas placed against the trench surface. Core is 8 cm wide. Photograph by P.A. Scholle. Reproduced with permission.

as in the Bahamas, with local teepees and cemented crusts that are reworked into flat pebble conglomerates during storms. Storm-deposited sediments (Figure 12.13a) are microbially laminated and have conspicuous fenestral pores (Figure 12.13b). Aragonite and dolomite layers as well as displacive gypsum laths are present within shallow subsurface sediment in the upper intertidal zone.

Supratidal environment

This 10-km-wide flat surface called a *sabkha* ("salt-encrusted flat" in Arabic), which is flooded during storms, is subject to intense evaporation. The sediment is well laminated but disrupted by desiccation, evaporite precipitation, and expansion polygons. The most obvious precipitate is displacive porphyroblastic anhydrite (Figure 12.13c, d) in the form of seams of coalesced nodules (enterolithic structures). These and less-abundant celestite and magnesite also precipitate from seawater that is swept onto the flats during storms.

Stratigraphy

One of the most important discoveries about muddy tidal flat dynamics was revealed in the 1960s by geologists from Imperial College and petroleum companies when trenches were cut into a supratidal flat in the Persian Gulf (Figure 12.14). The stratigraphic succession, a couple of meters thick, comprised (from the bottom up): (1) lagoonal muds such as those in

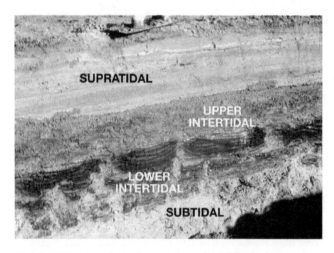

Figure 12.14 A trench cut through the supratidal zone in the United Arab Emirates illustrating the prograding succession that grades up from subtidal to supratidal facies. Trowel for scale, 15 cm long. Photograph by P.A. Scholle. Reproduced with permission.

Khor al Bazam; (2) a suite of spectacular microbial mats with desiccation cracks; and (3) laminated dolomitic muds and sands with layers of anhydrite. The section clearly indicated that this was a shallowing-upward succession that recorded the gradual transition from subtidal to intertidal, and finally supratidal environments. Careful coring and dating showed that this succession formed as a result of about 10 km of seaward progradation over the last 5000 years.

The shallowing-upward peritidal cycle

This meter-scale succession, such as that in the Persian Gulf, is common throughout the geological record. In most cases it comprises two parts: a lower relatively thin unit of coarse-grained reworked particles and a much thicker upper part that comprises the shallowing-upward portion. These segments have been interpreted as parasequences, the lower transgressive and upper regressive (progradational) parts of the cycle respectively (Figure 12.15).

A series of three different cycles can be generated from modern environments. The Bahamas is normal marine with a humid climate, Shark Bay is hypersaline with a semi-arid climate, and the Persian Gulf is nearly normal marine with an arid climate (Figure 12.16).

The discovery of the shallowing-upward succession has several important consequences for the geological record. Recall that the source of muddy peritidal sediment comes largely from the adjacent ocean. The whole platform cannot gradually shallow up to sea level because the source of sediment that feeds the tidal flats would be cut off long before that happened. It therefore follows that all shallowing-upward peritidal successions must be progradational (Figure 12.17). This is accomplished either by outward progradation from the hinterland or progradation around irregularities and islands on the platform.

In the case of simple progradation (Figure 12.17), muddy sediment is transported landward onto the tidal flat during large cyclonic storms. Such events sweep sediment-laden waters well onto the supratidal zone and so the whole flat receives a layer of carbonate mud. These sediments are then colonized by microbial mats, desiccated, and locally lithified during the long periods between storms such that the distinctive features of the supratidal, intertidal, and shallow subtidal zones, ponds, channels, marshes, or sabkhas are impressed on the deposits. The next storm then dumps a new layer of sediment across the flat and the process takes place all over again. At the same time, however, sediments are deposited along the marginal zone of the flat. Over the course of

METER-SCALE MUDDY PERITIDAL CYCLE

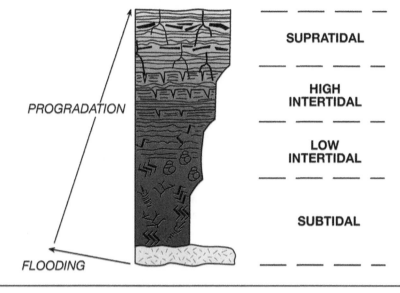

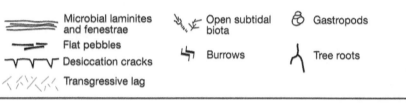

Figure 12.15 Sketch of a typical shallowing-upward meter-scale cycle comprising a thin basal flooding interval and a comparatively thick prograding shallowing-upward interval.

MODERN MUDDY PERITIDAL CYCLES

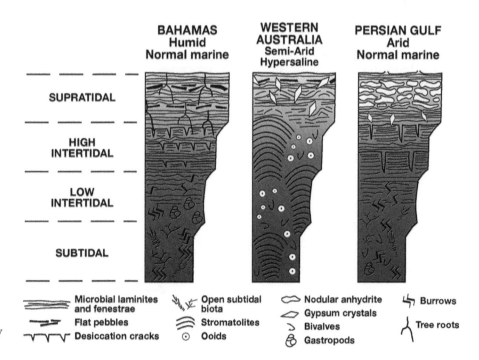

Figure 12.16 Sketch of three shallowing-upward meter-scale cycles where climate and hydrography produce different sedimentary motifs.

MUDDY TIDAL FLAT PROGRADATION

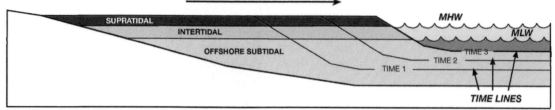

Figure 12.17 The genesis of a single cycle via progradation. Adapted from Hardie and Shinn (1986). Reproduced with permission of the Colorado School of Mines.

time the whole system gradually moves seaward, or progrades. The critical point here is that whereas the subtidal, intertidal, and supratidal environments will produce layered sediment, these layers are *diachronous* because the system is *progradational.*

In contrast to this shoreline system, islands in wide shallow lagoons or epeiric seas can be surrounded by muddy tidal flats (Figure 12.18). In such a system, islands shift laterally over time as do their associated peritidal deposits, creating an areally complex and largely unpredictable suite of interdigitating facies.

How do numerous peritidal cycles form?

Shallow-water successions in the carbonate rock record are usually characterized by numerous cycles that are stacked on top of one another (Figure 12.1). Such successions form either via *autogenic* processes where the carbonate factory simply overproduces sediment and sea level does not move, or *allogenic* processes whereby sea level fluctuates and the sediment factory responds.

Autogenic cycles

A shallowing-upward cycle can be generated during one period of sea-level stasis by continuous production in the carbonate factory (Figure 12.19). The offshore subtidal factory supplies sediment to the tidal flat that then progrades out onto the platform (ramp) as, for example, in the Persian Gulf. Progradation must eventually come to a halt because progradation of the tidal flat is accommodated by a reduction in the width of the carbonate factory. As the prograding flat nears the platform margin, the sediment source area becomes too small to supply the required sediment. Progradation therefore ceases and the whole system comes to a halt. If subsidence continues, however, the ocean will eventually flood the platform again, the carbonate factory will begin to produce sediment once more, and the whole depositional cycle begins anew.

Generation of the second cycle is not, however, immediate because the platform is shallow, swept by waves and swells, and not yet fully colonized by the carbonate-producing biota. There is therefore a *lag time* during which little sediment is produced on the platform and not much mud can be washed onto the tidal flats.

TIDAL FLAT ISLANDS

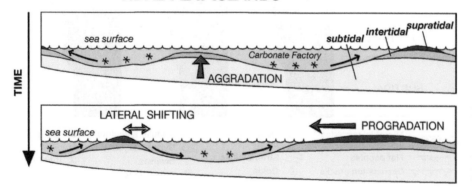

Figure 12.18 The genesis of meter-scale tidal flat cycles via the shifting tidal flat island concept. Source: Adapted from Pratt et al. (1992). Reproduced with permission of the Geological Association of Canada.

AUTOGENIC CYCLICITY

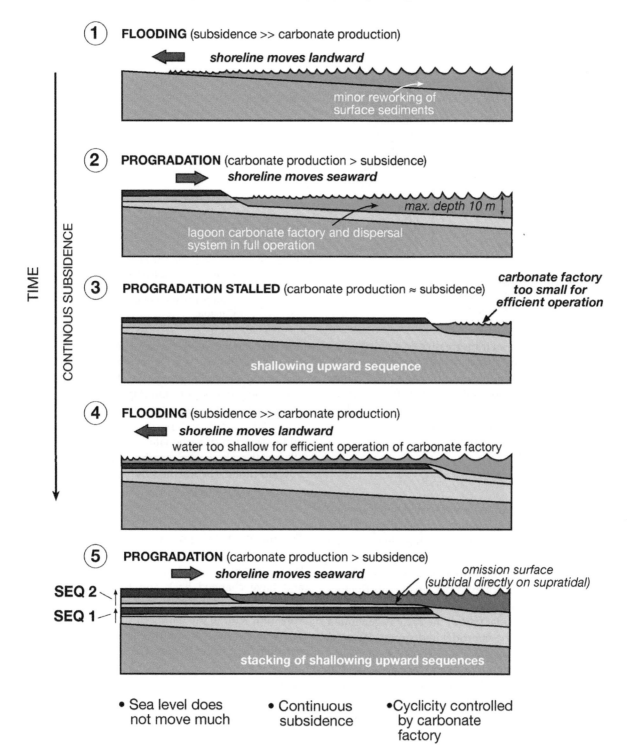

Figure 12.19 The generation of autogenic tidal flat cycles without changing sea level (after an unpublished diagram of R.N. Ginsburg, personal communication). Source: Adapted from Hardie and Shinn (1986). Reproduced with permission of the Colorado School of Mines.

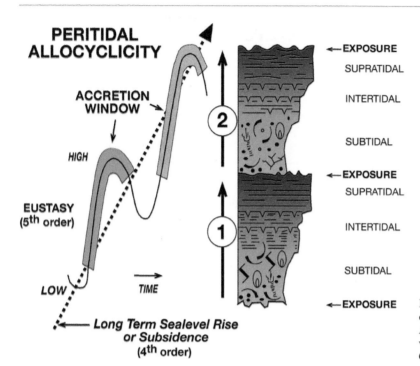

Figure 12.20 The generation of allogenic tidal flat cycles resulting from changing sea level. Source: Adapted from Pratt et al. (1992). Reproduced with permission of the Geological Association of Canada.

This is when the thin basal transgressive part of the cycle, composed of intraclasts and worn biofragments, is produced. Mud production does not begin, or become available for transportation onto the tidal flat, until sea level rises to the point where water is several meters deep and the rim is operational. Progradation then resumes and the next parasequence forms.

Allogenic cycles

Sea level has fluctuated on different spatial and temporal scales throughout geological history (see Chapter 20). Such eustatic changes at the meter-scale have the potential to generate shallowing-upward peritidal cycles that may operate in concert with the autogenic processes. This is especially evident when, for example, small-scale fluctuations (5th order) are superimposed on larger fluctuations (Figure 12.20). The subtidal part of the cycle is generated during sea-level rise. The intertidal and supratidal parts are developed as sea-level rise slows. The subsequent sea-level fall results in subaerial exposure and possible diagenesis or dolomitization. The next cycle is developed as sea-level rise floods this recently developed succession.

A caveat

The preceding arguments are elegant intellectual explanations for the numerous meter-scale muddy peritidal cycles in the geologic record. Coring of the Bahamian tidal flat has however revealed that, below the modern surface, there is not a classic shallowing-upward succession. Instead, there is a basal lag: a hiatus representing a long period of non-deposition and then rapid accumulation of pond sediments overlain by a variety of intertidal–supratidal deposits. Furthermore, analysis of a wide spectrum of carbonate successions reveals that not all meter-scale cycles exhibit the classic shallowing-upward motif. The detailed dynamics of this system still seem to be eluding us!

Temporal variations on the peritidal cycle theme

Whereas climate and oceanography are eternally variable and have changed back and forth through geological time, organism biology has changed in a unidirectional way and, with it, the nature of the marine and terrestrial realm. This has, of course, influenced the nature of peritidal deposition (Figure 12.21).

Proterozoic

Microbes were the only organisms in the world ocean for much of Earth's history; together with ooids, lime mud, and intraclasts, they dominated the Precambrian peritidal system. With this sediment factory of a somewhat limited biodiversity, it is more difficult to discern stressed subtidal facies than in younger rocks.

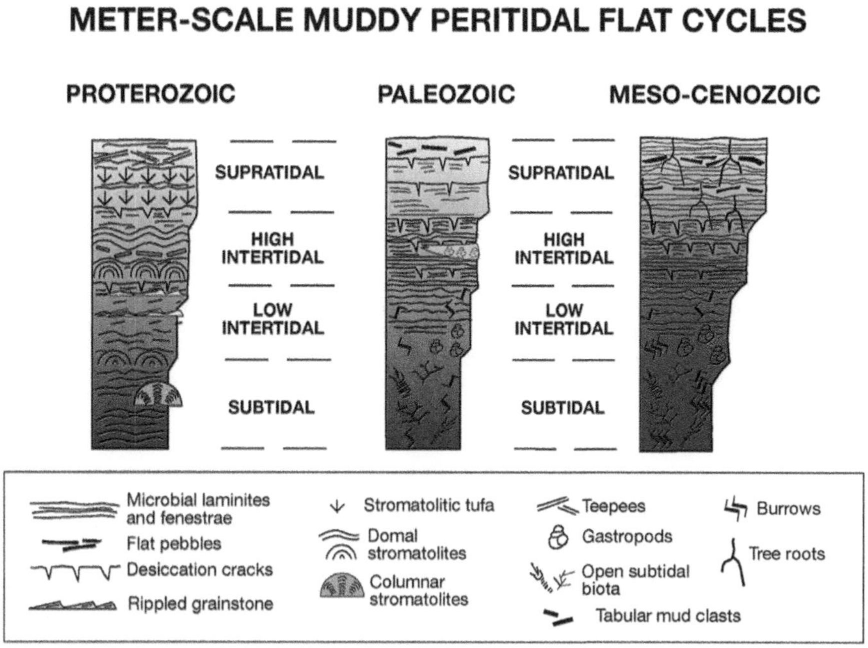

METER-SCALE MUDDY PERITIDAL FLAT CYCLES

PROTEROZOIC   **PALEOZOIC**   **MESO-CENOZOIC**

SUPRATIDAL
HIGH INTERTIDAL
LOW INTERTIDAL
SUBTIDAL

Microbial laminites and fenestrae — Stromatolitic tufa — Teepees — Burrows
Flat pebbles — Domal stromatolites — Gastropods — Tree roots
Desiccation cracks — Columnar stromatolites — Open subtidal biota
Rippled grainstone — Tabular mud clasts

Figure 12.21 A sketch of typical meter-scale peritidal cycles at different periods in geologic time.

Paleozoic

The Paleozoic ocean was the site of a wide diversity of invertebrates that generated a spectrum of skeletons and peloids that had a profound effect on peritidal sedimentation. Most important, perhaps, was the evolution of herbivorous gastropods in the late Cambrian whose grazing seems to have greatly restricted the distribution of microbial mats, largely confining them to upper intertidal and supratidal environments. The terrestrial realm was very Precambrian-like for much of the early Paleozoic, a generally barren landscape except for microbes and lichens. Appearance of land plants in the middle Paleozoic completely changed the supratidal part of the depositional spectrum, especially in humid climatic belts.

Mesozoic–Cenozoic

Evolution of angiosperm plants in the middle Mesozoic affected both the marine and terrestrial realms. In particular, marine grasses bound more muds in the subtidal zone, inhibiting transfer of fine sediment onto the tidal flats. Mangroves likewise anchored sediment, thus reducing the effect of cyclonic storms. The rapid diversification of crustaceans led to increased burrowing in intertidal and supratidal environments than before, resulting in obliteration of some sedimentary features.

Further reading

Berkeley, A. and Rankey, E.C. (2012) Progradational Holocene carbonate tidal flats of Crooked Island, south-east Bahamas; an alternative to the humid channeled belt model. *Sedimentology*, 59, 1902–1925.

Drummond, C.N. and Dugan, P.J. (1999) Self-organizing models of shallow-water carbonate accumulation. *Journal of Sedimentary Research*, 69, 939–946.

Ginsburg, R.N. (ed.) (1975) *Tidal Deposits, a Casebook of Recent Examples and Fossil Counterparts*. New York: Springer-Verlag.

Goldhammer, R.K., Lehmann, P.J., and Dunn, P.A. (1993) The origin of high-frequency platform carbonate cycles and third-order sequences (Lower Ordovician El Paso Group, west Texas): constraints from outcrop data and stratigraphic modeling. *Journal of Sedimentary Research*, 63, 318–359.

Hardie, L.A. (1977) *Sedimentation on the Modern Carbonate Tidal Flats of Northwest Andros Island, Bahamas*. Baltimore, MD: The Johns Hopkins University Studies in Geology, Johns Hopkins University Press.

Hardie, L.A. and Shinn, E.A. (1986) Carbonate depositional environments, modern and ancient. Part 3: Tidal flats. *Colorado School of Mines Quarterly*, 81, 1–74.

James, N.P. (1979) Facies models 10. Shallowing-upward sequences in carbonates. In: Walker R.G. (ed.) *Facies Models*. Toronto: Geological Association of Canada, Geoscience Canada Reprint Series 1, pp. 109–119.

Kendall, C.G.S.C. and Skipwith, S.P.A.D.E. (1969) Holocene shallow-water carbonate and evaporite sediments of Khor al Bazam, Abu Dhabi, southwest Persian Gulf. *AAPG Bulletin*, 53, 841–869.

Logan, B.W., Davies, G., Read, J.F., and Cebulski, D.E. (eds) (1970) *Carbonate Sedimentation and Environments, Shark Bay, Western Australia*. Tulsa, OK: American Association of Petroleum Geologists, Memoir no. 13.

Maloof, A.C. and Grotzinger, J.P. (2012) The Holocene shallowing-upward parasequence of north-west Andros Island, Bahamas. *Sedimentology*, 59, 1375–1407.

Playford, P.E., Cockbain, A.E., Berry, P.F., Roberts, A.P., Haines, P.W., and Brooke, B.P. (2013) *The Geology of Shark Bay*. Perth: Geological Survey of Western Australia, Bulletin no. 146.

Pratt, B.R. and James, N.P. (1986) The St. George Group (Lower Ordovician) of western Newfoundland: tidal flat island model for carbonate sedimentation in shallow epeiric seas. *Sedimentology*, 33, 313–343.

Pratt, B.R., James, N.P., and Cowan, C.A. (1992) Peritidal carbonates. In: Walker R.G. and James N.P. (eds.) *Facies Models: Response to Sea Level Change*. St John's, Newfoundland: Geological Association of Canada, GEOtext 6, pp. 303–322.

Rankey, E. and Berkeley, A. (2012) Holocene Carbonate Tidal Flats. In: Davis Jr R.A. and Dalrymple R.W. (eds) *Principles of Tidal Sedimentology*. Netherlands: Springer, pp. 507–535.

Shinn, E.A. (1983) Tidal flat environment. In: Scholle P.A., Bebout D.G., and Moore C.H. (eds) *Carbonate Depositional Environments*. Tulsa, OK: American Association of Petroleum Geologists, Memoir no. 33, pp. 171–210.

Wanless, H.R., Tyrrell, K.M., Tedesco, L.P., and Dravis, J.J. (1988) Tidal-flat sedimentation from Hurricane Kate, Caicos Platform, British West Indies. *Journal of Sedimentary Petrology*, 58, 724–738.

Wright, V.P. (1984) Peritidal carbonate facies models: a review. *Geological Journal*, 19, 309–325.

CHAPTER 13
NERITIC CARBONATE TIDAL SAND BODIES

Frontispiece Cross-bedded Oligocene cool-water biofragmental limestone in a paleoseaway complex, North Island, New Zealand. Note the person (circled) for scale.

Origin of Carbonate Sedimentary Rocks, First Edition. Noel P. James and Brian Jones.
© 2016 Noel P. James and Brian Jones. Published 2016 by John Wiley & Sons, Ltd.
Companion website: www.wiley.com/go/james/carbonaterocks

Introduction

Carbonate sand bodies (Frontispiece) are physically deposited high-energy grainstones (and associated packstones) composed of carbonate sand with lesser gravel- and mud-size grains. They occur in high-energy settings, where oceanographic processes such as tides, waves, swells, and storms prevent the deposition of mud. They are not bioconstructions (such as reefs; see Chapters 14, 15).

Carbonate sand bodies typically show a strong process–product relationship because of their intimate relation with energetic hydrodynamics (Figure 13.1). That is to say, the internal structure and external morphology of sand bodies are related to the process that controlled their deposition. In this sense, carbonate sand bodies share many similarities with siliciclastic deposits. Carbonate sands, however, still require a carbonate factory to source the sediment. Such sand bodies occur in both warm and cool marine depositional environments with particles generated by photozoan and heterozoan factories. Ooids and reef-associated biofragments characterize the photozoan factory, whereas most components in the heterozoan factory are skeletal remains.

This chapter describes carbonate sands that form in tidal-dominated settings along platform margins, in straits, in seaways, and in flooded incised valleys. Whereas tidal currents are generally dominant, storm-generated currents also play a significant role in deposition. Carbonate storm deposits in general are highlighted in Chapter 3.

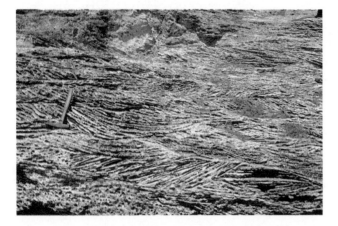

Figure 13.1 Outcrop of Pleistocene oolite with distinctive bidirectional cross-stratification, New Providence Island, Bahamas. Hammer for scale, 13 cm long.

Tides and tidal currents

An understanding of open water tidal depositional systems first requires an appreciation of tidal dynamics. Astronomic and oceanic forces drive tides in the marine realm. A tide is any periodic fluctuation in water level that is generated by the gravitational attraction of the Moon and Sun. Lunar tides represent the vector sum of the gravitational attraction of the Moon and the centrifugal force caused by the revolution of the Earth–Moon system about its common centre of mass. The water in the ocean therefore piles up in two bulges, one underneath the Moon (gravitational) and the other on the opposite side of the Earth (centrifugal). As a result of the Earth's rotation the bulges appear to travel around the Earth as two tidal waves, causing water levels to rise and fall regularly. The continents get in the way of these waves, so the movement is restricted to the ocean basins where the wave rotates about various amphidromic points. The height of the wave is lowest at the center and highest around the margins. The amplitude can be further enhanced or reduced by the geometry of the local ocean basin.

Rising water levels are called *flood* tides whereas falling water levels are called *ebb* tides. The tidal period is either semi-diurnal (12.4 hours) or diurnal (once a day). Semi-diurnal tides predominate in most places today. The tide is also a function of solar attraction. When the Sun and the Moon are aligned, the tides are higher than average (spring tides); when they are at right angles, they are smaller (neap tides). The tidal range varies dramatically on the Earth's surface from nearly zero to a maximum of ~16 m. As emphasized in the previous chapter, the ranges are classified as microtidal (0–2 m range), mesotidal (2–4 m range), and macrotidal (>4 m range).

The maximum tidal current speed at any location is determined by the volume of water that must pass that point in each tidal half-cycle, a quantity that is called a *tidal prism*; the tidal prism is a function of both the tidal amplitude and the size of the area being drained and flooded. What is important is the cross-sectional area of the passage through which the volume of water must pass. Strong tidal currents can occur near the continental shelf edge if a large volume of water has to move on and off the shelf during each tidal cycle, especially if the water is shallow. Flood currents are typically faster than ebb currents, so there is commonly a flood dominance of tidal currents in coastal seas. Finally, tidal height can be enhanced during storms when ocean water is blown onshore by strong winds and heightened by low atmospheric pressure (a storm surge; see Chapter 3).

Tidal sand bodies

Shallow-water carbonate sand bodies formed by tidal currents (Figure 13.2) generally occur in settings where tidal water flow is focused either by significant shallowing of the seafloor or by areal constriction. They are particularly common at or near platform margins or the outer fringes of inner ramps. Sand bodies in these locations can be a several tens to a few hundreds of kilometers in length. Those on the platform itself are generated in seaways and straits between islands or between islands and the mainland, so their size is constrained somewhat by the size of the channel. Tidal deposition along the shoreline is focused between barrier islands. Spectacular carbonate sands can also form when the ocean inundates incised valleys or tectonic embayments along the shoreline. Tidal sedimentation occurs in all of these settings because the cross-sectional area through which the tidal prism must flow is comparatively narrow; as a result, the tidal currents are amplified. These enhanced tidal currents are capable of transporting and depositing significant volumes of carbonate sand.

Sand body structure

Each sand body consists of a series of superimposed sediment accumulations of progressively larger size, namely: (1) small-scale ripples and other sedimentary structures; (2) subaqueous dunes; and (3) bars or ridges that are separated by channels. Bars (or barforms) are the most conspicuous features of modern marine tidal carbonate sand bodies. They are formed by large-scale flow patterns that are distinct from processes that form subaqueous dunes and ripples. Bars are the product of strong tidal water flow that comprises the whole water column and are generally developed parallel or subparallel to the directions of flow. They can grow to heights that are equal to the water depth, although this is not always the case. By contrast, subaqueous dunes and current ripples are formed by smaller-scale perturbations within the flow. These two latter bedforms are generally oriented normal or subnormal to the direction of tidal flow. Ripples are less than 5 cm in height, whereas dunes can grow to approximately 20% of the water depth. The result is that bars are almost always larger and are adorned by dunes.

Small-scale tidal sedimentary structures

The periodic changes in current speed and direction associated with tidal action produce diagnostic sedimentary structures that allow for characterization in modern deposits and their recognition in ancient rocks. Confirmation of tidal sedimentation is best achieved by the co-occurrence of features such as an abundance of

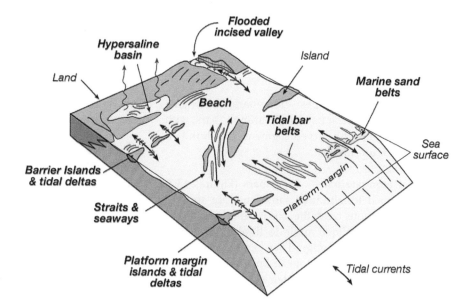

Figure 13.2 Isometric sketch indicating the different locations of carbonate sand body formation.

cross-bedding, flaser, wavy, or lenticular bedding, her-ringbone cross-stratification (Figure 13.1), reactivation surfaces, periodically spaced mud drapes, and landward-directed cross-beds. Herringbone cross-stratification is defined as adjacent cross-beds having opposed (bidirec-tional) dip directions. Not all of these features might be present in an individual tidal sand body because of the energetic hydrodynamic regimen and associated erosion during the subsequent tidal cycle. In addition, one of the two tidal currents can be stronger than the opposite current, thereby generating net unidirectional sediment transport. This is the reason why herringbone cross-stratification is not common. In two-dimensional out-crops, care must be taken not to misinterpret end-on views of troughs as herringbone cross-laminations.

Subaqueous dunes

The three main types of subaqueous dunes, with coun-terparts in siliciclastic tidal deposits, are: (1) two-dimensional (2D) simple dunes that have straight to sinuous crests and planar cross-stratification; (2) 3D simple dunes that have crests that bifurcate every few meters and internal trough cross-bedding; and (3) larger compound dunes that possess superimposed smaller bedforms. The largest dunes are called 2D compound dunes.

Compound dunes (previously called sand waves) are large to very large bedforms, upon which there are smaller simple dunes, and are common features of most tidal environments. They range from 1 to 15 m in height, can have wavelengths of tens to several hundred meters, and be continuous for more than 1 km. The dune is asymmet-rical with the steep (lee) face inclined in the direction of sediment transport and bedform migration.

Sand bars or ridges

There are three main bar types – flow-parallel elongate bars, flow transverse linear bars, and parabolic bars, with the latter restricted to the curved segments that link two elongate bars (Figure 13.3).

Flow-parallel elongate bars. These bars (Figures 13.3, 13.4a), also termed longitudinal sand bars, are straight with a variably sinuous crest; the axis is oriented subpar-allel to or at a low oblique angle to tidal flow, mostly almost normal to the shelf edge. They can occur in later-ally extensive groups, are up to 25 km long and with 3.5 km spacing (tending to broaden on-platform), and range from 3 to 6 m high. The shallowest parts can be

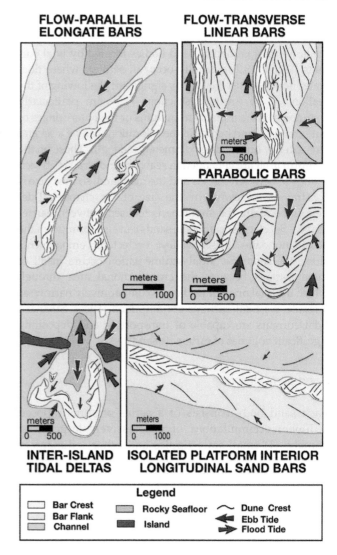

Figure 13.3 Sketch illustrating the different types of sand bodies in the Bahamas. Source: Adapted from Rankey and Reeder (2011). Reproduced with permission of the SEPM Society for Sedimentary Geology.

exposed at low tide. The crest in warm-water systems can have marine cementation, local corals, and even islands.

Flow transverse linear bars. These sand bodies (Figures 13.3, 13.4b, c), also called transverse shoulder bars, are asymmetric, straight with slightly sinuous crests, and oriented transverse to the currents. Each bar can be sev-eral meters high with a steep side and gentle side. They can exceed 30 km in length and 5 km breadth, and have up to 5 m bathymetric relief.

Parabolic bars. These arcuate structures (Figures 13.3, 13.4b, c) link two adjacent elongate linear bars and have

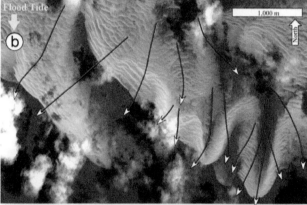

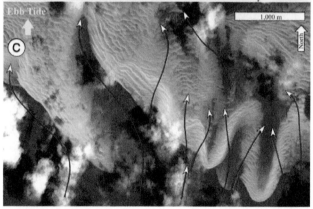

Figure 13.4 (a) Oblique aerial view of flow-parallel elongate bars, Schooner Cays, Bahamas. Photograph by P.A. Scholle. Reproduced with permission. (b and c) satellite images of flow tranverse bars and parabolic bars, Bahamas, with red arrows indicating the location of the strongest (b) and ebb (c) tidal flow. Images courtesy of G. Rankey. Reproduced with permission.

similar topographic relief. In most cases, their crests remain slightly submerged at low tide, but some aggrade to form small islands.

Bahamian platform ooid sand bodies

Marine ooid sands are unique to the warm-water photozoan depositional system (see Chapter 10). Although common in the geological record, they are not abundant in the modern ocean. Most Holocene deposits occur in the Caribbean Sea, especially on the Bahama Banks (Figure 13.5) and nearby carbonate banks. They also develop along the southern shores of the Persian Gulf, in the isolated gulfs of Shark Bay (Australia) and, as recently discovered, in some central Pacific atolls. They are closely associated with tidal currents and tidal bars are widespread in ooid-forming settings. Of these, the Bahamian ooid sands have been extensively studied as models for similar sand bodies in the geological record, and are therefore described in detail in this chapter.

The sediments are formed mainly of ooids and peloids: (1) ooid grainstone forms on high-energy current-swept rippled crests of sand bars; (2) ooid packstone accumulates on lower-energy stabilized burrowed bottoms; (3) peloid-packstone also collects on lower-energy stabilized and burrowed sand bottoms; (4) pellet wackestone forms on very low-energy restricted bottoms (lee of islands); and (5) lithoclast rudstone accumulates in high-energy active tidal channels.

Tidal ooid sand bodies

Most of these sand bodies have formed at the margin of platforms where flood tides are dominant. The tidal currents commonly approach $100\,cm\,s^{-1}$ ($\sim4\,km\,hr^{-1}$) but decrease as they sweep onto the platform due to flow expansion and frictional dissipation of energy. The best-formed ooids are found in platform margin areas where currents exceed $30\,cm\,s^{-1}$ ($\sim1\,km\,hr^{-1}$). Areas with the strongest currents are therefore characterized by the largest ooids and are associated with the largest and broadest ooid facies belts. Sand bodies form where reefs at the platform margin are discontinuous or non-existent. Here reef growth is inhibited because of the 'bad' nutrient-rich, saline, relatively hot waters flowing off the bank during ebb tides.

Tidal versus storm-dominated ooid sand bodies

Bahamian sand bodies are clearly tide-dominated but the location of the Bahamas, directly in the path of powerful westward-moving Atlantic tropical cyclones (hurricanes), has led to the proposition that hurricanes are involved in their genesis. According to this notion, hurricanes move large volumes of sediment and are responsible for much

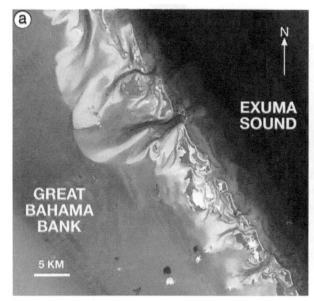

Figure 13.5 Tidal deltas in the Exuma Island Chain, Bahamas. (a) Satellite image of the Exuma Island chain illustrating the numerous inter-island ooid sand tidal deltas. Copyright © 2014 Esri, i–cube. Reproduced with permission. (b) Oblique aerial image showing the dominance of flood tidal deposition. Image width 10 km. Photograph by E. Hiatt. Reproduced with permission.

on-bank or off-bank transport of ooid sand: for example, the parabolic dunes are really storm spillover lobes/ deltas. This is an ongoing controversy, and revolves around the issue of whether or not infrequent intense storms move more sediment than the everyday tidal currents.

Types of Bahamian platform sand bodies

The extent and configuration of Bahamian tidal-domi-nated ooid sand bodies (Figure 13.6) is a function of platform geometry, location on the platform, degree of exposure, local bedrock topography, and orientation relative to the tidal currents, storms, and prevailing winds. With so many different controls involved, it is not surprising that the sand bodies are highly variable in shape and extent.

When viewed from the air, there are three recurring types: (1) those composed of separate, isolated, elongate sand ridges called *tidal bar* sand bodies; (2) those com-posed of a wide complex of sinuous sand shoals com-monly decorated with subaqueous dunes called *marine bar* sand bodies; and (3) those developed as inter-island *tidal delta* sand bodies (Figures 13.2, 13.5). Tidal bar sand bodies are distinguished by long linear sand ridges oriented roughly perpendicular to the shelf edge. Marine bar sand bodies are more massive and composed of parabolic bars or spillover lobes with numerous

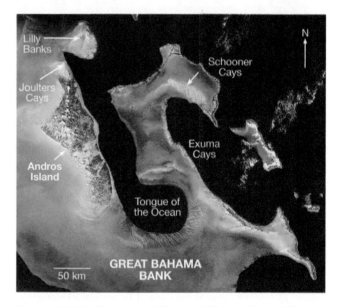

Figure 13.6 Satellite image of the Great Bahama Bank with sand body localities located by arrows. Copyright © 2014 Esri, i–cube. Reproduced with permission.

superimposed elongate bars. Some of these latter sand bodies have accreted to sea level such that the sand bodies are largely moribund. This three-fold classification has proven useful over time, and is used here with the caveat that all of these types grade into one another. We know most about the marine bar sand bodies.

Tidal deltas (Figures 13.3–13.5) are complex lobate features that are composed of both flood and ebb tidal deltas that occur at the ends of tidal inlets between islands. They are also called spillover lobes or washover lobes where there are no emergent islands. They have 2–4 m bathymetric relief and are generally submerged by 0.5 m at low tide, although some have aggraded to form islands. The deltas develop because a tidal current decelerates due to flow expansion, resulting in sediment deposition. Measured current velocities show no clear relation to bar form.

Some examples of Bahamian sand bodies

The following are a few selected examples that illustrate the spectrum of sand body complexes in the Bahamas. Some sand bodies comprise a whole suite of bar types, whereas only one or two types dominate other areas. Remember, however, that these sand bodies represent only ~6000 years of deposition, since Holocene sea-level rise first flooded the platform top.

Tidal sand bar belts

Schooner Cays complex. The Schooner Cays shoal complex (Figures 13.6, 13.7a, b), is a 16 × 60 km sand body constructed by the complete range of flow-parallel elongate sand ridges and parabolic bars. Individual bars are up to 13 km long and 1.5 km wide, with many being asymmetric in cross-section. Intervening 4–8-m-deep channels have rocky bottoms or are floored with skeletal sediment. Deeper parts are burrowed and covered with seagrass and muddy sediments. In all areas, the sediments become finer inboard from the platform margin. The bars have built to sea level in some places and resultant islands are now vegetated (Figure 13.7c). The rock record of the Schooners Cays complex, in its present state, would be a series of semi-isolated decimeter-thick linear lenses oriented subnormal to the shelf margin with thin skeletal sands between.

Tongue of the Ocean complex. This is the largest oolitic tidal bar belt system in the world and lies at the cul-de-sac end of the deep Tongue of the Ocean, through which the tidal currents are accentuated. The sand body (Figures 13.6–13.8b) stretches more than 130 km along the curved margin and has a variety of bar types. Parabolic bars are dominant in the east, whereas the sand body complex in the west is well known for its spectacular longitudinal current parallel sand bars. Individual bars

are >10 km long with widths of several hundreds of meters and are separated by channels that average 1.2 km wide and are over 8 m deep. If preserved like it is today, the sand body complex would comprise a series of elongate sand lenses normal to the shelf edge with thin sands between.

Marine sand bar belts

Cat Cay complex. The Cat Cay sand body (Figure 13.9a) runs along the western leeward margin of the Great Bahama Bank adjacent to the Straits of Florida for a distance of 14 km, and passes bankward into burrowed seagrass-covered muddy peloidal sand. The 2-km-wide complex is characterized by numerous spillover lobes (parabolic bars) that are ~1 km long, 500 m wide with heights of ~2 m, and oriented normal to the trend of the sand body. The lobes have a predominant axial channel that is flanked by shallower areas that represent elongate tidal bars. The shallow crests are textured by numerous subaqueous dunes at various angles, but mostly subparallel to the long axis of the sand belt. These dunes have ~3 m relief and spacing of ~600 m. The surface is covered with medium- to small-scale subaqueous simple dunes (1–10 cm amplitude and 50 m spacing) oriented perpendicular to the shelf edge. These dunes are in turn covered with ripples.

The sand body is ~4 m thick and overlies ~3–4 m of burrowed muddy ooid-peloidal sand (Figure 13.9b). Inclined laminations dip platformward on the eastern side and basinward on the western side into the Straits of Florida. Daily tides shift both the subaqueous dunes and smaller ripples. The lobes are variously interpreted to have formed either by tidal currents, during severe storms, or a combination of both. This complex would be preserved as a massive, somewhat muddy, sand body with a clean ooid sand fringe on the shelf edge.

Joulters Cays complex. The Joulters Cays complex (Figure 13.10) is located at the northern end of Andros Island (Pleistocene limestone), slightly inboard from the margin of an unrimmed shelf, and extends some 20 km onto the platform. The sand body, which is up to 7 m thick is largely a moribund sand flat bordered on the seaward margin by an active ooid sand belt with small ebb tidal deltas (Figure 13.10). The active shoal facies is 1–2 km wide and ~2–3 m thick (Figure 13.10b). Islands have formed along part of the seaward margin of the shoal due to aggradation, active longshore transport, accretion, and meteoric cementation. The sand flats on the western, bankward side of the islands are penetrated from the

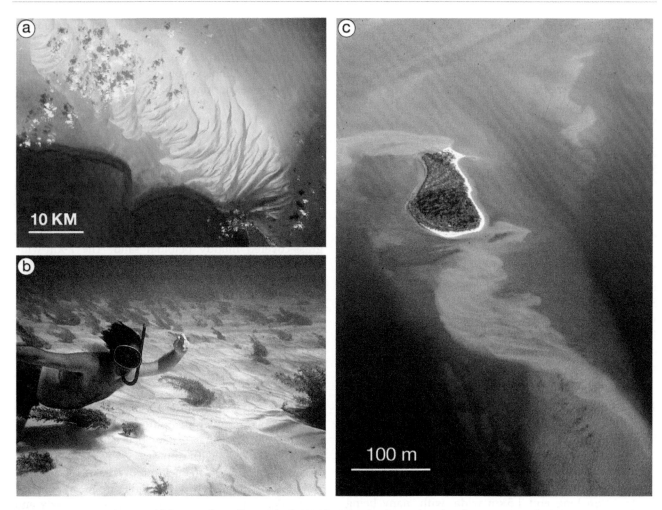

Figure 13.7 (a) Aerial view of Schooner Cays. Copyright © 2014 Esri, i–cube. Reproduced with permission. (b) Swimmer struggling against the tide across the top of an ooid shoal, Schooner Cays. (c) Holocene island developed on top of an ooid sand shoal, Bahamas.

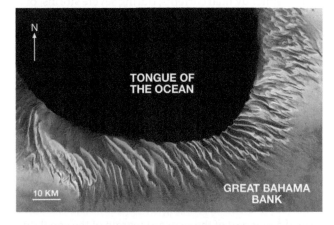

Figure 13.8 Spectacular linear ooid sand body complex at the southern end of the Tongue of the Ocean, Bahamas, the largest such deposit in the modern world. Photograph by S. Andrefouet. Reproduced with permission.

oceanward side by a number of active channels. Local mud layers up to 5 cm thick are the result of post-hurricane suspension settling locally on the floor of tidal channels. The channels are also sites of numerous intraclasts of cemented ooid sand. Much of the sand flat, which is stabilized by seagrass or algae, is intensively burrowed sediment.

This is an aggraded shoal where active ooid production has filled most of the available accommodation. Bedrock is directly overlain by muddy fine-grained peloidal sand in an area more than 30 km wide and 2.5 m in thickness. The upper mobile sand fringe grades bankward into burrowed muddy ooid sand and seaward into skeletal sand (Figure 13.10c). The bankward muddy sediments are stabilized by seagrass. The record of this complex would likely be a narrow marginal sand body with linear islands that graded bankward into somewhat muddy ooid sand.

CAT CAYS OOID BANK COMPLEX

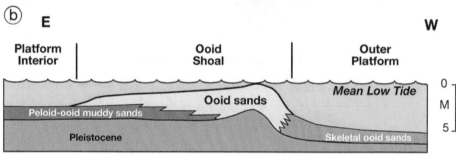

Figure 13.9 Cat Cays ooid shoal complex. (a) Oblique aerial image looking south showing the numerous parabolic bars and tidal deltas; and (b) generalized cross-section illustrating sedimentary facies. Source: Adapted from Ball (1967). Reproduced with permission of the SEPM Society for Sedimentary Geology.

It is entirely possible that many of the now-active shoal systems will eventually end up like this.

Inter-island tidal ooid sand bodies (tidal deltas)

Islands on the Bahamian Platform, both Pleistocene and Holocene, constrict tidal flow and therefore accelerate tidal currents. This situation results in the formation of tidal deltas similar to those in siliciclastic systems. These sand bodies are caused by flow expansion past the constriction point, resulting in a decrease in flow capacity and competence and thus deposition. Most of these sand bodies are composed of smaller elongate and parabolic bars.

The tidal currents are accelerated in the passes between islands to speeds exceeding 1.5 m s^{-1}, and decelerate to form a series of flood tide and ebb tide parabolic-shaped deltas, particularly between islands along the western margin of Exuma Sound (Figures 13.5, 13.6). The size of the tidal deltas varies with the availability of sediment and the size of the inter-island channel; the greater the volume of available sediment, the larger the channel or the bigger the delta. Some bars are exposed at low tide. The most bankward parts are bioturbated, covered in sea-grass, and have small reefs. Channels between the islands are hard-bottomed with sponges and corals and locally with subtidal stromatolites, but little sediment. The rock record of such deposits is difficult to predict, but isolated on-bank flood and off-bank ebb sand bodies are likely.

JOULTER CAYS OOID SAND BODY COMPLEX

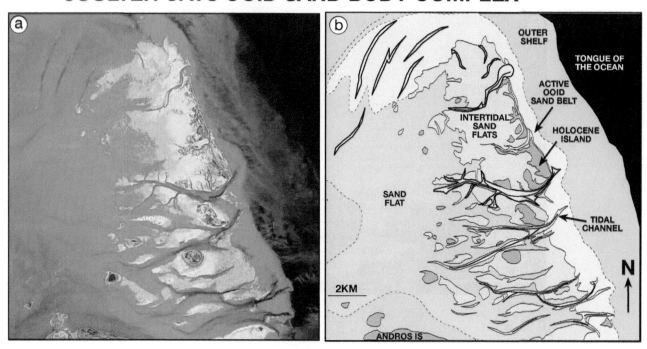

General Cross-section

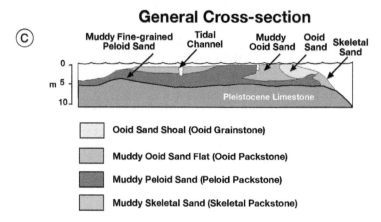

Figure 13.10 Joulters Cays ooid sand body complex, Bahamas. (a) Satellite image of the complex. Copyright © 2014 Esri, i–cube. Reproduced with permission. (b) Simplified facies map of the shoal complex. Adapted from Harris (1979). (c) Cross-section of the complex along line A–B. Source: Adapted from Major et al. (1996).

Platform interior Bahamian ooid sand bodies

Widespread sand sheets

As tidal currents are typically much slower across open platforms, ooid sands that are spread over platforms are typically influenced more by waves and storms than tides. A good example of these deposits is the ooid sands that currently cover much of Caicos Platform, a large carbonate bank in the southern Bahamas (Figure 13.11). Hurricanes average once every 5.5 years with category 4 to 5 storms (210 to >250 km hr^{-1}) every 50 years. Resulting strong storm surges and waves that sweep across the platform lead to deep scouring of sands and result in both off-bank and on-bank sediment transport across the southern rim. More importantly, the bank is constantly swept by brisk easterly winds that blow year-round at 30–45 km hr^{-1}. These trade winds promote good cross-bank circulation and persistent agitation that maintains seafloor sediment with a grainstone texture.

The platform, with a protective reef and an aeolianite island rim on the northern side, lies in a more arid

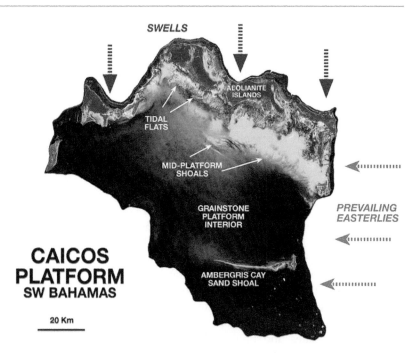

Figure 13.11 The Caicos Platform with a series of Pleistocene islands along the northern margin protecting muddy tidal flats. Most of the bank is swept by prevailing easterly trade winds and covered by subtidal ooid sands and sand shoals. Photograph by H. Wanless. Reproduced with permission.

climate than the northern Bahamas described above. Shallow seawater is therefore more saturated with respect to carbonates and ooids easily form across the open platform seafloor. The persistent but not constant agitation results in sheets of ooids that are more micro-bored and irregular in shape than those in the tidal-dominated northern Bahamas. This ooid sand body would, in the rock record, be a widespread, meter-thick ooid grainstone sand sheet.

Isolated elongate flow parallel sand bars

These linear ridges occur in the center of the bank seemingly "in the middle of nowhere," are up to 30 km long, have narrow (1 km wide) crests and no flanking channels, and contain discontinuous hardgrounds. For example, the linear Ambergris Cay ooid sand shoal on Caicos Bank is a 20-km-long isolated sand bar that is oriented roughly parallel to the predominant trade winds (Figure 13.11). The role of wind waves versus tides in the construction of such sand bodies is unresolved.

Carbonate ramp tidal ooid sand bodies

The carbonate depositional system in the southern Persian Gulf is a classic inner ramp. Ooid sands form there largely because of elevated seawater salinities of 42–44‰. The ooid-rich sediments are strikingly similar to those in the Bahamas with the addition of highly abraded thick-shelled bivalves. The system is a series of Holocene barrier islands (Figures 12.11, 12.12) composed of ooid sands separated by channels up to 10 m deep, each is flanked by a delta of ooid sand and filled by a complex series of banks and channels. Currents in the channels are 0.4–0.65 m s^{-1}. Ebb depositional systems dominate and the seafloor seaward of the islands is a vast spread of ooid sand with numerous Bahamian-like bars.

Carbonate sand bodies in straits and seaways

Straits and seaways are narrow marine passages that link larger bodies of water and are generally bordered by islands or shallowly submerged areas, between which tidal currents are amplified (Figure 13.2). The carbonate sediments can be either photozoan or heterozoan. Currents are produced either by a tidal phase difference between the basins and the open shelf at either end of the seaway, or by oceanic circulation that produces unidirectional flow. Current speeds can exceed 50 cm s^{-1}. Lateral flow-transverse dunes of various types (simple and compound dunes) and sizes are typical and can occur as isolated dunes or as sand sheets. If the seaways are too broad or too deep to generate much of a constriction, then open ocean storms and fair-weather waves are the dominant processes.

Modern ooid sand bodies

Elongate tidal bars are present in straits between islands and the mainland in Abu Dhabi and between Bahrain and the mainland of Saudi Arabia. These shallow straits are <5 m deep but can be up to 20 km wide with waters of 40–50‰ salinity. The ooid bars, parallel to the axis of the channel, are typically *en echelon* and can be up to 10 km in length and 5 m high.

Sub-recent skeletal sand bodies

A particularly well-documented example of these sand bodies is present in the Cenozoic carbonates that filled tectonic depressions on the North Island of New Zealand (Frontispiece). These interbedded siliciclastics and carbonates accumulated in seaways that were 50–80 km wide and up to 200 km long. The major facies are: (1) cross-bedded calcarenites; and (2) horizontal-bedded calcarenites. The carbonates and minor siliciclastic sediments were deposited in a tidal-current-dominated system that was periodically perturbed by storms and associated wave hydrodynamics.

The spectacular cross-bedded calcarenites comprise inclined beds 2–16 cm thick that are separated by much thinner 1–1.5-cm-thick fine-grained siliciclastic layers. These sands represent migration of large (5–7 m high) flow-transverse subaqueous dunes in paleowater depths >40 m with current speeds of 50–130 m s⁻¹. Minor grain fragmentation indicates that most sediment was generated in the dune fields proper. The 2–25-cm-thick associated wave- and storm-dominated, horizontally bedded units are separated by siliciclastic sands and silts 1–1.5 cm thick, contain rare HCS and wave ripples, and are conspicuously more bioturbated.

Carbonate sands in flooded incised valleys

Valleys cut into bedrock or created by faulting and then subsequently filled with marine sediments during later marine inundation are abundant in the geological record, but the fill is usually siliciclastic and typically estuarine. Carbonates are less common and could only accumulate in the valley when freshwater fluvial input was suppressed. Some of the most spectacular carbonates fill the complex sub-recent Neogene paleovalleys in the peri-Alpine terrane of southern France (Figure 13.12a).

These incised valleys are filled with stacked limestone sequences, each of which records a separate period of deposition. The biofragmental carbonates are

Figure 13.12 Miocene cool-water biofragmental limestone, Provence, France: (a) bimodal cross-bedded tidal sands; (b) bidirectional cross-bedded sands; and (c) bipolar cross-bedded sand with mud drapes (arrows). Scale in 10 cm increments.

parautochthonous, having been sourced from just outside the valley mouth and within the valley itself via factories along macroalgal-covered rock walls, in adjacent seagrass meadows that grew in areas with weaker currents, and between subaqueous dunes. Tidal compound cross-bedded calcarenites form units up to 8 m thick. Tidal bidirectional cross-bedding is ubiquitous (Figure 13.12b) with many units containing distinctive carbonate mud drapes (Figure 13.12c). Individual dunes can be 5–7 m high with large tidal point bars indicating postulated paleowater depths of 25 m for some of the units. Overall, this system is dominated by the flood tide. The cross-bedded grainy sediments were restricted to the narrow valley. Once the valley became filled, the sediments spread out across and drowned the interfluves with current speeds that were much lower because of the lack of flow constriction compared to that within the paleovalley.

Carbonate sands in hypersaline basins

There is a wide variability in carbonate sand bodies and they are not exclusively limited to the aforementioned environments. One example includes carbonate sands that are forming today in the marginally isolated basins of Shark Bay in Western Australia (Figures 12.9, 13.13), the same place as spectacular modern stromatolites (see Chapter 5).

Salinity is the major control with shallow metahaline environments (40–53‰) colonized by seagrasses, whereas similar hypersaline environments (56–70‰) are sites of ooid formation. The hypersaline basin of Hamelin Pool, Shark Bay (see Chapter 12), has a very low-diversity biota of mollusks, benthic foraminifera, and algae. Ooid sand bodies occur from low tide to about 2 m depth. They are localized to promontories along the shoreline as long elongate ribbons parallel to the shoreline, or prograding beach ridges.

The rock record of tidal ooid sands

The dynamics and resultant sand body morphologies of modern tide-dominated ooid systems are now mostly well known, but the geological record is less well understood. Predictions from the modern suggest that the shoal-channel system such as that in the Bahamas would partition any strike-parallel sand body into unpredictable strike-normal grainy ooid versus muddy peloid-biofragmental facies. The few complexes that have been studied in three dimensions reveal that complex feedback

Figure 13.13 Rippled ooid sands and stromatolites, Shark Bay, Western Australia. Note hammer (circled) for scale, 13 cm long.

processes dominate, emphasizing that each sand body is different.

These sand bodies are of great importance as hydrocarbon reservoirs. This is particularly the case in the Mississippian of North America, and in the Jurassic and Cretaceous of the Middle East and North America.

Flat-topped photozoan platform sand bodies

A recurring theme from the Holocene to the Pleistocene is that active bar systems in photozoan neritic environments inevitably shallow to sea level as the rate of sediment formation outpaces the rate of sea-level rise. The resultant formation of islands during these latter stages profoundly alters the tidal and therefore depositional dynamics; essentially from tidal bars to tidal deltas. This is nicely illustrated in the Holocene of Joulters Cays and Cat Cays in the Bahamas and the Pleistocene around Miami and in the lower Florida Keys. In short, if the present is any key to the past, ancient ooid sand bodies should be simple at the large scale but unpredictable at the small scale.

Ramp sand bodies

Ooid sand bodies in the inner ramp were more prevalent during specific times in Phanerozoic geologic history than in others. Whereas many of these sand bodies were oolitic, others were dominated by biofragments. Oolitic, cross-bedded grainy limestones with many of the attributes described above were especially numerous in the Mississippian and Jurassic. By contrast, carbonate sands composed of skeletal remains, crinoids in the early and middle Paleozoic and large benthic foraminifers in the Permian (e.g., fusilinids) and early Cenozoic (e.g., lepidocyclinids) also formed many distinctive and discrete inner ramp sand bodies.

Ancient sand body geometries

The geometry of carbonate sand bodies in straits, seaways, and flooded incised valleys is largely determined by the confining nature of surrounding pre-existing topography. Sand bodies on platform margins and inner ramps are not so constrained. To a first degree, they should be elongate and strike parallel. The internal structure will be complex but, just as the modern is typified by ridges and channels, so the rock body will be characterized by isopach thicks and thins; it will not be of uniform thickness! In other instances where deposition has reached sea level they will be more homogeneous but muddier and extensively bioturbated. Stacking of individual sand bodies through time could probably blur this simple picture, but will still impart a heterogeneous fabric to the deposit. Finally, ooid sands are widespread in the geological record but are not always in the form of lenticular, cross-bedded sand bodies. Many are meter-scale, thin-bedded, widespread ooid grainstone units, resembling the dispersed sands on Caicos Bank.

Further reading

Anastas, A.S., Dalrymple, R.W., James, N.P., and Nelson, C.S. (2006) Lithofacies and dynamics of a cool-water carbonate seaway; mid-Tertiary, Te Kuiti group, New Zealand. In: Pedley H.M. and Carannante G. (eds) *Cool-Water Carbonates; Depositional Systems and Paleoenvironmental Controls*. London: Geological Society, Special Publication no. 255, pp. 245–268.

Ball, M.M. (1967) Carbonate sand bodies of Florida and the Bahamas. *Journal of Sedimentary Research*, 37, 556–591.

Burchette, T.P., Wright, V.P., and Faulkner, T.J. (1990) Oolitic sandbody depositional models and geometries, Mississippian of southwest Britain: implications for petroleum exploration in carbonate ramp settings. *Sedimentary Geology*, 68, 87–116.

Harris, P.M. (1979) Facies anatomy and diagenesis of a Bahamian ooid shoal. In *Sedimenta*, University of Miami, Comparative Sedimentology Laboratory, pp. 1–163.

Harris, P.M., Purkis, S.J., and Ellis, J. (2011) Analyzing spatial patterns in modern carbonate sand bodies from Great Bahama Bank. *Journal of Sedimentary Research*, 81, 185–206.

James, N.P., Dalrymple, R.W., Seibel, R.M., Besson, D., and Parize, O. (2014) Warm-temperate, marine, carbonate sedimentation in an early Miocene, tide-dominated, incised valley; Provence, south-east France. *Sedimentology*, 61, 497–534.

Major, R.P., Bebout, D.G., and Harris, P.M. (1996) *Facies heterogeneity in a modern ooid sand shoal – an analog for hydrocarbon reservoirs*. Geological Circular 96-1, Bureau of Economic Geology, Austin, Texas, 30 pp.

Palermo, D., Aigner, T., Nardon, S., and Blendinger, W. (2010) Three-dimensional facies modeling of carbonate sand bodies: Outcrop analog study in an epicontinental basin (Triassic, southwest Germany). *AAPG Bulletin*, 94, 475–512.

Pomar, L., Obrador, A., and Westphal, H. (2002) Sub-wavebase cross-bedded grainstones on a distally steepened carbonate ramp, Upper Miocene, Menorca, Spain. *Sedimentology*, 49, 139–169.

Puga-Bernabéu, Á., Martín, J.M., Braga, J.C., and Sánchez-Almazo, I.M. (2010) Downslope-migrating sandwaves and platform-margin clinoforms in a current-dominated, distally steepened temperate-carbonate ramp (Guadix Basin, Southern Spain). *Sedimentology*, 57, 293–311.

Rankey, E. and Reeder, S.L. (2011) Holocene oolitic marine sand complexes of the Bahamas. *Journal of Sedimentary Research*, 81, 97–117.

Rankey, E. and Reeder, S.L. (2012) Tidal sands of the Bahamian Archipelago. In: Davis R.A.J. and Dalrymple R.W. (eds) *Principles of Tidal Sedimentology*. Dordrecht, Netherlands: Springer, pp. 537–565.

Reynaud, J.-Y. and James, N.P. (2012) The Miocene Sommières basin, SE France: Bioclastic carbonates in a tide-dominated depositional system. *Sedimentary Geology*, 282, 360–373.

Wanless, H.R. and Dravis, J.J. (1989) *Carbonate Environments and Sequences of Caicos Platform*. Washington DC: American Geophysical Union, 28th International Geological Congress, Field Trip Guidebook T374.

Wantland, K.F. and Pusey, W.C., III (eds) (1975) *Belize Shelf: Carbonate Sediments, Clastic Sediments, and Ecology*. Tulsa, OK: American Association of Petroleum Geologists, Studies in Geology no. 2.

CHAPTER 14
MODERN REEFS

Frontispiece A shallow water (~3 m deep) coral reef composed of a variety of corals and a trunkfish, Bermuda. Image width 1 m. Photograph by W. Martindale. Reproduced with permission.

Origin of Carbonate Sedimentary Rocks, First Edition. Noel P. James and Brian Jones.
© 2016 Noel P. James and Brian Jones. Published 2016 by John Wiley & Sons, Ltd.
Companion website: www.wiley.com/go/james/carbonaterocks

Introduction

Reefs have been part of the carbonate world since the Archean and today thrive in shallow- and deep-water environments worldwide. The most familiar are the dazzling coral reefs of the tropics (Figure 14.1) growing in crystal-clear waters and home to an amazing array of marine life. Reefs can, however, also grow in the cold, dark, and deep ocean. There is much to be learned from reefs that grow in the modern ocean, but translating this information to the interpretation of ancient reefs that were produced by now-extinct organisms is challenging. This chapter focuses on reefs in the modern world, whereas the next chapter (see Chapter 15) discusses reefs throughout geologic history.

Figure 14.1 Aerial view of a Pacific Ocean atoll reef, 20 km wide. Photograph by Dale Stokes. Reproduced with permission.

The reef mosaic

Reefs are *biologically constructed reliefs* that rise above the surrounding seafloor. Complex communities of calcareous and non-calcareous organisms, inorganic and organic carbonate precipitates, and sediment collectively form these topographically elevated features (Figure 14.2). The principal reef-builders today range from corals to microbes. Growth of the various organisms generally produces a highly porous structure that can be lithified early to form a hard feature (buildup) on the seafloor. The growing surface of the reef is, at the same time, continuously broken down by a variety of bioeroders.

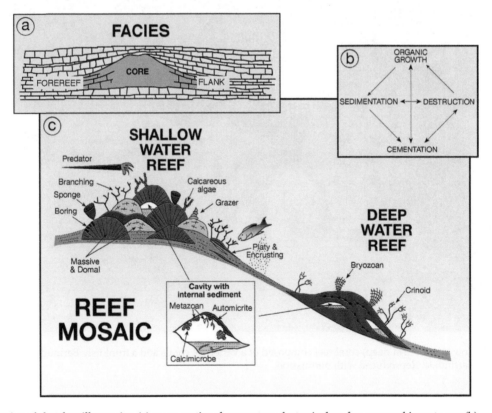

Figure 14.2 A series of sketches illustrating (a) cross-sectional geometry of a typical reef as exposed in outcrop; (b) complex interrelationship between processes that control reef composition; and (c) main attributes of shallow-water and deep-water reefs. Source: James and Wood (2010). Reproduced with permission of the Geological Association of Canada.

Corals

Today, the major reef-builders are scleractinian corals (Figure 14.3), cnidarians that evolved in the Triassic. These animals are mixotrophs, sessile micropredators whose tissues also house symbiotic photosynthetic cyanobacteria or microalgae. The most important symbionts are dinoflagellates called zooxanthellae. The coral ingests food whereas the symbionts produce carbohydrates and lipids via photosynthesis; these substances can be used by the coral. By providing almost limitless energy, zooxanthellae allow hermatypic (reef-building) corals to produce calcium carbonate several times faster than ahermatypic (non-reef-building) corals that lack zooxanthellae. For example, shallow-water corals (mixotrophs) grow at rates of ~0.2–1.0 cm a^{-1} in shallow water <15 mwd, and ~0.5 cm a^{-1} in water of depths >15 mwd. Zooxanthellae are limited to the photic zone and prefer low-nutrient (oligotrophic) environments. Luxuriant modern coral reefs are therefore limited to shallow-water nutrient-poor settings. Conversely, corals without symbionts are mostly confined to deep, dark nutrient-poor environments where they grow very slowly at rates of only 0.05–0.25 cm a^{-1}.

Microbes and algae

A significant amount of reefal carbonate precipitation, especially in the Proterozoic, is associated with microbes. These range from peloidal, clotted, or laminated micrite (automicrite) to calcified microbial sheaths (calcimicrobes),

Figure 14.3 (a) Large branching (~3 m across) coral *Acropora palmata* in ~2 m of water off Grand Cayman, Caribbean. (b) Two species of the hemispherical coral *Diploria* (*D. strigosa* and *D. labyrinthiformis*) in ~2 m of water off the Bermuda islands. Image width 3 m. (c) A cut slab of modern reef rock (^{14}C age 1500 years BP), Bermuda, composed of hydrozoans (H), encrusting foraminifers (F), and biofragmental sand cemented by magnesium calcite cement (S) and bored by sponges (Sp) (numerous holes). (d) A large *Acropora palmata* colony ~2 m high in ~2.5 m depth of water off Grand Cayman, that has been toppled by a recent hurricane.

with a spectrum of intermediate structures whose attributes depend upon seawater carbonate saturation and post-mortem preservation. Encrusting red algae are critically important in modern reef systems because they bind and stabilize reef frameworks and produce sediment. Branching and segmented green algae on modern reefs are, in contrast, mainly sediment producers.

Internal cavity systems

Inside any reef there is a surprising volume of open space because cavities are formed as a byproduct of the intricate metazoan-algal-microbe growth architecture. Cryptic organisms that encrust walls and hang down from cavity ceilings populate many of these voids (Figure 14.2c inset). They range from photic organisms near the openings to heterotrophs in the dark, lightless interiors. Cavities are also sites where fine-grained sediment (internal sediment) accumulates. This material is composed of tiny biofragments and mud-grade sediment that trickles into the holes from above or of skeletons that drop from the walls and ceiling after death. These particles collect on cavity floors and thereby form geopetal sediment. These tranquil crypts offer a stark contrast to the turbulent surface of the reef growth surface just above.

Synsedimentary lithification

Many reefs are lithified immediately below the living surface by a variety of calcite or aragonite cements (Figure 14.3c). Much of this cement is microcrystalline and therefore not obvious, but other cements are spectacular, precipitating as large botryoidal crystal arrays from the walls and ceilings of cavities or among sediment grains (Chapter 24). Cements help to make reefs rigid wave-resistant structures and can also fill much of the original pore space.

Bioerosion, grazing and predation

There are a host of organisms that erode the living reef structure via boring, grazing, and predation (see Chapter 3). These actions enhance organism diversity, weaken reef structure, and produce particulate sediment. Sponges, bivalves, and worms dominate the modern endolithic (boring) biota. Grazing (scraping and rasping) by herbivores, particularly gastropods, fish, and echinoderms, removes seaweed, algal and microbial coatings, and hard calcareous algae, thus significantly eroding the growing reef. More importantly, herbivores reduce the proliferation of fast-growing soft algae and seaweed and therefore create bare open space for the settlement and colonization of reef-building calcareous metazoans. Fish that consume the fleshy part of a reef organism and excrete the ingested skeletal fragments as feces curtail growth of calcareous reef-builders and generate significant amounts of sediment. Finally, weakening of the reef structure by generations of bioerosion makes larger components susceptible to fragmentation during major storms (Figure 14.3d).

The coral reef growth window

The growth window of a modern reef is determined by the combination of many factors that control the growth of coral, which are the dominant reef-building organisms today. These mixotrophs will only thrive if specific conditions of temperature, nutrients, and light exist (Figure 14.4).

Light

A single coral colony can house several different types of zooxanthellae, each adapted to specific local light conditions. Recent studies indicate that zooxanthellate corals, if stressed, either die or revert to being heterotrophs with significantly lower growth rates. Stress often causes the corals to expel the zooxanthellae, which results in the coral losing its color, a phenomenon known as "bleaching."

Light intensity decreases exponentially with depth. Today, the lower limit for hermatypic coral and calcareous green algal growth is 80–100 m. Vertical growth rates below ~15 m decrease in a non-linear, almost exponential fashion. Domal and branching forms of corals, with their encompassing tissue, are well adapted to relatively shallow water where light is refracted and comes from all directions. In deeper water, lamellar growth forms with thin subhorizontal plates maximize surface area relative to size and are particularly adapted to lower light levels, where all intercepted light is vertical. There is a conundrum, however, as flat, encrusting growth forms are also adapted to high-energy conditions at the reef crest and may also develop in shaded areas in shallow-water settings. Care must therefore be taken when interpreting environmental conditions based solely on coral growth form.

Nutrients

Modern coral reefs flourish in oligotrophic (nutrient-impoverished) oceanic regions and seem to have done so since the Triassic. This is due partly to the photosymbionts and partly because corals retain and recycle nutrients very

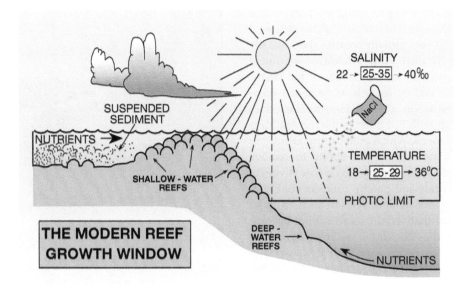

Figure 14.4 Sketch of the environmental parameters that define the growth window of modern mixotroph coral reefs. Source: Adapted from James and Wood (2010). Reproduced with permission of the Geological Association of Canada

efficiently, using inorganic nutrients from the water and waste products (ammonia, organic phosphates) from the reef animals. Even though the waters are low in nutrients, adequate amounts are supplied in energetic environments by high water flux across the reef.

Nutrient levels in the water range from oligotrophic (low nutrients) to mesotrophic to eutrophic (high levels) (Figure 4.3). Whereas corals grow best in oligotrophic waters, green calcareous algae are most prolific and contribute most sediment under near-mesotrophic to mesotrophic conditions. Growth of reef corals and algae is, however, suppressed under increasingly mesotrophic conditions and ceases above nutrient levels of ~2 mg m^{-3}. This has been called the 'reef turn on/turn off point' (Figure 4.3).

Increasing nutrient levels, from upwelling on the outer platform, fluvial runoff, or introduction of human waste (commonly associated with increased tourism), lead to dramatic changes in the reef. Phytoplankton growth impedes light from reaching the coral, while at the same time benthic algae can overgrow and crowd out the calcareous reef-building benthos. Other heterotrophic invertebrates, filamentous algae, fleshy algae, and small suspension-feeding animals (barnacles and bivalves) replace corals at intermediate nutrient levels. Reefs still grow under such conditions, but only when herbivores are present to graze back the algae. Here, a paradox exists because increased nutrient supply, by enabling other organisms to thrive, leads to modified or arrested reef growth. Such conditions are familiar to many tourists in the tropics: the reef directly offshore from the hotel is dead (e.g., western Barbados) and covered with green algae, because we send

out our untreated nutrient-rich waste directly from the hotel into the ocean, promoting its demise!

The above is an extreme case. When nutrients are less abundant they elicit a different response. If the reef is adjacent to land (e.g., volcanic islands such as Tahiti), the reefs today as well as those that grew during the Holocene sea-level rise contain prolific microbial crusts in addition to corals. Whereas all reefs, even in oligotrophic settings, have some microbial encrustations, it appears that abundant nutrients associated with runoff and groundwaters that emerge onto the seafloor fertilize the microbes. As a result microbial reef-building components grow more actively than they would otherwise, creating microbial encrustations.

Temperature and salinity

Zooxanthellate corals grow in waters between 18°C and 36°C (Figure 14.4), but are best adapted to construct reefs in waters between 25°C and 29°C. Periodic exposure is not necessarily lethal and some intertidal corals in the Pacific are exposed for many hours each day. The salinity window of modern reef-building corals ranges from 22‰ to 40‰, but most grow best in waters between 25‰ and 35‰.

Sedimentation

Most corals have a poor tolerance for suspended sediment. Coarse sediment in high-energy settings can cause abrasion, whereas fine-grained suspended sediment decreases light penetration and can cover and clog the polyps. It is difficult to decouple the effects of fine terrigenous sediment and nutrients on coral reef growth because they usually occur together. Corals deal with siliciclastic sediment fallout by polyp expansion, and enclosure of the coarse particles with mucous and by tentacle movement sweeping it away and by the action of cilia. Sediment alone does not arrest coral reef growth but it limits the reef-builders to sediment-tolerant corals, which either grow rapidly to rise above the substrate or have the ability to remove sediment as described above. Comparatively low-diversity exceptionally large corals and patchy coral-covered substrates are the response of modern reefs in such environments. This not to say that certain corals cannot become accustomed to periodic siliciclastic sedimentation, for example, in arid climates (Red Sea and Gulf of Aqaba) with intermittent fluvial outflow of coarse-grained material, but also in humid equatorial settings (Mahakam Delta and Amazon) where abundant mud is delivered to the ocean and corals appear adapted to turbid waters.

Shallow-water reefs

Coral reefs

The familiar beautiful reefs of modern shallow seas are dominated by animal skeletons, which form most of the rock volume. Growth forms and skeletal diversity are strongly controlled by hydrodynamic energy and available space. Diversity is lowest at the reef crest and in deep water with, as stressed above, both environments favoring sheet-like skeletal morphologies. The highest diversity of skeletons and shapes, and ultimately rock types, occur at intermediate depths. The shape of reefs in plan view varies from linear fringing reefs (Figure 14.5a) and barrier reefs (Figure 14.5b) to atoll reefs (Figure 14.1), subcircular cup reefs, and patch reefs (Figure 14.5c). Other reef types can develop at the same time in either quiet water settings across the platform, on ramps, or on the slope.

Zonation

The shallow parts of modern coral reefs commonly have a well-developed windward–leeward zonation. This is true whether they form the margins of large platforms or isolated structures inboard. Zonation is best developed in windward locations (Figures 14.6, 14.7). This zonation is particularly sensitive to perturbations of the growth window, especially a decrease in light, which leads to shallowing of all zones.

Figure 14.5 (a) Aerial image of the Ningaloo fringing reef off the coast of Western Australia, lagoon width ~4km. (b) Aerial image of the outer part of the Great Barrier Reef east of Cairns in the Coral Sea; note the dive boat of length 40m for scale (arrow). (c) A series of coral patch reefs in the Belize barrier reef lagoon on a very calm day. Circular reef in foreground (arrow) is 30m across. Photograph by W. Martindale. Reproduced with permission.

The *reef crest*, which can extend to a depth of ~15m, receives most of the wind and wave energy (Figure 14.8a). Only organisms that can encrust, generally sheet-like forms, are able to survive where wind and swell are

ZONATION OF A MODERN SKELETAL REEF

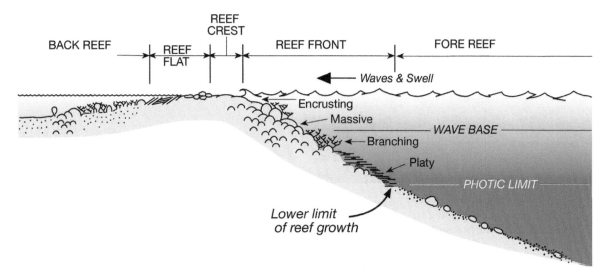

Figure 14.6 Sketch illustrating the facies of a modern mixotroph skeletal reef growing on the windward margin of a carbonate platform. Source: James and Wood (2010). Reproduced with permission of the Geological Association of Canada.

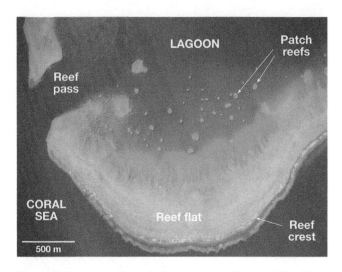

Figure 14.7 A vertical aerial photograph of ribbon reef No. 5 Great Barrier Reef off Cairns, Australia, with facies labeled. Source: Adapted from James and Bourque (1992). Reproduced with permission of the Geological Association of Canada.

constant and intense. Under the most energetic conditions of continuous pounding by waves and rolling swells, the crest is an *algal ridge* composed of encrusting corallines, encrusting foraminifers, vermetid gastropods, and hydrocorals that can, under exceptionally rough conditions, extend downward to 10 mwd. The seaward portion of this zone in living reefs is typically a series of seaward-trending ridges and intervening channels called 'spurs and grooves'.

If wave and swell intensity are more episodic or only moderate to strong, encrusting corals still dominate but can be bladed or have short, stubby branches. Where wave energy is moderate, robust branching corals with rapid growth rates proliferate. The large variety of branching corals on modern reefs that have the ability to grow comparatively quickly is a relatively recent (late Cenozoic) growth mode.

The *reef flat* (Figure 14.8b) varies from a pavement of cemented, large skeletal clasts with scattered rubble and coralline algal nodules in areas of intense waves and swell, to shoals of well-washed lime sand in areas of moderate wave energy. Most material comes from the reef crest and is swept back onto the pavement during cyclonic storms. The vagaries of wave refraction can pile the sands into cays and islands that in turn protect small, quiet water environments behind the reef crest (Figure 14.1).

The protected *back reef* (Figure 14.9a, b) is where much of the mud formed on the reef comes out of suspension. Low-energy conditions, coupled with the prolific growth of sand- and mud-producing bottom biota such as calcareous green algae, also increases the relative proportion of mud-rich sediment. Corals here are stubby and dendroid or large globular forms that extend above the substrate, hence able to withstand both episodic agitation and quiet times when mud-grade sediment settles.

The seaward *reef front* lies between about 10 m (the base of surface wave action) and 100 m. This is an environment of diverse reef-builders (Figure 14.9c) varying in shape

Figure 14.8 (a) Prolific coral and coralline algal growth among the surging swells at the reef crest of the Great Barrier Reef, Australia. Note people for scale (arrow). (b) Looking across the cemented reef flat pavement toward the reef margin at the outer edge of the Great Barrier Reef, Australia, at low tide. Circled hammer for scale.

Figure 14.9 (a) Exposed corals (mostly *Acropora* and *Porites* spp.) in the lagoon of Lady Elliot reef off eastern Australia. Note hammer, 15 cm length, for scale. (b) Prolific growth of the coral *Acropora cervicornis* in the lagoon of the Florida Reef Tract, USA. Coral in foreground is 1 m high. (c) Prolific coral growth on a reef in ~2 m of water off Heron Island, Great Barrier Reef, Australia. Image width ~3 m. Photograph by P. Pufahl. Reproduced with permission. (d) Growth of plate-like corals (*Montastraea cavernosa*) in ~40 m of water off northern Jamaica (Discovery Bay). Image width ~4 m.

from hemispherical to branching to columnar to dendroid to platy. Accessory organisms and various niche dwellers such as bivalves, gastropods, coralline algae, segmented calcareous algae, and sponges are common. Below about 30 m depth, wave intensity is lower, light is attenuated, and most reef corals are prone and plate-shaped (Figure 14.9d) or delicately branching. Pockets, streams, and chutes of skeletal sand, especially calcareous algal particles, accumulate seaward between the spurs and grooves.

The *fore-reef facies* below the zone of coral and algal growth is dominated by gravel and sand composed of whole or fragmented skeletal debris, blocks of reef limestone, and skeletons of reef-builders (Figure 14.10). Such deposits grade basinward into muds exhibiting many of the attributes described in the chapter on carbonate slopes (see Chapter 16).

Figure 14.10 The steeply dipping fore-reef slope in ~130 m depth of water off Tobacco Cay, Belize. Blocks of coral in foreground (arrow) ~1 m high.

Nutrient–sediment zonation

There is typically a cross-shelf zonation in reef composition that partly reflects the seaward decrease in fine sediment and nutrients.

Inner shelf reefs are characterized by: (1) quickly growing corals with a high tolerance for fine sediment and low salinity, and an inability to withstand turbulent waters; (2) large and abundant heterotrophic-only sponges; (3) low epifaunal diversity; (4) few soft corals; (5) abundant soft algae; and (6) few calcareous algae.

Outer shelf reefs (in areas of little upwelling) are distinguished by: (1) slow-growing mixotrophic corals that cannot withstand suspended sediment or low salinities, but are adapted to high-energy conditions; (2) reduced numbers of sponges, some of which contain photosynthetic symbionts; (3) common tridacnid bivalves in the Pacific; (4) high epifaunal diversity; and (5) prolific calcareous algae.

Energy zonation

The preceding section describes the nutrient–sediment zonation across the platform, but there is also a striking energy zonation. As waves decrease in intensity inboard, so different corals dominate reefs in a predictable succession (Figure 14.11). Domal corals and algae first replace the robust branching forms; for example, in the Caribbean *Montastraea* sp. and *Diploria* sp. replace *Acropora palmata* and coralline algae. As waters become calmer in the same settings, delicate branching types such as *Porites porites* replace these massive corals forming coral algal banks (see below).

Coralline algal reefs

Modern reefs built by coralline red algae (Figure 14.12) are up to 10 m high and a few tens of meters in diameter and grow on the seaward margins of platform. They are similar in composition to the intertidal to shallow subtidal coralline algal ridges described above, but grow in situations where water temperature or turbidity prevent the growth of the robust branching reef crest coral (*Acropora* spp.) community. In Bermuda, these well-cemented "cup reefs" have kept pace with rising sea level since atoll flooding at ~6 ka. Cup reefs form a discontinuous series of reefs around the atoll that are emergent at low tide. The reef surface is an almost continuous skin of coralline algae, embedded vermetid gastropods, and hydrozoans. Scattered domal corals grow on the reef surface below the intertidal zone. Extensive voids are inhabited by encrusting red foraminifers (*Homotrema* sp.) and are partially to completely filled with biofragmental sand. This algal-dominated community also grows in the intertidal zone along rocky coasts in the Mediterranean, but does not form reefs.

Other shallow-water reefs

There are other bioconstructions in modern shallow-water settings that barely qualify as reefs but have analogs in the rock record that have many reef attributes.

Coral-algal banks. The lowest-energy accumulations in the modern shallow-water reef spectrum are typified by nearshore banks that grow seaward of the Florida Keys,

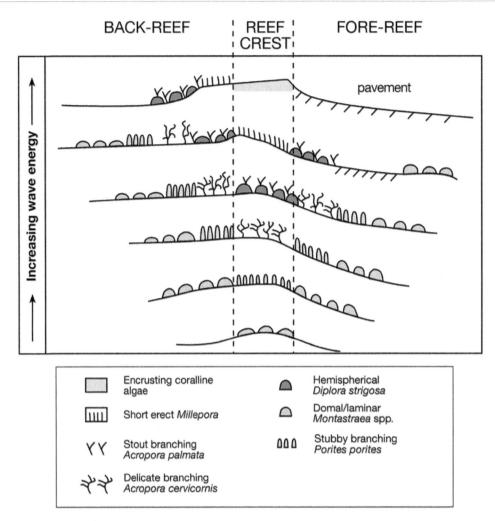

Figure 14.11 Sketch illustrating the change in modern coral zonation in response to increasing wave energy.

and are protected from open ocean waves by the Florida reef tract (Figure 14.13). These banks of delicate branching corals and a variety of calcareous algae are 1 to 3 km long, rise 2 to 4 m above the seafloor, and are exposed at spring low tide. They are zoned with a windward margin of branching coral and branching coralline algae that passes landward into seagrasses with a rich calcareous green algal flora and an infaunal bivalve and burrowing crustacean fauna. Coring reveals that the record of such banks is one of branching coral-algal sediment that passes laterally into bioturbated muddy sediment.

Halimeda (green algal) reef mounds.   These structures, rich in the green calcareous alga *Halimeda*, are found in the Java Sea, Timor Sea, the Great Barrier Reef, and the Caribbean. They all rise to within 40–15 m of the sea surface and do not occur in shallow water. They range

from living to relict, but are nevertheless Holocene in age. Structures have seafloor relief of 2 to 20 m, steep margins, and thicknesses of up to 50 m. They are dominated by *Halimeda* spp. but may have accessory encrusting sponges, bryozoans, and octocorals.

The sediment is mostly sand-grade *Halimeda* fragments with accessory coralline red algae, small corals, and ostracods in some places or mollusks, bryozoans, and large benthic foraminifers in others. The structures range from unlithified to cemented with many exhibiting extensive boring and micritization. The mound community appears to be a subtropical assemblage, and all interpretations point to the influence of cool, nutrient-rich upwelling waters that promote this style of reef mound growth.

Linear mud banks.   These elongate features rise 3 to 4 m above the surrounding seafloor in lagoonal settings (Florida Bay and Belize). Their origin is partly

Figure 14.12 Coralline algal reefs, Bermuda. (a) Numerous cup reefs along the shallow inshore, south coast (persons for scale, arrow). (b) Close-up of the surface geometry and water surging around the periphery (structure 3 m across). (c) Underwater view of a small cup reef ~5 m across illustrating the deep depressed center and lack of corals. Photograph by W. Martindale. Reproduced with permission. (d) Slabs of cup reef limestone (Bermuda) composed of coralline algae, vermetid gastropods, the encrusting red foraminifer *Homotrema*, and cemented carbonate sand that is bored by the bivalve *Lithophoga*, and filled by lithified sand. Image width 6 cm.

hydrodynamic and partly biogenic. Storms or tides sweep soft pellets into shoals as if they were sand grains. Roots of prolific seagrasses bind the sediments, preventing erosion. Since the pellets are not lithified, these structures would be preserved as mud in the rock record with no evidence of hydrodynamics except local winnowed shell lags.

Deep-water reefs

Modern deep-water reefs (Figure 14.14) that grow in dark, cold waters in areas such as Norway and New Zealand are constructed by azooxanthellate scleractinian corals, such as the branching form *Lophelia pertusa*, that lack photosymbionts. These reefs are most numerous between water depths of 250 and 1500 m where temperatures range from 4 to 12°C. Structures can be up to 5 km long and 40 m high, generally with steep sides. Many such reefs appear to thrive best on bathymetric highs such as seamounts, drowned glacial moraines, subaqueous dunes, and sediment drifts. They are also located in areas of elevated nutrients such as fronts between large-scale water masses or areas of upwelling. Others are sited on top of cold hydrocarbon seeps.

The corals range from small centimeter-scale cups to meter-scale dendroid bushes (Figure 14.14b). Other important organisms are soft corals with spicules

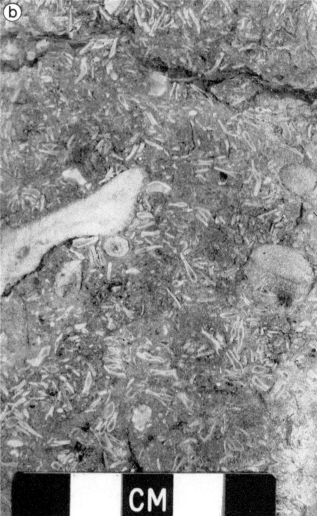

Figure 14.13 (a) Aerial view of Tavernier Key, a coral and algal bank topped by a small island just off the Florida Keys (in background). (b) A plastic impregnated core of sediment from just below the surface of Tavernier Key composed of the branching coral *Porites porites*, prolific green algal plates (*Halimeda* sp.), and carbonate mud.

Figure 14.14 (a) Sketch of a lithoherm in the Straits of Florida showing the geometry of and major organisms on the structure. Adapted from Messing et al. (1990). Reproduced with permission of SEPM Society for Sedimentary Geology. (b) Prolific growth of the coral *Lophelia pertusa* (it has no photosymbionts) on the up-current end of a lithoherm in ~600 m of water in the Straits of Florida. Photograph by A. C. Neumann. Reproduced with permission.

(alcyonarians), stylasterene hydrocorals, zoanthids (anemones), crinoids, spiculate and calcareous lithistid sponges. Most reefs are unlithified, with the exception of *lithoherms* in the Straits of Florida. The reefs are somewhat self-organized. Individual colonies coalesce to form thickets (an aggregate of closely spaced colonies), thickets merge to form coppices (piles of skeletons colonized by living corals), and coppices join to form banks or reefs. The coppices are usually intensively bioeroded, filled with muddy internal sediment, and act as a substrate for ancillary epibenthic growth leading to increased biotic diversity. A theme of all these deep-water reefs is one of numerous corals in a fine-grained matrix that can be either carbonate mud, siliciclastic mud, or both, containing numerous planktic foraminifers and pteropods.

Further reading

Birkeland, C. (ed.) (1997) *Life and Death of Coral Reefs*. New York, NY: Chapman and Hall.

Freiwald, A. and Roberts, J.M. (eds) (2005) *Cold-Water Corals and Ecosystems*. Berlin, New York: Springer-Verlag, Erlangen Earth Conference Series.

Ginsburg, R.N. and Schroeder, J.H. (1973) Growth and submarine fossilization of algal cup reefs, Bermuda. *Sedimentology*, 20, 575–614.

Hopley, D., Smithers, S.G., and Parnell, K.E. (2007) *The Geomorphology of the Great Barrier Reef: Development, Diversity and Change*. Cambridge: Cambridge University Press.

James, N.P. and Ginsburg, R.N. (1979) *The Seaward Margin of Belize Barrier and Atoll Reefs*. International Association of Sedimentologists, Special Publication no. 3.

James, N.P. and Bourque, P.-A. (1992) Chapter 17. Reefs and mounds. In: Walker, R.G. and James, N.P. (eds) *Facies Models: Response to Sea Level Change*. St John's, Canada: Geological Association of Canada, pp. 323–347.

James, N.P. and Wood, R. (2010) Reefs. In: James N.P. and Dalrymple R.W. (eds.) *Facies Models 4*. Geological Association of Canada, GEOtext 6, pp. 421–447.

Messing, C.G., Neumann, A.C., and Lang, J.C. (1990) Biozonation of deep-water lithoherms and associated hardgrounds in the northeastern Straits of Florida. *Palaios*, 5, 15–33.

Montaggioni, L.F. and Braithwaite, C.J.R. (eds) (2009) *Quaternary Coral Reef Systems: History, Development Processes and Controlling Factors*. Elsevier, Developments in Marine Geology, Vol. 5.

Roberts, H.H. and Macintyre, I.G. (eds) (1988) *Halimeda. Coral Reefs*, 6, 121–271.

Seard, C., Camoin, G., Yokoyama, Y., *et al.* (2011) Microbialite development patterns in the last deglacial reefs from Tahiti (French Polynesia; IODP Expedition #310): Implications on reef framework architecture. *Marine Geology*, 279, 63–86.

Wright, V.P. (1992) A revised classification of limestones. *Sedimentary Geology*, 76, 177–185.

CHAPTER 15
ANCIENT REEFS

Frontispiece Exhumed Neoproterozoic microbial reef ~300 m high, Mackenzie Mountains, Northwest Territories, Canada.

Origin of Carbonate Sedimentary Rocks, First Edition. Noel P. James and Brian Jones.
© 2016 Noel P. James and Brian Jones. Published 2016 by John Wiley & Sons, Ltd.
Companion website: www.wiley.com/go/james/carbonaterocks

Introduction

Fossil reefs (Frontispiece, Figure 15.1) are of both academic and economic importance. They contain records of evolving marine biology and changing oceanography, are major hydrocarbon reservoirs in many regions of the world, and play an important role in the economic geology as hosts of both base and precious metals. It is, however, a challenge to synthesize and decipher ancient reefs because their growth was governed as much by interactions within the evolving biosphere as by universal physical and chemical laws. This chapter is an integration of themes that run through geological history and characterize reefs of all ages.

Although a strict uniformitarian approach to understanding reefs provides insights into processes that governed their development, it is difficult to predict the products of those processes and translate these attributes into the rock record. This problem is that ancient reefs are not modern reefs just formed by different organisms; in many cases, they had a fundamentally different ecological structure and geological attributes that have no modern counterparts! The major differences are: (1) evolutionary biology–modern reef-building organisms are not related to fossil reef-builders and it is not known if the ancient invertebrates were mixotrophs or heterotrophs; (2)

calcimicrobes are generally not important in modern reefs, but they were abundant in many fossil reefs; (3) synsedimentary cement was much more prolific in the rock record than it is today; and (4) automicrite was pervasive in many ancient reef structures but is relatively rare today.

The reason for the abundance of benthic calcimicrobes and the ubiquity of marine cements in Paleozoic and many Mesozoic reefs is puzzling, but it is probably due to the absence of calcareous plankton prior to the middle Mesozoic (see Chapter 17). These protists and algae removed so much carbonate from the ocean that they substantially lowered the carbonate saturation state of seawater, thus making inorganic and microbial carbonate precipitation more difficult.

The ancient reef factory

Calcareous metazoans

Ancient reef-building calcareous metazoans came from many algal and invertebrate taxa, but the most common groups were calcareous algae, sponges, corals, bryozoans, and bivalves. These organisms could have been autotrophs, heterotrophs, or mixotrophs. Many reef metazoans were clonal and obligate calcifiers in the sense that the

Figure 15.1 A cross-section of the Upper Devonian reef complex in the Canning Basin, Western Australia. Photograph by P.W. Playford. Reproduced with permission.

organism's genetics controlled calcification (or biomineralization). Many of these robust invertebrates have been nicknamed "hypercalcifiers" because of their propensity to produce large skeletons (single coral colonies and stromatoporoid sponges could be up to 3 m across) that ranged in shape from domal-hemispherical, to laminar or encrusting, to delicate plate-like and branching (Figure 15.2). The other important invertebrates were bryozoans that, even though relatively small and delicate compared to the foregoing, were critical elements of many deepwater reefs and acted as scaffolds between which automicrite could form and cements precipitate.

The symbiont conundrum

For those reef-building organisms that did contain phototrophic symbionts, light and nutrients would have been important limiting factors. It is still unclear, in spite of exhaustive geochemistry, if Paleozoic corals and stromatoporoids contained photosymbionts. Tabulate and colonial rugose corals were a diverse group of organisms and some genera may have contained photosymbionts. Fossil sponges, both spiculate and soft bodied, and calcified sponges (such as stromatoporoids and archaeocyaths) are poorly understood and their association with photosymbionts is uncertain even though modern sponges on outer shelf reefs do contain cyanobacterial symbionts. If Paleozoic corals and calcified sponges such as stromatoporoids were not mixotrophs or were only marginally dependent on photosymbionts, then shallow-water skeletal reefs could have extended into much deeper waters than they do today. Consequently, there are two possible facies profiles for skeletal reefs: the mixotroph profile (Figure 14.6) and the heterotroph profile (Figure 15.3). This concept has important consequences because it means that Paleozoic corals and stromatoporoids were not necessarily indicative of shallow-water paleoenvironments. Rudist bivalves, which were mound-builders and massive sediment producers in Cretaceous seas, were similar but unrelated to the modern, mixotroph symbiont-bearing giant clams (*Tridacna*). Some of the most important rudist groups, however, do not appear to have had symbionts.

Microbes, calcimicrobes, and calcareous algae

These autotrophs were the original reef-builders and have been a continuing part of the reef spectrum throughout geologic time. During most of the Phanerozoic, however, they were largely restricted to stressed environments and dominated periods following biological catastrophes.

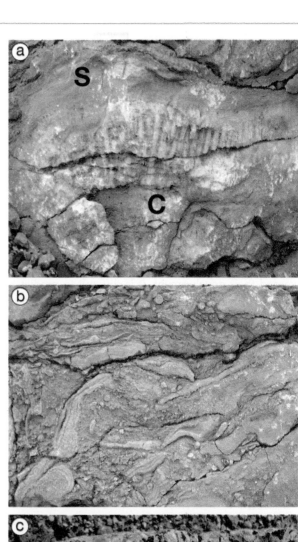

Figure 15.2 (a) Cross-section of a Middle Devonian hemispherical tabulate coral (C) encrusted by a stromatoporoid (S) in the Alexandra Reef Complex, Northwest Territories, Canada. Image width 40 cm. (b) Cross-section of plate-like stromatoporoids from the reef crest in the Middle Devonian Alexandra Reef Complex, Northwest Territories, Canada. Image width 35 cm. (c) A complex intergrowth of digitate stromatoporoids (*Amphipora* sp.) in lagoonal facies of an Upper Devonian reef complex, Norman Wells, Northwest Territories, Canada. Centimeter scale.

ZONATION OF A HETEROTROPH REEF

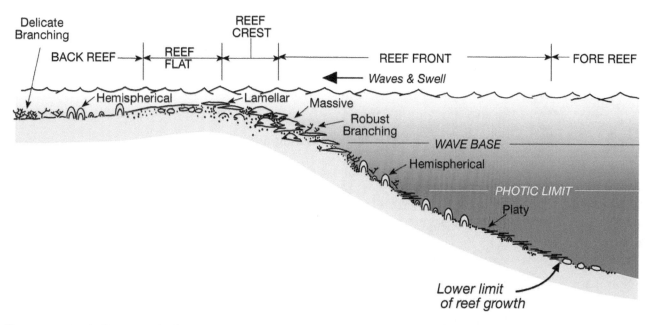

Figure 15.3 Sketch illustrating the facies of a skeletal reef composed mainly of heterotrophs that are not constrained by light, and can therefore grow into deeper water than their mixotroph counterparts. Source: James and Wood (2010). Reproduced with permission of the Geological Association of Canada

Stromatolites, composed of automicrite and synsedimentary cement, were the only reef-builders during the Precambrian but were augmented by thrombolites and calcimicrobes in the Neoproterozoic. They have continued to be a variable contributor to reef growth throughout the Phanerozoic.

Calcimicrobes, once established in the Neoproterozoic, continued to be a significant and sometimes dominant element of reefs throughout much of the Phanerozoic. The most common calcimicrobes are tubular (*Girvanella*, *Rothpletzella*), coccoid (*Renalcis*), or branching (*Epiphyton*) (see Chapter 5). Other similar microstructures such as *Tubiphytes* (a probable foraminifer-cyanobacteria consortium) have also been placed in this group of ancient reef components. Calcareous algae were most important during the late Paleozoic when several groups of green and red algae with large leaf-like fronds ("phylloid-algae") were particularly prolific.

Internal cavities

Just as in modern reefs, internal cavities were a hallmark of fossil reef structures (Figure 15.4a). These holes, which formed as a consequence of irregular growth of framework components, were typically the sites where cryptic organisms (algae, microbes, and invertebrates) encrusted the ceilings and walls, synsedimentary cement precipitated on a variety of surfaces, and geopetal sediment accumulated on the floor. Such sediment came from the skeletons of the cavity-dwelling organisms and from fine material that was pumped into the voids by waves and currents.

Lithification

Synsedimentary marine cements are a signature of ancient reefs (Figure 15.4b). These cements (see Chapter 24) are typically inclusion-rich, microcrystalline, fibrous, or botryoidal. Although most were originally aragonite or HMC, they have typically been diagenetically altered to dLMC. Cement crystals form rims around intergranular pores in sands and rinds that line and partially to completely fill large growth cavities. Such cements are typically interlayered with internal sediment. Many are less obvious precipitates in the form of micropeloids or irregular gravity-defying micritic laminations, both of which were probably formed by microbes.

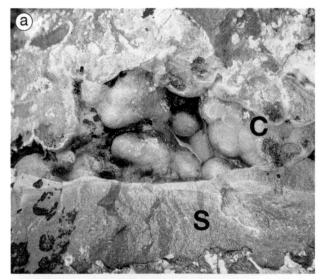

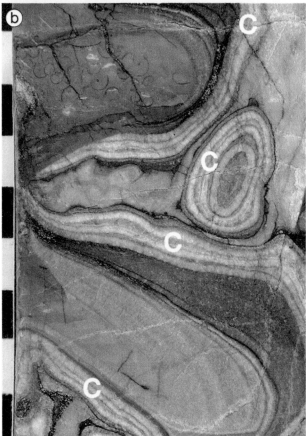

Figure 15.4 (a) A growth cavity in a lower Cambrian reef in southern Labrador, Canada that has been partly filled with geopetal internal sediment (S) and occluded by synsedimentary cement botryoids (C). Image is 20 cm wide. (b) Subsurface core of an Upper Devonian stromatoporoid reef in the Western Canadian Sedimentary Basin, with prolific fibrous banded synsedimentary cement (C). Centimeter scale. Photograph by W. Martindale. Reproduced with permission.

Stromatactis (see below), originally interpreted as a reef-building organism, is one of the most enigmatic cement-related fabrics in reef mounds. Visually it is a mass of fibrous calcite with digitate margins in a mud matrix. When examined in detail, it typically comprises a cavity that is filled with synsedimentary fibrous cement and geopetal internal sediment. Whereas the nature of the cement is clear, it is not known how the original cavity formed (especially in mudstones that have no supporting framework).

Boring and bioerosion

Microbes have been significant borers since the late Proterozoic and have become increasingly important ever since. Although present in the Paleozoic, grazing, rasping, bioerosion, and sediment production by invertebrates began in earnest in the Mesozoic and have increased dramatically since that time.

Reef stratigraphic nomenclature

For decades there has been an ongoing debate as to what constitutes an ancient reef. Regardless of the terms used in the literature, these features were bioconstructions, not rock bodies formed by sediments swept together by currents and waves, and not simply fossil-rich beds of limestone or dolostone. They had some or all of the features that characterize living reefs, namely: (1) in-place accumulation of biogenic carbonate; (2) internal cavities; (3) synsedimentary cement; and (4) bioerosion. Names for the most common rock types within these features are outlined later in this chapter. There are two different types of these bioconstructions: reefs and reef mounds (see below).

The terms bioherm and biostrome, which carry no genetic connotations regarding the composition of the bioconstruction (especially useful when dolomitization has obscured much of the original fabric), have long been used to define the overall geometry of ancient structures. A *bioherm* is a lens-shaped reef whereas a *biostrome* is a tabular rock body, usually a single bed of similar composition. It must be stressed that both of these features have the requisite reef features outlined above. Most of those described in the literature are bioherms. An *ecological bioherm* had clear depositional relief on the seafloor, as demonstrated by bedded limestone onlapping the structure. By contrast, *stratigraphic bioherms* are structures that had little seafloor relief (perhaps 1 m at most) when they grew, but stacked one on top of the

REEF TERMINOLOGY

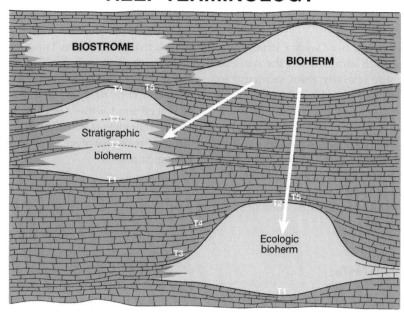

Figure 15.5 Sketch illustrating the stratigraphic attributes of a biostrome versus stratigraphic (very little depositional relief) and ecological (significant depositional relief) bioherms, T1 = time interval.

Table 15.1 Common terms for reef and reef-associated rocks

Feature	Description
Reef	A massive or layered, laterally restricted carbonate bioconstruction that formed *in situ*, possessed bathymetric relief, was built by the intergrowth of large, meter-scale calcareous organisms, might contain internal cavities and synsedimentary cement.
Reef mound	A laterally restricted carbonate bioconstruction that formed *in situ*, possessed bathymetric relief, but lacked large skeletal frame-builders and instead contained relatively small skeletal elements, might be rich in carbonate mud, internal cavities, and synsedimentary cement.
Bioherm	A lens, dome, or mound-shaped carbonate bioconstruction enclosed by sediments of a different lithology and composed mostly of the skeletal remains of *in situ* organisms and synsedimentary cement.
Biostrome	A laterally extensive, bedded, blanket-like carbonate bioconstruction composed of skeletal remains, microbial carbonate, calcimicrobes, *in situ* organisms and synsedimentary cement.
Stratigraphic reef	A thick, laterally restricted reef or reef mound that had little original relief above the seafloor at the time of growth.
Ecologic reef	A reef or reef mound generally formed during one specific time interval with considerable bathymetric relief during growth.
Carbonate buildup	A circumscribed body of carbonate rock that displays bathymetric relief above equivalent strata and differs in nature from typical thinner deposits of underlying and overlying rocks.
Reef complex	A major carbonate edifice consisting of reefs and associated sediments and usually divisible into several facies or zones (e.g., fore-reef, reef crest, back-reef, etc.)
Reef types	
Barrier reef	A reef-rimmed shelf margin with an inner lagoon belt bordering the shoreline of an older emergent area; the reef is relatively linear and can exceed 100 km in length.
Fringing reef	A reef growing near shore along the slopes of an older emergent area, essentially without an intermediate lagoonal belt.
Atoll reef	A ring-shaped reef surrounding a lagoon; the horizontal and vertical dimensions are usually within one order of magnitude (<10: 1).
Patch reef	An isolated reef generally occurring in lagoonal facies; horizontal dimension is typically <100 m and vertical dimension <20 m.
Cup reef	A small goblet-shaped reef composed principally of coralline algae, vermetid gastropods, and synsedimentary cement; generally <20 m high.
Pinnacle reef	An isolated reef or reef complex with horizontal and vertical dimensions within one order of magnitude (<10: 1); typically vertically exaggerated on subsurface stratigraphic cross-sections.

other through time. Swimming over such a structure, a diver would not notice much difference in the seafloor except that it would grade laterally from sediment to numerous corals and then back to sediment again. These have been called ecologic reefs and stratigraphic reefs, but perhaps a non-generic term is preferable until the exact nature of the structure is determined. These different structures are summarized in Figure 15.5 and Table 15.1. Another commonly used generic epithet with no compositional, size, or shape connotation is a *carbonate buildup*. These terms carry no implication of scale or composition (Figure 15.5).

The spectrum of ancient reefs

Ancient bioconstructions can be divided into reefs and reef mounds (Figure 15.6). Large skeletons that are generally more than 1 cm in size characterize reefs, whereas reef mounds, which have many of the same constructional attributes, are formed of skeletal components that

BIOCONSTRUCTIONS

REEFS

Skeletons
Mixotrophs

Skeletons   Microbes
Heterotrophs

. Large skeletons >10 cm - also stromatolites
. ± Internal cavities
. ± Synsedimentary cements
. ± Calcimicrobes
. Bioherms and platform margin complexes

REEF MOUNDS

Skeletons

Autotrophs   Mud
(algae)

. Small skeletons < 10 cm
. ± Internal cavities
. ± Synsedimentary cements (+ stromatactis)
. Mud (precipitated, detrital, microbial)
. Bioherms (shallow & deep water)

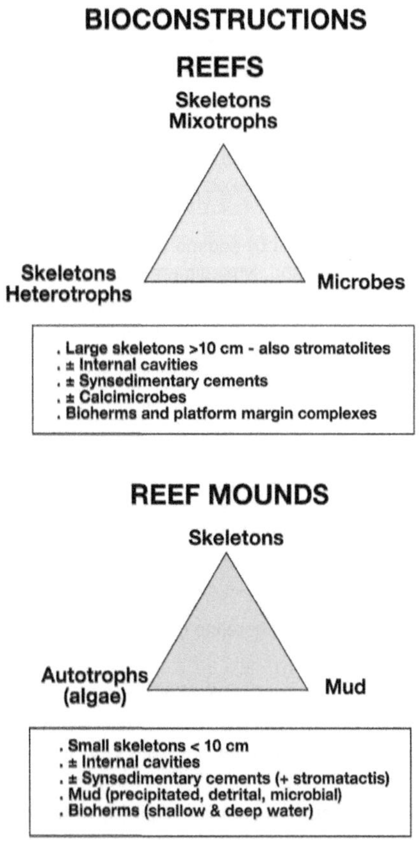

Figure 15.6 Attributes of reefs versus reef mounds.

are less than 1 cm in size. These are, however, merely end-members in a spectrum of biological reliefs that run the gamut from stacked large skeletons, to complex multi-organism constructions, to piles of *in situ* biogenically produced mud with masses of synsedimentary cement. The current view is that all such structures are reefs! These different reefs, however, have varied remarkably through geologic time.

Reefs

Considered against the background of geologic time, there are three types of reefs: (1) skeletal reefs composed mostly of relatively large calcareous invertebrates; (2) skeletal-microbial reefs constructed by the same metazoans but with the aid of numerous calcareous and non-calcareous microbes; and (3) microbial reefs built by microbes in the form of stromatolites and thrombolites.

Skeletal reefs

These structures, like those of today, were largely composed of algae and animal skeletons (Figure 15.7) and synsedimentary marine cement. They grew across the neritic environmental spectrum, but particularly along the platform rim. These reefs had the same basic attributes as modern reefs but with frameworks constructed by different skeletal invertebrates, the most prominent being rugose corals, tabulate corals, and stromatoporoids. Skeletons dominated the buildups and formed most of

Figure 15.7 Cross-section outcrop image of a Middle Devonian skeletal, framestone reef composed of stromatoporoids (S) and corals (C), Ellesmere Island, Arctic Canada. Note pencil (15 cm length) for scale.

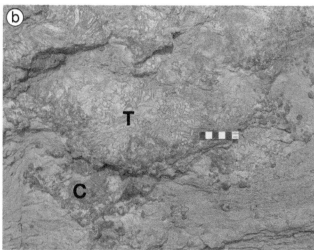

Figure 15.8 (a) Cross-section outcrop image of an Upper Ordovician coral-calcimicrobe bioherm Anticosti Island, Québec, Canada. Scale divisions on rod 10 cm. (b) Cross-section of a bioherm similar to that in (a) illustrating a halysitid tabulate coral (T) and calcimicrobe (C) intergrowth (*Wetheredella*, *Rothpletzella*, *Girvanella*) to form a bindstone. Centimeter scale.

the rock volume. Although synsedimentary cements, microbes, and automicrite were present, they were usually of subsidiary importance. Internal cavities were sites of a diverse cryptic biota, calcimicrobial growth, and internal geopetal sediment. Automicrite was present on, around, and between skeletons. Synsedimentary cement could be spectacular, with the largest amounts found in reefs at the platform margin.

Skeletal-calcimicrobial reefs

Such reefs were similar to skeletal reefs except that calcimicrobes formed a significant part of the buildup (Figure 15.8). Calcimicrobes in some reefs were volumetrically the most important constituents. Invertebrates were of decimeter-scale and larger, and grew in all platform environments including the rim.

Microbial reefs

These reefs were constructed entirely by microbes (not calcimicrobes) and were characterized by stromatolites and thrombolites (Figure 15.9). They were *the* reefs of the Archean and Proterozoic. Microbialites from centimeters to decimeters in size formed reefs across the depositional spectrum, from shallow to deep water. The reefs had many attributes of Phanerozoic structures, but changed in character from cement-dominated to microbe-dominated through time (see Chapter 18).

Reef mounds

Reef mounds are the most controversial of all bioconstructions! Although these structures are clearly bioherms or biostromes in terms of shape (Figure 15.6), they do not have the same large skeletons as modern reefs. Furthermore, even though they were recognized as reef-like by past researchers, there are no known modern analogs. In short, they are non-actualistic and they have therefore been referred to by a variety of names such as *mud mounds, lithoherms, algal reefs,* or *Waulsortian reefs*. These structures have recently been grouped under the terms *reef mounds* or *biogenic mounds*, but the term *reef mound* now seems most appropriate. They are all mud-rich, and this mud is in part depositional mud and in part mud-sized synsedimentary cement (see Chapter 24). Most have a distinct composition and stratigraphy (Figure 15.10).

Stromatactis is the name given to masses of crystalline calcite in numerous Paleozoic reef mounds, especially those that are very muddy. The feature was originally thought to be the recrystallized remains of the skeletal organism that formed the framework for such mounds, and so given a Latin name that we write in italics. It was subsequently realized that *Stromatactis* was a series of cavities filled with fibrous synsedimentary calcite cement and locally, internal sediment (Figure 24.13b). The scientific problem has now focused on the origin of the voids. Could they be: (1) holes that remained after dissolution of sponge tissue; (2) cementation of the muddy sediment leaving voids between lithified layers;

Figure 15.9 (a) Large Mesoproterozoic stromatolite reef complex ~100 m high (original depositional relief), Baffin Island, Arctic Canada. (b) Cross-section outcrop image of the reef illustrated in Figure 15.8a, composed of columnar stromatolite microbial framestone. Ruler is 10 cm long.

(3) microbial binding of the sediment leaving pores between; or (4) gas hydrates that dissolved forming pores? Regardless, the muddy reef mounds do not occur in post-Cretaceous carbonate rocks and so the origin of *Stromatactis* remains a puzzle.

Skeletal-calcimicrobial reef mounds

Skeletal-calcimicrobe mounds (Figure 15.6) were built by relatively small (centimeter-sized) skeletal invertebrates and calcimicrobes. The most common small skeletons range from small stromatoporoids and corals to bryozoans, bivalves, and sponges. The small size of the invertebrates prevented these mounds from growing at the platform margin and relegated them to protected inner ramp or deeper outer ramp to slope environments. Such reefs typically contain extensive framework cavities, with their own distinctive cryptic biotas and abundant synsedimentary cements. Whereas the reef-building skeletal organisms may have been heterotrophs, the presence of calcimicrobes implies that the mounds could have grown in the photic zone. Variably successful attempts have been made to link them with modern coral-algal banks, linear mud banks, or deep-water mounds.

These reef mounds also formed large slope-parallel biostromes that grew on the upper slope during the late Paleozoic (Mississippian–Permian). The biostromes were formed of isolated to amalgamated reef mounds. The shallowest water facies (estimated 0–30 mwd) were dominated by calcareous algae, calcimicrobes, and abundant mud, microbial sediment, and synsedimentary cement. Upper slope facies (50–350 mwd) were composed of domal to laminated microbial-synsedimentary cement boundstones characterized by overwhelming marine cements, as well as numerous bryozoans, crinoids, brachiopods, thin-walled bivalves, and local ostracods. These paleodepths would have been below the photic zone, suggesting that some of the microbes may have been chemoautotrophs. Mid-slope facies (250–750 mwd) were more massive with less cement and proportionally more skeletal components as layers alternating with sediment gravity-flow deposits mostly derived from the shallow-water slope facies.

Calcareous algal reef mounds

These reef mounds are dominated by autotrophs, particularly calcareous algae (Figure 15.11), with all of the same attributes of skeletal reefs that formed mounds

REEF MOUNDS

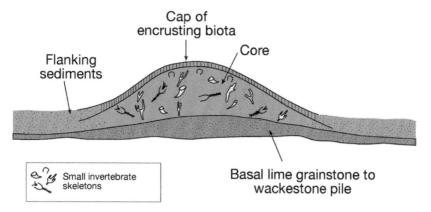

Figure 15.10 Sketch illustrating the general attributes of reef mounds. The structure typically nucleates on a previous bathymetric high or carbonate sand shoal, and is composed of small skeletons, muddy sediment, and synsedimentary cements. Some are capped by a thin crust of encrusting invertebrates or algae if the bioconstruction extended up into the photic zone. Source: James (1983). Reproduced with permission of the American Association of Petroleum Geologists.

Figure 15.11 Cross-section outcrop from the core of a Pennsylvanian phylloid algal reef mound composed of calcareous algae and bioclastic packstone-wackestone, New Mexico, USA. Finger for scale is 1 cm wide.

in protected environments at various times in the Phanerozoic. Particularly numerous in the late Paleozoic, they were constructed of leaf-like phylloid algae and grew as low-relief structures in shallow-water environments. Similar reefs composed of the green alga *Halimeda* are today found in illuminated nutrient-rich neritic settings and were also common in comparable but stressed Miocene environments (see Chapter 14).

Mud mounds

These reefs mounds (Figure 15.12) have no modern counterparts. Dominated by carbonate mud and a few delicate skeletons, they are usually found in outer ramp or downslope localities. They were large (commonly over 100 m high and 400 m wide), isolated, steep-sided structures that consisted of more than 80% mud. Mud-mound formation began in the Paleoproterozoic (probably Neoproterozoic) and extended until the Miocene, but they were predominantly a Paleozoic phenomenon. Mud mounds were particularly common in the early Cambrian, Early and Middle Ordovician, Late Devonian and the Mississippian.

Mud mounds are among the most puzzling of reefs. The micrite has both *in situ* (autochthonous) and detrital (allochthonous) attributes, but commonly shows accretionary structures constructed by successive phases of deposition. These polygenetic muds ("polymuds") formed on open surfaces and within semi-enclosed cavities. Such polymud fabrics produce complex, 3D accumulations that in turn form open frameworks that can subsequently be filled with mud or synsedimentary cement. Many mounds also display a rich epifaunal metazoan biota of crinoids, tabulate corals, brachiopods, trilobites, sponges, ostracods, and bryozoans.

The clotted, peloidal or laminated textures, the encrusting or frame-forming structure, and the inferred high-Mg

Figure 15.12 (a) Muleshoe Mound, a ~100-m-high (original depositional relief) Mississippian reef mound with fenestrate bryozoans, carbonate mud, and extensive synsedimentary cement, New Mexico, USA. (b) Cross-section outcrop image of cavity-filling white synsedimentary cement (*Stromatactis*) occluding depositional cavities in an upper Silurian reef mound, Gaspé, Québec. Centimeter scale.

mineralogy of the mud all argue for an *in situ* microbial or organomineralic origin augmented by rapid synsedimentary lithification. Abundant *Stromatactis* (Figure 15.12b) cavities that parallel the accretionary mound surface suggest an intimate relationship between mound formation, internal sediment filled voids, cementation, and carbonate production.

Reef geohistory

Reefs built by microbial stromatolites are first recognized in Archean rocks and extend throughout the Precambrian (Figure 15.13), with poorly understood but distinct temporal trends in their attributes. Such structures are detailed in Chapter 18. The history of Phanerozoic reefs is complex, largely because of biotic evolution and the changing chemistry of ancient oceans.

Early Paleozoic

Early Cambrian reefs (Figure 15.14) are astonishingly modern in structure with skeletal invertebrates (archaeocyathans), calcimicrobes, synsedimentary cements, internal cavities, and borings. These skeletal-calcimicrobial mounds disappeared at the end of the early Cambrian and the reef system regressed to a Precambrian style with middle and upper Cambrian buildups dominated by stromatolites and calcimicrobes. This gradually changed during the Lower Ordovician

with the evolution of corals, bryozoans, and stromatoporoids that locally constructed small skeletal-microbial mounds.

Middle Paleozoic

The Middle and Late Ordovician, Silurian, and Devonian represent the acme of Paleozoic reefs. Most were skeletal reefs and skeletal-calcimicrobial mounds. The prominent reef-builders were a consortium of corals, stromatoporoids, and calcimicrobes (Figures 15.1, 15.2).

Paleozoic stromatoporoid sponges, tabulate corals, and rugose corals sat on or were anchored in the sediment. Tabular stromatoporoids, locally bound together by calcimicrobes, automicrite, or cement and solidly rooted in sediment, inhabited high-energy zones. Domal, bulbous, and dendroid forms occupied quiet water zones, either below wave base or in sheltered areas of the back reef. The lagoon was populated by delicate stick-like amphiporids. These skeletons were locally reworked during cyclonic storms and redeposited as rubble. Stromatoporoids did not typically have an encrusting habit, so it is doubtful that they were successful builders in the surf zone.

This period of seemingly unrestrained reef growth was gradually brought to an end during the Late Devonian by the progressive extinction of various stromatoporoid groups. As a result, the latest Devonian buildups were largely downslope mounds dominated by calcimicrobes.

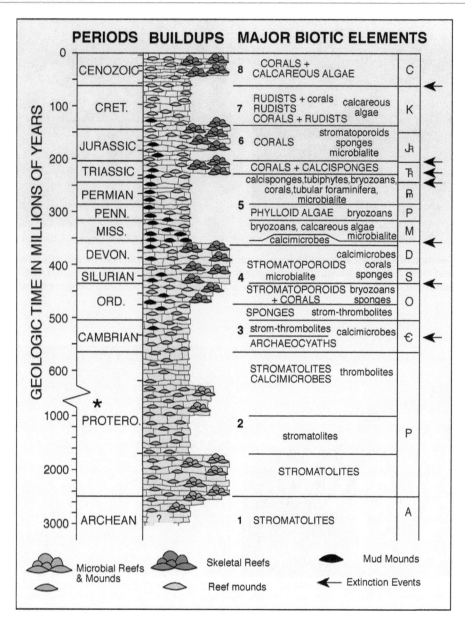

Figure 15.13 An idealized stratigraphic column representing geologic time, illustrating periods when there were only skeletal microbial reefs (biogenic mounds) and when there were both skeletal-microbial and skeletal reefs. Numbers indicate different associations of different reef- and mound-building taxa, arrows: significant extinction events; * scale change. Source: James and Wood (2010). Reproduced with permission of the Geological Association of Canada

Late Paleozoic

With the extinction of invertebrates capable of forming large calcareous skeletons, reefs during the late Paleozoic were mostly skeletal mounds, calcimicrobial-skeletal mounds, and mud mounds. The changing biota, however, meant that each period had a distinctive style. There were no reef-rimmed platforms during these times,

although some carbonate sand shoal-rimmed structures were present.

Shallow-water Mississippian reefs were constructed by various endemic communities that were constructed of laminated microbial mounds with a rich encrusting open surface and cryptic fauna dominated by algae, bryozoans, corals, and sponges. This time is better known for the development of spectacular outer ramp mud

Figure 15.14 (a) Cross-section of a lower Cambrian bioherm, southern Labrador, Canada. Staff increments 10 cm. (b) Slab of boundstone from the reef illustrated in (a) showing the intergrowth of archaeocyaths (coralline sponges) (A), the calcimicrobe *Renalcis* (R), red carbonate mud, and a growth cavity (C) partially filled with internal sediment. Centimeter scale.

Figure 15.15 (a) The exposed front of the Permian reef complex, New Mexico, USA illustrating the massive reef (R) and sloping fore-reef beds. Cliff is ~100 m high. (b) Outcrop image of reef facies dominated by calcareous sponges (S) and encrusting *Tubiphytes* (T), a possible calcimicrobe, and synsedimentary cement. Centimeter scale.

mounds, sometimes called *Waulsortian reefs*, that rose up to 150 m above the ancient seafloor. Most of these mud mounds were composed of polymuds with prolific marine cements (cementstones), sponges, and large fenestrate bryozoans. Crinoid-rich grainstone beds draped the mound slopes.

Although bryozoan-rich mud mounds continued to grow in deeper-water environments during the Pennsylvanian, newly evolved phylloid algae constructed small shallow-water skeletal mounds. These buildups,

which rose as much as 30 m above the seafloor, were replete with synsedimentary marine cements.

Widespread skeletal mounds and calcimicrobial mounds with a well-developed zonation characterized the middle–late Permian (Figure 15.15). Such mounds consisted, in part, of a primary framework of frondose bryozoans and calcified sponges (many of which were cavity-dwellers) that were bound by extensive crusts of laminated automicrite. *Tubiphytes* was abundant, together with various encrusting algae including *Archeolithoporella*.

Many of these reefs were volumetrically dominated by sediment and synsedimentary cement. This was, nevertheless, a time of significant carbonate platform growth and many of these structures had steep flanks. The platform margin and upper slope was a skeletal-calcimicrobe biostrome-like facies of calcimicrobes (*Donezella, Renalcis, Girvanella*, and *Ortonella*) as well as calcareous algae (phylloids and other types) with widespread synsedimentary cement, and extended across a gentle roll-over at the platform edge (below wave base) and downslope.

Figure 15.16 (a) Mountains composed of Upper Triassic reef and inter-reef limestones with patch reefs composed of scleractinian corals and calcareous sponges in the Alps. The mountains are ~200 m high. (b) Slab of Upper Triassic reef displaying numerous large dendritic scleractinian corals (*Retiophyllia* sp.) near Salzburg, Austria. Hammer handle is 15 cm long. Photographs by R. Martindale. Reproduced with permission.

Early Mesozoic

The early Mesozoic (Triassic–Jurassic) was a dramatic time in reef evolution because many organisms that had been instrumental in the development of late Paleozoic reefs became extinct and a new cadre of modern reef-builders emerged. The modern reef system has its beginnings at this time.

The first metazoan reefs to form after the end-Permian extinction were similar to late Permian buildups, with large framework cavities and large calcified microbial colonies. Middle and Upper Triassic carbonate platforms were among the largest in Earth's history. Middle Triassic skeletal reefs were built by new and improved stromatoporoids (unrelated to their Paleozoic ancestors), Permian-style calcareous sponges, and newly evolved colonial scleractinian corals with photosymbionts, together with the enigmatic encruster *Tubiphytes*. Upper Triassic reefs (Figure 15.16) were dominated by colonial corals, stromatoporoids, bryozoans, and calcareous algae with few Paleozoic holdovers. They were the first reefs to exhibit extensive effects of bioerosion by bivalves and algae. All Triassic reefs contained significant quantities of synsedimentary cement.

This reef biota was severely affected by the end-Triassic extinction and did not recover until the Middle Jurassic. After this time, reefs became progressively more complex with shallow-water reefs constructed by corals, stromatoporoids, red algae (*Solenopora*), and the most diverse and prolific suite of green calcareous algae in geologic history. Thrombolitic-stromatolitic columns or hemispheres constructed Upper Jurassic deep-water reefs that also included *Tubiphytes*, worms, and hexactinellid and lithistid sponges. The platy scleractinian coral *Microsolena*, which is thought to have been a mixotroph, grew in low-light settings such as shallow turbid or deep-water environments.

Late Mesozoic

Jurassic-style reefs persisted into the Early Cretaceous but shallow-water environments were gradually taken over by rudist bivalves (Figure 15.17). These extraordinarily large mollusks (often well over 1 m in length) dominated reef growth throughout the middle and Late Cretaceous. It is not certain if these large bivalves were mixotrophs like modern giant clams of the Pacific (*Tridacna*). The skeletal mounds that they constructed were widespread across platforms that rimmed the global Tethys Ocean; corals were restricted to deeper-water settings. Cretaceous rudist reefs did not form platform rim facies; that environment was usually occupied by benthic foraminiferal-rich sand shoals.

Cenozoic

The catastrophic end-Cretaceous extinction, together with the blossoming of calcareous plankton, changed the reef system forever and ushered in the modern reef archetype. Rudist bivalves disappeared, calcisponges (stromatoporoids) were decimated (today, only grow in deep and cryptic habitats), calcimicrobes became unimportant, and synsedimentary cements, although still critical, were never again so widespread; there are few cementstones in Cenozoic reefs. The diversity and numbers of scleractinian coral genera reached their maximum in the Oligocene. Although most Cenozoic reefs are modern in style (Figure 15.18), the

Figure 15.17 A reef mound of Cretaceous rudist bivalves, Oman. Centimetre scale.

Pliocene appearance of fast-growing branching corals such as the Acroporidae, Poritidae, and Seriatoporidae dramatically changed the reef crest community.

Reef rock classification

Folk and Dunham's limestone classifications (see Chapter 4) were not designed to deal with the different types of rocks that can result from the growth of reefs. As a result, other schemes have been developed to deal with these rocks. Such schemes include those developed by Embry and Klovan (1971) who expanded Dunham's scheme, Insalaco (1998) who modified Embry and Klovan's scheme, and Wright (1992) who developed an entirely different system.

Embry and Klovan's reef classification

This widely used 1971 scheme (Figure 15.19) builds on the Dunham classification and is applicable to reef rocks built by a variety of organisms. Terms for rocks, where the components are organically bound during deposition (i.e., boundstones), range from objective to subjective. *Framestones* are composed of large skeletons that are stacked on top of each other. *Bindstones* are formed by laminated or sheet-like anastomozing growth forms with cavities that can be filled with sediment or cement or both. This term is somewhat subjective because, in many cases, the subhorizontal skeletons are simply interlayered and not bound in

Figure 15.18 (a) The edge of a small uplifted Miocene carbonate platform along the Egyptian side of the Gulf of Suez, part of which is a scleractinian coral reef (R). Cliff is ~30 m high. (b) Outcrop image of the scleractinian coral framestone core of the reef illustrated in (a). Centimetre scale.

EMBRY AND KLOVAN REEF ROCK CLASSIFICATION

BAFFLESTONE   **BINDSTONE**   **FRAMESTONE**

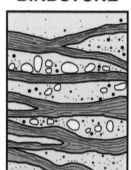

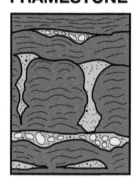

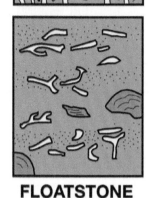

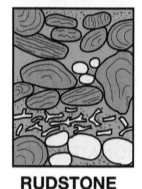

REEF LIMESTONE

FLOATSTONE   **RUDSTONE**

Figure 15.19 Classification of reef and reef-associated carbonate rocks proposed by Embry and Klovan (1971). Source: Embry and Klovan (1971). Reproduced with permission of the Canadian Society of Petroleum Geologists.

any way. *Bafflestones* comprise upright branching growth forms and reflect the idea that these seafloor forests baffled water movement, causing suspended sediment to fall and accumulate between the branches. In many cases there is simply no evidence for such processes, especially in deep-water muddy reefs and mounds, and so this is best used simply as a descriptive term.

Rudstones are skeletal conglomerates with the large particles touching, whereas the large skeletons in *floatstones* are suspended in finer sediment. Epithets are commonly used to differentiate types, for example, stromatoporoid framestone or coral floatstone with a trilobite wackestone matrix.

Insalaco's scleractinian reef classification

This 1998 classification (Figure 15.20) is a modification of the Embry and Klovan (1971) scheme and is specifically designed for post-Triassic scleractinian coral reefs but is potentially applicable to other reefs in the geological record. Rocks are divided into autochthonous (reef core) and allochthonous (reef margin particulate sediment) lithologies with each group divided into rock types. Skeletons in the autochthonous group comprise >60% of the reef core by volume. Rock types are differentiated by the dominant skeletal growth form with the particular fossils as epithets (e.g., coral sheetstone). Also important is the recognition of two end-member scleractinian coral growth fabrics, although they do not figure in the rock name. *Constratal growth* is where vertical growth is at about the same rate as sedimentation and so the skeleton is embedded in the sediment with only a small proportion (centimeters) peeking up above the sediment (biostromes). *Suprastratal growth* is when the skeletons project tens of centimeters to several meters above the seafloor, creating positive bathymetric relief (bioherms).

Wright's holistic classification

This classification (Wright 1992) is based on the concept that limestone textures result from the integration of deposition, biological activity, and diagenesis (Figure 15.21).

INSALACO SCLERACTINIAN REEF ROCK CLASSIFICATION

ALLOCHTHONOUS		AUTOCHTHONOUS						
More than 10% of fragments > 1 cm in size		Facies dominated by growth fabric of *in situ* and *in growth position* calcareous skeletons						
Matrix supported	Clast supported	Horizontal platy to tabular skeletons (width:height) 5:1 to 30:1	Horizontal platy to lamellar skeletons (width:height) > 30:1	Domal to irregular, massive skeletons	Skeletons with vertical growth attributes (tubular and branching)		No one skeletal growth form dominates
						PILLARSTONE		
FLOATSTONE	RUDSTONE	PLATESTONE	SHEETSTONE	DOMESTONE	*Sparse*	*Dense*	MIXSTONE

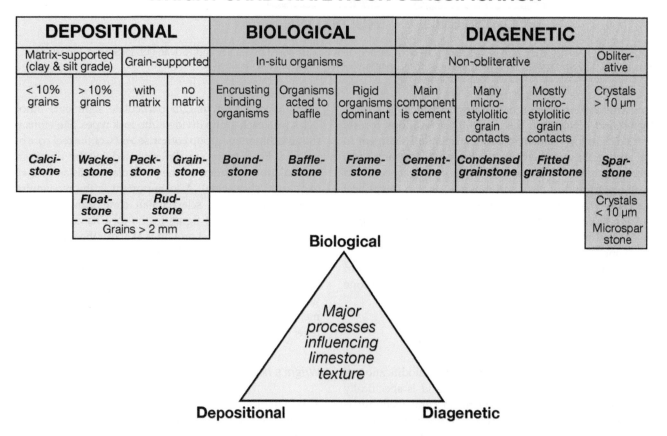

Figure 15.20 Insalaco's scleractinian reef classification. Source: Insalaco (1998). Reproduced with permission of Elsevier.

WRIGHT CARBONATE ROCK CLASSIFICATION

DEPOSITIONAL				BIOLOGICAL			DIAGENETIC			
Matrix-supported (clay & silt grade)		Grain-supported		In-situ organisms			Non-obliterative			Obliter-ative
< 10% grains	> 10% grains	with matrix	no matrix	Encrusting binding organisms	Organisms acted to baffle	Rigid organisms dominant	Main component is cement	Many micro-stylolitic grain contacts	Mostly micro-stylolitic grain contacts	Crystals > 10 μm
Calci-stone	*Wacke-stone*	*Pack-stone*	*Grain-stone*	*Bound-stone*	*Baffle-stone*	*Frame-stone*	*Cement-stone*	*Condensed grainstone*	*Fitted grainstone*	*Spar-stone*
	Float-stone	*Rud-stone*								Crystals < 10 μm Microspar stone
	Grains > 2 mm									

Biological

Major processes influencing limestone texture

Depositional   Diagenetic

Figure 15.21 Classification of all carbonate rocks proposed by Wright. Source: Wright (1992). Reproduced with permission of Elsevier.

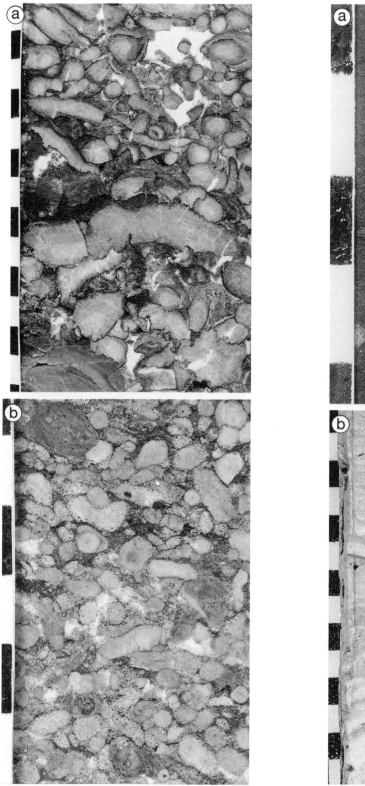

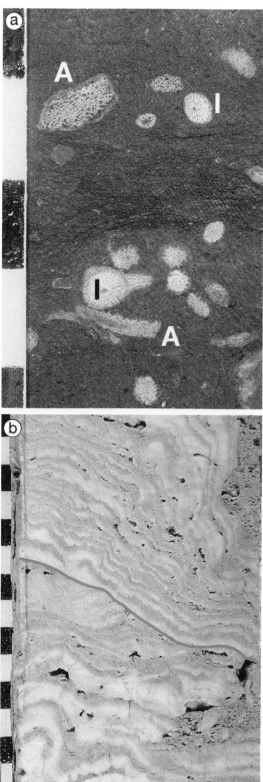

Figure 15.22 Devonian stromatoporoid rudstones in subsurface core from the Western Canadian Sedimentary Basin: (a) *Idiostroma* sp. rudstone; and (b) *Stachyodes* sp. rudstone. Centimeter scale. Photographs by W. Martindale. Reproduced with permission.

Figure 15.23 (a) Devonian stromatoporoid (A: *Amphipora* sp., I: *Idiostroma* sp.) floatstone in subsurface core from the Western Canadian Sedimentary Basin. Centimeter scale. (b) Devonian stromatoporoid framestone in subsurface core from the Western Canadian Sedimentary Basin. Blue line indicates the contact between different stromatoporoids growing on top of one another. Centimeter scale. Photographs by W. Martindale. Reproduced with permission.

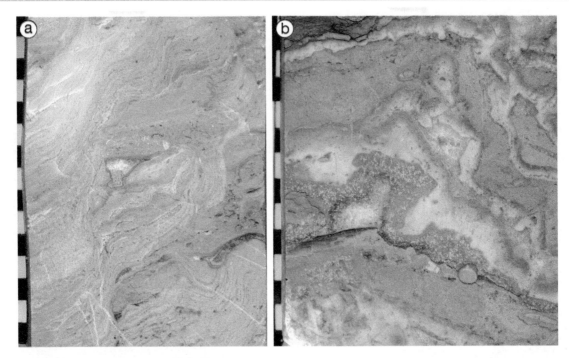

Figure 15.24 Devonian stromatoporoid boundstones in subsurface core from the Western Canadian Sedimentary Basin. Laminar stromatoporoids (a) with grey sediment between; and (b) locally attached to one another with cavities beneath partially filled with pendant *Renalcis* sp. (speckled material). Centimeter scale. Photographs by W. Martindale. Reproduced with permission.

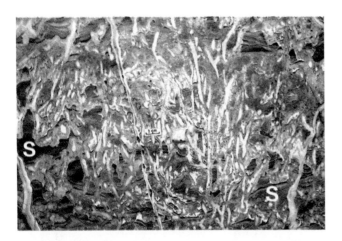

Figure 15.25 Vertical outcrop image of an Upper Devonian bafflestone from the Canning Basin, Western Australia, composed of upright (growth position) digitate stromatoporoids (*Stachyodes* sp.) surrounded by the calcimicrobe *Renalcis* sp. (speckled) and where the remaining pore spaces were filled with internal sediment (S). Image width 15 cm. Photograph by P.W. Playford. Reproduced with permission.

The depositional attributes come from Dunham's classification, the biological relationships from Embry and Klovan's scheme, and Wright added the diagenetic terms.

The diagenetic terms recognize contributions from different diagenetic realms. *Cementstone* recognizes the volumetric importance of synsedimentary cement in some reef limestones; cement volume commonly exceeds that of skeletal elements. *Condensed grainstone* and *fitted grainstone* refer to the importance of chemical compaction during burial, where the particles are no longer in depositional contact but are separated mostly or mainly by stylolites, respectively. Those rocks that are so altered as to have no recognizable components are called *sparstones* (crystals >10 μm) or *microsparstones* (crystals <10 μm).

The different lithologies of these classification schemes are illustrated in Figures 15.22–15.26.

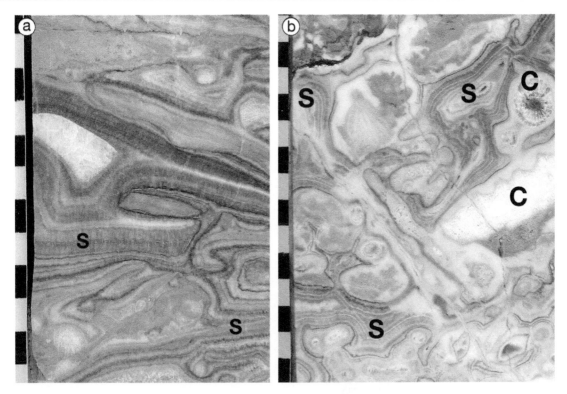

Figure 15.26 Devonian cementstone in subsurface core from the Western Canadian Sedimentary Basin. (a) Laminated synsedimentary fibrous cements (S) filling pore space between stromatoporoids. Centimetre scale. (b) Banded synsedimentary fibrous cements (S) filling pore space between colonial rugose corals (C) and stromatoporoids. Photographs by W. Martindale. Reproduced with permission.

Further reading

Embry, A.F. and Klovan, J.E. (1971) A Late Devonian reef tract on northeastern Banks Island, N.W.T. *Bulletin of Canadian Petroleum Geology*, 19, 730–781.

Fagerstrom, J.A. (1987) *The Evolution of Reef Communities*. New York: John Wiley and Sons.

Geldsetzer, H.H.J., James, N.P., and Tebbutt, G.E. (eds) (1988) *Reefs, Canada and Adjacent Areas*. Canadian Society of Petroleum Geologists, Memoir no. 13.

Insalaco, E. (1998) The descriptive nomenclature and classification of growth fabrics in fossil scleractinian reefs. *Sedimentary Geology*, 118, 159–186.

James, N.P. (1983) Reef environment. In: Scholle P.A., Bebout D.G., and Moore C.H. (eds) *Carbonate Depositional Environments*. Tulsa, OK: American Association of Petroleum Geologists, Memoir no. 33, pp. 345–440.

James, N.P. and Wood, R. (2010) Reefs. In: James N.P. and Dalrymple R.W. (eds) *Facies Models 4*, Geological Association of Canada, GEOtext vol. 6, pp. 421–447.

Keissling, W., Flügel, E., and Golonka, J. (eds) (2002) *Phanerozoic Reef Patterns*. SEPM (Society for Sedimentary Geology), Special Publication no. 72.

Monty, C.L.V., Bosence, D.W.J., Bridges, P.H., and Pratt, B.R. (eds) (1995) *Carbonate Mud-mounds: Their Origin and Evolution*. Oxford: Blackwell Science, International Association of Sedimentologists, Special Publication no. 23.

Stanley, G.D.J. (ed.) (2001) *The History and Sedimentology of Ancient Reef Systems*. New York, Boston, Dorchecht, London, Moscow: Kluwer Academic/Plenum Publishers, Topics in Geobiology.

Toomey, D.F. (ed.) (1981) *European Fossil Reef Models*. Tulsa, OK: Society of Economic Paleontologists and Mineralogists, Special Publication no. 30.

Wright, V.P. (1992) A revised classification of limestones. *Sedimentary Geology*, 76, 177–185.

CHAPTER 16
CARBONATE SLOPES

Frontispiece The submersible *Alvin* surfacing after a dive to 1500 m in the Tongue of the Ocean, Bahamas. Unlike shallow neritic environments, such manned and unmanned vehicles are the only way to investigate details of modern carbonate slope environments.

Origin of Carbonate Sedimentary Rocks, First Edition. Noel P. James and Brian Jones.
© 2016 Noel P. James and Brian Jones. Published 2016 by John Wiley & Sons, Ltd.
Companion website: www.wiley.com/go/james/carbonaterocks

Introduction

A carbonate slope is the region of the seafloor that stretches from the shallow, wave-swept, sunlit platform rim downward into progressively deeper, darker, and colder waters to finally merge with the basin floor, which can be hundreds to thousands of meters below sea level. Understanding these deposits poses a challenge. Unlike shallow-water settings, these environments cannot be reached easily by personal diving and, in the modern world, can only be studied using submersibles or remotely operated vehicles (Frontispiece). The seafloor on the slope is not a uniform environment, but progressively alters with increasing depth due to increasing hydrodynamic pressure, falling temperature, varying nutrients, and changing oxygen levels. Slope carbonates are therefore not one simple rock type but a suite of different lithologies. Ancient slope deposits are, with exceptions, exposed largely in orogenic belts and their complexities must be carefully unraveled in these complex tectonic settings. Finally, although locally important, slope deposits are not primary hydrocarbon exploration targets and have therefore been studied less than other more economically important carbonate rocks.

Depositional bathymetry

The seafloor along the edge of a rimmed carbonate shelf comprises the platform edge or margin, the slope (or foreslope), the toe-of-slope, and the basin. The shallow (<70 mwd) platform edge is either: (1) a photozoan factory of reefs or carbonate sand shoals and associated facies; or (2) a heterozoan factory of carbonate sands. Alternatively, the margin can be a gentle slope with the buildup of a deep (50 to >100 mwd) oligophotic structure that is marginally a photozoan but mostly a heterozoan carbonate factory of microbial carbonates.

The deposits

Sediments on the slope (Figure 16.1) are a mixture of fallout from overlying waters, erosional products from local reef-like biostromes growing on the upper slope, minor benthic production, and resedimented material episodically swept downslope from the adjacent platform and platform-margin reefs and shoals. The deposits are not generally disposed in submarine fan geometries because the origin of sediment is not a point source such as a submarine canyon. Instead, the platform margin is usually a linear complex (e.g., a barrier reef) that produces

CARBONATE SLOPE ENVIRONMENT

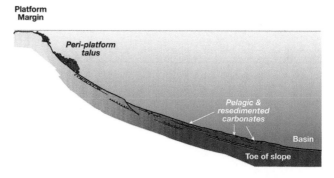

Figure 16.1 The major deposits on a carbonate slope, peri-platform talus, and a variety of resedimented and pelagic sediments that accumulate either on the slope itself or at the toe of slope where the gravity drive is removed.

sediment along its entire length. Thus, in most cases the slope deposits are a laterally continuous sediment apron seaward of the platform.

The deposits are various combinations of carbonate debris, grainy particles, carbonate muds, and shales (Figure 16.2). Local oceanographic conditions determine the character of the finer-grained sediments. Oxygenated settings have burrowed carbonates and red bioturbated shales; dysoxic environments comprise mainly structureless carbonates, and green shales, and dark, anoxic seafloors are typified by laminated, organic carbonates, and shales. The sediment dynamics are characterized by long quiescent time intervals of fine-grained sediment fallout, interrupted by episodic gravity flows or event beds that bring sediment ranging from sands to conglomerates downslope (Figure 16.3).

Muddy carbonates

Fine-grained sediment that mantles Phanerozoic slopes is predominantly derived from the adjacent platform (peri-platform ooze) (Figure 16.4a) and from the remains of pelagic calcareous phytoplankton and zooplankton (planktic ooze, middle Mesozoic–Holocene). When more than 25% of the sediment is siliciclastic (mainly silts and clays) the deposit is termed *hemipelagic*, even though it is still fine-grained. Although these sediments can remain soft for some time depending on conditions, they can also be lithified while on the seafloor

Peri-platform ooze. The neritic carbonate factory that operates across large areas of shallow platforms produces copious quantities of carbonate mud, especially in photozoan systems (see Chapter 3). This mud is easily

Figure 16.2 (a) The precipitous Upper Devonian reef margin in the Canning Basin, Western Australia, fronted by inclined carbonate slope deposits (beds are not rotated by tectonics). Total section is ~40 m thick. (b) The three basic elements of Paleozoic and early Mesozoic carbonate slope deposits – shales (S), ribbon lime mudstones (R), and carbonate debrites (D) – in the Lower Ordovician Cow Head Group, western Newfoundland, Canada. Circled backpack 25 cm high.

CARBONATE SLOPE SEDIMENTATION

Long periods of boredom alternating with short periods of terror

Figure 16.3 Sketch of the major sedimentation processes on carbonate slopes where long periods of fine sediment fallout alternate with short periods of chaotic deposition via sediment gravity flows.

put into suspension during strong cyclonic storms (Figure 16.4b), during which times the water on the platform looks like milk. From the air, it can be seen streaming out across the platform margin between sand shoals and reefs to be either transported elsewhere by surface currents or moved out to sea where, with time, the fine particles settle out of suspension.

These sediments in modern carbonate systems, often with floating grass blades (Figure 16.4c), are transported to the adjacent seafloor in several ways (Figure 16.4b). Simple fallout, where most of the material suspended in the water column settles onto the slope, is common. The amount of sediment deposited in this way decreases with increasing distance away from the platform margin. If the open water is density stratified, the fine sediment can fall

to the pycnocline and stay there, only to settle out later in somewhat more distal deep-water environments. Finally, if the sediment–water suspension is dense enough, it can move downslope as a sediment gravity flow and spread out across the lowermost slope and adjacent basin as a mud turbidite.

Planktic ooze. Muds generated by the disintegration of calcareous plankton are detailed in Chapter 17. They comprise the whole and partly dissolved skeletons of planktic foraminifers, the spheres, shields, and crystallites of coccolithophorids (single-celled green algae), and less-numerous shields and cones of pteropods (microscopic swimming gastropods) (Figure 16.4d). This material differs in two important ways from peri-platform ooze. It is produced in the upper part of the water column everywhere and is therefore not localized to the slope, but also forms basinal sediment above the CCD. It is, however, time-limited. Calcareous plankton did not evolve until the middle Mesozoic; before that time there was no significant source of pelagic carbonate and all carbonate muds came from the platform. Planktic carbonate is the major source of muds on modern temperate carbonate slopes because not much mud is produced on temperate platforms by benthic organisms.

Deposits. Rocks formed by lithification of planktic ooze and peri-platform ooze typically consist of clay-sized and silt-sized carbonate mudstones, chalks, and marls, with wackestone to muddy packstone textures. Beds are centimeter-scale with differential cementation, encrustation, and chert. Most are laminated to thinly bedded, and nodular bedding is common (Figure 16.5).

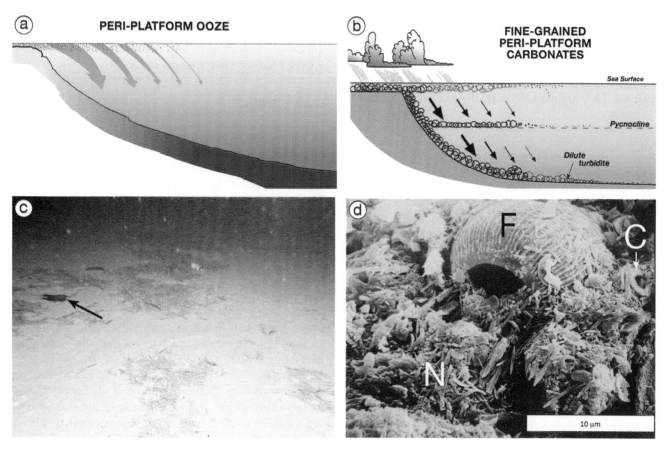

Figure 16.4 (a) Sketch illustrating the source of peri-platform ooze that diminishes in quantity away from the platform margin. Adapted from Coniglio and Dix (1992). (b) Sketch illustrating the various ways in which fine-grained neritic carbonate sediment is transported to the slope and basin. (c) Seafloor image of muddy sediment at ~900m depth in the Tongue of the Ocean, Bahamas, with patches of seagrass that have been eroded from the shallow platform top and settled down into deeper water in the same manner as the fine peri-platform mud. Animal in background (arrow) is 10cm long. (d) SEM image of peri-platform ooze from the Tongue of the Ocean, Bahamas composed of large planktic foraminifers (F), small coccolith disks (C), and fine benthic aragonite needles (N) from the shallow platform above.

Figure 16.5 Evenly bedded peri-platform lime mudstones separated by thin shales or argillaceous dolomites (ribbon or parted limestone) in the Lower Ordovician Cow Head Group, western Newfoundland, Canada. Hammer length 15cm.

Grainy carbonates

Origin. Grainy sediment on slopes, contributed entirely by sediment gravity flows, is derived for the most part from the platform margin complex. The material in tropical photozoan systems comes from ooid shoals, reefs, the upper slope, and from protected neritic environments behind them, and consists of ooid, skeletal and peloidal sands, and reef clasts. This shallow-water sediment is swept outward by strong tidal currents flowing through reef passes and across the shoals, and can reside for short periods on the uppermost slope. These particles are then episodically transported into deeper water by grain flows or turbidity currents (Figure 16.6).

Similar processes operate on temperate heterozoan open platforms, except that the source of sediment can be the entire outer platform. These sediments are dominated by skeletal remains and are also transported into deeper water by currents and storms.

Deposits.   The centimeter- to decimeter-scale beds are sand- and gravel-dominated, with grainstone to packstone textures. They are massive, normally graded, or inversely

Figure 16.6 Graded calciturbidite overlain by dolomitized grainstone, Permian, New Mexico, USA. Hammer 13 cm long. Photograph by P.A. Scholle. Reproduced with permission.

graded, with low-angle cross-laminations to planar laminations. Ta–Tb Bouma sequences, imbricated clasts, and grains are typical. Such successions are also commonly interbedded with hemipelagic sediments.

Coarse debris

Origin.   Coarse carbonate breccias and conglomerates interbedded with muds and sands are a hallmark of many slope deposits. They are sediment gravity-flow deposits or, more specifically, the deposits of coarse-grained turbidity currents and debris flows (Figure 16.7). Deposits are generally poorly sorted, massive, chaotic, ungraded sediments in the form of thick-bedded lenticular beds.

Clasts are *margin derived* from shallow-water neritic limestones and/or upper slope biostromes, or *slope-derived* from the fragmentation of lithified thin-bedded lime sands and muds (Figures 16.7, 16.8). Fragments range from cobble-sized to larger than a house. The largest, up to >100 m in size, are called *olistoliths* (Figure 16.8). Debrites that have an abundance of meter-scale or larger clasts that were generated by gravitational collapse or mass wasting of lithified carbonate from shallower water are termed *megabreccias*.

Fragments from shallow margin and adjacent facies are either calcarenites or reef rocks that were lithified by syn-sedimentary marine cements or calcite cements during

TYPES OF SLOPE CONGLOMERATE

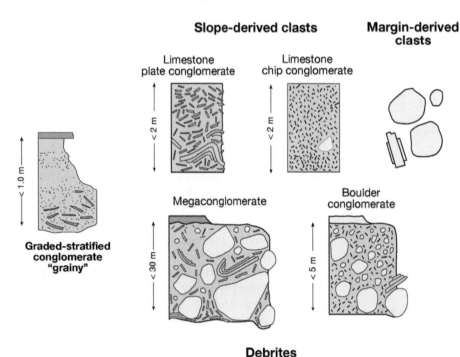

Figure 16.7 Sketch of the different types of slope conglomerates deposited from sediment gravity flows on carbonate slopes.

Figure 16.8 Olistolith (large isolated shallow-water limestone boulder (O)) in Triassic slope lime mudstones, northern Italy.

apart during transport because of the dispersive pressure of the fluidized mud matrix. This fluid pressure is only maintained by gravity and as long as the flow is on a slope. Once the material reaches the base of the slope, the gravity drive ceases, the flow stops, and deposition is almost instantaneous. A debrite (Figure 16.9) is a completely unsorted chaotic mixture of clasts and matrix. This simple picture can be complicated by the upward injection of underlying, locally semi-lithified lithologies into the mixture. On the other hand, the tops of the beds can have a thin, graded calcarenite or calcisiltite cap (turbidites).

Contourites

Once deposited, slope and deep-water sediments can be reworked by bottom currents. Among the most important of these are contour currents, semi-permanent currents that flow along slope, parallel to topographic contours. Because of the Earth's rotation and the Coriolis effect, contour currents are most intense along the western margins of ocean basins (western boundary currents). The currents usually have speeds of $10–20 \, cm \, s^{-1}$, which is the threshold for transporting fine sand. Such currents are generally deeper than ~300 m in the modern ocean and the associated deposits are called *contourites*. These deposits can be stacked to form slope-parallel *sediment drifts*, large elongate sediment bodies that can be tens to hundreds of kilometers long, a few tens of kilometers wide, and up to 1 km thick.

Contourites have been most extensively studied in siliciclastic systems and are among some of the most difficult deposits to identify in the rock record, largely because of slow depositional rates that allow for intensive bioturbation and thus destruction of physical sedimentary structures. Individual deposits can be either muddy or sandy depending on the prevailing current speed. Sediment texture commonly alternates because of variations in current speed driven by Milankovitch cyclicity. The alternations are usually 20–30 cm thick and comprise a basal coarsening upward interval, consisting of a muddy and sandy contourite overlain by a fining upward succession that terminates in a muddy contourite (Figure 16.10). Contourites are difficult to differentiate from turbidites with which they can be interbedded, but helpful attributes include: (1) local gradational bases of ripple cross-stratified units; (2) intercalated layers or drapes of mud within the cross-stratified unit; (3) common internal erosion surfaces; (4) lack of an ordered succession like turbidites; and (5) internal condensed horizons with Fe-Mn-PO_4 encrustations and hardground ichnofauna. The most diagnostic indicator is the presence of along-slope paleocurrent indicators. Physical sedimentary structures can, however, be masked or destroyed by the

periodic subaerial exposure. Those from upper slope boundstones have similar attributes but generally contain more microbial components. All such clasts can be somewhat rounded via clast-to-clast collision during transport (Figure 16.9a, b). Slope-derived limestone clasts, by contrast, are usually well-bedded carbonate mudstone (Figure 16.9c). Such slope lime mudstone clasts are generated by seismicity or over-steepening of the slope. They are in the form of small slabs (flat clast conglomerates), fragments of bedded lime mudstone, or, in exceptional circumstances, huge twisted rafts of slope muds. The fragments originate upslope as slumps and then can move downslope. In some cases, the sediments do not move very far downslope and therefore remain as slump features. In most cases, however, they leave subtle slump scars that in the rock record are referred to as "intraformational truncation surfaces" (Figure 16.9d).

Deposits. Debrites, the deposits of debris flows (Figure 16.7), are of a wide variety of clasts that are held

Figure 16.9 (a) Massive debrite in the Lower Ordovician Cow Head Group of western Newfoundland, Canada composed of platform margin boulders (B), twisted rafts of parted slope limestone (R), and argillaceous limestone matrix. Circled sledge hammer 10 cm long. (b) Massive Jurassic slope debrite in Oman (Musandam Peninsula), composed almost entirely of platform margin boulders and little matrix. (c) Debrite in the Lower Ordovician Cow Head Group composed entirely of broken carbonate slope mudstones (see Figure 16.2b, 16.5) that were deposited and lithified on the slope, and then eroded and redeposited in deeper water. Sledge hammer 10 cm long. (d) Intraformational truncation surface (arrows) in Lower Ordovician slope limestones, western Newfoundland, Canada.

CARBONATE CONTOURITE

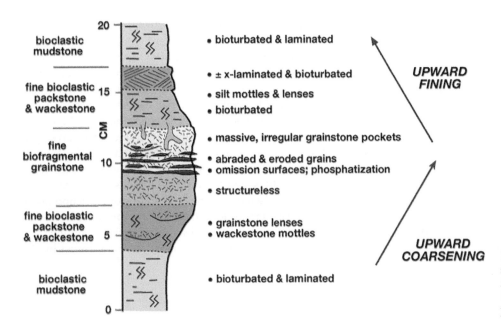

Figure 16.10 Conceptual cross-section of a contourite, synthesized from Robesco and Camerlenghi (2008). Reproduced with permission of Elsevier.

rain of calcareous plankton and increased abundance of infaunal burrowing in the Mesozoic and Cenozoic.

Slope types

Slopes can be divided into two main types that are largely dependent on the nature of the platform margin: (1) accretionary (depositional) slopes where much of the sediment accumulates on the slope itself; and (2) escarpment (bypass) slopes where the sediment flows across most of the slope and accumulates on the lower foreslope and in the adjacent basin (Figures 16.11, 6.12).

Accretionary slopes

Accretionary slope deposits (Figure 16.11) occur where the platform margin and slope interfinger and commonly have a prograding clinoform geometry. Deposits typically become thinner basinward. Excellent examples are found in the Permian of west Texas, USA and today along the leeward margin of the Great Bahama Banks in the Caribbean. Sediments on accretionary slopes are derived from either shallow reefs along the margin or from downslope biostromes.

Shallow reef-rimmed slopes. These slopes have a shallow sand shoal or reef rim that can extend from the intertidal zone to ~70 m depth. This region produces copious amounts of carbonate sand and mud. The slope is generally covered with a sediment apron that extends from the rim to the toe of slope as a series of stacked sheet-like units. The system, fed from numerous sources, can extend tens to hundreds of kilometers along-strike. The deposits pass from grain-dominated upslope to mudstone in the basin.

Upper slope biostromes. These slopes do not generally have a shallow reef rim, but instead are a series of bioclast-ooid-peloid sand shoals with a reef-like facies downslope. They were particularly common between the Late Devonian and the Early Triassic when large skeletal reef-builders were absent (see Chapter 15). The upper slope in the oligophotic zone, at estimated water depths of 50–300 m, was characterized by slope boundstones or biostromes rather than loose sediment.

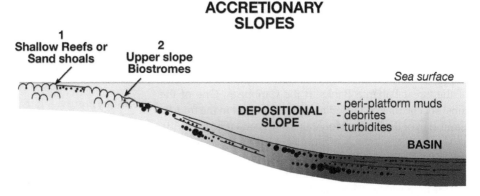

Figure 16.11 Sketch of the geometry and deposits on an accretionary carbonate slope with (1) shallow-water reefs or (2) upper slope biostrome. Adapted from James and Mountjoy (1983). Reproduced with permission of the SEPM Society for Sedimentary Geology.

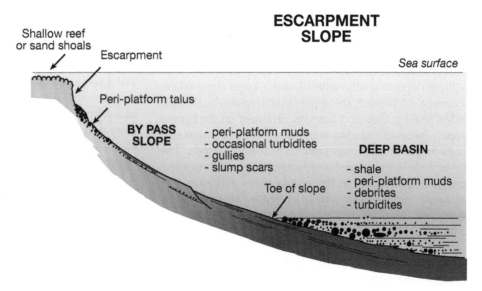

Figure 16.12 Sketch of the geometry and deposits on a large carbonate escarpment bypass slope. Adapted from James and Mountjoy (1983). Reproduced with permission of the SEPM Society for Sedimentary Geology.

Whereas a shallow reef is profoundly constrained by variations in accommodation, especially those produced by sea-level change, upper slope biostromes composed of reef mound facies are not. The upper slope environment is therefore one of continued growth, collapse, and production of debris. It is postulated that such biostromes grew in regions of upwelling and somewhat elevated trophic resources, and so their lateral growth exceeded that of shallow platform margin skeletal reefs. Such a situation promoted rapid lateral slope progradation. Most margins of this type are debrite-dominated aprons that extend from the deep-water biostromes to the base of slope. The stratigraphic package is characterized by numerous phases of gravitational collapse, scarps, scallops, and healed failure features.

Escarpment bypass slopes

Escarpment-type slopes (Figure 16.12) occur where the coeval slope and platform margin are physically disconnected by a surface of non-deposition. The shallow facies are either shallow platform margin reefs or upper slope biostromes. The escarpment itself can be *inherited* from earlier underlying platforms (possibly along faults) or it can be the result of differential platform *growth* where shallow-water facies accreted orders of magnitude faster than adjacent basinal facies. It can also form via margin collapse in the form of large submarine embayments. Examples in the rock record include the Late Devonian of the Canning Basin, Western Australia and today along the windward margins of the Great Bahama Banks in the Caribbean. One of the most spectacular collapsed margins is present in the middle Cambrian of the Rocky Mountains of Canada (Figure 16.13).

Inherited.   Inherited margins have a cliff zone directly seaward of the shallow platform rim that is derived from antecedent topography. They can be some of the largest slopes in the geological record with heights of >1 km, lengths of >10 km, and deposits extending many kilometers into the basin (e.g., the modern Bahamas). The system generally comprises an upper slope that is typified by net bypass where sediment gravity flows sweep across the slope without depositing any material, and a lower slope where sediment is deposited. The upper slope is typified by slumps, slides, small gullies, and canyons. Sediment on the bypass slope is generally muddy with conspicuous omission surfaces in the form of hardgrounds and firmgrounds. The lower slope is, by contrast, a sediment apron of debrites and grainstone-packstones in the form of sheets and lobes that are locally reworked by contour currents. Extensive debris sheets are interspersed through foreslope strata and can comprise up to one-quarter of the volume of these deposits.

Growth.   Synoptic relief of slopes is produced by rapid growth of the shallow-water platform factory through time

Figure 16.13 Cross-section outcrop of a middle Cambrian paleoescarpment slope (arrows) where part of the laterally extensive carbonate platform margin collapsed to form an escarpment that was later onlapped and filled with argillaceous limestones (this is the location of the famous Burgess Shale soft-bodied fossil site), Yoho National Park, front ranges of the Rocky Mountains, Canada. Cliff 400 m high. Photograph by D. Stewart. Reproduced with permission.

or progradation via upper slope biostromes. Differential relief can reach 1000 m or more and slope environments can be up to 5 km wide. The facies vary from mud- to grain-dominated, but debris is relatively minor because of margin stability through time.

Embayments.   Large-scale collapse of the margin (Figure 16.13) via seismic activity or over-steepening can generate large, irregular scalloped embayments many kilometers across. Flow is focused downslope and is characterized by channelization, sediment bypass, and point-source deposition that has many of the attributes of a lower-slope submarine fan.

Temporal and spatial variability

The geological record of these complex deposits is, as stated at the outset, an interplay of neritic, pelagic, and bathyal processes set against a background of continuing biological evolution. There is no one overriding system; each deposit is different and is the result of varying intrinsic and extrinsic influences.

Intrinsic controls

Secular geochemical and biotic variation would seem to be the most important intrinsic control. The most influential of these changes is the changing identity of the reef biota and the evolution of calcareous plankton.

Margin biotas. Archean and Proterozoic slope carbonate facies are surprisingly modern in aspect and include all of the attributes documented for Phanerozoic systems described above; the only difference is that the carbonate factory was completely microbial. Early Paleozoic slope facies are generally located at continental margins, yet contain some of the most spectacular deposits in the rock record. In many instances, their dynamics can only be partly understood because they are presently located in orogenic belts. The change to largely intracratonic and foreland basin carbonate deposition for the remainder of the Phanerozoic means that not only were basins shallower but that stratigraphic relationships were better preserved. Ordovician–Devonian slope facies were largely accretionary. The end-Devonian extinction of most large reef-building skeletons, however, changed the nature of slope deposition. The marginal facies ceased to be of importance and upper-slope biostromes dominated slope deposition until well into the Triassic. Reappearance of the reef biota in the Late Triassic and Jurassic led to a return of rim-dominated sedimentation, a feature that has prevailed to the present day.

Planktic carbonate. The Precambrian is an interesting case in slope deposition because pelagic carbonate mud seems to have been a significant part of the sedimentary record. The mud was either: (1) produced or influenced by microbes or organic matter in the water column; or (2) was precipitated inorganically because of high carbonate saturation levels in the ocean. By contrast, the early and middle Paleozoic was a time of no planktic carbonate production. Slope deposits of this age are therefore interbedded carbonates and shales; the only source of carbonate was from the platform. As a consequence, carbonates rapidly fade away basinward in Paleozoic slopes. The late Paleozoic had a few skeletonized planktic organisms, but they were not significant contributors to slope sediments. Much of the carbonate mud deposited on slopes was produced on and transported from platforms. Evolution of calcareous plankton in the Middle Jurassic had the result of not only adding to the mud component but also increasing the volume of carbonate slope deposits in general.

Extrinsic controls

The most important extrinsic controls are regional tectonics, local accommodation, oceanography, and external sediment input.

Tectonics. Particularly important here is tectonic setting. Slopes in regions of relatively quiescent tectonics are little affected by seismicity. Slopes in settings of active seismicity have numerous megabreccias and erosional embayments. Faulting, especially normal or growth faulting, often determines the character of escarpment margins either by generating strong bathymetric relief that is perpetuated through time or synsedimentary faulting during platform growth. Both types of faulting should be more common on newly rifted margins and in foreland basins and less common in intracratonic depocenters.

Accommodation. In simplest terms the carbonate platform factory exists in one of three states: exposed, flooded, or drowned (Figure 16.14). Accommodation is that space between the sea surface and the seafloor that is available for sediment to accumulate (see Chapter 19). At any one time, this space is an intricate balance between sea-level dynamics, subsidence rates, and sediment accumulation. When it is exposed, sea level lies below the platform rim, meteoric diagenesis and karst dissolution prevails on the platform top, and the carbonate factory is relegated to an ineffectual narrow platform on the slope; there is little to no slope deposition and it is either all pelagic carbonate or shale. When the platform is flooded, the neritic factory is operating at maximum output and carbonate sediment is delivered to the slope; slope deposition is at a maximum with the full spectrum of deposits. Upward platform growth or accretion dominates during times of high accommodation and comparatively little sediment is shed

**SEA LEVEL DYNAMICS
&
SLOPE/BASIN CARBONATE SEDIMENTATION**

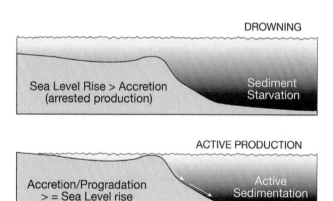

Figure 16.14 Sea-level dynamics and slope sedimentation. Slope sedimentation is most active when the neritic platform is relatively shallow and the platform is producing much sediment, a part of which is exported to the slope. When sea level falls below the platform edge, the factory is shut off; when it rises rapidly such that the factory falls below the photic zone, the slope is starved of sediment.

basinward. By contrast, phases of outward, basinward growth or progradation are periods of much slope deposition, especially debris, with the slope itself prone to collapse because of the comparatively rapid sedimentation.

When sea level rises so rapidly that the platform top is submerged below the photic zone, there is little sediment production and so the slope is starved of platform-derived sediment. In this situation, slope sedimentation is either totally pelagic carbonate or shale.

Oceanography.   A myriad of oceanographic factors can come into play and affect slope sedimentation. The most important seem to be wind direction, seafloor currents, and nutrient levels.

Wind direction has a profound effect on sedimentation, especially around offshore banks. Under an energetic wave climate, marginal neritic sands and gravels on the windward sides of platforms tend to be swept backward into the lagoon and hence the slopes are mostly mud-dominated. The leeward margin slopes on banks are sediment-poor because of the burying or poisoning of margin facies via off-bank transport.

Seafloor currents can winnow the slope sediment and promote whole-scale erosion of slope deposits. Slope-parallel contour currents, for example, can transport sediments laterally across and parallel to the slope for considerable distances.

Trophic resource levels mostly affect the carbonate-producing plankton with zones of upwelling, resulting in increased mud production that rains down on the slope. Similarly, increased nutrients promote growth of upper-slope biostromes. The slopes are consequently muddy and debris rich. Interbedded phosphates are possible if nutrients are abundant.

External sediment input.   The input of siliciclastic sediment has several effects, the main two being: (1) decreasing platform margin growth; and (2) adding fine sediment to the slope wedge, both of which tend to increase the likelihood of a mud-dominated slope.

Further reading

Coniglio, M. and Dix, G.R. (1992) Carbonate slopes. In: Walker R.G. and James N.P. (eds) *Facies Models: Response to Sea Level Change.* St John's, Newfoundland: Geological Association of Canada, pp. 349–373.

Cook, H.E., Hine, A.C., and Mullins, H.T. (eds) (1983) *Platform Margin and Deep Water Carbonates.* Tulsa, OK: SEPM (Society for Sedimentary Geology), Short Course no. 12.

Crevello, P.D. and Schlager, W. (1980) Carbonate debris sheets and turbidites, Exuma Sound, Bahamas. *Journal of Sedimentary Petrology,* 50, 1121–1148.

Della Porta, G., Kenter, J.A.M., and Bahamonde, J.R. (2004) Depositional facies and stratal geometry of an Upper Carboniferous prograding and aggrading high-relief carbonate platform (Cantabrian Mountains, N. Spain). *Sedimentology,* 51, 267–295.

George, A.D., Playford, P.E., and Powell, C.M. (1995) Platform-margin collapse during Famennian reef evolution, Canning Basin, Western Australia. *Geology,* 23, 691–694.

Grammer, G.M. and Ginsburg, R.N. (1992) Highstand versus lowstand deposition on carbonate platform margins: insight from Quaternary foreslope deposits in the Bahamas. *Marine Geology,* 103, 125–136.

Harris, M.T. (1994) The foreslope and toe-of-slope facies of the Middle Triassic Latemar buildup (Dolomites, northern Italy). *Journal of Sedimentary Research, Section B: Stratigraphy and Global Studies,* 64, 132–145.

James, N.P. and Mountjoy, E.W. (1983) Shelf-slope break in fossil carbonate platforms: an overview. In: Stanley D.J. and Moore G.T. (eds.) *The Shelf-break: Critical Interface on Continental Margins.* Tulsa, OK: SEPM Special Publication no. 33, pp. 189–206.

Kenter, J.A.M. (1990) Carbonate platform flanks: slope angle and sediment fabric. *Sedimentology,* 37, 777–794.

Payros, A., Pujalte, V., and Orue-Etxebarria, X. (2007) A point-sourced calciclastic submarine fan complex (Eocene Anotz Formation, western Pyrenees): facies architecture, evolution and controlling factors. *Sedimentology,* 54, 137–168.

Rebesco, M.A. and Camerlenghi, A. (eds) (2008) *Contourites.* Elsevier, Developments in Sedimentology no. 60.

Schlager, W. and Camber, O. (1986) Submarine slope angles, drowning unconformities, and self-erosion of limestone escarpments. *Geology,* 14, 762–765.

Spence, G.H. and Tucker, M.E. (1997) Genesis of limestone megabreccias and their significance in carbonate sequence stratigraphic models: a review. *Sedimentary Geology,* 112, 163–193.

CHAPTER 17
DEEP-WATER PELAGIC CARBONATES

Frontispiece A 40-m-high cliff of Upper Cretaceous chalk composed mainly of calcareous plankton, Birling Gap, southern England. Photograph by C.M. Reid. Reproduced with permission.

Origin of Carbonate Sedimentary Rocks, First Edition. Noel P. James and Brian Jones.
© 2016 Noel P. James and Brian Jones. Published 2016 by John Wiley & Sons, Ltd.
Companion website: www.wiley.com/go/james/carbonaterocks

Introduction

Pelagic carbonate (Frontispiece, Figure 17.1) accumulates largely below depths of ~200 m and is usually uniform over large areas. The modern deep open benthic carbonate factory produces sediment at a very slow rate because, in this twilight to perpetually dark and relatively cold environment, there are no algae or ooids, just sparse benthic invertebrates and skeletons of nektic animals (deepwater reefs are different and discussed in Chapter 14). Any carbonate sediments that accumulate here must come from the neritic zone via sediment gravity flows or from the planktic factory in the shallow part of the overlying water column. The deposits can be pure carbonate or associated with a variety of other deep-water sediment such as shale and chert.

These rocks are commonly called the *basinal carbonate facies*. They are important as source rocks and reservoir rocks for hydrocarbons. The deposits are strongly partitioned in space and time by tectonics, global sea level, geochemistry, and evolutionary biology (the evolution of calcareous organisms through time).

Universal controls

Global tectonics

Depositional environments for basinal carbonate facies (Figure 1.8) are either: (1) *cratonic*, on continental crust; or (2) *oceanic*, in the deep sea far removed from land, generally on oceanic crust. Cratonic locations are varied, but in general are epeiric seas, intracratonic basins and seaways, or continental margins. Oceanic settings range from abyssal plains to continental borderlands, volcanic seamounts, aseismic ridges, submerged plateaus, and mid-ocean ridges. Cratonic pelagic carbonate sedimentary rocks are generally well preserved as part of continental successions. In contrast, the oldest in-place oceanic pelagic carbonates are Late Jurassic. All Precambrian and Paleozoic and many younger pelagic carbonate successions are only preserved in tectonically complex areas of collisional and accretionary orogens.

Global sea level

Deep ocean basins can be repositories of pelagic sediment at any time, but cratons are depocenters only when sea level is exceptionally high such as during the Phanerozoic (e.g., Ordovician–Mississippian and Jurassic–Eocene). The Phanerozoic history of pelagic sedimentation can therefore be divided into *old* deepwater pelagic carbonate deposits and *young* deepwater pelagic carbonate deposits (Figure 17.2). Although pelagic carbonates accumulated

Figure 17.1 The Seven Sisters, a series of cliffs cut into Cretaceous chalk in south England along the northern coast of the English Channel.

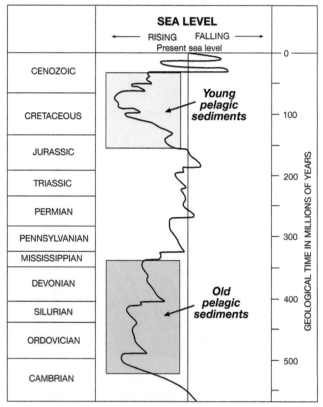

Figure 17.2 The Phanerozoic geological column highlighting the periods of old pelagic sediments and young pelagic sediments and their relationship to global sea-level highstands.

during other periods, they were largely confined to continental margin basins.

Evolutionary biology

The most important event in the history of pelagic sedimentation was the evolution of calcareous plankton in the middle Mesozoic. This event changed the nature of pelagic carbonate sedimentation forever. Precambrian pelagic carbonates might have formed in early oceans from highly carbonate-saturated seawater (see Chapter 18). Paleozoic pelagic deposits (Figure 17.3), the "old" pelagic carbonates, consist of condensed limestone sequences with shales or cherts and only accessory nektic and benthic macrofossils. By contrast, Mesozoic and younger pelagic carbonates are dominated by calcareous plankton. These microfossils are, for example, responsible for the widespread chalk deposits of the late Phanerozoic (Figure 17.1).

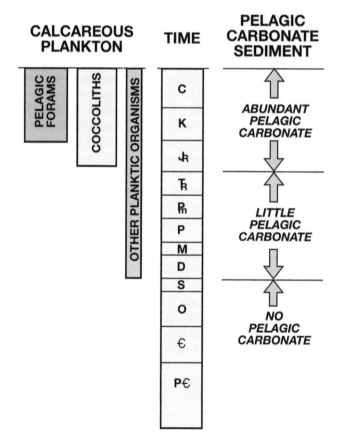

Figure 17.3 Diagram illustrating the relative abundance of pelagic carbonates and the evolution of coccoliths and pelagic foraminifers through time.

Depositional controls

Dissolution

Accumulation of calcareous sediment on the seafloor is a balance between the rate of supply versus the rate of dissolution. A major control on the preservation of pelagic carbonate is the location and depth of the carbonate compensation depth (CCD) and aragonite compensation depth (ACD). As outlined in Chapter 1, dissolution is a function of many variables including decreased temperature, increased hydrostatic pressure, and increased CO_2 content of the water with increasing depth. Even below these depths, dissolution may not occur immediately because most grains have organic coatings or are held in organic-rich fecal pellets. It is only after the organics have oxidized that the grains quickly dissolve. In the modern world, below the CCD where organic coatings have disappeared, there is no carbonate and the seafloor is covered with red clay and biosiliceous ooze. The deep-ocean hills that protrude up above the CCD today are mantled with white pelagic carbonate and have been called the "snow-capped mountains of the deep sea", with the snow being the white carbonate sediment. The lysocline, the depth where dissolution begins, is much shallower (Figure 2.10).

The positions of the CCD and ACD have varied in space and time. They are mostly affected by nutrient concentrations in the surface waters, which determine the rate of biogenic carbonate production and thus sediment supply, and the degree of undersaturation of the deep water. These controls are also affected by water temperature and salinity that are, in turn, controlled by plate tectonics. The CCD today is deep in areas of high pelagic productivity (e.g., the equator, where nutrients upwell), whereas it is shallow in areas of low productivity such as the nutrient-poor open ocean gyres. It is for these reasons that the positions of these critical interfaces are today different in the Atlantic and Pacific oceans.

Dissolution is also affected by the rate of organic matter breakdown, which releases CO_2. Deep water loses oxygen and becomes more CO_2 rich with increasing age because of biological consumption and organic matter decomposition. Older deep ocean waters that have been circulating for a long time, on the order of 1500 years, are therefore more chemically aggressive. The CCD also shallows close to continental margins where the input of organic matter is high and CO_2 levels are increased.

The position of the CCD has also varied with geological time and is tied to sea-level fluctuations; the CCD is relatively shallow during sea-level highstands because of prolific carbonate production on shelves, and deepens

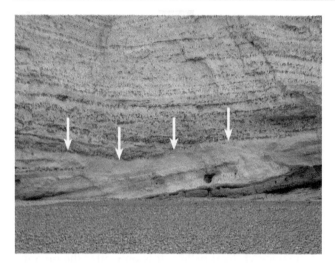

Figure 17.4 Major submarine erosion surface (arrows) in Cretaceous chalk, Normandy coast, France. Cliff with many chert nodules is 7 m high. Photograph by C.M. Reid. Reproduced with permission.

during times of lowered sea level because of decreased production on the relatively narrow shelves.

Submarine currents

The deep seafloor lies below the effects of tides and surface waves. The importance of internal waves is poorly understood but carbonate particles can be winnowed and sorted by other seafloor currents to produce grainy pelagic sediment (see Chapter 3). During icehouse times (see Chapter 21) and periods of invigorated oceanic circulation, bottom currents are highly active and can remove sediment, resulting in hiatuses in the form of hardgrounds or iron-manganese nodules. Contour currents along continental margins are particularly important in this regard. Today, contour currents along the eastern margin of the north Atlantic have velocities of 15–35 cm s^{-1}, and are easily capable of eroding, sorting, and transporting fine-grained pelagic sediment. Seafloor erosion can be significant even during greenhouse times and associated high sea levels (Figure 17.4).

Universal attributes

Phanerozoic pelagic carbonates are characterized by the following general attributes:

1. the presence of planktic or nektic macrofossil remains;
2. fine-grained, well-bedded sediments of great lateral extent with gradual lateral facies changes;

3. sediment with numerous fecal pellets, small-scale laminations, or centimeter- to meter-thick rhythms (e.g., chalk-marl cycles);
4. numerous carbonate nodules in otherwise argillaceous or marly sediments that formed through early synsedimentary lithification and subsequent pressure solution;
5. condensed sections (thin successions accumulating over long periods) that were deposited at slow rates and punctuated by numerous depositional breaks, typically in the form of multiple hardground horizons that mark submarine lithification events associated with hiatuses;
6. ripple marks but no large-scale bedforms;
7. burrow assemblages dominated by ichnogenera such as *Helminthoidea*, *Paleodictyon*, and *Chondrites* in deep-water facies and *Thalassinoides* and *Chondrites* in shallow-water settings; and
8. mineral-encrusted (glauconite, Mn-Fe oxides, phosphate) burrow linings and hardgrounds.

Old pelagic sediments

Paleozoic and early Mesozoic pelagic carbonates are mostly found in epicratonic settings because there was no biogenic pelagic carbonate production and so oceanic deep-water sediments were largely shale and biosiliceous chert. The epicratonic basins or depressions were, however, very large and extended over many hundreds of thousands of square kilometers. Some were so vast that they graded inboard into epeiric platforms (see Chapter 1). In most examples the carbonate successions are thin and condensed, illustrating repeated episodes of accumulation, hardground formation, and seafloor dissolution. These attributes reflect alternating periods of benthic productivity, cessation of accumulation and cement precipitation, or the intersection of the CCD, ACD, or lysoclines with the seafloor.

The muddy carbonate sediment that dominates Paleozoic and early Mesozoic successions is of uncertain origin. Possible sources of the lime mud include: (1) breakdown of macrofossils (maceration); (2) near-surface water column inorganic-biotic carbonate precipitation in the form of whitings; and (3) long-distance transport from neritic platforms or ramps. Although all of these origins are possible, we have not yet found a way to distinguish one from the other because they are now all dLMC.

The identity of Paleozoic nektic and benthic macrofossils varies, but the only pelagic contributors had siliceous, organic-walled, or phosphatic skeletons. The nektic and benthic components were somewhat partitioned by age. Cambro-Ordovician pelagic deposits were dominated by black shales and muddy limestones that locally contain abundant nautiloid cephalopods along with trilobites, echinoderms, mollusks, ostracods, bryozoans, brachiopods, and gastropods. By contrast, middle and late Paleozoic macrofossils included orthocone cephalopods, thin-shelled bivalves, conodonts, stylolinids, tentaculitids, and ostracods along with rare small brachiopods, gastropods, crinoid fragments, trilobites, solitary corals, goniatite ammonoids, thin-shelled ostracods, free-swimming bivalves (pteriids and pectens), fish, and calcareous algae in local shallow-water, photic environments. All of these components, however, were volumetrically limited. Bedded cherts were probably pelagic and devoid of macrofossils (dissolution). Although there are few Paleozoic pelagic carbonates, three examples highlight the nature of such deposition.

Ordovician limestones

These pelagic carbonates cover ~500,000 km² of the Baltic Shield and were deposited during a prolonged Early Ordovician sea-level highstand (Figure 17.5). The rocks, mostly lime mudstone with shaley intervals, are a condensed succession that is only 50 m thick. The succession contains numerous hardgrounds and discontinuity surfaces that are mineralized with glauconite, hematite, phosphate, and pyrite. Buckled beds and teepee structures resulting from seafloor cement precipitation and expansion of the surface layer are common. Nodular limestones are also abundant. Accumulation rates are estimated to have been on the order of 1 mm ky^{-1}.

Devonian Cephalopodenkalk (Germany) and Griotte (France and Spain)

These limestones with conspicuous cephalopods accumulated mainly on submarine topographic highs and in surrounding basins. The muddy carbonates are intensively burrowed and skeletons are bored. There are numerous hardgrounds, Fe-Mn coated clasts, neptunian dikes, and nodules. Condensed sections are best developed on paleo-highs (drowned reefs, volcanic seamounts, and basement highs). Similar facies are present in Morocco, where they are interbedded with very-shallow-water peritidal facies. All of these deposits have very low accumulation rates of 1–5 mm ky^{-1}.

Early Mesozoic (Triassic–Jurassic) Tethyan pelagic carbonates

This was a time of supercontinent breakup (Pangea) that resulted in the fragmentation and drowning of older platforms. Pelagic carbonates in the Mediterranean region such as the Ammonitico Rosso (Figure 17.6) accumulated on these foundering platforms. Sediments on the topographic highs include condensed limestones, nodular limestones interbedded with shales, and resedimented deposits in basins. Accumulation rates are estimated to have been between 0.5 and 2 mm ky^{-1}. Macrofossils were thin-shelled bivalves, small gastropods, foraminifers, radiolarians, sponge spicules, ostracods, and echinoderms, much the same as in Paleozoic pelagic carbonates.

Figure 17.5 Middle Ordovician Orthoceras Limestone, Sweden. (a) Quarry wall ~10 m high of highly condensed muddy limestone; and (b) close-up illustrating several hardgrounds (arrows) and cephalopod circular cross-sections (C). Lens caps 52 mm. Photographs by S. Stouge. Reproduced with permission.

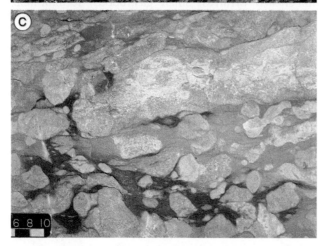

Figure 17.6 (a) Outcrop of Jurassic Ammonitico Rosso, Sicily. Note person for scale. (b) Surface outcrop of Jurassic Ammonitico Rosso illustrating the red nature of the limestone with numerous nodules, Sicily. Coin 1 cm diameter. (c) Polished slab of nodular Ammonitico Rosso, Sicily. Scale 2 cm intervals.

Nannofossils, including coccoliths, are present in the muddy matrix. $MnPO_4^-$ encrusted hardgrounds, poorly preserved ammonites because of shell dissolution, and corrosion on the seafloor all attest to dramatic synsedimentary diagenesis. The surprising presence of oncolites, stromatolites, and pelagic ooids sets these rocks apart from earlier pelagic deposits and suggests deep oligophotic environments. Local neptunian dikes are filled with pelagic sediment. These nodular- to thin-bedded white and red (e.g., Ammonitico Rosso) limestones are widespread throughout the Mediterranean region. The red color is caused by small amounts of Fe^{2+} (about 2%) and it is favored by low sedimentation rates, near-bottom waters rich in oxygen, the absence of sulfate-reducing bacteria, and well as the local Eh/pH potential within the sediment and at the sediment–water interface. Water depths are generally considered to have been ~50 to 100 m but stromatolites, if photosynthetic, suggest that they may have been even shallower.

Young pelagic sediments

The major turning point in pelagic carbonate sedimentation came in the middle Mesozoic with the evolution of coccoliths in the Jurassic and then planktic foraminifers in the latest Jurassic and Early Cretaceous. Pteropods, tiny swimming gastropods with aragonitic skeletons, another component of the pelagic fauna, evolved in the late Paleocene. Young pelagic carbonate rocks are relatively simple in composition; today they cover about 50% of the modern ocean floor and represent two-thirds of all modern carbonate production by volume. The water depth represented by Mesozoic–Cenozoic pelagic carbonates was from a hundred to a few hundred meters for cratonic chalks to perhaps thousands of meters for oceanic chalks.

The pelagic factory

Biogenic production

Calcareous and non-calcareous plankton are the base of the pelagic food chain. Whereas only calcareous plankton produce mineralized sediment, both types of plankton are food for heterozoans such as fish, crustaceans, mollusks, and other nektic animals. Planktic productivity is dependent upon nutrients, water temperature, light, and salinity, all of which vary with latitude, climate, and water circulation. Both organic productivity and carbonate productivity are highest in areas of upwelling along continental margins, equatorial regions, and at places where oceanic currents diverge (Figure 17.7). Most calcareous plankton production today is in warm surface waters at low latitudes. Siliceous plankton are concentrated in, but not confined to, cooler surface waters and areas of higher trophic resources.

Today, the community of planktic organisms in near-surface waters today comprises the autotrophic coccolithophorids and diatoms and the heterotrophic planktic foraminifers, pteropods, and radiolarians. The yellow-green "golden" unicellular algae known as coccolithophoridae produce spherical tests (microscopic skeletons) 10–100µm in diameter (coccospheres) that are formed of platelets 2–20µm wide (coccoliths) (Figure 17.8a). Coccoliths are plentiful, with 50,000–500,000L^{-1} of seawater. The less numerous but conspicuous pteropods are conical planktic gastropods up to a few millimeters long. The skeletons of calcareous plankton sink to the deep seafloor at various rates. Planktic foraminifers >50–60µm in size (Figure 17.8b) take only a few days to sink, whereas nannofossils such as coccoliths that are ~5µm in size take, on average, 100 years. If coccoliths are inside zooplankton fecal pellets (50–250µm in size can contain 10^5 coccoliths), then this time is reduced to only a couple of months.

Calcareous ooze

The most widespread and voluminous pelagic sediment in the modern ocean is *biogenic ooze*, a deposit of pelagic sediment of which at least 30% is skeletal. Early sailors who dredged this fine-grained, soft, soupy sediment from the deep seafloor worldwide coined the word "ooze."

The sediment grain size is polymodal, with peaks at 0.5µm (calcite laths from disintegrated coccoliths), 1–5µm (whole coccolith plates), 5–20µm (larger coccoliths, coccospheres, comminuted foraminifers, and other skeletal fragments), 25–64µm (whole foraminifers), and >64µm (foraminifers, bivalve prisms, and other macrofossil fragments). Accumulation rates in the modern ocean range from zero beneath open ocean gyres with low nutrient elements, to 1–5cm ky^{-1} in areas with abundant trophic resources.

The critical point in terms of subsequent diagenesis is that the dominant components, planktic foraminifers and coccoliths, are formed of LMC. In contrast, the pteropods are composed of aragonite, and are therefore more soluble and generally only seen in shallow tropical waters above the ACD.

Dilution of pelagic sediment

The slow accumulation rate of pelagic carbonate means that the deposit can be easily diluted by siliciclastic sediment (Figure 17.9). Dilution depends upon several factors such as the rate of pelagic fallout, distance from land, depth below sea level, local bathymetry, climate, and wind direction. Most of these controls affect the introduction of fine siliciclastics into the system via a variety of fluvial processes. Aeolian dust is also an important source of fine terrigenous sediment because it can be dispersed over such a wide area. Pelagic carbonates adjacent to platforms and ramps are both diluted by and interbedded with resedimented neritic carbonates.

Chalk

The most impressive sedimentary deposit of the late Mesozoic–Cenozoic is chalk (Figures 17.1, 17.10), from which the Cretaceous period gets its name. Chalk is pelagic carbonate sediment that is predominantly Late Cretaceous in age and is present across large cratonic areas of North America and Europe. Chalk accumulated during a greenhouse time when global sea level was 200–300m higher than it is today.

North America

Pelagic carbonates in North America (Figure 17.11) are restricted to the elongate, north–south Western Interior Seaway and, to a lesser extent, the Eastern Seaboard of the

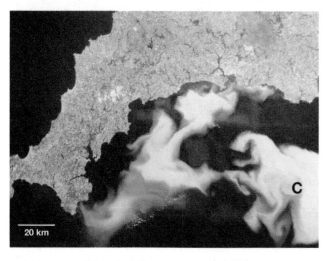

Figure 17.7 LandSat false color image (24 July 1999) of the southwest coast of the United Kingdom with an offshore coccolith bloom (C). Photograph courtesy of NERC Earth Observation Data Acquisition and Analysis Service, Plymouth, United Kingdom.

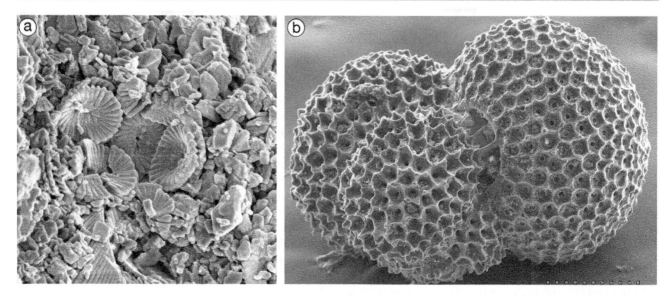

Figure 17.8 (a) Cretaceous chalk composed almost entirely of coccolith fragments, subsurface Cuba. Image width 9 μm. (b) SEM image of the planktic foraminifer *Globigerinoides ruber*. Image width 300 μm. Photograph by B. J. McCloskey. Reproduced with permission.

Figure 17.9 Chalk with abundant clay manifest as stylolite seams in a subsurface core from the North Sea. Image width 8 cm.

Figure 17.10 SEM image of Danian coccolith chalk from a core in the subsurface of the North Sea. Photograph by P.A. Scholle. Reproduced with permission.

continent. Their distribution in the Seaway was largely controlled by the influx of fluvial terrigenous sediment, especially from the rising Cordillera to the west. The influx of fresh water is also thought to have stressed the marine biota because the diversity of all organisms is relatively low. The macrobiota is mostly mollusks; there is a conspicuous lack of burrowers and hence excellent preservation of sedimentary structures, especially laminations. Hardgrounds are not well developed. Water depths were on the order of a few hundred meters. Lowstands are represented by siliciclastic sediments, especially shales.

Europe

Pelagic carbonates (Figure 17.12) accumulated over large areas of Europe, largely because there was little terrigenous input due to a somewhat arid climate. Successions are thicker than those in North America (generally ~200 m) and are up to 1000 m thick in the North Sea Basin. The ocean was more normal marine than in the Interior Seaway of North America, and there was a more diverse microbiota and macrobiota; the latter being dominated by siliceous sponges, brachiopods, mollusks, bryozoans, and echinoderms with lesser amounts of pelagic ammonites, belemnites, and crinoids. Shallow facies included a significant benthic fauna of echinoids, bivalves, sponges, brachiopods, and bryozoans. The sediments are intensively burrowed and contain much more chert (from sponges) (Figure 17.12) than the North American sediments. The carbonates also pass laterally into more diverse marginal-marine carbonate facies including greensands, phosphatic chalks, hardgrounds, and shelly chalks. Lowstand deposits are typically hardground omission surfaces. Spiculitic diagenetic

chert (flint) is a hallmark of the chalk (Figure 17.12) with many burrows replaced by chert nodules.

Overall, the chalk is evenly bedded and commonly burrowed (Figure 17.13). Nodular horizons, omission surfaces, and hardgrounds are present (Figure 17.14) at many levels. Hardgrounds are encrusted by bivalves, bryozoans, and serpulid worms and bored by fungi, algae, sponges, worms, and bivalves. Some hardgrounds are mineralized with glauconite and phosphate; these minerals penetrate the walls of burrows and surfaces (Figure 17.14c, d). Hardgrounds in the United Kingdom can be traced over many hundreds of square kilometers. Local facies include deep-water bryozoan mounds (Denmark) and spectacular large erosional channels (northern France).

Sedimentation rates

The accumulation rates of Mesozoic–Cenozoic pelagic sediments (uncompacted) were on the order of ~30 mm ky^{-1}. In the North Sea Basin they ranged from 150 to 250 mm ky^{-1}.

Diagenesis

Figure 17.11 Quarry exposure of Upper Cretaceous (Coniacian) Austin Chalk near Dallas Texas, USA, composed of shelf chalks and interbedded marly chalks with a low relief channel (arrows). Note person for scale. Photograph by P.A. Scholle. Reproduced with permission.

Diagenesis has been intensively investigated in the chalk because of its importance as a hydrocarbon reservoir. Much of the chalk is weakly lithified and has a very high microporosity. Alteration is nevertheless present in the form of conspicuous well-cemented and nodular layers. The softness of the chalk is thought to be due to: (1) the dominantly LMC original mineralogy of

Figure 17.12 (a) A 40-m-high Cretaceous chert-rich chalk cliff at Étretat, northern France, with bedding highlighted by layers of chert nodules; and (b) 30-m-high chalk cliffs, southern England, with conspicuous dark chert layers. Photographs by C.M. Reid. Reproduced with permission.

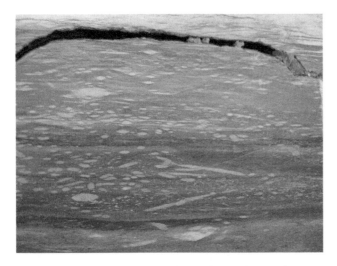

Figure 17.13 Heavily bioturbated (mainly *Zoophycos* sp.) argillaceous chalk in a core from the North Sea subsurface. Image width 6 cm.

the particles, with no aragonite to supply cement (see Chapter 25); and (2) the nature of the burial fluids that were of marine origin; there was no dissolution and cement precipitation from meteoric waters as with other limestones. Another factor is that they may not have been buried very deeply.

Seafloor lithification in the form of hardgrounds (Figure 17.14) was important because it led to the formation of subsurface aquicludes in otherwise porous stratigraphic sections. Cementation was principally by dHMC, aragonite, glauconite, and phosphate. There was a recurring progression of cementation associated with burrows: (1) burrow fillings were lithified; and then (2) lithification expanded to form calcareous nodules. In some situations, the nodules were then winnowed out to form clasts. Given their relatively simple composition, the diagenetic transformation of many pelagic sediments can be predicted and modeled. This is especially

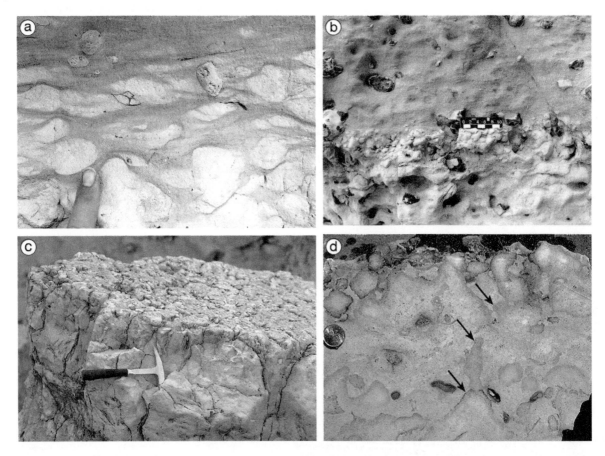

Figure 17.14 (a) Outcrop of nodular Cretaceous chalk in southern England with numerous argillaceous seams grading up to marl. Image width 16 cm. Photograph by C.M. Reid. Reproduced with permission. (b) Cross-section of nodular chalk with a nodular firmground–hardground overlain by marl, Normandy, France. Centimeter scale. Photograph by C.M. Reid. Reproduced with permission. (c) A well-developed planar hardground in Late Cretaceous shelf chalk south of Étretat, northern France. The greenish surface cast results from seafloor cementation by glauconite and phosphate in addition to Mg-calcite. Hammer 28 cm long. Photograph by P.A. Scholle. Reproduced with permission. (d) Polished slab of Late Cretaceous (Turonian) chalk showing multiple hardgrounds (arrows) with irregular, phosphatized (brownish) and glauconitized (greenish) surfaces. Note also cementation of thalassinoidean burrow systems and numerous reworked pebbles of hardened material, Kensworth Quarry, Bedfordshire, Great Britain. Coin scale 2.5 cm. Photograph by P.A. Scholle. Reproduced with permission.

important when considering such sediments as potential petroleum reservoirs.

Associated sediments

Pelagic lime mud can be interbedded or intermixed with non-carbonate materials, including siliceous, clayey, and other sediment components. Siliceous ooze is composed of pelagic radiolarians (Cambrian–Holocene) and diatoms (Cretaceous and younger), which have skeletons formed of opal-A. Maximum accumulation is in zones of high biological fertility, especially along ocean margins, beneath the equatorial divergence, and south of the Antarctic convergence. These sediments, along with siliceous sponge spicules, are the precursors of chert.

Red clays (really dark brown) consist of a variety of clay minerals that are coated with amorphous iron oxides. Minor components include volcanic ash, cosmic spherules, fish teeth and bones, and locally traces of carbonate and silica. They cover most of the deep sea-floor below the CCD (depths >4 km), with slow sedimentation rates of ~1 mm ky^{-1}. These fine materials are generally delivered to the ocean basins via sediment gravity flows from continental margins and by aeolian fallout.

Ocean anoxia

Global sea-level highstands during greenhouse periods (see Chapter 21) were times of warm climate, few icecaps, and stratified oceans with overall sluggish circulation. Such a situation also caused periodic cessation of deep-water transport from the tropics to the poles, leading to ocean stagnation. During such periods, oxygen was not supplied to the deep ocean floor, resulting in widespread anoxia. Sediments deposited during these *oceanic anoxic events* consist of organic-rich shales and pelagic carbonates and are now important hydrocarbon source beds. These units can be traced over vast areas and are interbedded with more normal pelagic carbonates. They were not only a Mesozoic phenomenon but are also present in the Paleozoic including, for example, the Upper

Figure 17.15 Boundary between shallow-water Upper Devonian limestones and black, phosphate-rich Late Devonian–Early Mississippian Exshaw anoxic shales in the Front Ranges of the Canadian Rockies. Note person for scale.

Devonian Kellwasser Shales in Europe and the Late Devonian–Early Mississippian Exshaw and Chattanooga shales (Figure 17.15) in North America.

Further reading

Bernoulli, D. and Jenkyns, H.C. (2009) Ancient oceans and continental margins of the Alpine-Mediterranean Tethys: deciphering clues from Mesozoic pelagic sediments and ophiolites. *Sedimentology*, 56, 149–190.

Ekdale, A.A. and Bromley, R.G. (2001) Bioerosional innovation for living in carbonate hardgrounds in Early Ordovician of Sweden. *Lethaia*, 34, 1–12.

Hsü, K.J. and Jenkyns, H.C. (eds) (1974) *Pelagic Sediments: On Land and Under the Sea*. Blackwell Scientific Publications, International Association of Sedimentologists, Special Publication no. 1.

Kennett, J.P. (1982) *Marine Geology*. Englewood Cliffs, USA: Prentice Hall.

Nieto, L.M., Reolid, M., Molina, J.M., Ruiz-Ortiz, P.A., Jiménez-Millán, J., and Rey, J. (2012) Evolution of pelagic swells from hardground analysis (Bathonian-Oxfordian, Eastern External Subbetic, southern Spain). *Facies*, 58, 389–414.

Quine, M. and Bosence, D. (1991) Stratal geometries, facies and sea-floor erosion in Upper Cretaceous Chalk, Normandy, France. *Sedimentology*, 38, 1113–1152.

CHAPTER 18
PRECAMBRIAN CARBONATES

Frontispiece Stacked stromatolites, Paleoproterozoic (1.9 Ga), Great Slave Lake, Northwest Territories, Canada. Image width 60 cm.

Origin of Carbonate Sedimentary Rocks, First Edition. Noel P. James and Brian Jones.
© 2016 Noel P. James and Brian Jones. Published 2016 by John Wiley & Sons, Ltd.
Companion website: www.wiley.com/go/james/carbonaterocks

Introduction

Precambrian limestones and dolostones that accumulated on the young Earth during Archean and Proterozoic times are among the most enigmatic of carbonates. They span more than 3 billion years of Earth history but have no real modern counterparts that we can use as actualistic models. The only possible analogs are in the hypersaline lagoons of Shark Bay, Western Australia, in the tidal race of the Bahamas, and in lakes. We must therefore work with these few recent environments that might be somewhat similar, and use our collective experience to determine how the Precambrian systems worked. It must also be remembered, however, that these carbonates have to be viewed against a world that changed from an Archean biosphere and ocean that is barely comprehensible to a Proterozoic one that, with time, gradually became more and more like the world we understand today. Before considering these carbonates, recall the Precambrian time scale, which is not familiar to most students of sedimentology (Figure 18.1).

Precambrian carbonate systems

Many very old carbonates are astonishingly well preserved and can be studied as easily as any Cenozoic limestone. The basic carbonate archetypes of rimmed platforms and hydrodynamically partitioned ramps (Figures 18.2, 18.3) contain all of the familiar facies such as peritidal mudflats, low-energy inner ramp or lagoon muds, sand shoals, storm deposits, reefs, and resedimented slope deposits. Although surprisingly similar to Phanerozoic systems in many ways, they are at the same time frustratingly foreign. This dichotomy exists because they also contain components, such as seafloor precipitates, giant ooids, molar tooth structures, and spectacular arrays of open marine stromatolites, that were not present in the Phanerozoic marine world.

The carbonate factory

The Precambrian carbonate factory lacked calcareous algae or invertebrates to produce sediment particles, boring organisms to break them down, or burrowing animals to disturb them once they had accumulated. As a result, unless diagenetically altered, the particles are well preserved and original sedimentary structures can look as though they were made yesterday. The dominant processes are microbial and abiogenic precipitation.

Microbial stromatolites

The microbial consortium, present in all conceivable environments, influenced Precambrian sedimentation by (1) trapping and binding sediment grains to form various microbial mat types; and (2) changing their immediate microenvironment, thereby inducing precipitation of carbonate in microbial mats. The surfaces of some microbes became calcified late in the Proterozoic to form thrombolites and calcimicrobes. Microbes were endolithic (boring) only at the very end of the Neoproterozoic.

Most Precambrian shallow-marine microbes are compared to modern photosynthetic microbes (some Precambrian taxa are still alive today) and so are assumed to be light-dependent. Slope to basin stromatolites are, however, not so readily interpreted and some may have been chemoautotrophs (living in perpetual darkness) although this possibility is not yet well resolved.

Stromatolites have almost every conceivable shape, and most classification schemes rely upon morphology as a primary attribute. There are two main schools of thought. One classification system, based largely on comparisons with modern examples, holds that the shape of the stromatolites was largely a function of the growth environment. The other classification system, based mostly on perceptions from the fossil record, holds that shape reflects various growth patterns that were engendered by different microbial communities. The latter

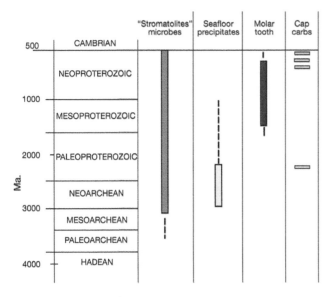

PRECAMBRIAN STYLE CARBONATES

Figure 18.1 Precambrian time scale with ranges of carbonate components that are somewhat unique to this period in time (except stromatolites).

PRECAMBRIAN RIMMED CARBONATE PLATFORM

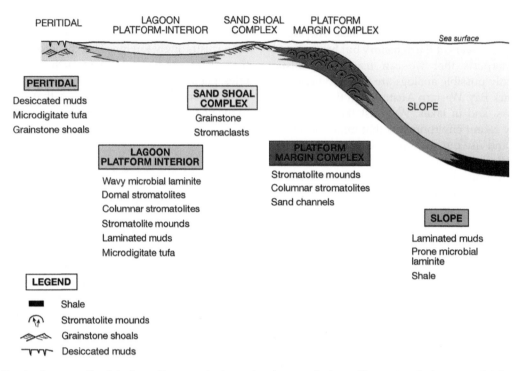

Figure 18.2 Sketch of a generalized Archean–Proterozoic rimmed carbonate platform, illustrating the location of different carbonate facies.

PRECAMBRIAN CARBONATE RAMP

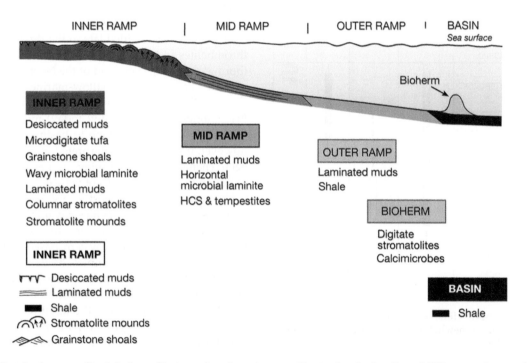

Figure 18.3 Sketch of a generalized Archean–Proterozoic carbonate ramp, illustrating the location of different carbonate facies.

classification system leads to these organo-sedimentary structures being given Latin generic and specific names. In reality, it seems that the form of these stromatolites was related to both the environment and the microbial constituents. A third possibility is that stromatolite form is due to emergent organism complexity, and is therefore non-deterministic by definition, resulting in simple biological and physical interactions. This means that it is not microbes and it is not the environment; it is simply the dynamics of accretion that determines form. Regardless, from an interpretive sedimentological point of view, shape is the most easily recorded attribute. One important aspect that is sometimes ignored when making paleoenvironmental interpretations is synoptic relief, which is the relief between the top of the growing stromatolite and the surrounding seafloor. Many spectacular columnar stromatolites, several meters in apparent height, were actually low-relief structures that continued to grow in the same place just above the surrounding seafloor over a long period of time. They are really "stratigraphic columnar stromatolites" (Figure 18.4).

Although the spectrum of shapes is impressive, microbial structures can really be divided into two basic types: (1) subhorizontal laminated structures with variable small domes and pustular protrusions; and (2) small and large columnar to hemispherical structures with centimeter- to meter-scale synoptic relief (Figure 18.5). These stromatolites can be stacked on top of one another in a semi-fractal arrangement to form spectacular reefs. If the stromatolites had meter-scale synoptic relief in high-energy environments, their bases could be eroded to form mushroom-shaped structures or the whole structure could be sculpted by waves and tidal currents as they grew into elongate, current-parallel shapes. Stromatolites growing in protected lagoonal or inner ramp settings were commonly small and *microdigitate* (Figure 18.6a) or spherical to elliptical *oncolites* (Figure 18.6b). Microdigitate stromatolites are also called *tufa* (unrelated to tufa in spring deposits and caves) or *microstromatolites*, and are found mostly in tranquil and peritidal locations. Since the term tufa is also used for spring precipitates through geologic time, care must be taken to determine context and setting when using this word. Another peculiar stromatolite growth form, usually restricted to deeper water facies, is *conoform*, meaning a series of stacked inverted cones (Figure 18.7). Small irregular stromatolites (Figure 18.8) can also be common in slope deposits that are formed of interbedded shale and carbonate mud.

Figure 18.4 Stromatolites from the Paleoproterozoic Pethei Group at Great Slave Lake, Northwest Territories, Canada. Although the microbial structures are columnar, many never had much synoptic relief above the seafloor (each lamina exhibits only 1–2 cm of relief from column to inter-column areas).

Figure 18.5 Gigantic stromatolite, Paleoproterozoic, Kuuvik Formation, Bathurst Inlet, Northwest Territories, Canada. Photograph by F. Campbell. Reproduced with permission.

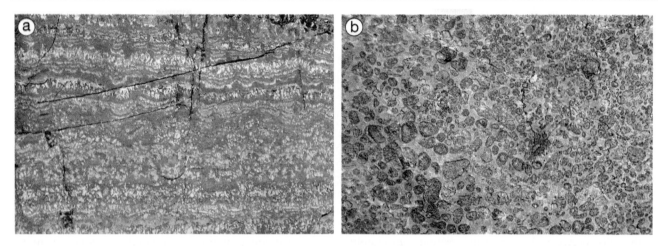

Figure 18.6 (a) Tidal flat tufa (microstromatolites) that appears to record the combined influence of benthic microbial communities, seafloor cement precipitation (both dark), and passive settling of carbonate mud (pale) from the water column. Late Mesoproterozoic, Hunting Formation, Somerset Island, Nunavut, Canada. Image width 7 cm. Photograph by E.C. Turner. Reproduced with permission. (b) Elliptical oncolites from the Pethei Group at Great Slave Lake, Northwest Territories, Canada. Image width 40 cm.

Figure 18.7 Axial zone of large Conophyton stromatolite, Paleoproterozoic Dismal Lake Group, Northwest Territories, Canada. Photograph by L. Kah. Reproduced with permission.

Figure 18.8 Interbedded slope carbonate mudstone and small stromatolites, Paleoproterozoic Pethei Group, Great Slave Lake, Northwest Territories, Canada. Image width 50 cm.

Familiar particles and structures

Grains. Carbonate grains in Precambrian rocks are not diverse and consist mostly of intraclasts, ooids, and peloids. Peloids were likely generated by microbes or formed as subspherical mud aggregates. They were probably similar to the modern micropeloids that are so commonly associated with synsedimentary cements and that have a microbial origin (see Chapter 24). Although Precambrian intraclasts are similar to intraclasts found throughout the geological record, many display stromatolite laminations and have therefore

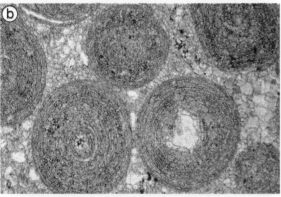

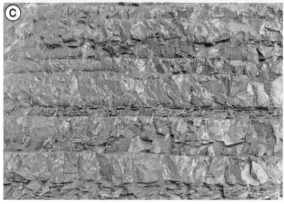

Figure 18.9 (a) Wave rippled bedding plane of ooid grainstone in the Paleoproterozoic Pethei Group, Great Slave Lake, Northwest Territories, Canada. Hammer 10 cm long. (b) Thin-section photomicrograph under plane-polarized light of neomorphosed ooids from Neoproterozoic limestones in the Mackenzie Mountains, Northwest Territories, Canada, where the original concentrically laminated aragonite cortex has been calcitized. Image width 3 mm. (c) Widespread, stratigraphically monotonous deep-water carbonate mudstone with no evidence of nearby shallow-water carbonate environment, implying that the carbonate mud formed in the water column ("pelagic carbonate") through abiogenic or biotically mediated processes, middle Neoproterozoic (~815 Ma), Mackenzie Mountains, Northwest Territories, Canada. Image width 60 cm. Photograph by E. C. Turner. Reproduced with permission.

been called *stromaclasts*; they probably "flaked off" stromatolites during high-energy events.

Most Precambrian ooids (Figure 18.9a, b) were either radial Mg-calcite or tangential aragonite originally, and were generally larger than their Phanerozoic counterparts. Large or "giant" ooids (pisoids) were present in many neritic early–middle Neoproterozoic environments, with some being 4 to 9 mm in diameter. The origin of these grains is still a puzzle, but it is suggested that low nucleation rates and increased agitation may have been be involved (the Moon was closer, so the tidal range was larger?).

Mud. Basinal carbonate mudstones are known from a range of Proterozoic successions, yet they cannot be explained by the mechanisms invoked for equivalent lithofacies in younger strata. Deep-water carbonate muds produced by calcareous plankton are an exclusively Mesozoic and younger phenomenon; no such pelagic carbonate factory existed prior to that time. Even though the Paleozoic lacked basinal carbonate muds, this lithofacies is present in older rocks (Figure 18.9c). An explanation of this conundrum lies in the probable existence of a Proterozoic pelagic carbonate factory in which mud-grade precipitates precipitated either spontaneously or under the influence of living and dead organic matter in the water column. This predominantly Precambrian way of forming carbonate mud could have been the result of the high levels of saturation in seawater prior to the evolutionary appearance of carbonate-precipitating skeletal benthos in the early Paleozoic, but may have a partial analog in the modern phenomenon known as whitings.

Teepees. Teepee structures, common in Phanerozoic muddy tidal flat and subtidal sediments, were also widespread. Caused by the rapid precipitation of carbonate cement resulting in lateral stress and upward buckling (Figure 18.10a), these were especially common in platform top and inner ramp locations.

Unfamiliar components

Seafloor precipitates. Late Archean, and to a lesser extent Paleoproterozoic and Mesoproterozoic, platforms are characterized not only by the stromatolites and sediments described above, but also spectacular seafloor precipitates (Figure 18.10b). Unlike synsedimentary cements in the Phanerozoic, which were mostly localized to internal cavities, these crusts of precipitated aragonite and Mg-calcite formed on the open seafloor. Now pseudomorphs preserved as calcite or dolomite, they precipitated as: (1) domes of aragonite (10–150 cm radius); (2) small microdigitate stromatolites; (3) isopachous aragonite crusts, micrometers–millimeters thick; (4) herringbone calcite; or (5) branching marine "tufas." They formed discrete beds up to 2 cm thick that can be traced over hundreds of square kilometers. Such precipitates are commonly associated with ooid sand shoals and ribbon limestones.

Herringbone calcite is massive cement with a characteristic repeated chevron style of fabric (interpreted to have been Mg-calcite originally) that is thought to have formed because of high Fe and Mn inhibition in the late Archean ocean. This is much the same as Mg inhibition preventing LMC precipitating in the ocean today.

Molar tooth structure.   This enigmatic structure is composed of vertical to ptygmatically crumpled sheets of finely crystalline calcite spar that originally precipitated in cracks that developed within dolomudstone or argillaceous lime mudstone on the seafloor (Figure 18.10c). The tiny crystals are pure, uniform, equant, polygonal, tightly packed blocky calcite 5 to 15 μm across, and are typically in sharp contact with the surrounding sediment. The structure is most abundant in Mesoproterozoic and Neoproterozoic carbonates. It is a subtidal phenomenon (shallow rimmed-platform or inner ramp) and is particularly common in the lower parts of shallowing-upward cycles. The synsedimentary nature of these crack-filling cement sheets is demonstrated by the fact that they were eroded from the softer

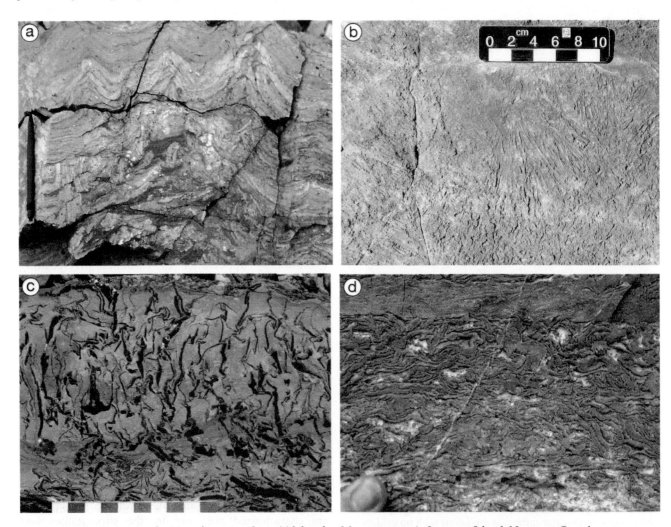

Figure 18.10 (a) Teepees at the tops of meter-scale peritidal cycles, Mesoproterozoic Somerset Island, Nunavut, Canada. Limestone beds 4 cm thick. Photograph by E.C. Turner. Reproduced with permission. (b) Small botryoids of former aragonite seafloor precipitates, now preserved as dolomite in Neoarchean carbonates at Steep Rock Lake, Ontario Canada. (c) Dark calcite molar tooth in dolomite from the Neoproterozoic Belt Supergroup, southern British Columbia, Canada. Centimeter scale. (d) A Neoproterozoic grainstone from Baffin Island, Arctic Canada, composed of the calcite spar fragments of molar tooth (the molar tooth calcite precipitated in soft sediment only to be eroded and redeposited as grains). Finger 1 cm wide.

surrounding sediment to form particles that were subsequently swept together to form calcite spar molar tooth intraclast grainstones (Figure 18.10d).

The origin of this feature is highly controversial. The origin of the cracks is variously thought to have been due to synaeresis, diastasis, earthquake-induced dewatering, evaporite replacement, or microbially induced gas bubble expansion. There are no modern or Phanerozoic counterparts. Laboratory experiments suggest that its formation may have been a two-step process in which biogenic gas created cracks and fissures within a meter or so of the sediment surface; the cracks were filled with H_2S, CO_2, or CH_4 that could not escape because of overlying microbial mats. Calcite spar then precipitated in the open voids. Geochemistry suggests, however, that there is no chemical or isotopic difference between the spar and surrounding carbonate sediment.

Cap carbonates. The Neoproterozoic is well known for its spectacular glacigene sediments of global extent (the three "snowball Earth" episodes) and the thin limestones and dolomites that directly overlie the diamictites of each glacial episode. These cap carbonates have a suite of features like no others in the geological record.

The carbonates (Figure 18.11) are typically pink to grey limestone or dolostone successions up to 20 m thick. They appear to be synchronous worldwide and coincide with initial marine flooding across the diamictites. Where best exposed, they are laminated, coarse peloid grainstones that are locally inversely graded or exhibit subtle hummocky cross-stratification. Spectacular meter-scale stacked teepee structures are common. Many such intervals contain layers of large centimeter- to decimeter-scale botryoids of calcitized aragonite. In some locations, the botryoids are stacked to form meter- to decimeter-scale biostromes and bioherms. The carbonates are present in both shallow- and deep-water environments, but the lack of standard sedimentary structures prevents their assignment to specific paleoenvironments.

Cap carbonates have a distinctive negative (–1 to –6‰) stable carbon isotope signature. Any interpretation of their origin must include an explanation of the geochemistry, their presence immediately following a profoundly cold period, and the obviously intensive carbonate precipitation. Models proposed for their origin include: (1) warm interglacial intervals that would have decreased the solubility of carbonate triggering precipitation; (2) turnover of a long-lived, previously stratified ocean that would have forced upwelling of anoxic, alkaline deep water; (3) renewal

Figure 18.11 (a) A section of Neoproterozoic stratigraphy exposed in the Mackenzie Mountains, Northwest Territories, Arctic Canada with the Marinoan cap carbonates forming a striking buff horizon (arrow) across the landscape. Total section is 120 m thick. (b) An outcrop section of classic Neoproterozoic cap carbonate (dolomite here) with teepees in the Mackenzie Mountains, Northwest Territories, Arctic Canada. This is a typical cap facies worldwide. Scale divisions 2 cm. (c) An outcrop cross-section of synsedimentary seafloor aragonite botryoids, now replaced by calcite in the Marinoan cap carbonate from the Mackenzie Mountains, Northwest Territories, Arctic Canada. Centimeter scale.

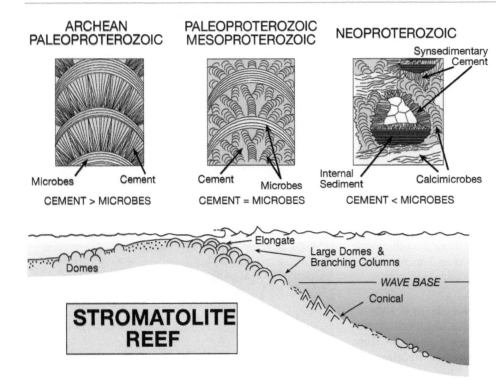

ARCHEAN PALEOPROTEROZOIC

PALEOPROTEROZOIC MESOPROTEROZOIC

NEOPROTEROZOIC

Synsedimentary Cement

Microbes   Cement

CEMENT > MICROBES

Cement   Microbes

CEMENT = MICROBES

Internal Sediment   Calcimicrobes

CEMENT < MICROBES

Elongate

Domes

Large Domes & Branching Columns

WAVE BASE

Conical

STROMATOLITE REEF

Figure 18.12 Sketch illustrating the different stromatolite morphotypes across an Archean–Proterozoic carbonate platform margin and how the relative proportion of synsedimentary cements versus microbes has changed through geologic time. Source: James and Wood (2010). Reproduced with permission of the Geological Association of Canada.

of organic productivity following glaciation; and (4) the buildup of CO_2 during glaciation followed by a post-glacial spike in continental weathering and delivery of solutes to the ocean.

Reefs

Although built by stromatolites, Precambrian reefs (Figure 18.12) have the same spectrum of bathymetrically delimited styles as Phanerozoic structures and range from patch reefs, to fringing and barrier reefs, to downslope and basinal buildups. Stromatolites varied from those that were dominantly abiogenic (effectively seafloor mineral crusts that look like stromatolites), to biogenic structures formed by microbially induced carbonate precipitation, to rare structures formed by the trapping and binding of sediment by microbial mats. Thrombolites were not important in reef formation until the Neoproterozoic.

Archean

Archean reefs were built primarily by stacked stromatolites that display a wide variety of morphologies (Figure 18.13). They have neither framework cavities nor inter-microbe synsedimentary cements. Many such reefs display a clear fractal or self-similar organization. Stromatolites can be delicate branching, lamellar,

hemispherical, or conical. These variants, as stressed above, are thought to be due to a combination of different microbial communities, hydrodynamics, and sedimentation rates.

Proterozoic

Shallow-water high-energy Proterozoic stromatolitic reefs (Figure 18.14) were constructed by stacked domes, linked domes, and columns. Stromatolite fragments (stromaclasts) were ripped off during storms and swept into cross-bedded gravels or, in the case of high-relief reefs, boulders and cobbles of reef rock. Reefs at platform margins show a strong lateral and vertical zonation of microbialite growth forms that resemble metazoan reef zonations (Figure 18.2). Isolated deep subtidal and slope reefs (probably below fair-weather wave base) were generally constructed by conical (conoform) stromatolites. Even though stromaclasts form carbonate sands, a part of the Phanerozoic reef system that is conspicuously absent from Precambrian reefs is the copious skeletal sand formed by reef-dwelling and bioeroding organisms.

Secular change

Archean and Paleoproterozoic carbonate rocks, although uniformly dominated by microbial factories, were inherently distinct carbonate systems. This difference reflects a

Figure 18.13 (a) A steeply dipping section of dolomitized Neoarchean stromatolites at Steep Rock Lake, Ontario, Canada. Note person (circled) for scale. (b) Closer view of the stromatolites illustrated in (a) showing the large scale domes. Centimeter scale circled. (c) Microstructure of stromatolites illustrated in (a, b) dominated by carbonate precipitates. Finger 1 cm wide.

change in the nature of the world ocean through Precambrian time.

The Archean was a time in Earth history where small volcanic-dominated continents were sites of carbonate sedimentation in a largely reducing ocean beneath an ultra-greenhouse atmosphere. The platforms, which remained small until the Neoarchean, were dominated by seafloor crystal fans and stromatolites, with rare carbonate muds and intraclasts. Seafloor precipitates were, however, present on virtually every known Archean platform.

Cratonization and formation of the first supercontinent (Kenorland) near the Archean–Proterozoic boundary (2.5 Ga) resulted in an abundance of diverse and stable shallow-marine environments available for carbonate deposition. There appears to have been a profound increase in the amount of free atmospheric oxygen at about this time. Before this great oxygenation event (GOE), microbial photosynthesis was producing molecular oxygen (O_2). The oxygen could not accumulate in the atmosphere because it was captured by organic matter, reduced minerals, and dissolved iron. The GOE was the point at which the supply of such substances became equal to or less than the rate of O_2 production, such that O_2 began to rapidly accumulate in the atmosphere and upper water column. This change was accompanied by a significant decrease in the volume and scale of seafloor carbonate precipitates (millimeter–centimeter in thickness but not larger). Most such precipitates were restricted to peritidal environments in the form of microdigitate stromatolites, botryoidal fans, and laminated precipitate-like crusts, which persisted in these settings through the Mesoproterozoic.

Ocean carbonate saturation

The decrease in seafloor precipitates outlined above points to a slow but unidirectional decrease in seawater carbonate saturation with time. The prolific Archean seafloor precipitates gradually decreased in abundance in the Paleoproterozoic and became rare thereafter (except in Neoproterozoic cap carbonates). This is consistent with observations that the atmospheric pCO_2 could have been greater early in Earth history and so the total alkalinity of seawater may have been much higher as a result. The decrease in the abundance and distribution of abiogenic precipitates may have been due to: (1) a permanent transfer of inorganic carbon from the atmosphere and ocean to the first supercontinent capable of preserving abundant carbonates; and (2) a decrease in reduced Fe and Mn known to be kinetic inhibitors of carbonate precipitation due to progressively increasing atmospheric and ocean O_2.

Herringbone calcite fabrics, for example, suggest that preferential growth of certain crystals faces was forced by the presence of an inhibitor, possibly Fe^{2+} or Mn^{2+}. Lower oxygen concentrations in seawater would therefore have led to greater solubility of iron and manganese, which in turn interfered with calcite precipitation to the extent that calcite was precipitated as highly distorted crystals.

This change in ocean chemistry is also recorded in the evolving reef system (Figure 18.14). *Paleoproterozoic* reefs were cement-rich with only accessory microbes and mostly hemispherical, columnar, laminated, and conical stromatolites, with most forming platforms or biostromes. *Mesoproterozoic* reefs had broadly similar stromatolite types, but the abundance of microbes increased in importance until they roughly equaled that of precipitates. Stromatolite morphology was more diverse overall, which may have been due to more varied paleoenvironments rather than to biotic evolution. *Neoproterozoic* reefs (Figure 18.15a) contained thrombolites and calcimicrobes for the first time. Microbes were volumetrically more important than cements and stromatolites, and conical stromatolites decreased in importance. Growth cavities with synsedimentary void-filling cement and internal sediment appeared for the first time (Figure 18.15b). The first skeletal metazoans populated late Neoproterozoic stromatolites and their winnowed fragments accumulated in channel lags between them; the most common are *Cloudina*, a small calcareous tube and *Namacalathus*, a goblet-like cup. Neoproterozoic reefs are the precursors to Phanerozoic metazoan-calcimicrobial reefs.

The dolomite conundrum

Although most Archean carbonates are limestone, many Proterozoic carbonates are dolostone. Many original depositional textures and fabrics are beautifully preserved in these older dolostones. These attributes have led to the proposition that dolomite was a primary marine carbonate precipitate during the Proterozoic (in place of aragonite and calcite). This suggestion has been challenged because of: (1) the evidence that mimetic dolomite can preserve texture at the nanoscale; (2) the clear original aragonite texture in dolomite and calcite pseudomorphs alike; and (3) the abundance of well-preserved limestones throughout the Precambrian. Doubt lingers however because preservation in many Proterozoic dolostones is so exquisite and thick formations are made of nothing but dolomite. Especially compelling are synsedimentary cements in some Neoproterozoic (Cryogenian) reefs. Whereas the components and very early marine cements were aragonite and HMC, most of the succeeding pervasive

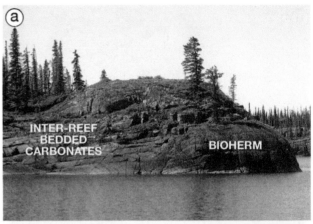

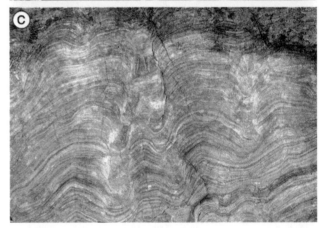

Figure 18.14 (a) Paleoproterozoic bioherm and surrounding inter-reef deposits Pethei Group, Great Slave Lake, Northwest Territories, Canada. Trees are ~8 m high. (b) Outcrop surface of Paleoproterozoic stromatolites in bioherms from the Pethei Group, Great Slave Lake, Northwest Territories, Canada. Image width foreground 3 m. (c) Cross-section of hemispherical stromatolites in bioherms from the Pethei Group, Great Slave Lake, Northwest Territories, Canada. Image width 1.5 m.

Figure 18.15 (a) Two Neoproterozoic reefs, the larger one of which is ~600 m wide and ~300 m high in the Mackenzie Mountains, Northwest Territories, Canada. (b) Thin-section photomicrograph of the core of the reef illustrated in (a) composed of calcimicrobes (mostly thrombolite clots, T) that form a framework with an internal cavity filled with internal sediment (I) and cement (C). Image width 2 cm. Photograph by E. C. Turner. Reproduced with permission.

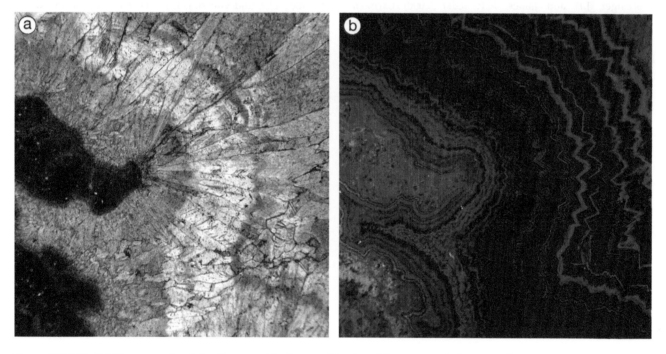

Figure 18.16 Radial fibrous marine dolomite cement in sheet cavities from Neoproterozoic Angepena Formation, Oodnaminta Reef Complex, Flinders Ranges, South Australia: (a) plane-polarized light; and (b) cathodoluminescence, field of view 4 mm. Photograph by M. Wallace. Reproduced with permission.

synsedimentary cements, on the basis of petrographic and CL evidence, were clearly fibrous, length-slow, fascicular dolomite (Figure 18.16). This may reflect the peculiar composition of a profound icehouse and an anoxic ocean. Furthermore, it implies that there may be other similar dolomite precipitates in very ancient rocks.

From the above, although it is clear from petrography and geochemistry that most Proterozoic dolomites are a replacement phenomenon, is it possible that there was some original dolomite sedimentation? As outlined in Chapter 2, modern seawater has a Mg:Ca ratio of 5:1 and is therefore chemically supersaturated with respect to dolomite. Dolomite should be precipitating today, but it is not because of kinetic constraints. If the Mg: Ca were higher in the Proterozoic, the seawater temperature was higher, the pCO_2 was higher, the SO_4 content was lower, and microbes were more involved, then these kinetic constraints might have been absent or reduced and it is entirely possible that dolomite could have precipitated directly from seawater. Given our rudimentary understanding of the Proterozoic ocean–atmosphere system, such a situation should not be discounted.

Further reading

Awramik, S.M. (1991) Archaen and Proterozoic stromatolites. In: Riding R. (ed.) *Calcareous Algae and Stromatolites*. Berlin: Springer-Verlag, pp. 289–304.

Grotzinger, J.P. and Rothman, D.H. (1996) An abiotic model for stromatolite morphogenesis. *Nature*, 383, 423–425.

Grotzinger, J.P. and Knoll, A.H. (1999) Stromatolites in Precambrian carbonates: evolutionary mileposts or environmental dipsticks? *Annual Review of Earth and Planetary Sciences*, 27, 313–358.

Grotzinger, J.P. and James, N.P. (eds) (2000) *Carbonate Sedimentation and Diagenesis in the Evolving Precambrian World*. SEPM (Society for Sedimentary Geology), Special Publication no. 67.

Hoffman, P.F. (2011) Strange bedfellows: glacial diamictite and cap carbonate from the Marinoan (635Ma) glaciation in Namibia. *Sedimentology*, 58, 57–119.

Hood, A.V.S. and Wallace, M.W. (2012) Synsedimentary diagenesis in a Cryogenian reef complex: Ubiquitous marine dolomite precipitation. *Sedimentary Geology*, 255–256, 56–71.

James, N.P. and Wood, R. (2010) Reefs. In: James N.P. and Dalrymple R.W. (eds) *Facies Models 4*. Geological Association of Canada, GEOtext 6, pp. 421–447.

James, N.P., Narbonne, G.M., and Sherman, A.G. (1998) Molar-tooth carbonates: shallow subtidal facies of the mid- to late Proterozoic. *Journal of Sedimentary Research*, 68, 716–722.

James, N.P., Narbonne, G.M., and Kyser, T.K. (2001) Late Neoproterozoic cap carbonates; Mackenzie Mountains, northwestern Canada; precipitation and global glacial meltdown. *Revue Canadienne des Sciences de la Terre (Canadian Journal of Earth Sciences)*, 38, 1229–1262.

Riding, R. (2008) Abiogenic microbial and hybrid authigenic carbonate crusts: components of Precambrian stromatolites. *Geologia Croatica*, 61, 73–103.

Sami, T.T. and James, N.P. (1994) Peritidal carbonate platform growth and cyclicity in an early Proterozoic foreland basin, upper Pethei Group, Northwest Canada. *Journal of Sedimentary Research, Section B: Stratigraphy and Global Studies*, 64, 111–131.

Sumner, D.Y. and Grotzinger, J.P. (1993) Numerical modeling of ooid size and the problem of Neoproterozoic giant ooids. *Journal of Sedimentary Petrology*, 63, 974–982.

Sumner, D.Y. and Grotzinger, J.P. (1996) Herringbone calcite; petrography and environmental significance. *Journal of Sedimentary Research*, 66, 419–429.

Turner, E.C., James, N.P., and Narbonne, G.M. (1997) Growth dynamics of Neoproterozoic calcimicrobial reefs, Mackenzie Mountains, Northwest Canada. *Journal of Sedimentary Research*, 67, 437–450.

CHAPTER 19
CARBONATE SEQUENCE STRATIGRAPHY

Frontispiece 800-m-high cliffs of massive Devonian carbonates (lower right) overlain by well-bedded Mississippian limestones and minor shales, Big Bend, Main Ranges of the Rocky Mountains, western Canada.

Origin of Carbonate Sedimentary Rocks, First Edition. Noel P. James and Brian Jones.
© 2016 Noel P. James and Brian Jones. Published 2016 by John Wiley & Sons, Ltd.
Companion website: www.wiley.com/go/james/carbonaterocks

Introduction

Sequence stratigraphy is an important unifying concept in sedimentary geology because the idea not only links sedimentology, stratigraphy, and the seismic expression of rock bodies, but it also allows the interpreter to place the carbonates being studied into an understandable dynamic context. More specifically, *sequence stratigraphy* is an analysis of the sedimentary response to changes in relative sea level and the depositional trends that emerge from the interplay of accommodation (space available for sediments to fill) and sedimentation.

The stratigraphic record of this interplay is transgression and regression. Transgression occurs when accommodation is not limiting and is manifest as either aggradation or retrogradation (Figure 19.1). *Aggradation* is generally vertical or subvertical, whereas *retrogradation* is lateral or sublateral with facies moving progressively shoreward or inboard. *Regression* takes place when accommodation is limiting and is visible as progradation.

The basic unit of sequence stratigraphy is the *depositional sequence* (Figure 19.2). This unit is framed above and below by *sequence boundaries*, which are subaerial unconformities in shallow water and their correlative conformities in deeper water. The sequence is independent of temporal or spatial scale.

Each depositional sequence is a series of different *systems tracts* (Figure 19.3) that develop in response to the combined effects of accommodation change and sedimentation. Systems tracts are linked contemporaneous depositional systems (e.g., peritidal + lagoon + reef margin).

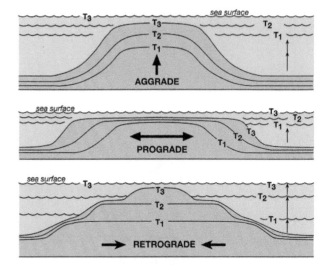

CARBONATE STRATIGRAPHIC PATTERNS

T_1 = Position of sea surface or platform at Time One.

Figure 19.1 Sketch illustrating the stratigraphic response of a carbonate factory to situations when the rate of sea-level rise matches (aggrade), is slower than (prograde), or is faster than (retrograde) the rate of carbonate production and accumulation.

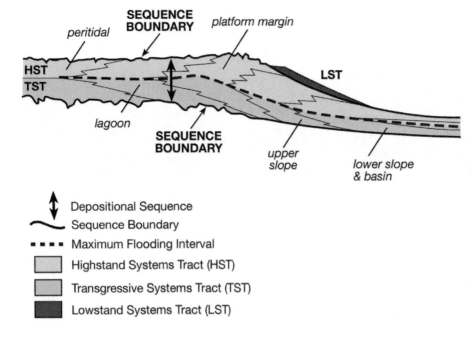

CARBONATE DEPOSITIONAL SEQUENCE

↕ Depositional Sequence
∼ Sequence Boundary
▪▪▪▪ Maximum Flooding Interval
▢ Highstand Systems Tract (HST)
▢ Transgressive Systems Tract (TST)
▆ Lowstand Systems Tract (LST)

Figure 19.2 Attributes of a generic carbonate depositional sequence. The depositional unit is bounded above and below by sequence boundaries in the form of unconformities. The transgressive systems tract (TST) comprises a genetically linked suite of depositional facies that move inboard with time in concert with rising relative sea level. The highstand systems tract (HST) is composed of a suite of genetically linked depositional facies that move basinward with time at the same time as relative sea level is rising slowly or static. The change from TST to HST is marked by the maximum flooding interval. The lowstand systems tract (LST) is formed by those facies that accumulate during the lowest point of sea level fall.

SYSTEMS TRACTS

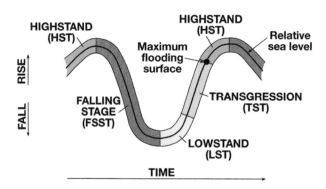

Figure 19.3 Sketch illustrating the different systems tracts during a cycle of sea level.

CYCLE HIERARCHY

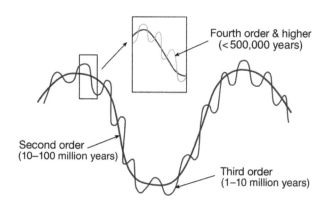

Figure 19.4 Sketch highlighting the hierarchy of different sedimentary cycles and their approximate duration.

A series of superimposed systems tracts are recognized when the depositional sequence is placed against a curve of relative sea-level change (Figure 19.3). The *transgressive systems tract* (TST) develops when relative sea-level rise is so rapid that sedimentation has trouble keeping pace and the facies therefore move shoreward or retrograde (Figure 19.3). In many carbonate systems, however, sedimentation can be sufficiently active as to keep pace with relative sea-level rise and the succession will therefore aggrade (Figure 19.3). This situation continues until the rate of sea-level rise slows, and sedimentation rate becomes greater than the rate at which the ocean surface is rising. Accommodation quickly decreases. From this point onward, because there is little accommodation, the sedimentary system is regressive and moves seaward or progrades. This regressive and prograding system is manifested as the *highstand systems tract* (HST) because it generally occurs just prior to the apex of sea-level rise.

The point between the transgressive and highstand systems tracts or, more specifically, the point at which the system changes from transgressive to regressive, is called the *maximum flooding surface* (Figure 19.3). This change can be rapid across a small stratigraphic interval or somewhat prolonged, in which case it is termed a *maximum flooding interval*.

Regression can be either normal or forced. *Normal regression* occurs during the HST when relative sea level is either rising slowly or is static and sediments gradually prograde. In carbonate sequence stratigraphy, it is at times useful to differentiate between different stages of normal regression. Initial or early highstand (EHST) is when accommodation is still being filled and so progradation is slow. Late highstand (LHST) is that period just prior to relative sea-level fall when sea level is nearly stationary and so progradation is rapid. *Forced regression* occurs

when sea level falls because of eustasy or tectonics and sediments rapidly prograde; the resultant rock body is termed a *falling stage systems tract* (FSST; Fig. 19.3). The *lowstand systems tract* (LST) develops when relative sea level is at its lowest point.

Systems tracts can be divided into relatively conformable successions of genetically related beds or bedsets bounded by flooding surfaces called *parasequences*. Parasequences in carbonate systems can be composed of meter-scale shallowing-upward cycles, even though flooding surfaces may not bound the units.

Sequences are either *allogenic*, where their character is a function of tectonic subsidence and eustatic (global) sea-level change, or *autogenic*, where their composition is controlled by sediment supply and local climate. Relative sea level is the space between the sea surface and the surface of the sediment pile, regardless of the cause. This datum is a function of tectonic subsidence, sediment compaction, and uplift. There is a hierarchy of stratigraphic cycles. Long-term cycles (first- and second-order) are driven by long-term sea-level changes (10–400 million years). Shorter third-order cycles (1–10 million years duration) are generally ascribed to glacioeustatic change. The shortest-term fourth-order and higher cycles could be driven by climate or glacial eustasy (Figure 19.4).

Carbonate sequence stratigraphy

The factors that are especially important for carbonate systems are climate, *in situ* sediment production, the presence or absence of reefs, and shelf margin cementation. Carbonate systems tracts mostly reflect the interaction between accommodation and sediment production.

Resultant carbonate rock bodies depend on the creation and destruction of accommodation as determined by the changes in relative sea level, caused by the interaction between tectonics and eustasy but also, most importantly, evolving marine biology.

The concept of accommodation has been usefully modified for carbonate systems into two types. *Physical accommodation* occurs when the system is dominated by hydrodynamic processes (i.e., there are no reefs). The loose sediment moved by waves and currents tends to build up to base level (either the sea surface or wave base, that is, the base of wave abrasion); once that is exceeded, it is exported to the slope and basin. This surface is commonly below sea level. *Ecological accommodation*, by contrast, is the ability of organisms to adhere to one another and build reefs and hence accumulate above the physical accommodation surface.

Marine carbonate deposition requires that there be space for the deposits to form and to accumulate. This seemingly obvious prerequisite is a combination of the position of the sea surface and regional subsidence. In some situations, carbonate sediment production can be so rapid that the seafloor aggrades to sea level and can rapidly reduce accommodation to nothing. This is called a *shallowing-upward motif* and typifies much of carbonate stratigraphy.

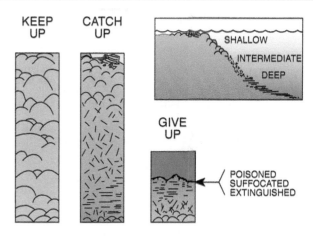

Figure 19.5 Growth strategies illustrated by reefs during a period of rising sea level. (Inset) The shallow, intermediate, and deep zones of reef growth; columns indicate the response of these zones to different rates of sea-level change. *Keep-up* reefs track sea level and have uniform facies throughout. *Catch-up* reefs can have initial shallow water facies, then lag behind sea-level rise for a short time only to grow upwards through progressively shallower waters to reach sea level. *Give-up* reefs start to grow, but are killed off by excessive nutrients (poisoned), abundant fine sediment (suffocated), or rapid subsidence below the photic zone (extinguished). Source: Adapted from James and Bourque (1992). Reproduced with permission of the Geological Association of Canada.

Shallow-water reef sequence stratigraphy

Reef nucleation

The initiation of reef growth can occur: (1) during a rise of sea level after subaerial exposure; (2) during sea-level fall when a suitable seafloor drops to within the reef growth window; (3) when reef growth creates an elevated seafloor for a new type of reef community; (4) when any local factor hostile to reef growth is eliminated; and (5) during a change in local tectonics. Where the initial reef grows will be determined by the ecological demands of the community.

Keep-up, catch-up, and give-up

The relationship between accommodation and sediment production, especially reef growth, can be encapsulated in three simple situations (Figure 19.5).

The carbonate factory can *keep-up* with sea-level rise, in which case the stratigraphic record is one of uniform facies throughout. If the rate of sea-level rise is relatively slow, the carbonate factory can *catch-up* to rising sea level. The

stratigraphic record in this case is one of progressively shallowing facies. Alternatively, if sea-level rise is just too rapid for the carbonate factory production to keep pace, then the stratigraphic succession deepens upward, finally ceases production, and *gives up*. In this latter case, the moment of surrender is usually represented by a bored hardground with a mineralized (P, N, Fe) surface; it was either poisoned by high levels of nutrients, extinguished by subsiding below the photic zone, or suffocated by siliciclastic mud.

Reef body geometry and internal structure

The development of any particular stratigraphy is determined by changing accommodation and the ability of the community to respond to that change. Responses are hierarchical and can be superimposed.

Aggrading reefs. An aggrading reef shows continuous near-vertical growth as the shallow reef system constantly keeps-up with sea-level rise and remains submerged (Figure 19.6). This growth is dominant when the amplitude of sea-level fluctuations is large and the period short, and when rates of production barely match those of accommodation space increase. Transgressive systems tract (TST) reefs can be narrow but thick accumulations, with near-vertical margins and limited peri-reefal sediment. They aggrade in a keep-up mode, but continuously lag behind sea-level rise and the reef community grows at an intermediate depth. Early highstand systems tract (EHST) reefs are keep-up to catch-up, evolving from intermediate-depth to shallow-water structures. Late highstand systems tract (LHST) reefs are catch-up and may show localized exposure surfaces and minor progradation. As long as there is sufficient underlying sediment to act as a foundation, lowstand systems tract (LST) reefs will prograde.

Successive phases of reef growth result in stacked structures during subsequent sea-level cycles (Figure 19.6, upper left). Relief above surrounding sediments depends upon rates of inter-reef sedimentation during lowstands. Alternatively, if long-term sea-level rise is rapid, then reefs will nucleate successively upslope, producing backstepping geometries (Figure 19.6, upper right).

Compound reefs. Compound reefs show aggradation followed by progradation. This situation can develop when the amplitude and period of sea-level fluctuation was moderate or when reef growth could easily match the fastest rates of sea-level rise (Figure 19.7). TST reefs with plenty of accommodation were constantly in catch-up or keep-up mode, and this part of the reef is normally but narrowest. EHST reefs grow up to sea level (catch-up) and form wide flat facies with growth focused on the oceanward side. LHST reefs have limited accommodation and so exhibit marked progradation and have wide reef flat facies. LST reefs are narrow fringing structures downslope, but on variably cemented fore-reef sediments they may have been prone to slumping. Successive reef phases of reef growth (Figure 19.7, inset) have a pronounced overall prograding geometry.

Progradational reefs. Progradation occurs when sea-level change is minimal and of long duration and the reef community growth rate can exceed rates of sea-level change. These reefs grow into shallow water and are often exposed; facies expand laterally due to restricted accommodation (Figure 19.8). Progradation is often established soon after reef initiation if growth is in shallow waters with minimal accommodation. TST reefs are typically zoned structures but the style of EHST and LHST reefs depend on basin relief. If the relief is large, then progradation can continue and wide reef flats will develop. If relief is minimal then reef geometry will flatten and much fore-reef debris would be created.

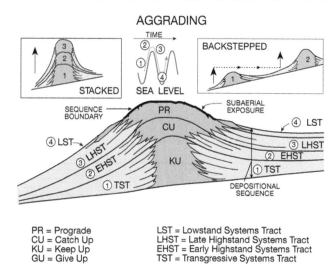

Figure 19.6 A sketch illustrating aggrading reef growth during a rise and fall in sea level when fluctuations were large and of short period (icehouse times), or the reef-builders could barely match rates of sea-level rise. In relatively shallow water aggradation is mainly during TST, EHST, and LHST phases. Such reefs may be stacked to form large complexes or, if sea-level rise is very rapid, may backstep upslope (insets). Source: James and Wood (2010). Reproduced with permission of the Geological Association of Canada.

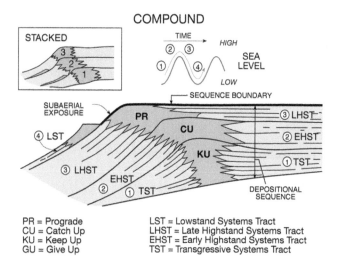

Figure 19.7 A sketch illustrating compound reef growth patterns during a rise and fall in sea level when the scale and period of sea-level fluctuations were intermediate, or reef-builders could match the fastest rates of sea-level rise. The motif is one of aggradation followed by progradation. TST reefs are mostly keep-up; EHST reefs are catch-up and LHST reefs are prograding. Inset shows stacking of reefs as the result of growth during several major fluctuations in sea level. Source: James and Wood (2010). Reproduced with permission of the Geological Association of Canada.

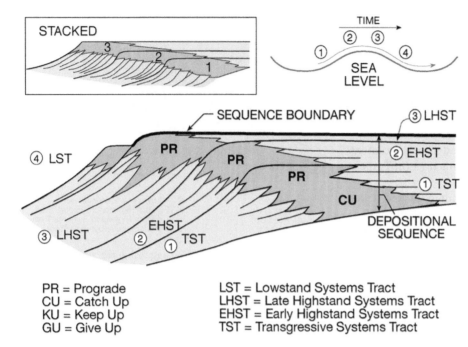

Figure 19.8 A sketch illustrating prograding reef growth during a rise and fall in sea level when changes in sea level were small and of long duration (greenhouse times), or the reef-builders could easily exceed the rates of sea-level change. If platform-basin relief is large, progradation prevails during LHST and LST; if small, then biostromes or sand shoals can develop. Inset shows the result of repeated progradation during several sea-level cycles. Source: James and Wood (2010). Reproduced with permission of the Geological Association of Canada.

Successive phases of reef growth then result in relatively thin structures with a pronounced progradational geometry. LST development will be represented by a slight downshift in facies.

Retrogradational reefs. This situation occurs when relative sea-level rise outpaces reef growth and the whole system is in give-up mode and backsteps (Figure 19.9). Once the system catches up during a highstand phase, progradation dominates.

Photozoan rimmed platforms

Depositional conditions

As outlined in Chapter 10, these structures are largely tropical and have a relatively narrow rim (kilometer-wide) and a shallow-water carbonate factory (<100 mwd). The shallow margin is either reefs with abundant marine cements or carbonate sand bodies (particularly ooids) with hardgrounds. The margin can accrete vertically or prograde laterally. This rim can, in fact, produce sediment much more rapidly than the interior lagoon. Large amounts of sediment can be shed onto the slope and into the basin because the factory produces much more sediment than can be stored on top. Such platforms can be attached to land or grow as unattached offshore banks.

Stratigraphy

Transgressive systems tract. The underlying older platform is progressively flooded. The rate of accommodation creation is usually greater than the rate of sediment creation except for the rim, and this can result in rimmed platforms with deep lagoons (Figures 19.10, 19.11). These are truly "bucket" systems with sediments accumulating in a depression surrounded by an elevated rim. Lagoons can become euxinic because of water stratification or develop hardgrounds due to low sedimentation rates. Alternatively, under arid climates the lagoon can become filled with evaporites. Ooid and peloid facies are abundant, especially during times when large reef-builders are absent and reefs are located downslope. This is the time of maximum vertical accretion, and there is not much slope or basin sedimentation. Stratigraphic sections can illustrate either keep-up or catch-up motifs.

Maximum flooding interval. The rate of accommodation creation is largely matched by the rate of sediment production. The switchover is reflected in stratigraphy by a noticeably high proportion of shallowing-upward facies.

Highstand systems tract. The rate of sediment production and accumulation is now greater than the creation of accommodation. Sand bodies and reefs at the margin are

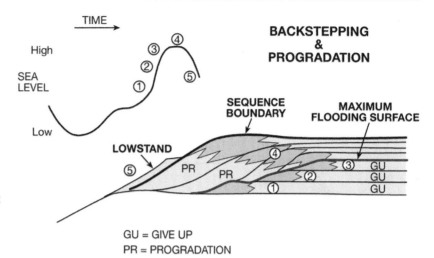

Figure 19.9 A sketch illustrating retrogradational reef growth during rapid sea-level rise and fall. Reefs backstep and transgress during the TST mainly in give-up mode. The situation quickly turns around during the highstand, with mainly catch-up followed by progradation as relative sea level drops.

PHOTOZOAN RIMMED PLATFORM

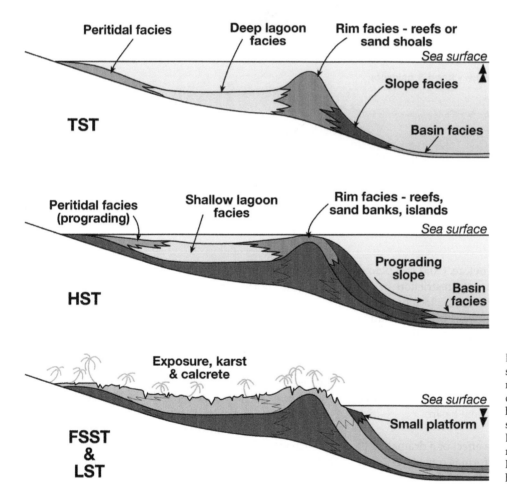

Figure 19.10 The sequence stratigraphy of an attached rimmed carbonate platform during a sea-level rise (TST), highstand (HST), and falling stage to lowstand (FSST and LST). Arrows at right denote relative movement of sea level. Prograding slope is the result of highstand shedding.

PHOTOZOAN OFFSHORE BANK

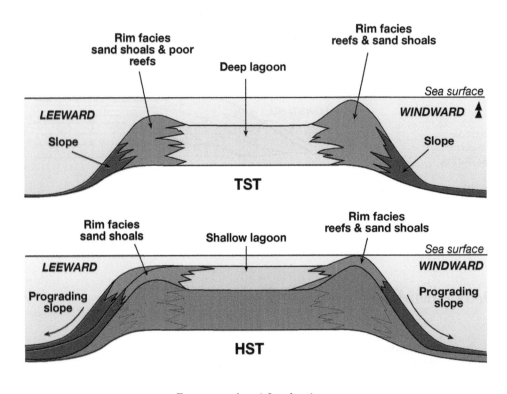

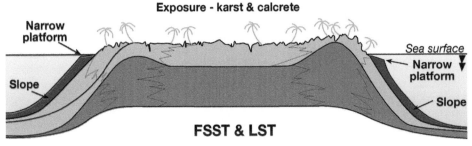

Figure 19.11 The sequence stratigraphy of a rimmed offshore bank during a sea-level rise (TST), highstand (HST), and falling stage to lowstand (FSST and LST). Arrows at right denote relative movement of sea level.

wider and more continuous than before. Ooid and peloid facies are widespread. The degree of restriction in the lagoon increases, so sedimentation can be locally evaporitic. Numerous facies shallow-upward and the relative proportion of mud increases, resulting in wide prograding peritidal mudflats.

This is a time of *highstand shedding*; because of low accommodation on the platform, large amounts of sediment are exported to the slope and basin and the whole margin can prograde into the adjacent basin.

Falling stage systems tract. The effect of a dramatic fall in sea level is profound. The platform is quickly exposed

with erosional processes at the shoreline as sea level falls, leading to short-term active downslope redeposition.

Lowstand systems tract. The whole shallow platform is drained and subaerially exposed; the carbonate factory dies and production shifts to the steep slope where a small narrow factory exists but is not capable of producing much sediment to be deposited downslope. There is no source of carbonate sediment to be swept onto the slope and into the basin and, as a result, platform-generated deep-water carbonate sedimentation stops (see Chapter 17). In the Paleozoic when there was no pelagic carbonate factory, slopes and adjacent basins were carbonate-starved during this time and deposits were either shales in open marine settings or evaporites in saline basins.

Bounding surfaces

Sequence boundaries in platforms are of two main types: (1) meteoric exposure unconformities; and (2) drowning unconformities.

Meteoric exposure unconformities. These surfaces are produced during the exposure of marine carbonate sediments to atmospheric conditions and fresh waters (see Chapter 25) during FSST and LST. If exposure is prolonged then, depending upon climate, there can be extensive lithification or development of surface karst or calcrete (see Chapter 26). Slope and basin sedimentation is arrested. This is unlike siliciclastic systems where the underlying deposits are eroded and transported basinward.

Drowning unconformities. The complete opposite effect occurs when, generally for tectonic reasons, relative sea level rises abruptly and the depositional surface falls below the window of active production, which is usually defined by the photic zone. This effect is generally called *drowning*. Flat-topped platforms suffer most with neritic deposits being abruptly overlain by omission surfaces, commonly hardgrounds, and then either pelagic carbonates (Mesozoic and younger) or shales (Paleozoic). Once again, slope and basin sedimentation is significantly reduced. The concept that this is actually a sequence boundary is currently controversial.

Evaporites and siliciclastics

Evaporites

Evaporite deposition takes place in an overall hot and arid climate. There are several possible situations in which evaporites can be precipitated during the typical sea-level cycle associated with a rimmed platform.

The TST is not usually a time of evaporite deposition because basin connection improves and circulation of open ocean waters across the platform is good. Muddy tidal flat deposition is typically regressive. During the HST the basin is deep and largely open marine, whereas the platform, with low accommodation, can be the site of lagoonal evaporites, peritidal evaporites, local salinas, or marginal marine lakes. If the LST platform is within a basin or rims a large basin, then sea-level drawdown can result in restricted basin circulation and precipitation of basin centre subaqueous evaporites, both sulfates and halites. These deposits can be either deepwater accumulations or, if drawdown is severe, shallow-water deposits in a deep basin.

Siliciclastics

When a platform is attached to land, there is the possibility of siliciclastic sediment being swept out onto and across the platform, especially under humid or semiarid climates.

The shoreline is continuously pushed inboard during the TST and siliciclastics are sequestered along the landward-moving shoreline. With accommodation reduced to a minimum during the HST, siliciclastics can then prograde out over the platform, but not usually to the edge. Terrigenous clastic sediments can be swept onto the platform during the LST, covering the karst surface and out over the edge into deep water, especially when the shoreline is at the shelf edge.

The overall result is a phenomenon known as *reciprocal sedimentation* where LST sedimentation is characterized by thin platform and thick basin sandstone deposits, whereas TST and HST deposits are carbonate (Figure 19.12). It is the logical extension of highstand shedding of carbonates and lowstand shedding of siliciclastics.

Heterozoan unrimmed carbonate platforms

Introduction

These platforms are generally cool-water in the modern world but could have been tropical during times of few reef-builders, especially along passive continental margins (Figure 19.13). This discussion will focus on the cool-water systems, which we know best.

The carbonate sediments are entirely biofragmental and the production window is wide (many tens of km), deep (<200 mwd), and almost wholly depth independent. There is no protective rim. The rate of outboard sedimentation is much lower than in tropical systems, whereas inboard the rates can be roughly equivalent. Sediments are mostly grainy, with mud typically localized inboard and in deep water.

EVAPORITES & SILICICLASTICS

ARID CLIMATE

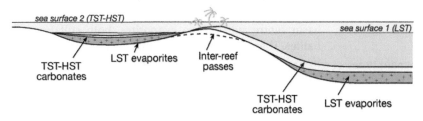

HUMID CLIMATE
(Reciprocal sedimentation)

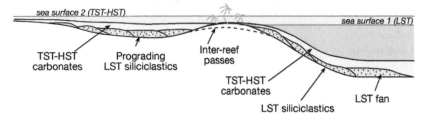

Figure 19.12 Sequence stratigraphy of an attached rimmed platform under a humid and arid climate. Arid conditions (upper panel) favor evaporite deposition on the platform and in the adjacent basin (if subject to restricted marine circulation) during lowstands. Carbonate deposition returns when the platform is covered by the ocean. Humid conditions (lower panel) promote weathering and can result in fluvial input of siliciclastic sediment that is swept across the platform during lowstands. Carbonate deposition returns when the platform is covered by the ocean. The resultant alternating siliciclastic–carbonate stratigraphic motif is called "reciprocal sedimentation."

HETEROZOAN OPEN (UNRIMMED) PLATFORM

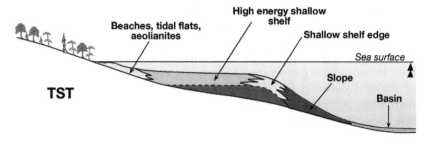

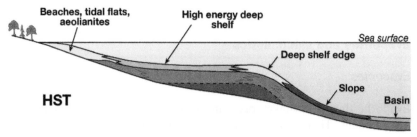

Figure 19.13 Sequence stratigraphy of heterozoan unrimmed (open) attached carbonate platform during a sea-level rise (TST), highstand (HST), and falling stage to lowstand (FSST and LST). Arrows at right denote relative movement of sea level.

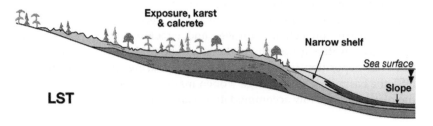

Depositional conditions

The most important attributes that separate cool-water from warm-water carbonates in terms of sequence stratigraphy are: (1) the lack of shallow-water reef-builders; (2) the lower rates of carbonate production; (3) the paucity of synsedimentary cementation; (4) the hydrodynamic redistribution of sediments; and (5) the lack of cementation when exposed to fresh waters. As a result, cool-water carbonate platforms lack an elevated rim, are typically underfilled, aggradation is limited, and the platform top beds are thin. They are, therefore, almost completely molded by allogenic forces, particularly sea-level change, which moves the photic zone, the thermocline, and energy levels up and down and hence back and forth across the platform.

They are like warm-water equivalents in that the sediment factory is *in situ* but hydrodynamic forces are much more important in redistribution of the material. Sedimentation below FWWB is slow but constant. Unlike tropical platforms, a fall in sea level does not shut down the system; it merely moves it seaward albeit to a narrower situation. Lack of cementation during exposure means that the loose sediment is extensively reworked during flooding, mixing both old and new sediments and transporting material inboard and outboard, thus somewhat blurring the interpretable stratigraphic record. Low rates of sedimentation mean numerous omission surfaces and mostly subtidal as opposed to peritidal cycles. The zone of mud is always below storm wave base.

Stratigraphy

The surf zone and zone of waves move across the shelf during the TST, eroding unlithified sediment and mixing new and older particles to generate palimpsest deposits. Sediment is moved basinward where it covers the underlying wedge and smothers reef mounds. Inboard facies (seagrasses today) accrete rapidly and prograde seaward. Waves sweep across the entire shelf reworking sediments above swell wave base. Sediment is also moved both landward and seaward. The shelf edge is a zone of prolific bryozoan and sponge growth. These conditions exist into the HST.

As sea level begins to fall, erosion dominates and sediments are reworked and mixed. There is considerable downwasting and sediment gravity flows into deeper water. During the LST, most of the shelf is exposed and the relatively narrow depositional system lies over the old gently sloping seaward edge and on the underlying slope,

creating a lowstand wedge. Reef mounds, generally rich in bryozoans, grow on a slope with much spiculitic sediment.

These combined processes result in an outer shelf and upper slope depositional system that is a series of aggrading and prograding wedges intercalated with transgressive and highstand sediment blankets. The middle shelf is a zone of omission-bounded subtidal cycles with low accumulation rates, whereas the inner shelf can be a zone of aggradation and progradation.

Subaerial surfaces

Karst surfaces are less common compared to those in tropical systems because lithification of the mainly calcitic sediments is slower. Surfaces formed by forced regression are therefore not as easily identified because the relatively soft sediments are easily reworked during subsequent transgression. Differentiation between exposure and flooding and confirmation of sequence boundaries is, as a consequence, somewhat more difficult.

Evaporites and siliciclastics

The relatively cold ocean waters preclude extensive evaporite deposits in the basin, but adjacent semi-arid terrestrial climates can result in peritidal evaporites in sabkha-like mud flats. By contrast, humid climates that are common in temperate latitudes can be devastating for carbonate deposition. If fluvial runoff is active, then carbonate accumulation is relatively slow because it will shut off carbonate sedimentation altogether. On the other hand, if riverine input is low or attenuated by a semi-arid climate, then a situation similar to reciprocal sedimentation can develop.

Ramps

Ramps are usually simple inclines where facies migrate up and down the slope in concert with sea-level change. When sea level falls, the carbonate factory simply stays in the same water depth but shifts downslope (Figure 19.14). Whereas the upslope abandoned environments are subjected to meteoric diagenesis, the factory does not die; it merely moves. With sea-level downshift, the record is one of an updip unconformity and downdip shallow-water facies overlying deeper-water facies. Accommodation is generally not a problem because, once it is filled, facies simply prograde basinward down the ramp.

RAMPS

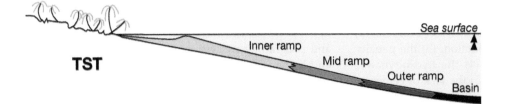

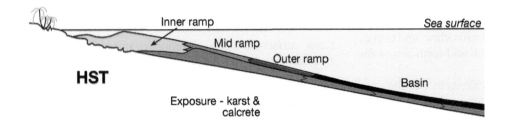

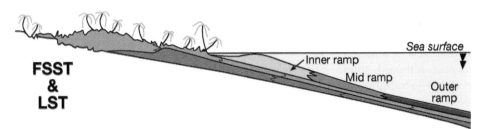

Figure 19.14 Sequence stratigraphy of a carbonate ramp during a sea-level transgression (TST), highstand (HST), and falling stage to lowstand (FSST and LST). Arrows at right denote relative movement of sea level. The main sedimentary response is for facies to move up or down the ramp in response to sea-level change.

During a relative sea-level rise, shallow ramp facies will onlap the bounding unconformity of the previous sequence in the same way as marginal sand bodies (inner ramp). As water deepens over the outer ramp, carbonate production will decrease there and shaley, organic-rich sediments could accumulate in these outboard environments. Under arid conditions, hypersalinity can cause density stratification with high potential for organic matter preservation. As the TST approaches the maximum flooding surface, coastal onlap reaches its most landward position and ramp facies begin to turn around and prograde seaward on top of and downlap onto outer ramp facies, initiating highstand sedimentation. The HST is characterized by an offlapping narrow lagoonal or sabkha complex. These facies prograde into deeper water and the ramp can become distally steepened with a distinctive slope break. The entire ramp sequence is generally exposed during the LST. If the climate is humid, siliciclastic sediments can be transported across the exposed surface by a variety of fluvial processes; if the climate is arid, the siliciclastics can move across the exposed surface as aeolian dunes. LST prograding complexes are common features of ramps because of the low depositional angles. These LST deposits usually have siliciclastics at the base and high-energy grainstones at the top. Under arid climates the shallow basin in front of the ramp can even become hypersaline with subaqueous evaporite deposition. Finally, during drowning, the factory merely shifts upslope but all previous deposits are buried either by basinal shales or pelagic carbonates.

Higher-order cycles (parasequences)

Most sequences described above are the result of second-order sea-level changes. Superimposed on these relatively long-term sea-level oscillations are higher-, third- and fourth-order frequency fluctuations (Figure 19.4) that commonly result in smaller depositional packages called *parasequences* or *cycles*. Carbonate parasequences range

SHALLOWING-UPWARD CARBONATE CYCLES

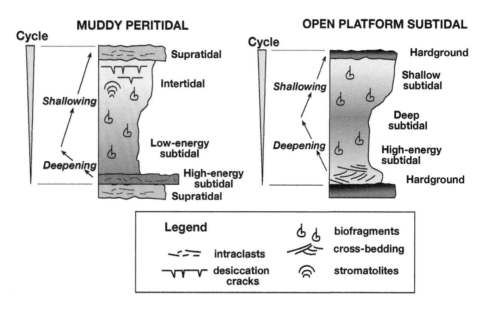

Figure 19.15 Two common recurring meter-scale shallowing-upward stratigraphic motifs that result from sea-level fluctuation. The "muddy peritidal cycle" records an initial thin deepening unit with transgression and high energy and then shallows-upward via progradation of the tidal flat wedge. The cycle top is a subaerial exposure surface. The "open platform subtidal cycle" records initial transgression and upward shallowing via accretion until the seafloor intersects fair-weather wave base, when sediment is swept away and hardgrounds typically develop.

from a meter to tens of meters in thickness. It is these higher-order changes that control the classic bedding patterns of carbonate stratigraphy.

Depositional cycles

Accommodation is framed by two different bounding surfaces – the sea surface (upper) and fair-weather wave base (lower) – thus producing two different cycle types (Figure 19.15). Cycles are largely differentiated by the nature of the uppermost capping unit.

Peritidal cycles

Carbonate sediments obviously cannot accumulate above the sea surface, except perhaps on the inner parts of muddy tidal flats. Peritidal cycles are therefore limited by and develop a predictable cycle developed on rimmed platforms and inner ramps. The cycle comprises a thin basal flooding unit, a thicker overlying subtidal unit, and a capping muddy or grainy peritidal unit.

Subtidal cycles

If swept by ocean waves, as on unrimmed platforms or mid ramps, the sediments cannot accumulate above the energetics of FWWB or they would be swept away. The

depositional cycle or parasequence under such conditions comprises a thin transgressive unit, a thicker subtidal unit, and a thin cap, usually in the form of a subtidal hardground. This hardground and associated hiatus is a submarine unconformity.

Further reading

Bova, J.A. and Read, J.F. (1987) Incipiently drowned facies within a cyclic peritidal ramp sequence, Early Ordovician Chepultepec interval, Virginia Appalachians. *Geological Society of America Bulletin*, 98, 714–727.

Catuneanu, O., Abreu, V., Bhattacharya, J.P., *et al.* (2009) Towards the standardization of sequence stratigraphy. *Earth-Science Reviews*, 92, 1–33.

Goldhammer, R.K., Dunn, P.A., and Hardie, L.A. (1990) Depositional cycles, composite sea-level changes, cycle stacking patterns, and the hierarchy of stratigraphic forcing: examples from Alpine Triassic platform carbonates. *Geological Society of America Bulletin*, 102, 535–562.

Harris, P.M., Saller, A.H., and Simo, J.A. (eds) (1999) *Advances in Carbonate Sequence Stratigraphy; Application to Reservoirs, Outcrops and Models*. Tulsa, OK: Society for Sedimentary Geology (SEPM), Special Publication no. 63.

James, N.P. and Bourque, P.-A. (1992) Reefs and mounds. In: Walker R.G. and James N.P. (eds.) *Facies Models: Response to Sea Level Change*. St John's, Newfoundland: Geological Association of Canada, pp. 323–347.

James, N.P. and Wood, R. (2010) Reefs. In: James N.P. and Dalrymple R.W. (eds.) *Facies Models 4*. St John's, Newfoundland: Geological Association of Canada, GEOtext 6, pp. 421–447.

Loucks, R.G. and Sarg, J.F. (1993) *Carbonate Sequence Stratigraphy; Recent Developments and Applications.* Tulsa: American Association of Petroleum Geologists, Memoir no. 57.

Lukasik, J.J. and James, N.P. (2003) Deepening-upward subtidal cycles, Murray Basin, South Australia. *Journal of Sedimentary Research*, 73, 653–671.

Neumann, A.C. and Macintyre, I. (1985) Reef response to sea level rise: keep-up, catch-up or give-up. *Proceedings of the Fifth International Coral Reef Congress, Tahiti*, 3, 105–110.

Osleger, D. (1991) Subtidal carbonate cycles: implications for allocyclic vs. autocyclic controls. *Geology*, 19, 917–920.

Osleger, D.A. (1998) Sequence architecture and sea-level dynamics of Upper Permian shelfal facies, Guadalupe Mountains, southern New Mexico. *Journal of Sedimentary Research*, 68, 327–346.

Pomar, L. and Ward, W.C. (1994) Response of a late Miocene Mediterranean reef platform to high-frequency eustasy. *Geology*, 22, 131–134.

Tucker, M.E. (1993) Carbonate diagenesis and sequence stratigraphy. In: Wright, V.P. (ed.) *Sedimentology Review*, Blackwell Publishing Ltd., Vol. 1, pp. 51–72.

CHAPTER 20
THE TIME MACHINE

Frontispiece A mountainside of Cambrian recessive weathering shales and resistant limestones (trees ~3 m high) in the foreground and a wide saline lake with Holocene carbonate deposition in the background; The Great Basin, Utah, USA.

Origin of Carbonate Sedimentary Rocks, First Edition. Noel P. James and Brian Jones.
© 2016 Noel P. James and Brian Jones. Published 2016 by John Wiley & Sons, Ltd.
Companion website: www.wiley.com/go/james/carbonaterocks

Change is the only thing that is constant in our world.

Introduction

The preceding chapters have laid out the principles of carbonate sedimentation by describing the processes operating in modern marine and terrestrial settings and then applying those principles in a general way to the rock record. A clear theme that emerges from this approach is that the rock record of such processes is dynamic and complex and the modern is only a guide to interpreting the past. The focus of this chapter is to place some of the themes developed in previous chapters into an understandable time perspective.

The three main controls that determine the nature of limestone formation at any given time are global tectonics, biological evolution, and very-long-term secular geochemical changes. Global tectonics is the long-term driver of all geological processes on the planet, except biology. Organisms have in turn modified our world from the very beginning to the modern day and continue to do so. Cyanobacteria, for example, were directly responsible for the change in the Earth's environments from reducing to oxidizing in the Archean; without them there would be no oxygen! It is estimated that all available planetary oxygen passes through organisms every 2000 years; all available CO_2 passes through organisms every 300 years; and all H_2O in the ocean is broken down and reformed by organisms every 2 million years.

Carbonates and plate tectonics

By determining the spatial position of the cratons and global sea level, plate tectonics control where carbonate platforms grow. Plate movement dynamics decide the rates of seawater movement through the mid-ocean ridges, a process that in turn influences the chemistry of seawater and therefore the mineralogy of marine carbonate precipitates. The disposition of plates controls the patterns of seawater circulation that, in turn, influence climate.

The supercontinent cycle of rifting, seafloor expansion, contraction, and collision (the Wilson Cycle) has played itself out at least four times in Earth history (Figure 20.1). Global tectonics was particularly active during each Wilson Cycle, with extrusive ocean volcanism displacing seawater, resulting in flooding of the cratons. In contrast, tectonic quiescence during periods of supercontinents led to relatively high-standing continents and narrow continental margins.

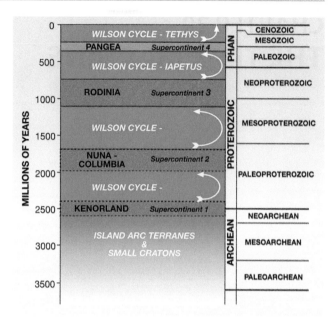

Figure 20.1 A conceptual diagram of alternating periods of active plate tectonic movement (Wilson Cycle) and periods of supercontinents (e.g., Rodinia).

The Archean is somewhat enigmatic in this context because it is not certain if plate tectonics, as we now know it, were operational at that time. Regardless, Archean carbonates were present, mostly in the form of narrow, poorly developed platforms adjacent to small volcanic-dominated cratons. Cratonization at the beginning of the Proterozoic provided wide shallow-marine environments that were quickly occupied by vast platforms that persisted through the first two supercycles.

The Phanerozoic has a much clearer geotectonic-carbonate history. Following Neoproterozoic rifting of the supercontinent Rodinia and widening of oceans such as the Iapetus Ocean, carbonate platforms were first sited along continental margins. Using North America as a well-studied example, this situation persisted along the western margin of the craton adjacent to the Proto-Pacific until well into the Ordovician, whereas in the east, recurring collision of microcontinents focused carbonate deposition into foreland basins. The global highstands of sea level during the early and middle Paleozoic led to flooding of the cratons and the appearance of vast epeiric seas (Figure 20.2). Most middle and late Paleozoic carbonates developed in intracratonic or foreland basins. Carbonate deposition during the late Permian and Early Triassic was relatively minor and confined to intracratonic basins on the supercontinent Pangea (Figure 20.3).

Early Mesozoic breakup of Pangea resulted first in Triassic rift-basins and then in elongate Jurassic seaways

Late Ordovician
(450 Ma)

Figure 20.2 Interpreted disposition of continents in the Late Ordovician. North America was undergoing tectonism along the eastern margin while most of the continent was covered by an epeiric sea (E) resulting from globally high sea level. Image courtesy of Ron Blakey and Colorado Plateau Geosystems, Inc.

with extensive continental margin platforms at low latitudes (Figure 20.4). Continued widening during the Cretaceous led to circumglobal circulation and continental margin carbonate platforms that encircled the globe in tropical realms. Globally high sea level also led to extensive deposition on the cratons (Figure 20.5). Closure of the Mediterranean and the Isthmus of Panama in the Cenozoic resulted in more meridonal (longitude-parallel) circulation; carbonate platforms, although extensive, were more partitioned and not as laterally extensive.

Epeiric seas

The conundrum. Although the geological axiom that "the present is the key to the past" is largely true, it fails when confronted with the record of carbonate deposition in ancient epeiric seas. Epeiric environments developed where the surfaces of cratons and continents were flooded by marine water to form vast shallow epicontinental marine seas many thousands of kilometers across (Figure 1.8). The term was first used to explain the origin of perplexing shallow-water carbonate rocks that extended for vast distances across cratons. Such an extent implies that depositional gradients were either flat or inclined at an extremely low angle. The problem is that

Late Permian
(260 Ma)

Figure 20.3 Interpreted disposition of continents in the late Permian to form the supercontinent Pangea with shallow seas confined to the supercontinent margin. Image courtesy of Ron Blakey and Colorado Plateau Geosystems, Inc.

there are no modern equivalents! Understanding how these depositional systems worked is critical, because this

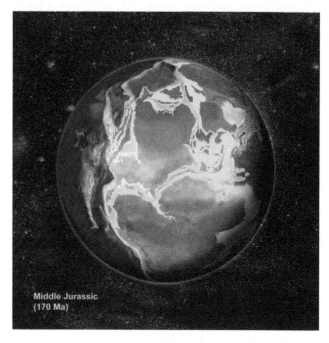

Middle Jurassic
(170 Ma)

Figure 20.4 Interpreted disposition of continents in the Middle Jurassic during fragmentation of Pangea, with shallow seas localized to recent rift zones. Image courtesy of Ron Blakey and Colorado Plateau Geosystems, Inc.

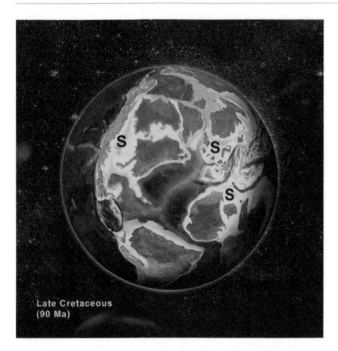

Figure 20.5 Interpreted disposition of continents in the Late Cretaceous, a time of extraordinarily high sea level with much of North America and other continents covered by shallow seas (S). Image courtesy of Ron Blakey and Colorado Plateau Geosystems, Inc.

is where the many fossil "open marine subtidal" carbonates are preserved and where we get much of our information about the ancient marine world.

The most extensive epeiric seas veneered cratons during global highstands of sea level (greenhouse periods; see below) with the most spectacular being during the early–middle Paleozoic and the Cretaceous, times of the very different old pelagic limestones and young pelagic limestones (see Chapter 17). Classic epeiric seas were, as indicated by the facies and biota, generally shallow marine throughout and could extend thousands of kilometers across the craton (e.g., lower Paleozoic carbonates of North America extended from New York to California with shallow water depths throughout; Figure 20.2).

The oceanography and the controls on deposition are difficult to comprehend, especially when shallow-water environments were 10–200 m deep and extended for hundreds of kilometers over a flat seafloor with little change in the biota. The lack of modern analogs means that the oceanic conditions in these long-vanished seas must be deduced from the rocks themselves, hydrodynamic modeling, and experience from modern settings. The major problems concern the oceanography, particularly the role of temperature, salinity, tides, and storms.

Oceanography. The continuity of normal marine benthos across vast epeiric seafloors demands that the seawater itself was of near-normal salinity throughout. This situation suggests that if there was extensive evaporation, then it must have been balanced somewhat by rainfall, much like parts of Indonesia today. If so, there must have been fluvial input, but it appears that most such sediments were trapped along the shoreline. The paleogeography and shallowness of the water predicts that photozoan taxa should predominate. Furthermore, such shallow environments would have promoted early seafloor lithification in the form of firmgrounds and hardgrounds (see Chapter 24).

Today, the idea of strong tidal currents in this system being damped by bottom friction is hotly contested. Consensus, backed up by modeling, suggests that many epeiric seas were microtidal to virtually tideless due to their relative isolation from the open ocean tidal bulge (see Chapter 13). Such damping, however, depended on how well the epeiric sea was connected to the open ocean.

These epeiric seas, in the absence of tidal mixing and their microtidal regimes, would have been susceptible to stratification, oxygen depletion, mass mortality, and organic carbon preservation. Since friction would have damped the amplitude of tides, the influence of fair-weather waves and storms would have dominated. If modern shallow, somewhat isolated, seas and large lakes offer even partial analogs, wind-driven waves and storms would have induced partial mixing. Such waves can today reach 60 mwd and models predict that they could have disturbed sediments to 100 mwd. The presence of muddy tidal flats in epeiric seas may therefore attest to the importance of storm-driven flooding rather than the action of tides.

The existence of cross-laminated carbonate sand bodies well away from the open ocean indicates that strong currents were present, as least, locally. These are ascribed to the presence of islands that served to funnel tidal flow and increase bottom stress.

Regardless of the above constraints, the combined effects of temperature and an overall microtidal system would have resulted in a strongly stratified water column. Attendant sluggish water circulation would probably have resulted in dysoxia and anoxia below the thermocline. Biological production would have been localized to shallow waters with organic-rich shales accumulating in deeper waters.

Models. *Epeiric platforms* (Figure 20.6) were formed by carbonate deposition in the shallow waters that covered the cratons. These monotonous environments stretched from ocean to ocean and passed insensibly into proximal

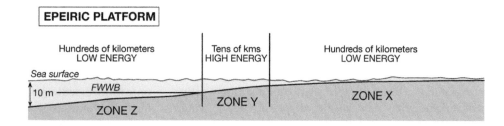

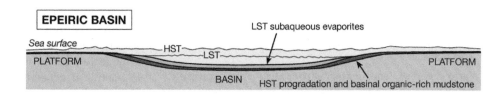

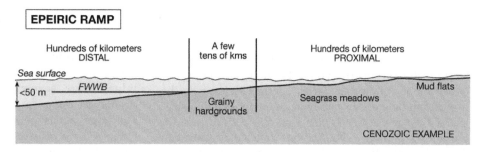

Figure 20.6 Three different aspects of epeiric sea carbonate sedimentation. An epeiric platform utilizing the original terminology, the location of an epeiric basin or depression on the epeiric platform, and an epeiric ramp fringing an epeiric basin. FWWB: fair-weather wave base. Source: Lukasik et al. 2000. Reproduced with permission of John Wiley & Sons.

low-lying land (e.g., early Paleozoic of North America; Triassic and Cenozoic of the Middle East). Depositional environments were assigned to different zones by early workers: Zone Z was the open ocean largely below wave base; Zone Y was that narrow zone under tidal influence (peritidal); and Zone X was that part of the platform inboard and largely beyond the effects of normal tidal exchange, but flooded during large storms. The environment was not necessarily everywhere subtidal and the shallow sea may have been dotted with a series of low-lying islands. Fossiliferous limestones graded landward into ooid shoals or peloidal grainstones that gradually passed into dolomudstone and, finally, evaporite rocks.

Epeiric basins were local, shallow depressions on the platforms (e.g., middle Paleozoic basins of North America and Australia; Mesozoic of the Middle East). Such structural basins or sags were either rimmed or open, but the presence of a rim did little to affect the shallow-water carbonate factory. The seafloor was similar to that on epeiric platforms.

Epeiric ramps were those benthic settings with imperceptible gradients that rimmed intracratonic depressions. Shallow, wide, epeiric ramps in Cenozoic temperate water paleoenvironments (where the biota is largely modern) were divided into proximal and distal facies separated by

a narrow band (tens of kilometers wide) of grainy deposits or submarine hardgrounds that marked the impingement of fair-weather waves on the seafloor. The grainy facies graded shoreward into extensive mollusk-rich grass banks that, in turn, pass shoreward into extensive tidal flat facies similar to those in epeiric platforms. Normal fair-weather waves were dissipated on the ramp by friction. Since the entire seafloor was above storm wave base, episodic storms affected the entire ramp and, together with tidal flat processes, were the dominant agents of sedimentation. Much of their record was, however, erased by burrowing.

Paleoclimate and paleoceanography

Greenhouse and icehouse climatic modes

Global climate and, by association, oceanography, has been divided into different intervals that were controlled by global tectonics. Times of active plate movement and relatively high CO_2 levels coincided with globally warm temperatures (*greenhouse mode*), warm seas, sluggish marine circulation, and high sea levels. Times of relative tectonic quiescence were the opposite, with globally cooler temperatures (*icehouse mode*), local and continental glaciation, cooler oceans overall, invigorated ocean circulation, and relatively low sea levels. Consensus suggests that these modes were the result of variations in plate tectonics. Greenhouse modes correspond to times of rapid seafloor spreading – active volcanism when new basalts displaced ocean water, resulting in continent flooding and possibly elevated levels of atmospheric CO_2 (much of the Wilson Cycle; Figure 20.1). Icehouse modes match those periods of megacontinent formation with much-reduced oceanic volcanism and lowered levels of atmospheric CO_2.

Warm- versus cool-water carbonates. The complex response of carbonate deposition to seawater temperature is clearly reflected in the photozoan versus heterozoan carbonate record. During greenhouse times, the neritic photozoan factory and associated depositional environments expanded away from the tropics into higher latitudes, resulting in spectacular and widespread warm-water carbonates. When the globe was in an icehouse mode, photozoan facies contracted towards the tropics and heterozoan carbonates were more universal.

Carbonate cyclicity. Fluctuating continental glaciation during icehouse times resulted in major excursions of sea level because much water was extracted from the ocean to form vast ice sheets. The amplitude of sea-level fluctuations during such times was on the order of tens of meters and of relatively short duration (100 ky; Figure 20.7). Peritidal and shallow marine cycles in icehouse periods were typically thick because of large-amplitude sea-level fluctuations and resultant significant accommodation. Most of these cycles

were, however, typically amputated because rapid sea-level fall prevented completion of progradation. Muddy tidal flats were not extensive because of the rapid frequency of sea-level fluctuations, so there was not much time for accumulation. Meteoric diagenesis was extensive because, once formed, sediment was exposed to meteoric effects during long intervals of lowered sea level.

In contrast, during greenhouse times, there was little continental glaciation and sea level fluctuated only by a few tens of meters at the most. Carbonate cycles from greenhouse periods are thin because of low accommodation, typically muddy because of extensive shallow subtidal carbonate factories, and subject to extensive early diagenesis. Most of this diagenesis was seawater-related in the form of dolomitization.

Calcite and aragonite seas

Mid-ocean ridges are giant seawater pumping systems. All of the water in the ocean is circulated through the ridges every several million years, but the rate of circulation depends on the rate of seafloor spreading. The ridges act as a huge rock-fluid ion exchange system where Ca^{2+} is released to the fluid while Mg^{2+} is consumed by the alteration of oceanic basalt by hot seawater. Intervals characterized by low spreading rates (equivalent to times of supercontinents) are periods of low hydrothermal brine flux and should lead to elevated Mg:Ca ratios in open ocean seawater. If this ratio rose above ~2 for warm surface water, then aragonite ± HMC would precipitate. Conversely, high spreading rates or times of high hydrothermal brine flux should lower the Mg:Ca

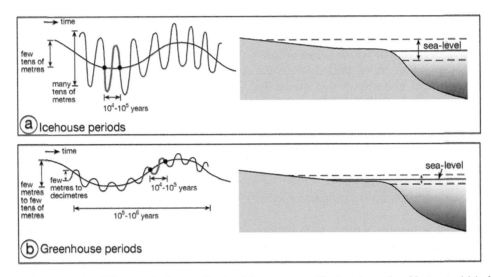

Figure 20.7 Sketch illustrating the differences in the amplitude of short-term oscillations in sea level between (a) icehouse and (b) greenhouse conditions. Source: Skelton (2003). Reproduced with permission of Cambridge University Press.

ratio; if this ratio dropped below ~2 then LMC should precipitate. These tectonically controlled times of LMC versus HMC-aragonite have been termed "calcite seas" and "aragonite seas," respectively (Figure 20.8). Recent work, however, suggests that, because of temperature considerations, aragonite could still precipitate in warm, shallow, tropical platforms in calcite sea times.

Although the understanding of the Precambrian is still rudimentary, these changes can be clearly seen in Phanerozoic supercontinent cycles (Figure 20.8): (1) active

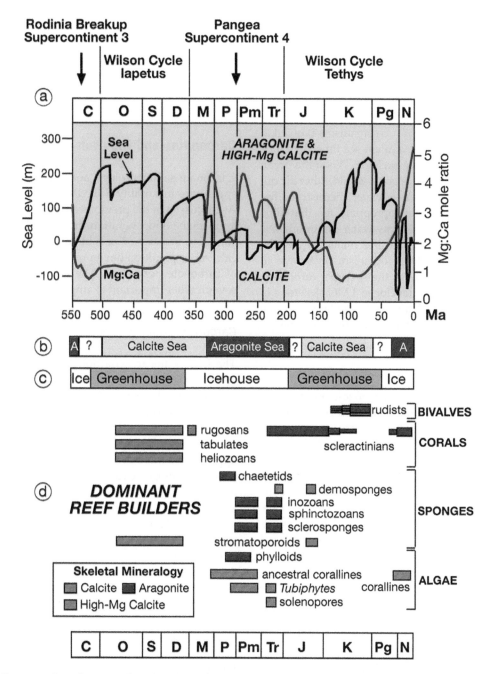

Figure 20.8 (a) Correspondence between changing ocean chemistry and carbonate mineralogy through time as a function of the Mg:Ca ratio in seawater and general global sea-level change. The boundary between fields of calcite (<4 mole % $MgCO_3$), high-Mg calcite (>4 mole % $MgCO_3$), and aragonite is the horizontal line at Mg:Ca = 2. (b) The different dominant non-skeletal mineralogy precipitated in seawater through time. (c) The different global climatic and oceanographic periods. (d) Mineralogy of different reef-building organisms. Source: Adapted from James and Wood (2010). Reproduced with permission of the Geological Association of Canada

seafloor spreading in the early and middle Paleozoic resulted in calcite seas; (2) slow or negligible seafloor spreading at the time of the supercontinent Pangea (Mississippian–Jurassic) resulted in aragonite seas; (3) active spreading (Cretaceous–early Cenozoic) resulted in calcite seas; and (4) slowed spreading (late Cenozoic) resulted in aragonite seas (as today).

Hypercalcifiers

Plate tectonics may have even affected the carbonate biosphere. Hypercalcifiers (Figure 20.8) are those algae, sponges, and corals that are relatively unsophisticated carbonate secreters. It seems that if such organisms evolved in a particular time period, say in a calcite sea time, then they would have produced a calcite skeleton. When the ocean changed to an aragonite sea, this taxon would either die out or have a hard time with the new water chemistry; other taxa with mineralogies more appropriate for that time would go on to dominate shallow-water environments. The conspicuously alternating original mineralogy of the major Phanerozoic reef-building organisms is shown in Figure 20.8. Similarly, among the pelagic carbonate-producers coccoliths, which developed LMC skeletons in the Cretaceous, had a difficult time in the modern aragonite sea, probably due to relatively high Mg concentrations.

Biogenic calcification

The effect of Wilson Cycles on Phanerozoic ocean chemistry is especially important for various biogenic components and has been studied in the laboratory by growing organisms in seawater in which the Mg:Ca ratio has been modified. Coccoliths, whose skeletons are LMC, easily precipitate LMC when the Mg:Ca ratio is low, and this is probably the reason for extensive Middle–Late Cretaceous chalk deposits when the Mg:Ca ratio was close to one. Organisms that secrete Mg-calcite today, such as echinoids, serpulid worms, and coralline algae, changed their mineralogy in concert with changing seawater chemistry. As the Mg:Ca ratio was lowered, for example, the Mg content in skeletal HMC decreased. This same relationship is present in microbial biofilms.

Today, organisms that secrete aragonite under high Mg:Ca seawater ratios, such as corals and calcareous green algae, respond somewhat differently in laboratory experiments under lowered Mg:Ca ratios. Corals changed part of their skeletal mineralogy to LMC, but still retained some aragonite. Green algae remained aragonitic but their calcification rate, growth rate, and primary production was reduced. These reactions suggest that, under conditions of

depleted Mg:Ca ratios in ancient oceans, the algae would have been smaller, less abundant, less competitive for space on the seafloor, and less resistant to grazing pressure.

These results go a long way towards explaining the changing biogenic nature of carbonate sedimentation from the calcite seas of the Cretaceous–early Cenozoic, which were characterized by prolific coccoliths, echinoids, coralline algae, and serpulids, to the resurgence of corals and calcareous green algae in the post-Eocene aragonite ocean once the Mg:Ca ratio of seawater began to rise. Such controls were doubtlessly present in Precambrian seas, but the clues are more difficult to discern.

Carbonates and the evolving biosphere

Carbonate sedimentology and the evolving marine biosphere are inexorably linked through Phanerozoic time. As discussed in the previous chapters, aspects such as the appearance and evolution of skeletal invertebrates, the changing identity of reef biotas, and the appearance of calcareous plankton have profoundly affected the nature of carbonate deposition. The following sections outline several other important examples of this relationship.

Carbonate sediment partitioning

The largely linear evolution of the biosphere stands in contrast to the cyclic nature of global tectonics. Organisms have modified our world, beginning with the generation of oxygen, and have, at the same time, profoundly affected the carbonate system. Their first effect concerns partitioning of marine carbonate deposition through time. There appear to be three periods: (1) the Precambrian, a time of profound abiogenic carbonate precipitation (both benthic and pelagic); (2) the Cambrian–Jurassic, a time of global skeletal invertebrate accumulation; and (3) the Cretaceous–Holocene, a time dominated by pelagic sedimentation (Figure 20.9). Precambrian seas were populated by microbes, and little else. The abundance of synsedimentary cements, as seafloor crusts and major components of stromatolites, suggests that seawater was much more saturated with respect to carbonate than it is today. This saturation state was diminished somewhat by the appearance of calcimicrobes, calcareous algae, and calcareous invertebrates in the Neoproterozoic, because they extracted large amounts of carbonate from the world oceans. The clear evidence of this is the abundance of interparticle and void-filling marine cements in Paleozoic and early Mesozoic reefs and mounds, but the complete absence of synsedimentary seafloor cement crusts. The

PARTITIONING OF CARBONATE DEPOSITION THROUGH GEOLOGIC TIME

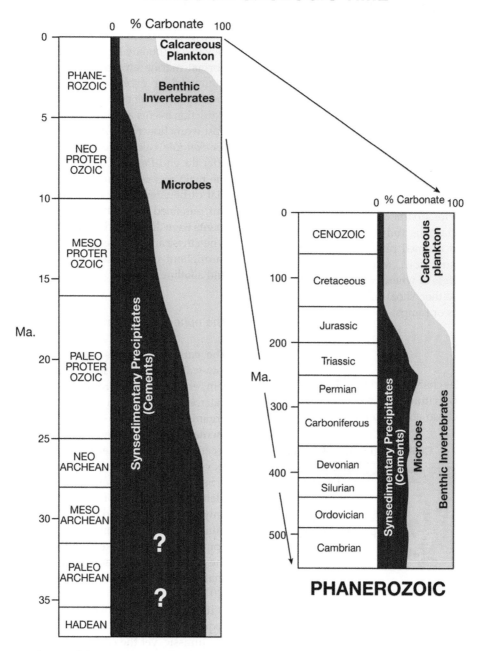

Figure 20.9 A conceptual view of the partitioning of carbonate deposition through geologic time.

evolution of calcareous plankton in the Jurassic changed the dynamics of carbonate sedimentation once again. As emphasized in Chapter 18, calcareous plankton extract enormous amounts of carbonate from the ocean, such that two-thirds of all carbonates today are from calcareous plankton. This dramatic pelagic extraction of seawater carbonate since the Jurassic may go a long way towards explaining the dramatic reduction or disappearance of calcimicrobes, automicrites, and *Stromatactis* at about the same time. The period immediately after the end-Permian extinction is particularly instructive; coincident with the disappearance of skeletal invertebrates, there was an increase in inorganic seafloor carbonate precipitates, but only for a short time until diverse skeletal taxa reappeared in the Early Triassic.

Platform and ramps

A suite of hypercalcifiers such as stromatoporoids or corals is generally needed to build a rimmed platform. When these organisms are absent, carbonate edifices exist as open platforms and ramps (Figure 20.10). This relationship between biological evolution and reef-building has profound implications for not only carbonate structures, but also for facies disposition. Phanerozoic times characterized by rimmed platforms are therefore Middle Ordovician–Late Devonian, Middle Triassic–Early Cretaceous, and various parts of the Cenozoic. There is some debate if Cretaceous rudist bivalves formed true reefs and generated rimmed platforms. During most of the remaining Phanerozoic, when large reef-builders were absent, regardless of where carbonate systems were located they were either shoal-rimmed platforms with downslope buildups or ramps.

Microbes

Skeletal invertebrates characterized most Phanerozoic carbonate depositional environments. Nevertheless, microbes that dominated Archean and Proterozoic seascapes were

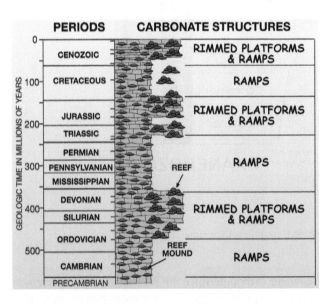

Figure 20.10 A stratigraphic column of Phanerozoic time illustrating the nature of reefs and reef mounds; the position of rimmed platforms and ramps is highlighted. Source: Adapted from James and Wood (2010). Reproduced with permission of the Geological Association of Canada

waiting in the wings to re-emerge as important constituents. Throughout geologic time, microbes and calcimicrobes were always present in stressed paleoenvironments, such as saline lagoons and tidal flats. They were particularly prolific in high-nutrient settings and at times of elevated carbonate saturation. They also quickly occupied normal marine environments when, for whatever environmental or evolutionary reasons, calcareous invertebrates were absent. This is particularly the case after mass extinction events and is repeatedly evident in the nature of post-extinction reefs. After the end-Devonian extinction, for example, calcimicrobes became the main reef-builders; after the end-Permian extinction, other microbialites were the main reef constituents. In short, microbes and microbial carbonates did not disappear after the Precambrian, but remained in the background when marine environments were dominated by invertebrates, only to re-emerge when these animals were removed. They also were present throughout in non-marine settings such as fresh, saline, and alkaline lakes.

The ephemeral biota

The feature that strikes anyone who swims in marine shallow-water carbonate settings today, regardless of water temperature, is the prolific vegetation such as seagrass, macroalgae, and even salt-tolerant trees. Although these organisms all have a profound effect on modern sedimentation (see Chapters 3 and 4), they are comparatively recent additions to neritic carbonate environments and leave little or no good fossil record.

Seagrasses are flowering plants that adapted to a completely aquatic life in the marine environment. There are only a few confirmed fossil seagrass occurrences, and much of our understanding comes from their calcareous epiphytes. Most workers believe that they originated in the Middle–Late Cretaceous and were well established in shallow photic marine environments by the Eocene (40 Ma).

It is generally assumed, largely on the basis of DNA sequencing, that modern macroalgae (kelps) evolved in the Jurassic, yet the earliest-known fossils that can confidently be assigned to the phaeophytes are Miocene. Fossils comparable in morphology are, however, known from strata as old as Late Ordovician, but their taxonomic affinity is far from certain. Fossils termed fucoids (from their similarity to the modern seaweed genus *Fucus*), and others with strong resemblance to the order *Laminariales* (kelps), were present in the middle Paleozoic. Today, brown macroalgae are limited largely to hard substrates, but it is not certain if this was the case in the past.

Most evidence suggests that mangroves (salt-tolerant swamp trees) evolved in the Late Cretaceous and consensus is that they have been widely distributed since the early Eocene. These plants possess a unique combination of morphological and physiological adaptations for living in the tidal environment.

It is intriguing to speculate what shallow-marine environments would be like without these plants and algae.

Ocean acidification

The preceding text has focused on the past, but what about the future of carbonate sedimentation? It seems as though the amount of anthropogenic CO_2 production might be increasing over the last several hundred years. Roughly one-third of this gas goes into the ocean, so is lowering seawater carbonate ion concentration by decreasing pH and thus the saturation state (Ω) of $CaCO_3$. This is referred to as *ocean acidification*, but the oceans are still quite alkaline.

Under experimental conditions, many $CaCO_3$-secreting organisms exhibit reduced calcification with elevated CO_2, decreasing pH, and lowered CO_3^{2-}. Although the implications would seem to be clear, the exact responses of individual taxa are not yet well known because most of our understanding is based on short-term experiments under laboratory conditions. Regardless, the organisms most affected are planktic coccoliths and pteropods, together with benthic mollusks, echinoderms, reef corals, and coralline algae. Although aragonite is soluble under these circumstances, HMC is even more so; coralline algae and echinoderms therefore appear to be the most vulnerable benthic organisms in this context.

Laboratory results of acidification experiments are, however, complex. For example, invertebrate species with different mineralogies, such as corals, clams, echinoids, calcareous worms, gastropods, and bivalves, exhibited decreased calcification levels with increasing pCO_2 (decreased carbonate saturation) and some even displayed dissolution at the highest pCO_2 levels. By contrast, other species such as some echinoids, gastropods, green calcareous algae, and coralline red algae exhibited increased calcification at intermediate pCO_2 levels but decreased rates at the highest levels. Finally, one species, a blue mussel, displayed no response whatsoever to elevated pCO_2.

Several factors might explain these variations: (1) the organism's ability to regulate pH at the site of calcification; (2) the extent of organic layer coverage of the external shell; (3) the organism's biomineral solubility; and (4) whether the organism utilizes photosynthesis. Regardless,

if these experiments are any guide, then it is clear that the short-term effects will be highly variable and not straightforward. Finally, other experimental data on bivalves, gastropods, and echinoderms indicate that acidification of internal body fluids leads to the dissolution of skeletal carbonate to compensate.

Although certain algae and invertebrates would seem to be losers with increasing CO_2 in seawater, seagrasses and their ecosystem will be winners in nutrient-rich waters because of increased rates of photosynthesis. By contrast, it seems that endolithic microbes will bore more deeply into lithic and reef substrates, therefore resulting in greater rates of bioerosion.

Finally, coral reef calcification depends on the saturation state of aragonite in surface waters; if the saturation levels decline, then reef-building capacity also decreases. Some predictions suggest that, by the middle of the next century, atmospheric carbon dioxide concentrations will rapidly rise to double pre-industrial levels. This could be catastrophic for reefs and the modern carbonate system overall because the time scales on which natural feedbacks operate are far longer than the rate of recent CO_2 increase. Experimental work and modeling suggest that an increased concentration of carbon dioxide will lower pH and decrease the aragonite saturation state in the tropics by 30% and biogenic aragonite precipitation by 14–30%. Biological calcification rates are already known to be 10–20% lower than under pre-industrial conditions.

It is early days in this fascinating quest to understand the possible future of the carbonate system, yet it is a critical mission.

Further reading

Allison, P.A. and Wright, V.P. (2005) Switching off the carbonate factory. A: tidality, stratification and brackish wedges in epeiric seas. *Sedimentary Geology*, 179, 175–184.

Allison, P.A. and Wells, M.R. (2006) Circulation in large ancient epicontinental seas: what was different and why? *Palaios*, 21, 513–515.

Doney, S.C., Fabry, V.J., Feely, R.A., and Kleypas, J.A. (2009) Ocean acidification: The other CO2 problem. *Annual Review of Marine Science*, 1, 169–192.

Fischer, A.G. (1981) Climatic oscillations in the biosphere. In: Nitecki M.H. (ed.) *Biotic Crises in Geological and Evolutionary Time*. New York: Academic Press, pp. 103–133.

Hönisch, B., Ridgwell, A., Schmidt, D.N., *et al.* (2012) The geological record of ocean acidification. *Science*, 335, 1058–1063.

James, N.P. and Wood, R. (2010) Reefs. In: James N.P. and Dalrymple R.W. (eds) *Facies Models 4*. Geological Association of Canada, GEOtext 6, pp. 421–447.

Lukasik, J.J., James, N.P., McGowran, B., and Bone, Y. (2000) An epeiric ramp: low-energy, cool-water carbonate facies in a Tertiary inland sea, Murray Basin, South Australia. *Sedimentology*, 47, 851–881.

Mitchell, A.J., Allison, P.A., Gorman, G.J., Piggott, M.D., and Pain, C.C. (2011) Tidal circulation in an ancient epicontinental sea: The Early Jurassic Laurasian Seaway. *Geology*, 39, 207–210.

Pratt, B.R. and Holmden, C. (eds) (2008) *Dynamics of Epeiric Seas*. Geological Association of Canada, Special Paper no. 48.

Ries, J.B., Cohen, A.L., and McCorkle, D.C. (2009) Marine calcifiers exhibit mixed responses to CO2-induced ocean acidification. *Geology*, 37, 1131–1134.

Sandberg, P.A. (1983) An oscillating trend in Phanerozoic non-skeletal carbonate mineralogy. *Nature*, 305, 19–22.

Skelton, P.W. (2003) Changing climate and biota; the marine record. In: Skelton P.W., Spicer R.A., Kelley S.P., and Gilmour I. (eds) *The Cretaceous World*. UK: Cambridge University Press and The Open University, pp. 163–184.

Stanley, S.M. and Hardie, L.A. (1998) Secular oscillations in the carbonate mineralogy of reef-building and sediment-producing organisms driven by tectonically forced shifts in seawater chemistry. *Palaeogeography, Palaeoclimatology, Palaeoecology*, 144, 3–19.

PART III
CARBONATE DIAGENESIS: AN OVERVIEW

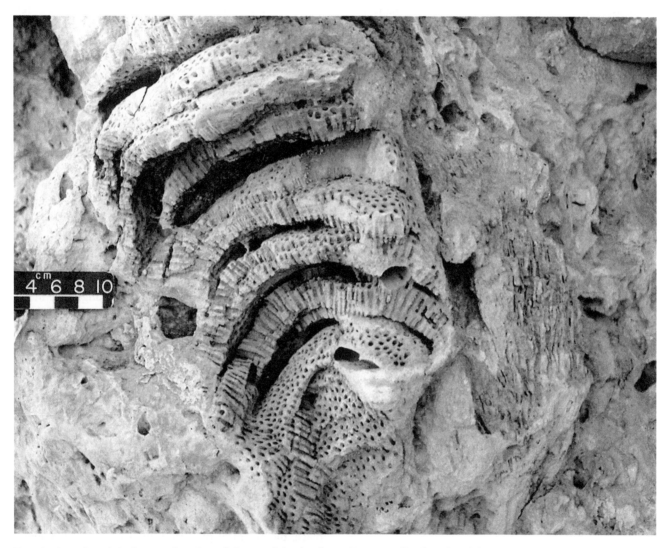

Frontispiece An original aragonite scleractinian coral that has been diagenetically altered to dolomite but retained part of its original fabric, Miocene, Gulf of Suez, Egypt. Centimeter scale.

Origin of Carbonate Sedimentary Rocks, First Edition. Noel P. James and Brian Jones.
© 2016 Noel P. James and Brian Jones. Published 2016 by John Wiley & Sons, Ltd.
Companion website: www.wiley.com/go/james/carbonaterocks

Introduction

The diagenesis of carbonate sediments and rocks, because they are so soluble in natural waters, can be surprisingly rapid or infinitely prolonged, but in all cases the changes are profound. Seafloor sediments can be transformed into hard limestone (a hardground) in mere years. These same deposits can be changed into dolostone such that all hints of original texture and fabric are obliterated. Dissolution can also alter once-bedded limestone into galleries and intricate passages of underground caves. At the same time, two-billion-year-old limestones are so beautifully preserved that they look as though they were made yesterday. Such is the nature of carbonate diagenesis! It is the ultimate intellectual challenge to interpret the products of all these changes in an understandable and defensible way. To do this, it is imperative that all of the depositional, petrographic, and geochemical attributes be integrated. The following chapters are devoted to understanding how all of these alterations take place and how we can recognize them in the geological record.

The approach is to first look at the processes of diagenesis in a general way, the analytical methods commonly used to unravel those processes, and the basic geochemical principles of such diagenesis. With this background, we assess the changes that occur as the carbonate sediments pass from the seafloor to the freshwater meteoric and finally the burial realms. But that is not all. An additional complexity is that these deposits can be changed to dolomite at any time. Such alteration can occur while the sediments are being deposited, in shallow subsurface aquifers, and later during burial or interaction with hot saline subsurface brines.

Documentation is fine, but all such processes and products are dynamic and need to be placed against the framework of sequence stratigraphy. Finally, whereas it is critical to understand carbonate petrogenesis, it is the evolution of porosity that is economically paramount. If it were not for extensive porosity and permeability, there would be no hydrocarbon or base metal accumulations in carbonate rocks. The pathway to understanding diagenesis can be quite challenging – good luck!

What drives diagenetic change?

The principal driver of diagenesis is chemical disequilibrium between different mineral phases and the waters that surround them. Before delving into specific diagenetic environments it is important to recognize the different drivers that promote diagenesis, principally the fluids, mainly water, and the minerals. Microbes, together with lithostatic and hydrostatic pressure, are also drivers in many systems.

The most important control is the interaction between different minerals and waters. Carbonates are always formed in aqueous environments, either freshwater (lacustrine) or seawater (marine). Once deposited, these sediments can later be flushed with meteoric waters. These fresh waters are either from local sources or from distant recharge areas. They can be mixed with seawater in the shallow subsurface and have varying CO_2 and salinity concentrations, resulting in universal carbonate dissolution and local carbonate precipitation. The carbonate minerals precipitated in seawater have different solubilities in fresh water. As a result, aragonite, HMC, and, to a lesser extent, LMC components all alter to LMC or dolomite because they are out of chemical equilibrium with fresh waters. This change, however, occurs at different rates and in different ways.

Although the above are the most important drivers, the role of microbes and chemical compaction are also important. Microbe-driven diagenesis happens when microbes or associated biopolymers change equilibrium conditions and promote carbonate precipitation or dissolution. Compaction-driven diagenesis is the result of increased stress during burial, generally due to overburden pressure. The changes that take place have no relationship to original sediment or rock composition, be it limestone or dolomite.

Even though most of the products brought about by these drivers are sequential, many are also concurrent and so several different processes can be taking place at the same time. Thinking about carbonate diagenesis is therefore a lot like trying to keep several concepts in the mind all at the same time; it is truly an intellectual juggling act.

Carbonate geochemistry

Before attempting to unravel the details of carbonate diagenesis, it is imperative to fully understand the way in which the minerals and waters interact. More important perhaps is the question: which chemical techniques can we use to work out past diagenetic processes? To resolve this problem, we *must* know the chemical signature of marine and terrestrial components before any change took place, the chemical processes involved in such changes, and the chemical attributes of the resultant components.

Among the most useful tools in this regard are trace elements and stable isotopes. The incorporation of trace elements into calcite or dolomite depends on mineral crystallography. Fe, Mn, and Sr are the most helpful elements in this respect because these bivalent cations substitute easily for Ca and Mg in the carbonate crystal structure. The most useful stable isotopes are those of oxygen and carbon. Their relative abundance is mostly dependent on organic processes, fluid temperature, and water salinity. If we understand how they are incorporated into the carbonate minerals, we can interpret the processes that led to their formation.

Limestones

In its simplest form, limestone diagenesis is the interaction between carbonate minerals and waters of different compositions (see Chapter 2):

$$CaCO_3 + H_2O + CO_2 \leftrightharpoons Ca^{2+} + 2HCO_3^-$$

Such changes involve chemical, mineralogical, and biological processes and all take place in the presence of water; there are virtually no solid-state reactions in carbonate diagenesis. Water temperatures are mostly near-surface and less than ~50°C, but can be as high as 150°C or more in deep burial situations.

The fluids are generally marine, meteoric, or deep subsurface in origin but can be substantially modified. With time, as carbonates are formed, deposited, buried, eroded, exposed, and reburied, they interact with these fluids in different ways. Such interactions result in the spectrum of carbonate rocks that we see in the geological record. Since each stage of alteration takes place in waters of somewhat different composition, each setting can be thought of as a separate diagenetic environment much in the same way that we think of depositional environments (see Chapter 2).

Dolomites and dolostones

Dolomite is precipitated in shallow sediments on the seafloor and in marginal marine environments, and replaces carbonates in contact with shallow groundwaters, during deep burial, and from hydrothermal fluids.

$$2CaCO_3 + Mg^{2+} \leftrightharpoons CaMg(CO_3)_2 + Ca^{2+}$$

Although dolomite is not abundant on the modern seafloor and is difficult to precipitate in the laboratory, under near-surface (<50°C) abiotic conditions, it does form in a variety of different diagenetic settings. The trick for precipitation seems to be to overcome the numerous chemical and biochemical factors that inhibit formation.

The dynamics of diagenesis

The processes and products of diagenesis change through time, both over the short term with oscillations in sea level and over the long term with changing seawater chemistry and biotic evolution. The evolution of marine organisms is important because the nature and mineralogy of aquatic calcareous algae and invertebrates that are the basic components of carbonate rocks, has changed profoundly through geologic time.

The types of fluids that percolate through the rock or sediment are largely dependent upon the position of sea level. If sea level is low, then the rock or newly formed sediment is rained upon and flushed with fresh water. If sea level is high, then the carbonate deposit is flooded by the ocean and filled with seawater. On the basis of the foregoing chapters, we can therefore predict which processes should take place and what products should result at different times during any one sea-level cycle.

By contrast, the long-term changes that affect diagenetic processes are largely the product of changes in seawater chemistry brought about by plate tectonics. Times with active plate motion and marine volcanism are characterized by ocean water with a low Mg:Ca ratio because Mg is stripped from seawater during serpentinization at mid-ocean ridges. By contrast, during periods of global tectonic quiescence, the ratio is quite high. Since the Mg:Ca ratio determines which carbonate minerals are precipitated from seawater (see Chapter 20), such changes in ocean chemistry can have a profound effect.

Porosity and permeability

At any given time, the relative porosity and permeability of carbonate rocks controls the way in which fluids flow through the rocks and thus determines diagenetic patterns. More important from a human standpoint, however, is that the nature of the holes and their connectivity also decides whether the rocks will be aquifers, potential hosts for hydrocarbons, locales for mineral deposition, or dense and impermeable.

Further reading

These books or series of articles are great sources of information for readers who want a more in-depth understanding of carbonate diagenesis. References at the end of each chapter also address more specific topics.

Allan, J.R. and Wiggins, W.D. (1993) *Dolomite Reservoirs: Geochemical Techniques for Evaluating Origin and Distribution.* Tulsa, OK: American Association of Petroleum Geologists, Continuing Education Course Note Series 36. *A series of succinct articles reviewing principles of dolomitization and their applicability to the petroleum industry.*

Choquette, P.W. and James, N.P. (1990) Limestone: The burial diagenetic environment. In: McIlreath I. and Morrow D. (eds) *Diagenesis.* St John's, ND, Canada: Geological Association of Canada, Reprint Series 4, pp. 75–111.

James, N.P. and Choquette, P.W. (1990) Limestone: The meteoric diagenetic environment. In: McIlreath I. and Morrow D. (eds) *Sediment Diagenesis.* St John's, ND, Canada: Geological Association of Canada, Reprint Series 4, pp. 35–74.

James, N.P. and Choquette, P.W. (1990) Limestone: The sea floor diagenetic environment. In: McIlreath I. and Morrow D. (eds) *Diagenesis.* St John's, ND, Canada: Geological Association of Canada, Reprint Series 4, pp. 13–34. *A series of papers summarizing all aspects of carbonate diagenesis up to the early 1900s. Much of the thinking in the following chapters is an outgrowth of this compilation.*

Moore, C.H. and Wade, W.J. (2013) *Carbonate Reservoirs: Porosity and Diagenesis in a Sequence Stratigraphic Framework.* Amsterdam: Elsevier, Developments in Sedimentology no. 67. *A complete overview of modern concepts that govern the development and diagenesis of porosity in carbonate rocks and much more; a must-reference work on the topic.*

Morrow, D.W. (1990) Dolomite. Part 1: The chemistry of dolomitization and dolomite precipitation. In: McIlreath I. and Morrow D. (eds) *Diagenesis.* St John's, ND, Canada: Geological Association of Canada, Reprint Series 4, pp. 113–123.

Morrow, D.W. (1990) Dolomite. Part 2: Dolomitization models and ancient dolostones. In: McIlreath I. and Morrow D. (eds) *Diagenesis.* St John's, ND, Canada: Geological Association of Canada, Reprint Series 4, pp. 125–139. *Two very readable summaries of the basic principles of formation and numerous examples of dolomites. An excellent place to start.*

Purser, B.H., Tucker, M., and Zenger, D. (eds) (1994) *Dolomites.* The International Association of Sedimentologists, Special Publication no. 21. *A collection of 24 papers on all aspects of dolomitization.*

Saller, A.H. (1986) Radiaxial calcite in Lower Miocene strata, subsurface Enewetak Atoll. *Journal of Sedimentary Petrology,* 56, 743–762.

Tucker, M.E. (ed.) (1988) *Techniques in Sedimentology.* Palo Alto, CA: Blackwell Scientific Publications. *An extremely useful series of chapters on the petrography of carbonate diagenesis.*

Tucker, M.E. (1990) Chapter 7. Diagenetic processes, products, and environments. In: Tucker M.E. and Wright V.P. (eds) *Carbonate Sedimentology.* Oxford: Blackwell Scientific Publications, pp. 314–364. *The most recent overview of carbonate diagenesis prior to this volume. A scholarly and in-depth treatment of the subject.*

Zenger, D.H. and Dunham, J.B. (1980) Concepts and models of dolomitization: an introduction. In: Zenger D.H., Dunham J.B., and Ethington R.L. (eds) *Concepts and Models of Dolomitization.* Tulsa, OK: Society for Sedimentary Geology, Special Publication no. 28, pp. 1–10. *A series of papers by different authors summarizing understanding of dolomites.*

CHAPTER 21
THE PROCESSES AND ENVIRONMENTS OF DIAGENESIS

Frontispiece Slab of limestone (right) with numerous molds of the bivalve *Pterotrigonia* sp. with the original aragonite shell being held at left. Upper Cretaceous, Mississippi, USA.

Origin of Carbonate Sedimentary Rocks, First Edition. Noel P. James and Brian Jones.
© 2016 Noel P. James and Brian Jones. Published 2016 by John Wiley & Sons, Ltd.
Companion website: www.wiley.com/go/james/carbonaterocks

Introduction to the processes

Carbonate sediments and limestones are altered relatively easily in natural waters even at low temperatures, because they are so soluble. This is manifest as partial to complete dissolution and precipitation of a variety of minerals in the ocean, on the land surface, and underground. The original particles and cements can be altered at any stage in their diagenetic history (paragenesis) with their original character retained, leached to form molds, or completely destroyed.

Carbonate dissolution

Dissolution (Figure 21.1) can take place on the seafloor, throughout the meteoric zone, and during deep burial. The meteoric zone is where all holes in the sediment or rock can be filled with fresh water. Components dissolve when the surrounding pore waters become undersaturated with respect to any particular mineral phase, including LMC, and this situation can be brought about in several ways. The simplest method is to create carbonic acid by adding CO_2 to the waters, either by microbial activity, oxidation or decay of organic matter, diffusion from atmospheric air, increasing hydrostatic pressure, or decreasing water temperature. This is where it gets tricky! Undersaturation can also occur during the mixing of waters with different CO_2 contents or with different

salinities. Finally, dissolution can take place when burial or tectonic stress and resultant focused pressure is applied. Burial stress is usually vertical, whereas tectonic stress can be at any angle to bedding but is mostly subhorizontal.

The most important aspect is if the dissolution is fabric-specific or non-fabric-specific. *Fabric-specific* is where each component of the rock or sediment is affected in a different way. This is different from *non-fabric-specific* dissolution, which cuts across all previous particles and cements, creating a spectrum of different holes.

Carbonate precipitation

Carbonate minerals, including dolomite, can form in any open space in sediment, limestone, or dolostone. Such holes occur across the spectrum of these deposits. They range from: (1) small voids inside skeletons (intraparticle); to (2) larger cavities inside reefs (framework); to (3) spaces between grains of all sizes (interparticle); to (4) spaces between crystals; to (5) dissolution voids (molds and vugs); to (6) fenestral pores (small irregular voids called "birdseyes" in older literature) in tidal flat muds; to (7) caves (see Chapter 32). The minerals can also precipitate on the ground surface within tufa springs.

Interparticle cement precipitation, which results in lithification, is arguably the most important process that takes place during diagenesis. There is a spectrum of such precipitates and, although their attributes are the result of a combination of factors, water composition and crystallization rate are the most important. The critical aspect regarding water composition is ionic strength, especially seawater versus fresh water. Crystal size and purity appear to be largely determined by the precipitation rate.

There are several rules of thumb that can be used when interpreting the origin of cement precipitates (Figure 21.2). Dirty, or inclusion-rich cement (Figure 21.3) can be precipitated from saline water (especially seawater), is microbially influenced, or both. Clear (limpid) inclusion-free crystals (Figure 21.4) are generally precipitated from fresh waters. Fibrous or microcrystalline cement crystals are generally (but not always) formed rapidly in saline waters whereas columnar crystals, some of which have pyramidal terminations, precipitate more slowly from fresh waters.

Interparticle and intraparticle cement shape is also illuminating (Figure 21.2). Those cements that are a rind of uniform thickness around a grain (Figure 21.5a) or lining the surface of a void (isopachous) must have precipitated in a fluid-filled pore (*phreatic*, i.e., below the water table or

DIAGENETIC PROCESSES

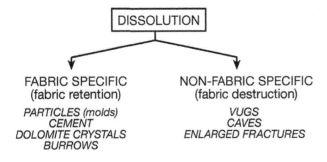

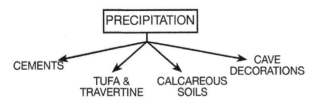

Figure 21.1 The main diagenetic processes, involving dissolution and precipitation.

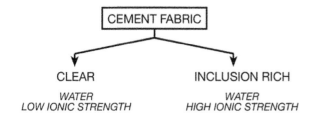

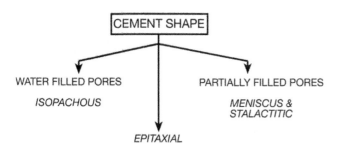

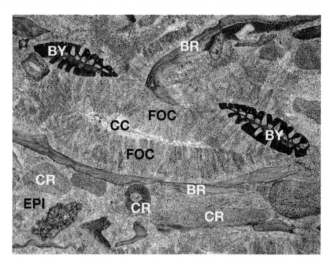

Figure 21.3 Thin-section image (PPL) of early Silurian interparticle inclusion-rich fascicular optic calcite cement (FOC) surrounding bryozoans (BY), brachiopods (BR), and epitaxial cements (EPI) around crinoids (CR). Late-stage calcite cement (CC) is clear. Image width 5 mm. Anticosti Island, Québec, Canada.

Figure 21.2 The different carbonate cement fabrics and shapes.

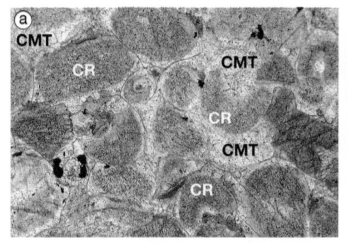

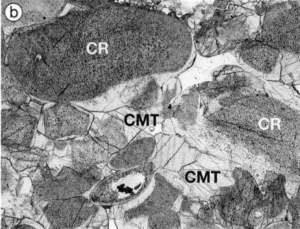

Figure 21.4 Thin-section images (PPL) of early Silurian crinoid grainstones and epitaxial cements from Anticosti Island, Canada. (a) Crinoids (CR) surrounded by inclusion-rich (synsedimentary marine) cement (CMT). Image width 4 mm. (b) Crinoids (CR) surrounded by clear, inclusion-free (burial) cement (CMT). Image width 2.5 mm.

on the open seafloor) with little inhibition. Those cements that have a meniscus (Figure 21.5b) or stalactitic morphology, however, must have precipitated against a fluid surface (air–water interface) and therefore have no crystal terminations; these are *vadose* (above the water table) in origin. Such phreatic and vadose conditions are present in both marginal marine seafloor and meteoric diagenetic environments. By contrast, epitaxial cements (crystals precipitated in optical continuity with the host grain, also called syntaxial cements) have no such constraints because they precipitate on grains that are already single

crystals or single-crystal mosaics (e.g., an echinoderm (Figure 21.4), a planktic foraminifer, or a prismatic bivalve).

The identification and discrimination of cement from recrystallization products is sometimes difficult but the following relationships and criteria can help (Figure 21.6). Cement, resulting from *competitive growth*, is where crystals grow outward from a free surface but interfere with each other as they grow. Crystals that have their c-axes oriented normal to the surface have the growth advantage. The result is a gradual reduction in the number of crystals and increase in size of the more successful crystals

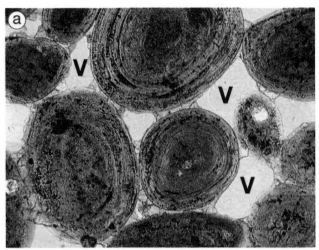
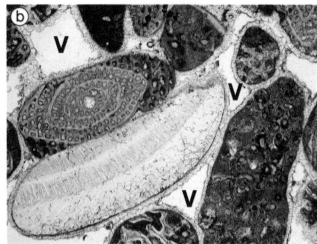

Figure 21.5 Meteoric cements. (a) Holocene aragonite ooids from Joulters Cays, Bahamas, partially lithified by LMC meniscus cement. Image width 1.5 mm. Photograph by W. Martindale. Reproduced with permission. (b) Pleistocene biofragmental grainstone, Bermuda, cemented by isopachous dLMC (V: void). Image width 2.2 mm.

CARBONATE CEMENT FABRICS

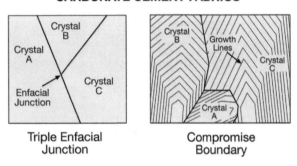

Triple Enfacial Junction   Compromise Boundary

Figure 21.6 The geometric crystal relationships at a triple junction (left) where one boundary is 180° (enfacial junction) and a compromise boundary (right) resulting from competitive crystal growth into a void where there is no junction at 180°. Source: Adapted from Bathurst (1975). Reproduced with permission of Elsevier.

outward. This is called "drusy cement". A *compromise boundary* is where adjacent crystals meet and interrupt the freedom of each other's growth but they continue to grow along a straight contact. A *triple junction* is the meeting place of three intercrystalline boundaries (Figure 21.6). An *enfacial junction* is a triple junction where one angle is 180°. All of these boundary conditions are encountered in cement crystals.

Neomorphism

In spite of the forgoing, not all clear sparry calcite is cement. Limestones are easily altered to a variety of crystal mosaics mosaics because of their solubility. The term *neomorphism* (new form) is commonly applied to these altered carbonates. More specifically, neomorphism is defined as "all transformations between one mineral and itself or a polymorph." This term embraces both polymorphic transformation (also called "inversion," for example, the aragonite to dLMC transformation) and recrystallization (alteration where the mineralogy remains unchanged, for example, dLMC crystal enlargement). Differentiation of cement from such crystal mosaics or neomorphic spar is one of the vexing problems of carbonate petrology. It is a relatively simple task when all or part of an obvious grain or mass of micrite is altered to clear calcite spar. It is much more difficult when most of the rock is altered to a mosaic of different calcite crystals.

There are two recurring neomorphic calcite crystal types: microspar and neomorphic spar. *Microspar* (Figure 21.7) is a limestone composed of small subequidimensional crystals with a relatively uniform size range commonly 5–6 μm, but up to 50 μm. It is typically an alteration product of micrite. *Neomorphic spar*, as emphasized above, is more difficult to tell from cement, but has the following attributes: (1) crystal size ranges of 4–100 μm; (2) crystal size varies greatly and is patchy (unlike microspar which is uniform or drusy cement which increases in size away from the host surface); (3) it can have a substellate (star-like) fabric that radiates away from a micrite center; (4) crystal boundaries are curved and wavy (unlike cement boundaries which are typically planar); (5) crystals margins are typically embayed; and (6) there are very few enfacial junctions (unlike cements where there are many). Cathodoluminescence (CL) is also useful in differentiation where cement zones are well defined and parallel to crystal boundaries. Neomorphic spar, in contrast, is generally poorly zoned.

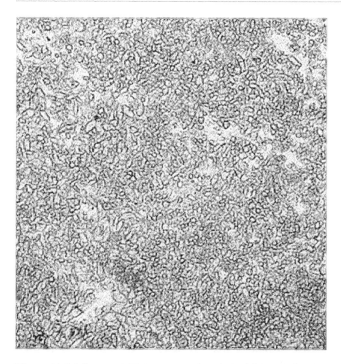

Figure 21.7 Microspar, Pleistocene calcrete, Barbados. Image width 300 μm.

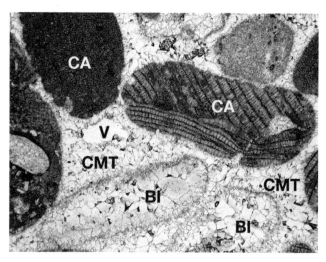

Figure 21.8 Thin-section image of late Pleistocene dLMC grainstone from Bermuda that has been intermittently exposed to meteoric waters for ~500,000 years. The coralline algae particles (CA), originally Mg-calcite, are well preserved but the aragonite bivalves (BI) are molds filled with dLMC cement (CMT) that also partially fills interparticle pore spaces with only a few voids (V) remaining. Image width 2.3 mm.

These neomorphic fabrics are present in the youngest limestones and not just confined to very old carbonates. Finally, there is a caveat: the terms microspar and neospar are commonly used today for cement-sizes or crystals with uncertain origins, especially those associated with microbial carbonates and not strictly neomorphic fabrics. Care is therefore needed when reading about such crystals.

Component alteration

The most universal alteration process involves the alteration of aragonite and Mg-calcite components to diagenetic low-magnesium calcite (dLMC). The process is mineral-driven and involves: (1) dissolution of the more soluble phases such as aragonite and Mg-calcite; (2) resultant oversaturation of the water with respect to low-Mg calcite as a consequence; and (3) precipitation of new dLMC. The critical point is that primary non-diagenetic LMC is not dissolved because oversaturation is achieved and maintained by dissolution of aragonite components.

The aragonite–dLMC transformation, because it entails change from an orthorhombic to a hexagonal crystal, involves macroscopic dissolution of the aragonite and possible precipitation of calcite, with generally little or no textural preservation. When there is minor fabric preservation, the process is called calcitization. By contrast, HMC–dLMC transformation does not entail crystallographic change. Instead, the process typically involves microscopic dissolution of the Mg-calcite and immediate precipitation of dLMC. As a consequence, there is excellent textural preservation (Figure 21.8). The microscopic texture of Mg-calcite components is therefore usually preserved even though they are altered to dLMC, whereas aragonite components either dissolve or are altered to dLMC with only vague preservation of original fabric (*calcitized*). The eventual fabric of any limestone is therefore primarily a function of its original mineralogical composition.

The environments

In its simplest form as outlined above, diagenesis is the interaction between carbonate minerals and waters of different compositions. Three diagenetic environments are recognized (Figure 21.9). The *synsedimentary marine diagenetic environment* includes the seafloor and sediments directly below that are bathed in normal or evaporated seawater. Diagenetic alteration is driven by changes in water chemistry and microbial action. The *meteoric diagenetic environment* includes environments above and below the water table as well as underlying mixed freshwater–marine environments. Diagenetic alterations result from

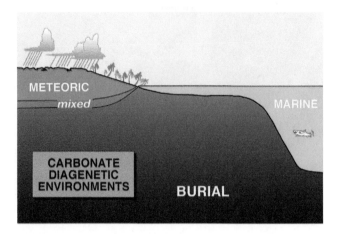

Figure 21.9 The spatial relationships between different diagenetic environments.

different mineral solubilities and microbial processes. The *burial diagenetic environment* is where pores are filled with waters that could have been marine, evaporative, or fresh, but have been modified by interaction with other diagenetic fluids and minerals under conditions of increased temperature and pressure in the burial realm. The processes here generally result from overburden compaction and water–mineral interactions.

Synsedimentary marine diagenetic environment

Carbonate sediment particles are precipitated from seawater, either directly or under biogenic influence. Intuitively, it would seem that they should remain there unaltered because they are largely in chemical equilibrium with their birth fluid seawater. This is, however, not the case. As emphasized in Chapter 2, due to petrifyingly low temperatures and unimaginably high hydrostatic pressures, the deep parts of the ocean are undersaturated with respect to carbonate and so all particles are dissolved. By contrast, the neritic zone, which is bathed in tropical, temperate, or polar waters is generally perceived as a place of carbonate precipitation and particle formation. Whether or not this is true, preservation is largely, but not wholly, dependent on seawater temperature. In the tropics, carbonate is precipitated in a variety of forms ranging from grains such as ooids to pore-filling cements to whitings of carbonate mud. Such precipitates are not common in temperate waters, but cements are present at hardgrounds. There are few carbonate precipitates in polar seas.

Given this situation, dissolution should not occur in warm neritic environments. The fact that it does occur further emphasizes the fact that carbonates never seem to obey the rules! Dissolution takes place just below the seafloor

in both tropical and temperate neritic environments. Dissolution in the Arctic and Antarctic is common on the seafloor even at depths <200 mwd.

Meteoric diagenetic environment

Introduction

Carbonate sediments can accumulate layer upon layer on the seafloor to form a thick sedimentary package. Given the vicissitudes of sea level fluctuations with time or because of tectonic processes, it is common for these deposits to be periodically exposed to terrestrial alteration. Sediments above and below ground are flushed with fresh meteoric waters, unlike the seawater in which they were born, and profound mineralogical, chemical, textural, and fabric changes are the result. This is the zone of action!

Mg-calcites and aragonites are not chemically stable when bathed in fresh waters, so typically change their character. These processes happen relatively rapidly and persist until all of the deposit is changed to dLMC or dolomite. This is, however, not the only process. Combined water- and microbial-driven diagenesis result in the formation of caves and other karst features, local precipitation of spelean carbonates, and soil carbonate development. These processes continue as long as the carbonate is exposed or re-exposed to meteoric water flow. Finally, significant terrestrial carbonate deposits can also form in lake, swamp, and spring environments.

The bottom line is that, when recently deposited carbonate sediments or ancient limestones of any age are exposed to meteoric waters, there is wholesale dissolution and export of carbonate from the system. Although precipitation does occur and can be spectacular, it is comparatively minor in the larger scale of things.

The surface meteoric environment

Dissolution.   All carbonates exposed to the open air are subject to dissolution because rainwater contains trace amounts of CO_2 and so is a weak acid. The effects range from relatively minor on bare rock, if short term, to profound if beneath a soil or over long periods of time.

Precipitation.   Most lacustrine carbonates precipitate from relatively dilute fresh waters and so are, as expected, LMC. As with marine systems, if the Mg:Ca ratio of the water is elevated, HMC and aragonite can precipitate. Dolomite is also a widespread mineral in such systems.

Iron carbonates in the form of siderite (FeCO$_3$) and ankerite (Ca[Mg$_x$Fe$_{1-x}$]CO$_3$) are also common. Sodium carbonate and bicarbonate minerals such as trona (NaHCO$_3$.Na$_2$CO$_3$.2H$_2$O), nacholite (NaHCO$_3$), and natron (Na$_2$CO$_3$.10H$_2$0) can occur in saline lake systems, usually under conditions of extreme brine concentration.

Carbonate spring systems precipitate tufa and travertine whose mineralogy is either LMC or aragonite in a wide variety of crystal forms. Soil carbonates, usually called calcrete, are almost always LMC except near the ocean, where seawater is blown inland under very arid conditions to provide the source of carbonate and form dolomite (dolocretes).

The shallow subsurface meteoric environment

The meteoric setting is divided into vadose and phreatic zones separated by the water table (Figure 21.10). The water table is the level at which hydrostatic pressure is equal to atmospheric pressure.

The vadose zone. Rainwater, meltwater, or runoff enters the subsurface either directly via bare rock or by percolating through a calcrete or soil zone. The *zone of infiltration* is a region of complex carbonate–water–microbe interactions. Pores filled with water, air, both water and air, or organically produced gas, characterize the *zone of gravity percolation*. Under conditions of vadose seepage, water trickles down through the limestone via a series of small fractures or interconnected pores. The *zone of vadose flow*, by contrast, is where water moves rapidly downward through joints, large fissures, or sinkholes directly to the water table.

The phreatic zone. This zone, where water occupies all available pore spaces, is one of active water flow. Such movement is horizontal or subhorizontal toward the local base level (a spring, a river, or a lake) or the ocean. This movement is locally so active that tongues of water can extend beyond the shore and out beneath the continental shelf.

The lenticular zone. A common situation, especially near coasts, is where the fresh water filling interparticle pores has a lenticular lens-like geometry and "floats" on more dense seawater beneath. In unconfined aquifers, the depth to which the lenticular zone extends below sea level is about 40 times the height of the water table above sea level (the Ghyben–Herzberg Principle; Figure 21.11). The precise nature of each lens is different and depends upon the rock permeability and the rate of groundwater recharge. The important point is that only a slight elevation of the water table can lead to deep penetration of meteoric waters into the sediment or rock below. This has profound implications for diagenesis.

The transition or mixing zone. This is a zone of physical and diffusive mixing. It is thickest in more permeable strata where mixing processes are most effective and near the coast. Waters are fastest moving and most dilute in the upper part of the zone. The zone can be quite thick compared to the freshwater lens above it, especially in permeable carbonates of oceanic islands.

Deep phreatic zone. Water in the deep phreatic can be slow-moving or almost stagnant. In the case of young exposed carbonates it can be seawater because the freshwater lens itself is relatively thin.

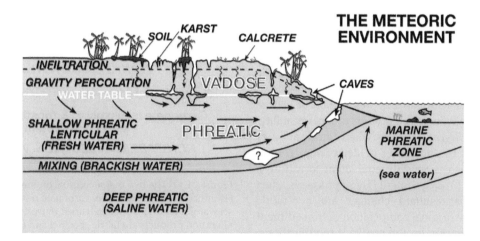

Figure 21.10 The main zones of the meteoric diagenetic environment. Source: Adapted from James and Choquette (1984) . Reproduced with permission of the Geological Association of Canada.

SMALL SAND CAY OR OCEANIC ATOLL

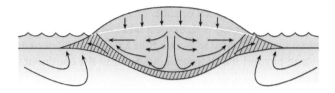

LARGE CARBONATE PLATFORM

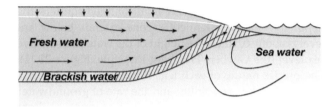

Figure 21.11 Different scales and configurations of freshwater lenses. Source: Adapted from James and Choquette (1984). Reproduced with permission of the Geological Association of Canada

Carbonate sediment and limestone alteration in the meteoric zone

There are two processes occurring simultaneously when carbonate sediments are first exposed to percolating fresh waters, mineral-driven alteration, and water-driven alteration (Figure 21.12). Once all of the metastable minerals (HMC and aragonite) have changed to dLMC, then alteration continues but it is only water-driven.

Mineral-controlled diagenesis takes place when a suite of minerals is no longer in equilibrium with the waters from which they precipitated. Aragonite- and Mg-calcite-rich sediments, for example, will be prone to alteration when flushed with fresh water. Resultant changes are fabric-specific, that is, each component reacts differently and so is preserved in a different way. Such changes in the meteoric zone result in the alteration of all carbonate minerals to dLMC (or dolomite) with variable preservation of texture and fabric but typically increased porosity and permeability.

Water-controlled diagenesis occurs when composition of the water changes by increasing or decreasing: (1) ion concentration (which is roughly equivalent to salinity); (2) temperature; or (3) CO_2 content. These processes affect all carbonates and resultant changes are non-fabric selective, that is, the original composition of the sediment or rock has little bearing on the resultant modifications. Changes in the meteoric zone are manifest as extensive surface and subsurface karst, formation of calcareous

soils, and precipitation of cave speleothems. Although these are the most important drivers, the role of microbes is important throughout. This is especially so for soils and cave precipitates.

Burial diagenetic environment

Almost all ancient limestones have at one time been buried beneath a pile of younger sedimentary rocks and have therefore been compacted to varying degrees under overburden pressure. Such compaction is both physical and chemical. Physical (or mechanical) compaction takes place first in the upper several hundred meters of burial. It is here that, if they are not already cemented, grain fracturing and porosity reduction take place by squeezing and closer grain packing. Chemical compaction can also take place in this zone, but is usually in the form of dissolution at grain contacts. Widespread chemical dissolution and accompanying cement precipitation does not generally occur until major physical compaction is finished at depths of several hundreds to thousands of meters. The predominant features are stylolites (Figure 21.13) and solution seams, together with precipitation of burial cements. Although compaction-driven diagenesis dominates, water composition and dissolution by organic acids is important.

METEORIC DIAGENESIS
ALTERATION IN FRESH WATER

---

1. MINERAL CONTROLLED

HIGH MG-CALCITE ⟶ LOW MG-CALCITE

ARAGONITE ⟶ LOW MG-CALCITE

LOW MG-CALCITE ⤏ LOW MG-CALCITE

FABRIC SELECTIVE

---

2. WATER CONTROLLED

$$CaCO_3 + H_2O + CO_2 \rightleftharpoons Ca^{2+} + 2HCO_3^-$$

Carbonate is dissolved & mostly transported out of the system

NON-FABRIC SELECTIVE = VUGS & KARST

Figure 21.12 The two major controls on meteoric diagenesis: (1) mineral controlled, where carbonate minerals stable in seawater alter to LMC when bathed in fresh water and resultant alteration products are fabric specific; and (2) water controlled, where all components regardless of mineralogy are dissolved and locally reprecipitated as dLMC and resultant alteration is in the form of karst and caves and so non-fabric specific.

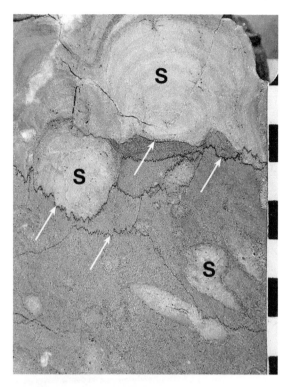

Figure 21.13 A subsurface Upper Devonian limestone core from the Western Canadian Sedimentary Basin showing large stromatoporoids (S) and numerous stylolites (arrows). Centimeter scale.

Dolomite and dolostone

Dolomite (Figure 2.4) is being precipitated today within shallow sediments on the seafloor, in marginal marine environments, in meteoric groundwaters, during deep burial, and from hydrothermal fluids. It also replaces carbonate sediments and rocks across the geological spectrum. The terms *dolomite* and *dolostone* are variably used for the rock but, strictly speaking, dolostone is the proper term.

Unlike limestone, where we can visit most sites of diagenesis or easily duplicate most processes in the laboratory, dolomite and dolomitization are completely different. The problem is simple but the answer is multifaceted. Shallow seawater is oversaturated with respect to dolomite and the Mg:Ca ratio is on the order of 3:1; yet the mineral is only locally being precipitated in the modern ocean. Furthermore, as much as half of all carbonate rocks are dolomite and yet comparatively few seem to be forming today.

Our understanding of such diagenesis comes only partly from the modern world. This is because some systems are long lasting and involve depositional systems larger than those present today (especially evaporites). Many diagenetic processes are therefore formulated as models from recurring stratigraphic situations in the rock record. Interpretations of diagenesis are therefore usually part observation, part measurement, and part deduction.

Determining the nature of diagenesis in dolostone is, however, of inordinate importance because much porosity and permeability in carbonate rock successions is in dolostones as opposed to limestones. These porous dolostones hold a disproportionate amount of global hydrocarbon resources and are typically associated with base metal deposits.

Further reading

Arvidson, R.S., Collier, M., Davis, K.J., Vinson, M.D., Amonette, J.E., and Lüttge, A. (2006) Magnesium inhibition of calcite dissolution kinetics. *Geochimica et Cosmochimica Acta*, 70, 583–594.

Bathurst, R.G.C. (1975) *Carbonate Sediments and their Diagenesis*. Amsterdam: Elsevier Science, Developments in Sedimentology.

Bricker, O.P. (ed.) (1971) *Carbonate Cements*. Baltimore, Maryland: John Hopkins Press, John Hopkins Studies in Geology.

Broecker, W.S. (2003) The oceanic $CaCO_3$ cycle. In: Elderfield H. (ed.) *The Oceans and Marine Geochemistry*. Amsterdam: Elsevier, Treatise in Geochemistry Vol 6, pp. 529–549.

Given, R.K. and Wilkinson, B.H. (1985) Kinetic control of morphology, composition, and mineralogy of abiotic sedimentary carbonates. *Journal of Sedimentary Petrology*, 55, 109-119.

James, N.P. and Choquette, P.W. (1990) Limestone: The meteoric diagenetic environment. In: McIlreath I. and Morrow D. (eds) *Sediment Diagenesis*. St John's, ND, Canada: Geological Association of Canada, Reprint Series 4, pp. 35–74.

Meldrum, F.C. (2003) Calcium carbonate in biomineralisation and biomimetic chemistry. *International Materials Review*, 48, 187–224.

Morse, J.W. (1985) Kinetic control of morphology, composition and mineralogy of abiotic sedimentary carbonates. *Journal of Sedimentary Petrology*, 55, 919–934.

Morse, J.W. and Arvidson, R.S. (2002) The dissolution kinetics of major sedimentary carbonate minerals. *Earth-Science Reviews*, 58, 51–84.

Pierson, B.J. (1981) The control of cathodoluminescence in dolomite by iron and manganese. *Sedimentology*, 28, 601–610.

Reeder, R.J. (ed.) (1983) *Carbonates: Mineralogy and Chemistry*. Mineralogy Society of America, Reviews in Mineralogy Vol. 11.

Schneidermann, N. and Harris, P.M. (1985) *Carbonate Cements*. Society of Economic Paleontologists and Mineralogists, Special Publication no. 36.

CHAPTER 22
ANALYTICAL METHODS

Frontispiece Students examining the microscopic attributes of modern carbonate sediments that they have just collected; Bermuda Institute for Ocean Sciences.

Origin of Carbonate Sedimentary Rocks, First Edition. Noel P. James and Brian Jones.
© 2016 Noel P. James and Brian Jones. Published 2016 by John Wiley & Sons, Ltd.
Companion website: www.wiley.com/go/james/carbonaterocks

Introduction

Interpretation of the depositional and diagenetic history of carbonate sediments and rocks relies on the acquisition of many different types of data. Prior to any investigation, it is critical to determine how the rocks will be analyzed and what information is needed to formulate answers to the questions being asked. Whereas the array of components and fabrics gleaned from outcrop and core examination can usually provide comprehensive insights into the original depositional settings, an understanding of diagenesis usually depends on mineral and chemical information.

This chapter provides an overview of the techniques that are most commonly used in the analysis of carbonate rocks. Forward planning is essential because it will determine the type of samples collected and the methods by which they will be analyzed. Such planning is also important because the acquisition of data can be time-intensive and costs can be high.

Background considerations

The last 10–15 years have been marked by rapid technological changes that have, in general, meant that more data can be collected with greater precision more rapidly than ever before. Potentially, data for the analysis of carbonate samples can come from fieldwork, analysis of cores and hand samples, thin-section petrography, X-ray diffraction (XRD) analysis, scanning electron microscopy (SEM), electron microprobe analysis, stable isotope analyses, and radiogenic isotope analyses (Figure 22.1). Planning is needed because each kind of analysis requires a different type of sample preparation.

The following points are critical to any dataset and the interpretation of those data.

- Scale is critical because the assumption that underpins all analyses is that the sample being considered is representative of the rock layer from which the sample was first collected. Downsizing starts in the field as hand samples are collected from the strata being studied. Thin-sections are a $3 \times 2\,cm$ slice of the rock; XRD samples are typically about $1\,g$, SEM samples are typically less than $1\,cm^3$, and samples for isotope analysis many be only a few milligrams. These differences in scale should be factored into any comparison of datasets from different types of analyses.
- All data should be thoroughly checked to make sure that the numbers are robust and without error. If possible, data should be checked by analyzing duplicate samples, running different types of analyses, or by having some samples processed by different laboratories.
- It is important that the accuracy, limitations, and errors be fully understood for all data. Comparison of datasets produced in different laboratories should also include comparisons of the methods used, because even subtle differences in sample processing can lead to significant differences in the resultant data.
- In an ideal world, the interpretation of samples should utilize as many lines of evidence as possible with as many different types of data as possible. In reality, however, time and budget constraints will dictate the type and volume of data that can be assembled.

Collectively, this means that it is critical to know the limitations of the analytical techniques that were used to obtain the data, know the limitations of the data produced, and know the limitations of the interpretative capabilities. Nevertheless, it is also important to view some datasets with a degree of imagination; thinking

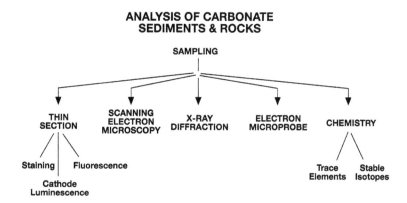

Figure 22.1 The various common techniques used to interpret the diagenesis of carbonate rocks.

outside of the box commonly leads to significant advances in our understanding of natural systems that produce carbonate sediments and rocks.

Sample acquisition

Fieldwork underpins all analyses because it provides the information that is needed to establish the large-scale setting of the strata being considered, which in turn creates the framework for all subsequent interpretations. Data collected during this phase should provide a complete description of all aspects of the rocks being studied. This can, for example, include information on facies succession, nature of the bedding, and recognition of unconformities, as well as specific information about the composition of the facies, the fossils that they contain, and any other features of note. At this stage, comprehensive notes describing the rocks are required and photographs of key features are essential. The use of 10% HCl is invaluable in providing initial differentiation of calcite and dolomite in the rocks being examined. Careful sampling of the main rock types (facies) is essential and great care must be taken to systematically label the samples with an indication of their precise location and "way-up". This aspect is particularly important for analysis of diagenesis. The key here is to label the samples so that there can be no misunderstandings as to their location once they are in the laboratory.

Core analysis

Cores offer the only source of information in areas where the rocks are not exposed at surface. Core analysis based on a detailed examination of the core and the associated petrophysical data sets (e.g., gamma-ray logs) therefore provides the information that is needed. For many cores there are also comprehensive analyses of the porosity, permeability, bulk densities, and grain densities that provide valuable insights into diagenesis. Such analyses are essentially the same as fieldwork because they allow assessment of the sedimentological successions, facies, fossil distributions, and various other features. Sampling can, however, be restricted by regulations associated with the storage, availability, and confidentiality of the core. As with samples from the field, core samples must be systematically labeled with the well location, depth in the well, and a "way-up" arrow. Core depths should always be checked against log depths, because the latter is regarded as "true" depth and some adjustment to the core depth will probably be required.

Petrography

General considerations

Thin-section petrography is critical to the analysis of carbonate rocks because it provides information about the mineralogy, crystal morphology, depositional fabrics, fossil components, and diagenetic components. Many different techniques are used in the thin-section analysis of carbonate rocks and it is vital that careful planning is exercised when getting the sections made.

For carbonate rocks, thin-sections may be small (typically 3×2 cm) or large (7×5 cm), with the larger ones being used to establish the large-scale fabric relationships of the rocks being considered. Making these thin-sections is a highly skilled, time-consuming, and expensive process. The process involves: (1) cutting the sample to the desired size; (2) polishing one surface and then mounting onto a glass slide using epoxy; and (3) grinding the sample down to a thickness of 30 μm or less depending on the density and grain size of the material being examined. Carbonate rocks are commonly friable and grinding will commonly cause small pieces to be "plucked" from the sample, thereby creating false porosity. Carbonate samples are therefore usually impregnated with blue epoxy before grinding so that the original porosity (blue), as opposed to false porosity created by "plucking" (white), is highlighted. Cover slips are typically applied to the thin-section as a finishing step. With carbonates, however, this step is commonly delayed or omitted altogether. This is because it is often necessary to apply different techniques to the thin-sections in order to highlight various features in the rock.

Staining

In many cases, it can be difficult to distinguish between the different mineralogical components of a carbonate rock, especially if it is finely crystalline. With carbonates, this problem can be overcome by staining the rock using various chemicals. These stains provide an easy and inexpensive way of obtaining mineralogical data. Although numerous different stains have been used, the most commonly used stains are as follows:

- Alizarin Red S solution (ARS) stains calcite red while leaving dolomite unstained (Figure 22.2).
- Potassium ferricyanide imparts a blue to purple color to Fe-rich calcite or dolomite (Figure 22.2b). Experimental work has shown that the stain takes if there is more than 2% Fe in the calcite or dolomite.

Samples that do not stain are typically referred to as being non-ferroan, even though they may contain up to 2% Fe. This stain can be: (1) applied by itself; or (2) mixed with Alizarin Red S solution. The latter process yields combined colors that allow separation of calcite from dolomite and also highlights if the components are Fe-rich or Fe-poor. Only freshly made solutions should be used because the potassium ferrocyanide deteriorates rapidly and quickly becomes useless.

• Fiegel's solution stains aragonite black, but is difficult to apply and use.

• Clayton yellow, a histological stain, is specific for Mg (Figure 22.3) and the rock becomes progressively more red with increasing $MgCO_3$ content. The stain is useful for differentiating aragonite, which remains white, from HMC in young limestones. It fades with time and must be fixed with NaOH.

All descriptions of thin-sections should be fully supported by photomicrographs that readily illustrate the main components and features of the rock being studied. Estimating the percentage contribution of each

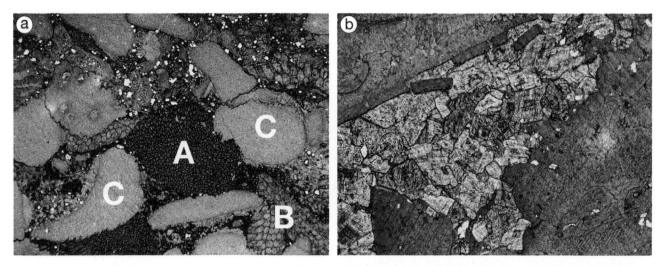

Figure 22.2 (a) Bryozoan (B), crinoid (C), and algal (A) grainstone in which all grains are calcite and so stained red with Alizarin Red-S. Middle Ordovician, Ontario, Canada. Image width 1.5 cm. (b) Calcite grainstone (stained red with Alizarin Red S) and Fe-rich dolomite (stained blue with potassium ferricyanide), lower Cambrian, Labrador, Canada. Image width 1 cm.

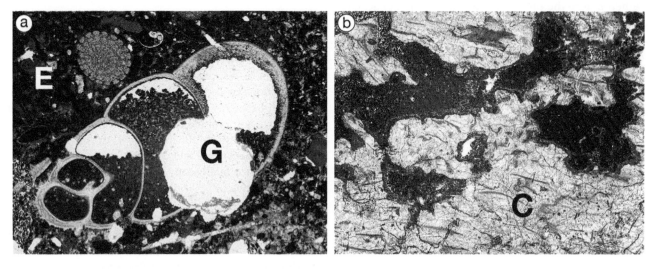

Figure 22.3 Thin-section images (PPL), stained with Clayton yellow where aragonite remains unstained and Mg-calcite stains a vivid red. Synsedimentary cemented sediment, shallow modern Belize reef crest. (a) Microgastropod (G: aragonite unstained) filled with micropeloidal Mg-calcite sediment and cement (red) with Mg-calcite echinoid spine (E) in surrounding sediment. Image width 2 mm. (b) Aragonitic coral (C) surrounded by fine Mg-calcite rich sediment and cement (red). Image width 4 mm.

component is a critical part of most studies because this provides a basis for classification and provides valuable insights into the origin of the rock. Numbers have traditionally been obtained by: (1) point counting, where the percentage composition is derived from the components found at various points (typically 100–300 points) throughout the thin-section; or (2) visual estimates. With the advent of digital photographs, image analysis is commonly used to determine the composition of the rock. This type of analysis, however, relies on components that are readily identifiable on the basis of their color. Porosity that has been highlighted by blue epoxy, for example, can easily be determined using this technique.

Acetate peels

Acetate peels, commonly known as the "poor-person's thin-section," offer an easy and inexpensive way of making a replica of an etched surface. The process involves: (1) etching of a cut and polished surface; (2) covering the surface with acetone; and (3) applying a thin acetate sheet onto the surface. The solvent softens the lower side of the acetate sheet that then settles into the irregularities of the surface. Once dried, the sheet is peeled off the sample and quickly sandwiched between two glass slides to prevent curling. Peels are best suited to low-porosity rocks because highly porous rocks or rocks with large vugs allows the solvent (acetone) to drain away before it can interact with the acetate sheet. Although easy to produce, experience is needed before perfect peels can be made routinely. As with thin-sections, the rock can be stained before the peel is made. Examination of acetate peels on a microscope can provide almost as much information as a standard thin-section. The exception is that a peel cannot be used to determine mineralogy using cross-polarized light.

White card technique

Organic matter is revealed in exquisite detail when a white card is slipped beneath the thin-section and observation is made by oblique reflected light. The method is an alternative to fluorescence microscopy for highly matured organic material.

Cathodoluminescence microscopy

Traditionally, cathodoluminescence equipment has been attached to petrographic microscopes so that information pertaining to the luminescence of the rock components can be established. In recent years, this approach has been supplemented by the attachment of cathodoluminescence units to scanning electron microscopes and electron microprobes.

In most cases, thin-sections are used for this type of analysis. They should not have a cover slip because the electron beam needs to react with the trace elements that are in the carbonate minerals being viewed. In calcite and dolomite, some trace elements are activators (e.g., Mn^{2+}) and emit photons that have wavelengths in the visible light spectrum as a result of electron excitation (Figure 22.4). Conversely, other elements (e.g., Fe^{2+}) are emission inhibitors. With this technique, the rock components can be described as being brightly luminescent, dully luminescent, or non-luminescent. In some cases, these descriptors also reflect the color of the luminescence (e.g., bright red, dark orange). This technique is particularly useful for delineating compositional zoning in calcite or dolomite cements that are found in many carbonates.

It is often difficult to precisely determine what elements are responsible for different luminescent colors. This is especially true when the analysis is based solely on observations made with a petrographic microscope. This issue can, however, be resolved using an electron microprobe because it is then possible to obtain precise compositional information for areas that are characterized by different luminescent responses. An alternative approach is to use a spectrometer attachment because this instrument can often determine activators that are below detection limits of the microprobe.

Fluorescence microscopy

Fluorescence is a type of luminescence whereby a material emits light in response to excitation by ultraviolet light. Thin-sections (without cover slips) used for this purpose need not be highly polished. These can be viewed on a standard petrographic microscope that has had a fluorescent adaptor attached to it or on specifically designed fluorescence microscopes.

Fluorescence microscopy can be used in two different ways as follows:

1. On standard thin-sections it is used to highlight: (1) ghost structures that may be present in recrystallized limestones or dolostones; and (2) zoning in cements. Although the exact cause of fluorescence in carbonate rocks is poorly understood, it is widely suspected that fluorescent images reflect the residual organic matter that is trapped in the altered limestone or dolomite. It is

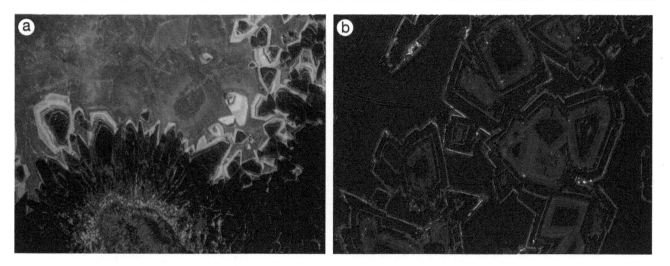

Figure 22.4 Cathodoluminescent thin-section images. (a) Cements precipitated into a void. First stage black = no Mn (meteoric vadose); second stage bright yellow = abundant Mn (meteoric phreatic); and third stage orange = Mn but quenched with iron (burial). Lower Cambrian Labrador, Canada. Image width 1 cm. (b) Strongly zoned dolomite crystals where the bright zones are rich in Mn, whereas the black zones have little or no Mn. Oligocene, South Australia. Image width 0.5 mm.

important to note, however, that not all carbonate rocks will fluoresce.

2. Autofluorescent dyes such as rhodamine-B can be added to the epoxy that is used to impregnate the carbonates before a thin-section is made. Under normal light, the epoxy will appear clear with no evidence of the dye being present. Under fluorescent light, however, the rock components will become black whereas the epoxy will appear red. This approach is very useful for highlighting the porosity that is present in a carbonate rock and making maps of its distribution.

Fluid inclusions

Fluid inclusions, ranging in size from <1 μm to 1 mm, are present in most carbonate crystals. They generally contain a fluid that is a sample of waters trapped when the mineral was precipitated. A gas or solid phase that separated during later cooling may also be present. Petrographic study using freezing and heating stages is used to determine the composition and original rates and temperature histories of the fluids involved in crystal formation. Heating is utilized to ascertain temperatures at the time of cement precipitation; freezing point determinations on the inclusion can establish salinities and the approximate composition of the waters present at the time of inclusion entrapment. The composition of the fluid can also be analyzed directly by ion microprobe or mass spectrometry. The techniques are tricky; if you are

planning to attempt this method, be sure to read one of the books written on the topic.

X-ray diffraction analysis

XRD is used to identify carbonate minerals, determine the percentages of various carbonate minerals in the sediment or rock, yield information about their chemical composition, and determine the ordering of dolomite crystals. This information is typically derived from a powder sample (usually ~1 g) that has been prepared from the material being investigated. It is important that: (1) the powder is uniform in terms of its grain size; and (2) the minerals are randomly distributed throughout the powder. A smear mount should only be used if qualitative and not quantitative results are needed, because certain minerals will be preferentially oriented and therefore overrepresented in the diffraction pattern.

X-ray diffractometers produce X-rays that are collimated into a beam that is focused onto the sample, which is being rotated at a constant speed specified in degrees per minute. When the mineral planes attain an appropriate angle, they diffract the X-ray according to Bragg's Law. The diffraction angle also depends upon the composition of the X-ray tube (usually copper or cobalt) that yields Kα radiation. Output is typically in the form of a strip graph that shows the intensity of the refraction (y axis) plotted against degrees 2θ, which is the angle of diffraction (Figure 22.5). The latter can easily be converted

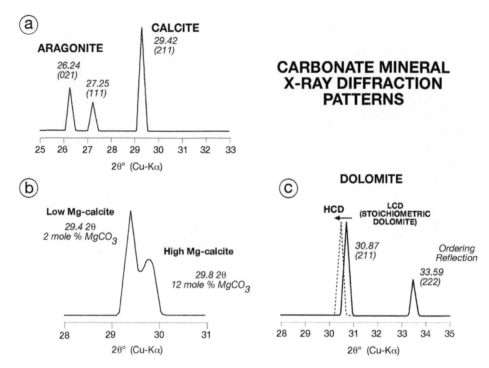

Figure 22.5 Synthetic diffractograms produced using Cu-Kα radiation, illustrating: (a) location of the main aragonite and calcite peaks; (b) the displacement of the main calcite peak with increasing amounts of Mg in HMC; and (c) the main dolomite peak and the peak shift expected with increasing amounts of Ca in the lattice (HCD = high calcium dolomite; LCD = low calcium dolomite), together with one of the ordering peaks.

to *d*, which is the lattice spacing that is then used to identify the minerals. With most diffractometers, the attached computer automatically performs mineral identification or gives a choice of possible minerals in a degree of certainty. Although all peaks must be scanned for correct mineral identification, the region between ~25° 2θ and 35° 2θ is where the strongest and most useful carbonate peaks occur (Figure 22.5).

XRD analysis of carbonates can be used in the following ways:

1. Identification of the minerals (e.g., aragonite, calcite, dolomite) in a sample (Figure 22.5): such analyses can also provide semi-quantitative analyses that provide approximations of the amount of each mineral found in the sample.
2. Quantitative estimates of mineral percentages: these must be based on peak areas and not peak heights (Reitveld Analysis), the latter being mostly a function of crystallinity.
3. Determination of mole % $MgCO_3$ in HMC (Figures 22.5b, 22.6): there will be a second main calcite peak because the Mg has changed the cell size. Increasing amounts of Mg in the calcite lattice shrinks the cell size and therefore shifts this second peak to a higher 2θ. If HMC and LMC

are both present then the peaks will overlap, one on the shoulder of the other (Figure 22.5b). Since the shift of this peak is directly proportional to $MgCO_3$ content, the amount of Mg in the HMC can be accurately determined by simply measuring the peak displacement.

4. Determination of mole % $CaCO_3$ in dolomite (Figure 22.5c): as the amount of $CaCO_3$ increases, the main dolomite peak on the diffractogram is displaced towards the calcite peak or lower 2θ. This displacement is directly proportional to the mole % $CaCO_3$ content and can therefore be used to accurately determine the amount of excess Ca in the dolomite.
5. XRD of dolomites gives information on the ordering of the crystals: the segregation of cations into sheets in dolomites creates a set of suprastructure or ordering reflections. The reflection at *d*222 is the easiest to see because it is near the main dolomite peak (Figure 22.5c).

Scanning electron microscopy

Developed in the 1960s, the scanning electron microscope (SEM) is fundamentally designed to produce high-resolution, high-magnification, three-dimensional images of an object (Figure 22.7). The addition of various attachments,

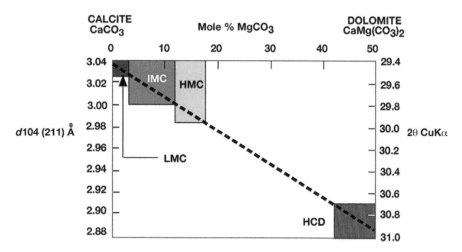

X-RAY POWDER DIFFRACTION
MgCO₃ in Calcite & Dolomite

Figure 22.6 The most commonly used graphic plot illustrating displacement of the calcite d_{104} peak with increasing $MgCO_3$ to dolomite. LMC: low-magnesium calcite; IMC: intermediate-magnesium calcite; HMC: high-magnesium calcite; HCD: high-calcium (calcian) dolomite.

Figure 22.7 Two different techniques using the SEM for carbonate analysis. (a) Photomicrograph showing bushes formed of aragonite crystals with some being rooted on rhombic calcite crystals. Modern spring deposit, Jifei, China. Image width 0.2 mm. (b) A polished and slightly acid-etched sample of Cenozoic subsurface dolomite from Grand Cayman, where the core has been dissolved and etching has highlighted crystal boundaries. Image width 70 μm.

however, means that additional information can be obtained from the samples being examined.

Various types of samples can be examined on the SEM. The most common type is the "fracture sample," up to a maximum size of about 1 cm³ that is simply broken off the parent material. The fractured surface must be clean and devoid of dust. The sample is then mounted on a metal SEM stub using double-sided carbon tape or conductive glue before being sputter-coated with carbon, gold, chromium, or any other conductive substance that will allow electrons to "drain off" the surface of the sample. The sample is then placed in the SEM sample chamber where air is extracted, creating a vacuum. An electron gun, mounted at the top of the chamber, then fires a constant stream of electrons onto the sample. The electrons that reflect from the surface of the sample are collected and used to construct an image of the surface being examined (secondary electron imaging). The image is built on a line-by-line basis, from the top to the bottom of the area being imaged. Each line records the topography of the surface from the electrons that are reflected from that surface; in effect, it produces a graph that depicts the topography of the surface along that line. The computer that controls the SEM then integrates all of the lines to produce a single "image" that shows the surface topography of the entire area being viewed. The image will always be black and white.

An SEM is typically used to image micro-scaled features that cannot be resolved with a thin-section. It is particularly useful for imaging microfossils (e.g., conodonts, foraminifera, coccoliths, diatoms), microbes (e.g., bacteria, cyanobacteria, fungi), zoning in cements, and microscale fabric variations. Be warned, however, that the SEM provides a view into the microworld that includes many features that we do not yet fully understand.

Samples can be treated in various ways to accentuate different features. Acid etching of a polished surface of calcite or dolomite, for example, can highlight microscale variations that may reflect growth defects or subtle differences in the internal structures of crystals. The ideal amount of etching is, however, a matter of trial and error.

Various attachments on a SEM can be used to obtain additional information about a sample. The use of an energy-dispersive X-ray (EDX) analyzer, for example, will allow determination of the elements that are present in a specific spot on a sample. Such analyses are typically based on a sample volume of about $1\,\mu m^3$. This information, when combined with the physical appearance of a crystal, allows accurate identification of the mineral being examined. It must be stressed, however, that EDX analyses are only semi-qualitative in nature.

Electron microprobe analysis

An electron microprobe, like a scanning electron microscope, operates by bombarding a sample with an electron beam. This produces X-rays with wavelengths that are indicative of the elements being analyzed and allows determination of the abundance of each element in small sample areas of the substrate (usually about $10–30\,\mu m^2$). An electron microprobe is used primarily for obtaining precise compositional measurements of any substance that is being examined. Elemental data as low as 100 ppm

can be determined in this way. Such analyses are far more accurate than those obtained using EDX analysis on a SEM. Conversely, images obtained from an electron microprobe are generally poorer than those obtained from a SEM.

Thin-sections used for such analysis must be uncovered, highly polished, and coated with carbon. From the perspective of carbonate rocks, an electron microprobe is used for the following purposes:

• Precise compositional information (to ppm level) for selected spots on a sample. Such analyses can be extended so that the spatial variation in composition along selected transects can be graphed. Compositional maps of selected areas can also be generated. Such transects and maps are especially useful for identifying zoned crystals and establishing the elements responsible for the definition of those zones.
• Electron backscattered imaging produces images that highlight spatial differences in the average atomic weight of the constituents in the field of view (Figure 22.8). This powerful technique can be used to highlight fabrics and zoned crystals. The electron microprobe can then be used to determine the composition of each zone that was highlighted by this technique.
• Some electron microprobes are equipped with attachments that can provide cathodoluminescent images. If these images show features of interest, such as zoning, then compositional transects can be obtained in the hope that variations in the composition can be correlated to the zones.

Acquiring compositional analyses on the electron microprobe is a time-consuming exercise that increases as the number of elements being analyzed and the number of points being sampled increases. Today, for example, this problem can be overcome by programming the microprobe so that such analyses can be obtained without an operator being present. The microprobe may therefore be allowed to run overnight so that the operator simply has to collect the data the following morning.

Chemical analyses

General considerations

Chemical analysis of carbonate rocks involves: (1) collection and preparation of samples; (2) analysis of the samples; and (3) analysis of the data derived from the samples. Most of these analyses are performed using a mass

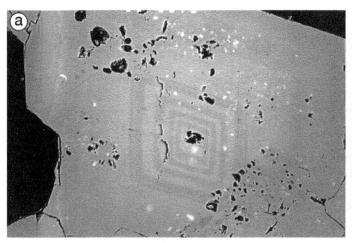 

Figure 22.8 Back-scattered electron microprobe images of dolomite crystals: (a) zoned crystal with the Ca-rich zones showing as light zones. Oligocene, South Australia. Image width 300 μm. and (b) similar crystal but it is a detrital sand grain from the seafloor. Holocene, South Australia Shelf.

spectrometer (MS) that measures the spectra emitted from a sample, allowing the elemental composition and the chemical structures of molecules to be determined. The instrument works by ionizing samples and then measuring their mass-to-charge ratios. Many different techniques, which depend on the nature of the sample (solid, liquid, gas) are used to ionize the sample, with the mass spectrometers generally being named after the technique used. This includes, for example, ICPMS (inductively coupled plasma mass spectrometer), TIMS (thermal ionization mass spectrometer), and SIMS (secondary ion mass spectrometer). The techniques are explained below but the applicability of the results is detailed in Chapter 23.

Ideally, sampling should follow a pattern that is designed to answer the questions that are being posed with respect to a suite of carbonate rocks. Sampling for geochemical analyses is usually of two types:

1. "Whole rock" samples where a sample is analyzed without regard to the different components in the rock. Data produced by this method are averages of all the components. This approach is typically used to established large-scale trends throughout a geological succession.
2. "Selective analysis" where specific parts of a sample are analyzed. This can involve separation of different components of a carbonate, for example, separation of dolomite from calcite in a calcareous dolostone. In some cases, microsampling of zoned cements can be undertaken by microdrilling specific textural elements.

Carbonate samples can be analyzed for different geochemical characteristics including trace element (e.g., Fe,

Mn, Sr) concentrations, stable isotopes (e.g., O and C isotopes), radiogenic isotopes (e.g., $^{87}Sr{:}^{86}Sr$ ratio), and rare Earth elements (REE). Decisions as to the type of analysis required depend on the information that is needed for the interpretation of the rocks being considered. The potential application of each of these types of geochemical data is discussed in various chapters throughout the book.

In the simplest sense, data acquisition is straightforward: a sample is fed into a machine and that machine produces numbers that characterize that sample. Although the machine will always produce data, many other factors ultimately control the fidelity and robustness of those data. Here, the old adage of "garbage in = garbage out" is highly appropriate. Potential problems can arise in the following ways:

- Sample preparation is the most fundamental and critical step in the process because it can significantly impact the analytical results. Such problems can, for example, include: (1) mixing of different generations or components of a rock that could have formed under different conditions and at different times; (2) the introduction of contaminants because the tools used to produce the samples are not kept clean; and (3) incomplete sample preparation, for example, poorly controlled acid dissolution during sample preparation for stable isotope analysis.
- Instrumental errors can be difficult to detect and rectify, especially if the potential for such errors is ignored. Replicate analyses or including routine analysis of "standard samples" (i.e., samples of known and well-established compositions) are usually

undertaken in order to monitor this issue. Inclusion of these types of samples allows calculation of the accuracy, precision, and reproducibility of the data from a given machine.

Geological studies commonly rely on the acquisition of data from a large number of samples. This can mean that the samples are analyzed at the same laboratory but at different times (e.g., following each fieldwork season), or that the samples are sent to different laboratories in order to assemble all of the data in a timely fashion. Even subtle differences in the analytical protocols can lead to different results. The inclusion of specific reference samples in each batch of samples allows direct comparison between different sample batches.

Further reading

Barker, C.E. and Kopp, O.C. (eds) (1991) *Luminescence Microscopy and Spectroscopy: Qualitative and Quantitative Applications*. Dallas, Texas USA: SEPM (Society for Sedimentary Geology), Short Course no. 25.

Burruss, R.C., Cercone, K.R., and Harris, P.M. (1983) Fluid inclusions petrology and tectonic-burial history of the Al Ali No. 2 well: evidence for the timing of diagenesis and oil migration, northern Oman foredeep. *Geology*, 11, 567–570.

Dickson, J.A.D. (1966) Carbonate identification and genesis as revealed by staining. *Journal of Sedimentary Petrology*, 36, 491–505.

Folk, R.L. (1987) Detection of organic matter in thin sections of carbonate rocks using a white card. *Sedimentary Geology*, 54, 193–200.

Frank, J.R., Carpenter, A.B., and Oglesby, T.W. (1982) Cathodoluminescence and composition of calcite cement in Taum Sauk Limestone (Upper Cambrian), southeast Missouri. *Journal of Sedimentary Petrology*, 52, 631–638.

Goldstein, R.H. (1986) Reequilibration of fluid inclusions in low-temperature calcium-carbonate cement. *Geology*, 14, 792–795.

Jones, B., Luth, R.W., and MacNeil, A.J. (2001) Powder X-ray diffraction analysis of homogeneous and heterogeneous sedimentary dolostones. *Journal of Sedimentary Research*, 71, 790–799.

Major, R.P., Halley, R.B., and Lucas, K.J. (1988) Cathodoluminescent bimineralic ooids from the Pleistocene of the Florida continental shelf. *Sedimentology*, 35, 842–856.

Scholle, P.A. and Ulmer-Scholle, D.A. (eds) (2003) *A Color Guide to the Petrography of Carbonate Rocks: Grains, Textures, Porosity, Diagenesis*. Tulsa, Oklahoma: American Association of Petroleum Geologist, Memoir no. 77.

Tucker, M.E. (ed.) (1988) *Techniques in Sedimentology*. Palo Alto, CA: Blackwell Scientific Publications.

CHAPTER 23
THE CHEMISTRY OF CARBONATE DIAGENESIS

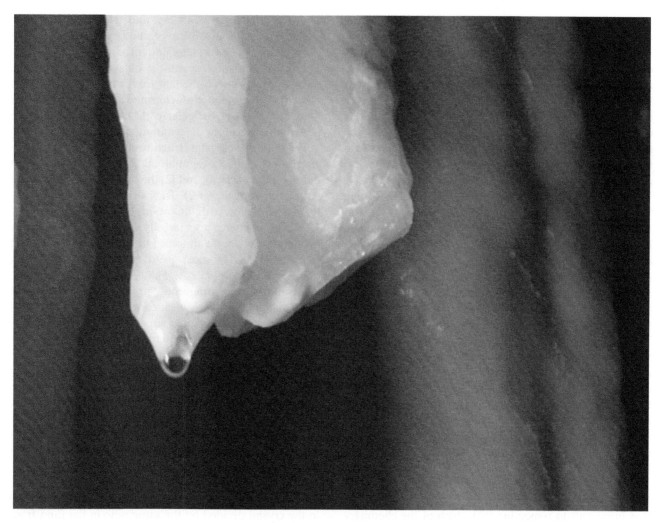

Frontispiece Stalactite from a Bermuda cave with a droplet of fresh water at the tip, illustrating the intimate geochemical relationship between carbonate diagenesis and water. Image width 3 cm.

Origin of Carbonate Sedimentary Rocks, First Edition. Noel P. James and Brian Jones.
© 2016 Noel P. James and Brian Jones. Published 2016 by John Wiley & Sons, Ltd.
Companion website: www.wiley.com/go/james/carbonaterocks

Introduction

To truly understand the origin of carbonate rocks is to understand their geochemistry. Specifically, the geochemical signatures of the rocks serve as proxies that allow interpretations beyond those obtained from fieldwork or petrographic observations alone. Such data allow us to: (1) confirm the origin of biogenic versus inorganic carbonate precipitates, both marine and terrestrial, and the way they were formed; (2) understand the processes operative during early marine and meteoric diagenesis; (3) interpret the nature of paleofluid flow; (4) understand burial diagenesis; and (5) reconstruct paleoceanography.

This is possible because of the nature of the minerals involved. The carbonate molecules, either $CaCO_3$ or $CaMg(CO_3)_2$, are composed of bivalent metal cations and the CO_3 anion complex. Other bivalent metal cations (e.g., Mn, Fe, and Sr) can substitute for Ca and Mg in these minerals depending on the character of the water from which they are derived and the crystallinity of the carbonate mineral. The CO_3, composed of C and O, is especially useful because each of these elements has stable isotopes that are systematically incorporated into the carbonate structure according to universal physical and biological controls.

When interpreting the trace element and stable isotope values in diagenetically altered carbonates, it is also critical to ascertain where the oxygen, carbon, and trace elements have come from in the new minerals. Natural fluids such as water with minor solutes are rich in oxygen, but they generally have far less carbon and trace elements than the rocks or sediments through which they pass. If the waters are flowing freely through the system, then the oxygen in the altered carbonate will come mostly from the waters. The carbon and trace elements will come from both the rock and the water. This situation is envisaged as an *open diagenetic system* or one with a *low rock-water ratio*. Alternatively, if the fluid system is largely isolated or fluid flow is sluggish, then the oxygen will still come mostly from the water but the carbon and trace elements will be mostly derived from the rock or sediment. This situation is called a *closed diagenetic system* or one with a *high rock-water ratio*.

This chapter focuses on the use of these aspects in geochemistry for determining the origin of carbonate rocks. It is important to remember that much of the chemistry discussed is experimental and inorganic, but we are dealing with a system that is a complex mixture of largely biogenic components that have been changed through geologic time and that may have been subject to alteration by a variety of different fluids.

Trace elements and element ratios

Component mineralogies

Carbonate mineralogies are both biogenic and abiogenic. Biogenic components comprise microbial precipitates, algae, and invertebrates, whereas abiotic components are generally ooids and synsedimentary and burial cements. Most components are aragonite or Mg-calcite with compositions ranging from ~4 to ~18 mole % $MgCO_3$ (Figure 23.1) and a few are low Mg-calcite.

Important trace elements

The trace elements Sr, Fe, and Mn are useful in understanding the nature of modern carbonates and interpreting the processes and products of diagenesis. The trace element contents of biogenic and abiogenic carbonates, however, deviate from the true equilibrium values because of kinetic factors and biological mediation. Element ratios, particularly Mg:Ca and Sr:Ca are potentially useful for determining paleotemperatures.

The important crystal structures that need to be considered are: (1) the hexagonal calcite structure (of which dolomite is a variant); and (2) the orthorhombic aragonite structure (see Chapter 2). The calcite structure has a six-fold coordination with the cations bonded to six oxygens. The trace elements most easily incorporated are those bivalent cations with an ionic radius that is less than that of Ca (0.99 Å). The most useful radii for understanding diagenesis are Mg (0.72 Å), Fe (0.78 Å), and Mn (0.83 Å). By contrast, the aragonite structure with nine-fold coordination most easily incorporates bivalent metal cations with an ionic radius larger than 0.99 Å. Sr (1.35 Å) is the most useful trace element found in aragonite.

Although crystal chemistry tells us which metal cations are most easily incorporated into a given crystal lattice, it does not provide information about how many cations can be included. This number comes from the experimentally determined *partition coefficient* (K_D) (Table 23.1). This physical constant predicts how much of a trace element in the fluid can be included in a crystal that precipitates from that fluid. The partition coefficient $(K_{DSr})_{aragonite}$, for example, defines how much Sr can be incorporated into aragonite. K_D is dependent on the molar ratio of Sr to Ca (mSr^{2+}/mCa^{2+}) in the crystal divided by the same ratio in the fluid from which the precipitation is occurring. Since the partition coefficient $(K_{DSr})_{aragonite}$ is close to 1 (precisely 1.12), the amount of Sr in the crystal will be roughly the same as that in the water. By contrast, $(K_{Sr})_{calcite}$ is 0.14, which is about 10% of that for aragonite. Accordingly, much less Sr can be

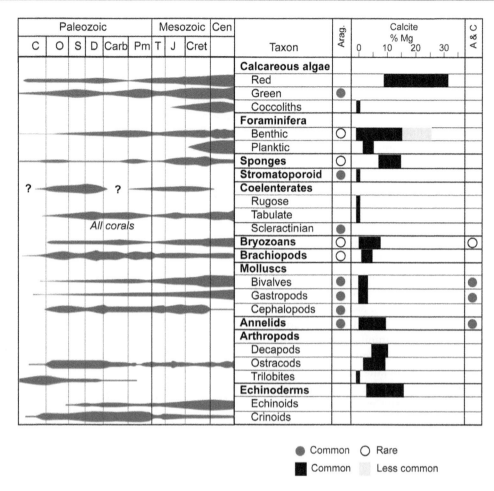

Figure 23.1 Geological ranges and mineralogical composition of skeletal organisms that contribute to carbonate sediment formation. Source: Adapted from Scholle and Ulmer-Scholle (2003) and Jones (2010).

Table 23.1 CaCO₃–H₂O trace element partition coefficients K for calcite and aragonite.

	Hexagonal, calcite	Orthorhombic, aragonite
Sr	0.30–0.13	0.9–1.2
Ba	0.1–0.4	1.0–2.0
Mg	0.013–0.06	0.0006
Mn	5.4–3.0	0.86
Na	0.00002	0.00014
Fe	1–20	? unconstrained
Cd	8–30	1.5–7.0

Table 23.2 Trace element contents (ppm) in the carbonate minerals calcite, aragonite, and dolomite.

	Calcite	Aragonite	Dolomite
Sr	1000	7000–9000	470–550
Na	200–300	1500	<110–160
Mg	16,300–75,400	750–6	130–400
Fe	2–39	–	3–50
Mn	1	4	1

incorporated into a calcite crystal that is precipitated from the same fluid as the aragonite (Table 23.2).

The partition for aragonite is:

$$(K_{DSr})_{aragonite} = (mSr^{2+}/mCa^{2+})\, solution = 1.12 \pm 0.04$$

where m is molalities (moles of solute divided by the kg of solvent), measured in moles kg⁻¹. By contrast, the partition coefficient of calcite is:

$$(K_{DSr})_{calcite} = 0.14 \pm 0.02.$$

Biogenic carbonate precipitation

Aragonite, HMC, or LMC produced by biomineralization do not necessarily obey the rules of thermodynamics or experimentation. There are two main processes of biogenic carbonate formation: biologically induced and biologically mediated.

Biologically induced precipitates result from the reaction between metabolic byproducts and ions in the external environment and are therefore chemically similar to inorganically precipitated crystals. Such minerals are most generally produced as a result of photosynthesis that is mediated by bacteria and algae. Crystal orientations are generally random. Trace element partitioning is generally close to that predicted from experimental results.

Biologically mediated precipitation is different. Every crystal or crystallite is enveloped by organic material with the mineral species and its crystallographic orientation directed by the host organism. Some skeletons have the experimentally predicted trace element concentrations, whereas others do not. Pelagic foraminifers and coccoliths, for example, precipitate LMC skeletons from seawater with high Mg:Ca ratios that predict the mineral should be either HMC or aragonite; their LMC skeletons should be forbidden! Some of these organisms even seem to precipitate amorphous calcium carbonate, which subsequently changes into aragonite or calcite.

Modern biogenic carbonates

Theoretically, the Sr^{2+} values in aragonite precipitated from tropical seawater should be ~8200 ppm (Figure 23.2). Ooids and precipitates are close to this value, generally containing between 8400 and 10,000 ppm (Figure 23.2). By contrast, corals are slightly lower at ~7900 ppm. Biogenic control of trace elements is most obvious in aragonite gastropod shells that contain only ~2000 ppm Sr^{2+}.

Limestones

The most useful cations for tracking early meteoric diagenesis are Sr, Mn, Fe, and Mg. Theoretically, those cations with $K_{cal} < 1$ (e.g., Sr) will decrease during early diagenesis, whereas those with $K_{cal} > 1$ (e.g., Fe, Mn, Mg) will increase because they fit more easily into the LMC crystal structure.

Strontium. Sr should decrease during diagenesis in meteoric water, but the amount incorporated into dLMC will depend on the rock-water ratio.

Iron and manganese. If present in the precipitating fluid, these cations will be easily incorporated into the dLMC crystal. Nevertheless, the inclusion of Fe and Mn in the dLMC lattice also depends on the oxidation state (Eh) of the fluid because these elements can only be included if they are in the reduced state (Fe^{2+}, Mn^{2+}). Waters in the vadose zone (above the freshwater table) are typically oxidizing and the dLMC is therefore typically non-ferroan and contains only small amounts of Mn. By contrast, most phreatic and burial fluids (below the water table) are reducing and ferroan (iron-rich) and Mn-rich dLMC cements are therefore common.

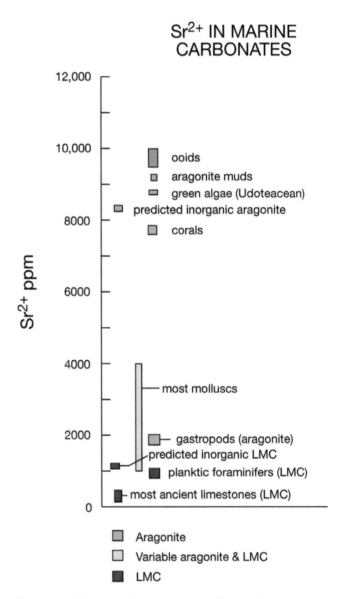

Figure 23.2 The strontium content of various modern carbonate skeletons and precipitates. Source: Adapted from Bathurst (1975). Reproduced with permission of Elsevier.

Element ratios. The Mg:Ca ratio in the calcitic tests of planktic and benthic foraminifers is, theoretically, a function of ambient seawater temperature; more Mg is incorporated into the test as the seawater temperature rises. This relationship has long been used to interpret seawater paleotemperatures. Given that different taxa have different ecological preferences, the best results are obtained when similar species are compared because each species has its own calibration curve. The fact that the Mg:Ca ratio of seawater has changed through geological time somewhat hinders the use of this ratio.

Likewise, the Sr:Ca in coral aragonite shows the same but inverse correlation with seawater temperature during biomineralization; the higher the temperature, the lower the Sr in the skeleton.

Stable isotopes

The nucleus of an element contains a set number of protons, but isotopes of that element have a different number of neutrons. Isotopes of the same element have slight differences in mass and energy and therefore form bonds resulting in slightly different physical and chemical properties. Such differences are highest for elements with low atomic numbers. When a molecule has two isotopes, the isotope with the smaller mass has bonds that are weaker than those isotopes with a higher mass. As a result, the lighter isotope is more reactive than the heavier isotope. The oxygen and carbon isotopes, which are commonly used in the interpretation of carbonate rocks, are reported in accordance with well-established standards.

Oxygen isotopes

Introduction

Oxygen has three isotopes comprising, in order of abundance: $^{18}O = 99.76\%$, $^{16}O = 0.20$ %, and $^{17}O = 0.04\%$ (Figure 23.3). Although determination of the concentration of each isotope is difficult, the ratio $^{18}O{:}^{16}O$ is relatively easy to measure with an isotope ratio mass spectrometer. $^{18}O{:}^{16}O$ is expressed as a ratio against a standard (V-PDB for $CaCO_3$ and V-SMOW for water). V-PDB represents a belemnite from the Cretaceous Pee Dee Formation, which has been chosen as the international standard against which all values are standardized. V-SMOW represents Standard Mean Ocean Water. The V stands for Vienna which is home of the International Atomic Energy Agency, the body responsible for setting the international standards.

OXYGEN STABLE ISOTOPES

- $^{16}O = 0.20\%$; $^{17}O = 0.04\%$; $^{18}O = 99.76\%$;
- easiest to measure ratio $^{18}O{:}^{16}O$
- measured in per mil (‰)
- $^{18}O{:}^{16}O$ plotted as a ratio against a standard (V-PDB - for $CaCO_3$)

$$\delta^{18}O = (R_x - R_{std}) / R_{std} \times 1000$$

R_x = value of unknown
R_{std} = value of standard

- *Most comes from water*

Figure 23.3 Chemical information concerning oxygen stable isotopes.

Delta ^{18}O ($\delta^{18}O$) is calculated:

$$\delta^{18}O = \frac{R_x - R_{std}}{R_{std}} \times 1000$$

where R_x is the ratio of the sample being analyzed ($= {}^{18}O/{}^{16}O$), and R_{std} is the ratio of the standard. $\delta^{18}O$ is usually expressed in units per mil (‰).

In the $CaCO_3$–H_2O–CO_2 system, most of the oxygen comes from the water (Figure 23.3). Seawater (officially designated as SMOW) has a $\delta^{18}O_{V\text{-}SMOW} = 0$. Marine carbonates should have a value of ~29‰ on the V-SMOW scale and –2‰ on the V-PDB scale. The relationship between $\delta^{18}O_{V\text{-}PDB}$ and $\delta^{18}O_{V\text{-}SMOW}$ is as follows:

$$\delta^{18}O\,(\text{calcite V} - \text{SMOW}) = 1.03086\,\delta^{18}O\,(\text{calcite V} - \text{PDB}) + 30.86$$

$PDB_{calcite}$ ($\delta^{18}O$ V– PDB = 0) is 30.86 on the SMOW scale

As stressed above, most of the oxygen in carbonate minerals comes from the fluid in which the crystal precipitates. The values are fractionated in various ways and it is these fractionations that are most useful in carbonate diagenesis. In terms of oxygen *fractionation*, it is the change in the ratio of two oxygen isotopes during a physicochemical process. In carbonates there is both equilibrium and non-equilibrium fractionation.

Equilibrium fractionation

Equilibrium fractionation refers to the situation where oxygen isotopes in the carbonate–water–bicarbonate system obey the rules of kinetic fractionation. The relationship between fractionation and fluid temperature for a given isotopic composition of water water is critical in this situation. (Figure 23.4); there is a known relationship between $\delta^{18}O$ V-PDB and water temperature. This relationship is fundamental in estimating paleomarine water temperatures throughout the Phanerozoic in both marine and terrestrial systems. Aragonite, LMC, HMC, and dolomite all have different fractionation values with the seawater that vary with water temperature.

Non-equilibrium fractionation

This situation occurs when organisms interfere with direct precipitation or when the crystallization rate is rapid. Biogenic carbonate precipitates are particularly disobedient because different organisms exert different physiological controls over isotopic fractionation. As a result, oxygen isotopes can be several parts per mil different from equilibrium fractionation. As a result, different biogenic particles in the same environment can have quite different isotopic compositions. Non-equilibrium fractionation is also affected by precipitation rate; the faster the precipitation, the further out of equilibrium the isotopes will be.

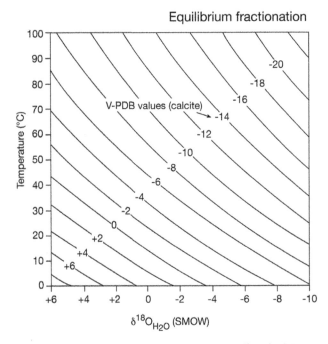

Figure 23.4 Equilibrium relationship between $\delta^{18}O$ of calcite, temperature and the $\delta^{18}O$ of water. The x axis represents $\delta^{18}O$ value of water (SMOW) and the y axis represents temperature between 0 and 100°C. The curved lines represent constant $\delta^{18}O$ values (PDB) for calcite calculated from expression in text.

What do the numbers mean?

Carbonates precipitated from tropical seawater at about 25°C should have a $\delta^{18}O$ value of ~0‰ on the V-PDB scale (Figures 23.5, 23.6). If the marine waters are colder, the values will be positive (e.g., $\delta^{18}O$ = +4‰), whereas if they are warmer they will be negative (e.g., $\delta^{18}O$ = –7‰). Conversely, if the waters have been evaporated and lost the lighter isotope, the remaining saline waters have more ^{18}O and their $\delta^{18}O$ values are positive (e.g., $\delta^{18}O$ = +5‰ for carbonate precipitates on the V-PDB scale). Advected into the atmosphere, the lighter isotope will eventually fall elsewhere as rain; meteoric water therefore has negative values (e.g., $\delta^{18}O_{V-SMOW}$ ≈ –4‰ for tropical environments). These meteoric waters become further depleted with increasing distance from the ocean, altitude, and latitude.

Paleowater temperature calculations

The ^{18}O in a carbonate crystal comes from the precipitating fluid, and partitioning is partly temperature controlled. This offers the possibility that the temperature of the paleofluid (ancient seawater, meteoric water, burial water) can be calculated from the $\delta^{18}O$ of the carbonate.

The chemical relationship between $\delta^{18}O$ and LMC precipitation has long been known:

$$T = 16.9 - 4.2(\delta c - \delta w) + 0.13(\delta c - \delta w)^2$$

where T is temperature (°C), δc is the $\delta^{18}O$ value for the solid carbonate and δw is $\delta^{18}O$ value for the water (both PDB).

The problem with this equation is that it includes one known value (the $\delta^{18}O$ of the precipitate) and two unknown values ($\delta^{18}O$ seawater and water temperature). In order to determine the precipitation temperature, the $\delta^{18}O$ value of the water (δw) must be assumed. One school of thought holds that the $\delta^{18}O$ value of seawater has always been 0‰, whereas the other cadre assumes that seawater has changed in composition through time. This controversy is examined later in this chapter.

Similar equations have been formulated for aragonite and Mg-calcite. Aragonite foraminifers are, for example, enriched by 0.6‰ compared to LMC foraminifers. Mg-calcite is enriched by 0.06‰ per mole $MgCO_3$ relative to LMC.

Clumped isotopes

Whereas most stable isotopes in carbonate skeletons, cements, and rocks are randomly distributed, a small fraction of the heavier isotopes (^{13}C, ^{18}O) have a tendency

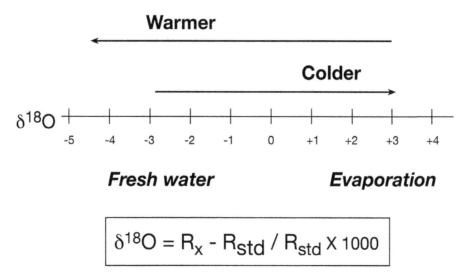

Figure 23.5 Diagram illustrating the general relationships between δ¹⁸O values and variations in water temperature and evaporation.

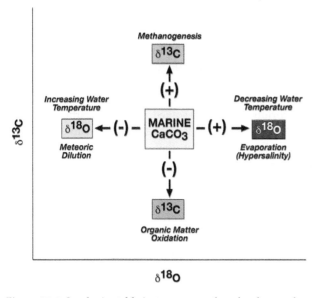

Figure 23.6 Synthetic stable isotope cross-plot of carbon and oxygen in carbonates and the main controls on their composition.

to bond with one another or form clumps (^{13}C, ^{18}O, ^{16}O) instead of being randomly distributed within the crystal lattice. Their enrichment relative to a random isotopic distribution is denoted Δ_{47}. The degree of clumping is a function of the temperature of precipitation and decreases with increasing temperature. This technique does not assume an original isotopic composition for the water as traditional oxygen isotope analysis does, and so holds great promise as a potential paleothermometer. Care must be taken because it has been shown that

disequilibrium occurs for certain speleothem precipitates, corals, and mollusks. The best present approach seems to be to use both traditional and clumped stable isotope analyses.

Carbon Isotopes

Introduction

Carbon has three isotopes, including the two stable forms $^{12}C = 98.89\%$ and $^{13}C = 1.11\%$ and the radiogenic isotope ^{14}C, which is used for dating. The easiest number to determine is the $^{13}C{:}^{12}C$ ratio. As for oxygen, this is calculated in parts per mil (‰) using the relationship:

$$\delta^{13}C = \frac{R_x - R_{std}}{R_{std}} \times 1000$$

(Figure 23.7). Since carbon is a fundamental building block of all living things, the marine organisms, terrestrial microbes, and organic molecules in the deep subsurface profoundly influence the isotopes via fractionation. Unlike the oxygen isotopes, most carbon does not come from the water but instead comes from the sediments, the rock, or the gases produced by microbes.

Most of the carbon in carbonate minerals comes from atmospheric CO_2 and metabolic processes within the organism, or from organic matter donated to the water by microbes. There is a mixture of effects. Organisms can profoundly fractionate carbon isotopes, thus giving a distinctive isotopic signature to associated precipitates. On the other

CARBON STABLE ISOTOPES

- ^{12}C = 98.89%; ^{13}C = 1.11%
- easiest to measure ratio ^{13}C:^{12}C
- measured in per mil (‰)
- ^{13}C:^{12}C plotted as a ratio against a standard (PDB- for $CaCO_3$)

$$\delta^{13}C = (R_x - R_{std}) / R_{std} \times 1000$$

R_x = value of unknown
R_{std} = value of standard (V-PDB)

- ***Most comes from rock & gas***
- ***Affected by metabolism***

Figure 23.7 Chemical information concerning carbon stable isotopes.

hand, many organisms precipitate their shells in equilibrium with $\delta^{13}C$ of dissolved CO_2 in their ambient fluid.

More specifically, departures from equilibrium fractionation in carbon isotopes are mostly due to a combination of carbonate precipitation rates and vital effects. The slower the precipitation rate, the closer to equilibrium fractionation the $\delta^{13}C$ values will be. The biochemical controls, often called "vital effects," are due to a mixture between dissolved inorganic carbonate and respiratory CO_2, the latter having low $\delta^{13}C$ values.

What do the numbers mean?

Most modern marine skeletons have $\delta^{13}C$ values of ~0 to +2‰ V-PDB, whereas cements precipitated from fluids enriched in soil gas (e.g., calcretes, speleothems) have negative values (about –8 to –12‰ V-PDB) (Figure 23.6). This is because vegetation results in ^{12}C-rich carbon that is incorporated into the calcite precipitate. Methanogenesis in the subsurface can result in carbonate precipitates with values in excess of +6‰.

Values for microbial and thermal alteration of organic matter that is incorporated into subsurface formation waters can be significant (Figures 23.6, 23.7). Bacterial oxidation or sulfate reduction will yield $\delta^{13}C$ values as low as –6 to –15‰ V-PDB. Conversely, bacterial fermentation will result in values as high as +25‰ V-PDB. Thermal decarboxylation in the deep surface can have values of –10 to –25‰ V-PDB.

A carbonate mineral, LMC, HMC, or aragonite that incorporates carbon from 100% marine organic or soil gas sources will have a $\delta^{13}C$ value close to –15‰. If the mineral incorporates carbon from a 50:50 mixture of organic and host-rock-sourced carbon with a value near 0, then its $\delta^{13}C$

will be close to –5‰. If the source of organic carbon in the pore fluid is low, then the value will be close to the $\delta^{13}C$ of host-rock inorganic carbon (which is the usual case).

Stable isotope values for modern biogenic carbonates

The $\delta^{13}C$–$\delta^{18}O$ cross-plot

One of the most useful ways to illustrate the isotopic composition of carbonates is to express the relationship between their $\delta^{13}C$ and $\delta^{18}O$ values as a cross-plot. This is a quick way to show the values and illustrate changes in isotopic composition of carbonate. Remember, however, that whereas $\delta^{13}C$ values are largely a function of organic processes, the $\delta^{18}O$ values are a function of both the isotopic value of the seawater and seawater temperature. These cross-plots are used extensively in the following chapters.

In broad terms, oxygen isotope equilibrium is quite common, whereas carbon isotope equilibrium is not. Carbon isotope disequilibrium is the rule in biogenic carbonate precipitation (Figure 23.8). There is usually a vital effect such that the $\delta^{13}C$ values observed are much more negative than the $\delta^{13}C$ values calculated for equilibrium with dissolved organic carbon in the surrounding water. This vital effect results in enrichment in ^{12}C, that is, carbon that is involved in metabolic processes.

Modern carbonate stable isotopes

Carbonate components. The $\delta^{18}O$ and $\delta^{13}C$ values of green calcareous algae are not in equilibrium with surrounding seawater (Figures 23.8, 23.9a). Mollusks, mainly bivalves and gastropods, grow in a wide range of marine and terrestrial environments from fresh to marine to hypersaline settings and their $\delta^{18}O$ values are variable (Figure 23.9b). The mineralogy of bivalves, ranging from aragonite to calcite to mixed aragonite and calcite, has an effect on both isotopes. In general, however, the $\delta^{13}C$ of bivalves is near isotopic equilibrium with the waters in which they grow. Gastropods, with most having aragonite shells, are usually not in equilibrium with the ambient water because they rapidly precipitate their shells during their short lives.

Foraminifers are variable (Figures 23.8, 23.9c). Numerous studies of LMC planktic protists confirm that their LMC tests are mostly in isotopic equilibrium with ambient seawater. As a result, they are commonly used

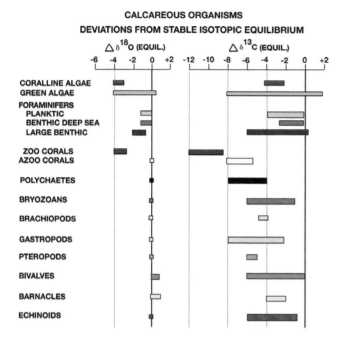

CALCAREOUS ORGANISMS
DEVIATIONS FROM STABLE ISOTOPIC EQUILIBRIUM

Figure 23.8 The differences of $\delta^{18}O$ and $\delta^{13}C$ from equilibrium values for a variety of important modern carbonate producing organisms. Source: Adapted from Wefer and Berger (1991). Reproduced with permission of Elsevier.

for paleoceanographic studies and paleotemperature estimates. Small benthic foraminifers are also generally in isotopic equilibrium. Large tropical benthic foraminifers, however, exert a strong vital effect on their $\delta^{13}C$ and $\delta^{18}O$ values because the photosymbionts in their tissue preferentially utilize ^{12}C. An increase in photosynthesis rates results in preferential incorporation of the light isotope into the test.

Reef-building corals (Figures 23.8, 23.9d), like the large benthic foraminifers, have photosymbionts (dinoflagellates) and therefore exert a strong vital effect that results in low to very low $\delta^{13}C$ values. The rapid growth of reef coral skeletons further increases the disequilibrium for both C and O isotopes with a strong seasonal bias (more pronounced in summer). $\delta^{18}O$ values, however, also reflect the tropical seawater in which they grow. Deep and cold-water corals have low $\delta^{13}C$ values, likely because of the nature of the dissolved organic carbon in their environment and food supply to the deep ocean. $\delta^{18}O$ values reflect the cold waters in which they live.

Bryozoans are among the dominant sediment producers in cool-temperate waters. Examples from southern Australia show their $\delta^{13}C$ values close to equilibrium values in surrounding seawater, whereas $\delta^{18}O$ values reflect the cool waters in which they thrive (Figures 23.8, 23.9e).

Echinoids use both metabolic- and seawater-derived CO_2 during calcification and their isotope values therefore

reflect a vital effect (Figures 23.8, 23.9e). Barnacles, by contrast, incorporate both carbon and oxygen in isotopic equilibrium with surrounding seawater.

Carbonate sediments

Neritic carbonate sediments have been analyzed from a variety of environments. Results of some recent studies (Figure 23.9f) show that overall the isotope fields for the sediments are smaller than those of the individual components. Although somewhat surprising, it is thought to be because extreme values are cancelled out in a mixture of grain types. Most of these data are from tropical neritic environments, and this is reflected by the relatively narrow range of $\delta^{18}O$ values (–1 to +3‰ V-PDB). Values from temperate carbonate environments of southern Australia and New Zealand have values that reflect the overall cooler marine waters (0 to +4‰ V-PDB).

By contrast, the $\delta^{13}C$ sediment values, although more variable, are similar overall with most values between 0 and +6‰. These data indicate that those calcareous organisms that do not metabolically fractionate carbon isotopes are producing most of the sediment. Recall, for example, that although corals build reefs, they do not produce much sediment. It is this information from sediments that should perhaps be utilized as a starting point in studies of diagenesis rather than the geochemistry of the individual organisms.

Carbonate stable isotope values through geologic time

Oxygen

It is important to know the $\delta^{18}O$ value of seawater from the past for paleoceanographic studies and for the interpretation of diagenetic isotopic signatures. The problem is that this value may not have been constant through time. For this discussion, it is best to begin with the Pleistocene and work backwards through time.

Pleistocene. The Pleistocene was characterized by periods of repeated continental glaciation. The water needed for these vast ice sheets came from the ocean via evaporation and meteoric water precipitation. With large volumes of meteoric water tied up in the ice with a low $\delta^{18}O$ signature, the $\delta^{18}O$ composition of seawater during the periods of maximum glaciations was about +1‰ and not 0‰ as it is today. An interesting question is: was this true for the Neoproterozoic Snowball Earth and the Late Paleozoic Ice Age?

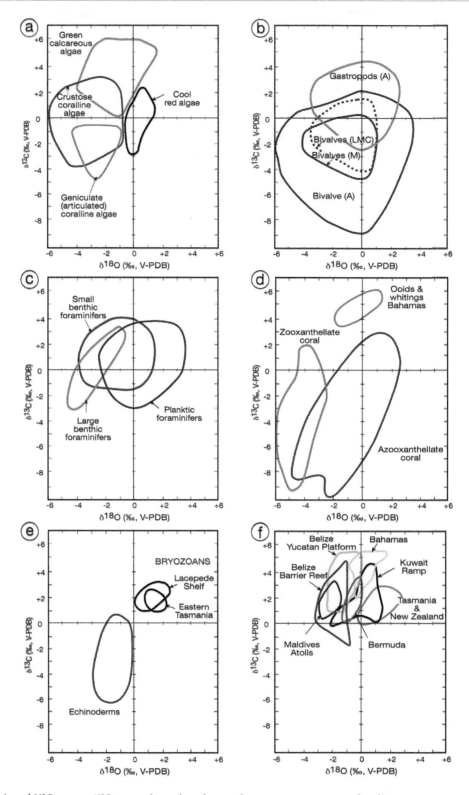

Figure 23.9 A series of $\delta^{13}C$ versus $\delta^{18}O$ cross-plots of modern carbonate components and sediments constructed from a variety of sources, principally Gross (1964), Milliman (1974), Morrison and Brand (1986), James and Choquette (1990) and references therein, Wefer and Berger (1991), Rao (1993), Rahimpour-Bobab et al. (1997), Bone and James (1997), Swart et al. (2009), and Duguid et al. (2010).

δ¹⁸O in carbonates through Phanerozoic time. It appears that the $\delta^{18}O$ of seawater may have changed by as much as 5.5‰ through geologic time (Figure 23.10). This is critical when trying to determine which $\delta^{18}O$ values of seawater to use when interpreting ancient limestones.

The original $\delta^{18}O$ of past seawater can be determined from the $\delta^{18}O$ values of invertebrate shells, providing they were originally formed of LMC and have not been diagenetically modified. The most common invertebrates with LMC skeletons are brachiopods and planktic foraminifers. Articulate brachiopods, which have lived in neritic environments since the early Cambrian, can have pores in their skeletons (punctae) that are filled with later cements. Great care must therefore be taken to ensure that only the original skeletal LMC is analyzed and not any pore-filling cements. Planktic foraminifers are, sadly, a Mesozoic invention and are therefore only useful for Cretaceous and younger limestones. Nevertheless, in this relatively recent time frame, they are superb proxies for ancient seawater.

Carbon

Carbon isotopes in Phanerozoic LMC shells (Figure 23.11) range from ~0 to +6‰ V-PDB. The $\delta^{13}C$ of mantle carbon is ~ −5‰ V-PDB, and in the absence of life this should be the $\delta^{13}C$ value of seawater. Life with photosynthetic capabilities has changed this value to ~0‰. Life is

therefore largely responsible for the carbon isotopic composition of seawater.

Strontium isotopes

Introduction

The $^{87}Sr:^{86}Sr$ ratio in a carbonate precipitate is the same as that in the water from which it crystallizes because there is no fractionation. The $^{87}Sr:^{86}Sr$ ratio in the water is related to the rocks and minerals that have donated Sr to the fluid. The value in seawater is largely controlled by input from continental crust or ocean crust (Figure 23.12). The Sr signature of weathered continental rocks, whose detritus is carried to the ocean, is ~0.7160. By contrast, the ocean crust that is affected by hydrothermal activity and submarine alteration of basalts has lower values of 0.7030–0.7050. These end-member values are buffered somewhat by marine carbonates and weathering of older carbonate rocks, such that most carbonate sediments and rocks are ~0.7070–0.7095. The relative proportions that come from each source are controlled by global tectonics and the values of ocean water, as recorded in carbonates, and therefore vary through time. The oceanic signature dominates during periods of rapid seafloor spreading, oceanic volcanism, high sea levels, continental flooding, and warm climates. By contrast, the continental signature

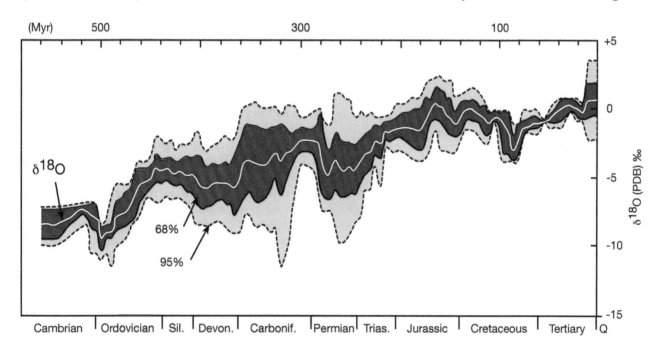

Figure 23.10 Phanerozoic $\delta^{18}O$ trend for LMC shells. Shaded areas around running mean include 68% ± 1σ and 95% ± 2σ of all data. Source: Adapted from Veizer and MacKenzie (2005). Reproduced with permission of Elsevier.

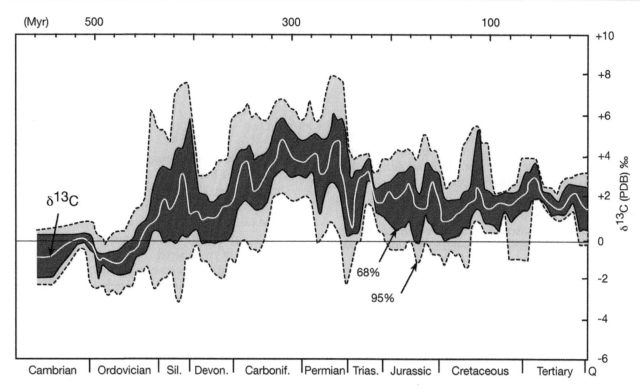

Figure 23.11 Phanerozoic trend of $\delta^{13}C$ for LMC shells. Shaded areas around running mean include 68% ± 1σ and 95% ± 2σ of all data. Source: Adapted from Veizer and MacKenzie (2005). Reproduced with permission of Elsevier.

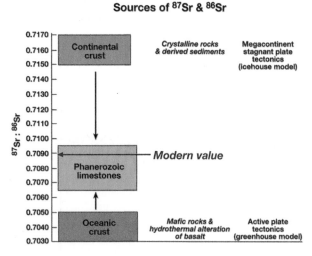

Figure 23.12 The main sources of Sr. The $^{87}Sr{:}^{86}Sr$ values from continental lithosphere versus oceanic lithosphere are demonstrably different, whereas most limestones have contributions from both sources.

prevails in times when plate movement is arrested, megacontinents are present, sea level is low, climates are cooler, and continental erosion is rampant. The values recorded in the least-altered limestones and fossils

have also varied in a consistent way throughout the Phanerozoic (Figure 23.13).

Utility in carbonates

Limestones, fossils, and dolostones that are proven to be unaltered since formation can, as a result of the above, be dated. Perhaps more important, however, is that if the age of the rock is known with certainty from other techniques (fossils, radiometric techniques), then much information can be extracted about diagenesis. For example, if the rock is unequivocally Ordovician but strontium isotope values from cements indicate ratios that are consistent with a Jurassic age, then such information places constraints on paragenesis.

The curve showing the relationship between the $^{87}Sr{:}^{86}Sr$ ratio and time is characterized by numerous oscillations (Figure 23.13). Whereas the late Mesozoic and Cenozoic curve is relatively smooth, the curve for the Paleozoic is more difficult to decipher. Finally, if the subsurface waters have passed through siliciclastic rocks then they will have picked up ^{87}Sr and have elevated values relative to ambient seawater, therefore influencing any paragenetic interpretations. In short, these isotopes are potentially useful but they must be employed together with all other available petrographic and geochemical information.

STRONTIUM ISOTOPES

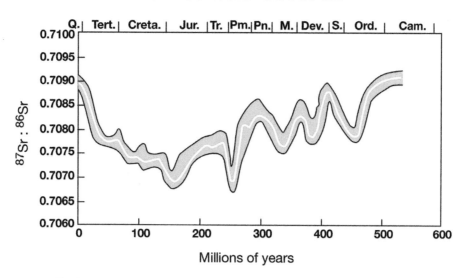

Figure 23.13 The variation of ^{87}Sr:^{86}Sr ratio as recorded in brachiopods through Phanerozoic time. Source: Adapted from Koepnick et al. (1985) and Veizer et al. (1999).

Further reading

Anderson, T.F. and Arthur, M.A. (1983) Stable isotopes of oxygen and carbon and their application to sedimentologic and paleoenvironmental problems. In: Arthur M.A. (ed.) *Stable Isotopes in Sedimentary Geology*. Tulsa, OK: SEPM (Society for Petroleum Geology), Short Course No. 10, pp. 1-1–1-151.

Bathurst, R.G.C. (1975) *Carbonate Sediments and their Diagenesis*. Amsterdam: Elsevier Science, Developments in Sedimentology.

Berner, R.A. (2004) *The Phanerozoic Carbon Cycle: CO$_2$ and O$_2$*. New York, NY: Oxford University Press.

Bone, Y. and James, N.P. (1997) Bryozoan stable isotope survey from the cool-water Lacepede shelf, southern Australia. In: James N.P. and Clarke J.A.D. (eds) *Cool Water Carbonates*. Tulsa, OK: SEPM (Society for Sedimentary Geology), Special Publication no. 56, pp. 93–105.

Burton, E.A. and Walter, L.M. (1987) Relative precipitation rates of aragonite and Mg calcite from seawater: temperature or carbonate ion control? *Geology*, 15, 111–114.

Duguid, S.M.A., Kyser, T.K., James, N.P., and Rankey, E.C. (2010) Microbes and Ooids. *Journal of Sedimentary Research*, 80, 236–251.

Eiler, J.M. (2007) "Clumped-isotope" geochemistry: The study of naturally-occurring, multiply-substituted isotopologues. *Earth and Planetary Science Letters*, 262, 309–327.

Ghosh, P., Adkins, J., Affek, H., et al. (2006) ^{13}C-^{18}O bonds in carbonate minerals: A new kind of paleothermometer. *Geochimica et Cosmochimica Acta*, 70, 1439–1456.

Given, R.K. and Wilkinson, B.H. (1985) Kinetic control of morphology, composition, and mineralogy of abiotic sedimentary carbonates. *Journal of Sedimentary Petrology*, 55, 109–119.

Gross, M.G. (1964) Variations in the O^{18}/O^{16} and C^{13}/C^{12} ratios of diagenetically altered limestones in the Bermuda Islands. *Journal of Geology*, 72, 170–194.

Hoefs, J. (2009) *Stable Isotope Geochemistry*, sixth edition. Berlin Heidelberg: Springer-Verlag.

James, N.P. and Choquette, P.W. (1990) Limestone: The meteoric diagenetic environment. In: McIlreath I. and Morrow D. (eds) *Sediment Diagenesis*. St John's, ND, Canada: Geological Association of Canada, Reprint Series no. 4, pp. 35–74.

Jones, B. (2010) Warm-water neritic carbonates. In: James N.P. and Dalrymple R.W. (eds) *Facies Models 4*. Geological Association of Canada, GEOtext 6, pp. 341–369.

Koepnick, R.B., Burke, W.H., Denison, R.E., *et al.* (1985) Construction of the seawater ^{87}Sr/^{86}Sr curve for the Cenozoic and Cretaceous: supporting data. *Chemical Geology (Isotope Geoscience Section)*, 58, 55–81.

Milliman, J.D. (1974) *Marine Carbonates. Part 1 Recent Sedimentary Carbonates*. New York: Springer-Verlag.

Morrison, J.O. and Brand, U. (1986) Geochemistry of recent marine invertebrates. *Geoscience Canada*, 13, 237–254.

Morse, J.W. (2005) Formation and diagenesis of carbonate sediments. In: Mackenzie F.T. (ed.) *Sediments, Diagenesis, and Sedimentary Rocks*. Amsterdam: Elsevier, Treatise on Geochemistry no. 7, pp. 67–85.

Rahimpour-Bonab, H., Bone, Y., and Moussavi-Harami, R. (1997) Stable isotope aspects of modern molluscs, brachiopods, and marine cements from cool-water carbonates, Lacepede Shelf, South Australia. *Geochimica et Cosmochimica Acta*, 61, 207–218.

Rao, C.P. (1993) Carbonate minerals, oxygen and carbon isotopes in modern temperate Bryozoa, eastern Tasmania, Australia. *Sedimentary Geology*, 88, 123–135.

Scholle, P.A. (1978) *A Color Illustrated Guide to Carbonate Rock Constituents, Textures, Cements, and Porosities*. Tulsa, OK: American Association of Petroleum Geologists, Memoir no. 27.

Scholle, P.A. and Ulmer-Scholle, D.A. (eds) (2003) *A Color Guide to the Petrography of Carbonate Rocks: Grains, Textures, Porosity, Diagenesis*. Tulsa, Oklahoma: American Association of Petroleum Geologist, Memoir no. 77.

Swart, P.K., Reijmer, J.J.G., and Otto, R. (2009) A re-evaluation of facies on Great Bahama Bank II: variations in the δ13C, δ18O and mineralogy of surface sediments. In: Swart, P.K., Eberli, G.P., and McKenzie, J.A. (eds) *Perspectives in Carbonate Geology*. IAS Special Publication no.41, pp. 47–59.

Veizer, J. (1983) Chemical diagenesis of carbonates: theory and application of trace element technique. In: Arthur M.A. (ed.) *Stable Isotopes in Sedimentary Geology*. Tulsa, OK: SEPM (Society for sedimentary Geology), Short Course No. 10, pp. 3-1–3-100.

Veizer, J. and Mackenzie, F.T. (2005) Evolution of sedimentary rocks. In: Mackenzie F.T. (ed.) *Sediments, Diagenesis, and Sedimentary Rocks*. Amsterdam: Elsevier, Treatise on Geochemistry no. 7, pp. 369–408.

Veizer, J., Ala, D., Azmy, K., *et al.* (1999) $^{87}Sr/^{86}Sr$, $d^{13}C$ and $d^{18}O$ evolution of Phanerozoic seawater. *Chemical Geology*, 161, 59–88.

Wefer, G. and Berger, W.H. (1991) Isotope paleontology: growth and composition of extent calcareous species. *Marine Geology*, 100, 207–248.

CHAPTER 24
LIMESTONE: THE SYNSEDIMENTARY MARINE DIAGENETIC ENVIRONMENT

Frontispiece Seaward-dipping beachrock along the southern coast of Grand Cayman, Caribbean Sea.

Origin of Carbonate Sedimentary Rocks, First Edition. Noel P. James and Brian Jones.
© 2016 Noel P. James and Brian Jones. Published 2016 by John Wiley & Sons, Ltd.
Companion website: www.wiley.com/go/james/carbonaterocks

Introduction

The marine environment, where carbonate sediments are born, is also one where important but localized diagenesis begins as soon as a particle is deposited on the seafloor. The diagenetic processes operative on and just below the seafloor are mostly dissolution and cementation and, to a lesser extent, neomorphism. Physiochemical and biological processes commonly act together to produce a wide variety of distinctive textures and fabrics. Such processes, operating on scales that range from individual particles to entire facies, can radically alter the composition of the sediment and, in some cases, create hardgrounds. Synsedimentary diagenesis (authigenesis) will generally reduce porosity and permeability (Figure 24.1). Increasing water depth, cooling temperatures, and increasing hydrostatic pressure in the ocean can result in local neomorphism and, at greater depths, complete dissolution (see Chapter 2).

The setting

This environment is the seafloor, the sediments on the seafloor, and the sediments to depths of several tens of meters below the seafloor. Although the environment ranges from the strandline across shallow lagoons and platform reefs to the slope and deep basin, the focus of this chapter is on neritic settings (from the strandline to the shelf edge, usually a maximum water depth of 200 m).

The calcareous skeletons of invertebrates and algal hard parts form in equilibrium with the metabolic fluids from which they precipitate. Since seawater is supersaturated with respect to these minerals in both tropical and temperate shelf environments, there is usually little physiochemical-driven mineral or water-driven alteration once they become sediment particles. There can be, however, extensive carbonate precipitation in the form of cement or as microbially influenced dissolution.

From the perspective of synsedimentary diagenesis, there are two realms: (1) the neritic realm where seawater is mostly oversaturated with respect to $CaCO_3$; and (2) the slope and basin where the seafloor can intersect the lysocline for various minerals and, in deeper waters, the calcite compensation depth (CCD).

Dissolution

Neritic environments

Today, the shallow seawater on shelves, banks, and ramps is generally oversaturated with respect to both aragonite and Mg-calcite. It is therefore surprising that recent research has shown that dissolution seems to be occurring in the sediments just below the seafloor, both in warm tropical and cool temperate environments. The impact of this is small in tropical systems because the sediments are accumulating so rapidly. In temperate environments, however, the effect can be significant because the rate of sediment accumulation is considerably slower. The dissolution appears to be associated with microbial reduction of organic matter in the shallow subsurface that leads to the production of acids that, in turn, lower the alkalinity and promote dissolution within the sediment. This microbially driven dissolution is mineral-specific as it focuses on the dissolution of aragonite rather than calcite. In cool-water systems, this process can potentially remove all aragonite from the near-surface sediment in as little as 30,000 years (Figure 24.2). This situation means that these sediments may become entirely calcitic even before they are exported to the next diagenetic environment.

Deep-water environments

Diagenetic processes in these environments are largely water-driven. Waters below the aragonite or Mg-calcite lysoclines are undersaturated with respect to aragonite and calcite. As calcareous grains sink below the aragonite and calcite lysocline, dissolution will start and continue until, once below the compensation depths, the grains will be completely dissolved. Here, it is important to remember that: (1) the depth of the lysoclines and compensation depths vary from ocean to ocean; and (2)

HIGHLIGHTS
SYNSEDIMENTARY SEAFLOOR DIAGENESIS

MINERALOGY	*NO CHANGE*
PROCESSES	*Boring*
	CEMENTATION
	Dissolution
CONTROLS	*WATER DEPTH*
	POSITION ON PLATFORM
	SEDIMENTATION RATE
	SUBSTRATE STABILITY
POROSITY & PERMEABILITY	*REDUCED*

Figure 24.1 The main diagenetic processes and products in the marine synsedimentary diagenetic environment.

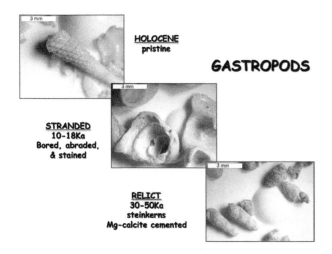

Figure 24.2 Images of gastropods of different ages from surface neritic carbonate sediment in the Great Australian Bight ranging from pristine Holocene (0–6 ka) aragonite to bored, abraded and stained skeletons (10–18 ka), to totally dissolved shells with only internal HMC sediment (steinkerns) remaining (30–50 ka); none had been exposed to meteoric waters. Photograph by J. Rivers. Reproduced with permission.

the depth of the aragonite lysocline, as in the Pacific Ocean today, can be as shallow as 500 m (see Chapter 2).

Precipitation

Neritic environments

Today, precipitation is largely restricted to neritic environments where it is manifest as aragonite or Mg-calcite cements that are precipitated within grains and skeletons, inside biological borings, between grains, and in reef cavities. Although precipitation does occur in cool temperate systems, it is far more intense in warm waters.

Cements in the modern ocean are fibrous, micropeloidal, or micritic (Figure 24.3) and typically isopachous (Figure 24.4). Cements in reef cavities can be spectacular botryoidal arrays that are not necessarily isopachous (Figure 24.5). Cement rinds can be multi-generation and in large cavities are commonly intercalated with internal sediments. Micropeloids (Figures 24.5d, 24.6) that precipitate in the water column and fall to the cavity floor can be layered. Micritic cement is typically automicrite (Figure 3.12), as evidenced by its formation on cavity ceilings and microbial fabrics. Cement inside grains is perhaps the most common type of precipitate, occurring in grains that are still rolling around on the seafloor. Conditions for precipitation in these intraskeletal microenvironments can be quite different from the surrounding seawater. Grains can therefore contain intraparticle

cements even though they are not cemented together on the seafloor.

Seawater precipitates

Modern marine cements, precipitated from seawater with a high Mg:Ca ratio, are mostly aragonite or Mg-calcite. They are usually iron-poor and non-luminescent. Stable isotopes of modern cements clearly indicate that there is some involvement of organic processes; they are therefore not simply the product of abiogenic precipitation from supersaturated seawater.

Mg-calcite cements. Calcite cements can contain between 4 and 19 mole % $MgCO_3$ where the magnesium is distributed somewhat randomly throughout the crystal lattice. The crystals are trigonal and precipitate either as rhombs less than 4 μm long (micrite-size) or as fibers a few tens of micrometers long.

• *Micrite-size cement crystals*: This is the most common type of modern Mg-calcite cement and occurs in a wide variety of sediments. Although commonly developed as a thin (tens of microns thick) isopachous rind around particles and algal filaments, it can also completely fill intergranular pores. It is easy to recognize a cement when it is a coating around grains. If located between grains, however, it can be difficult to distinguish this

MODERN SHALLOW MARINE CEMENTS

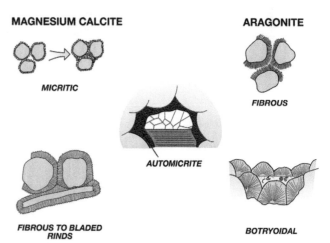

Figure 24.3 The fabric of synsedimentary carbonate cements from the modern seafloor: microcrystalline and fibrous magnesium calcite, microbially influenced magnesium calcite automicrite, and fibrous to botryoidal aragonite. Source: James and Choquette (1983). Reproduced with permission of the Geological Association of Canada

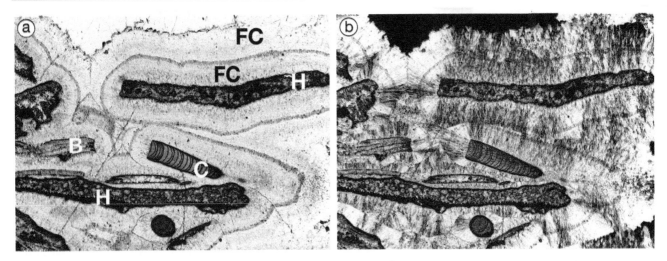

Figure 24.4 Fibrous isopachous synsedimentary cement, Pliocene Hope Gate Formation, Jamaica. (a) Biofragmental grainstone composed of Halimeda (H), coralline algae (C), and bryozoans (B), with all grains surrounded by two layers of fibrous high-magnesium calcite cement (FC). (b) Same image under partially polarized light, illustrating the fibrous nature of the cement. Image width 6.5 mm. Photographs by T. Frank. Reproduced with permission.

Figure 24.5 Holocene deep reef limestones, Belize. (a) Margin of a reef growth cavity with numerous aragonite botryoids that partially fill the synsedimentary void. Image width 6 cm. (b) Thin-section image under XPL, stained with Clayton yellow, of an aragonite botryoid filling a small reef void (HMC cement in red). Image width 9 mm. (c) SEM image of botryoidal aragonite needles. Image width 1 mm. (d) Microcrystalline aragonite cement (clear) and scattered HMC micropeloids (red). Image width 1.5 mm.

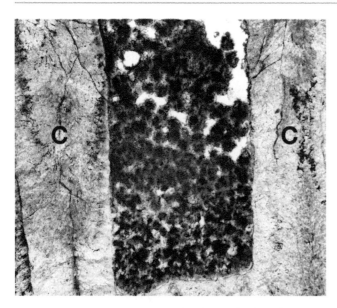

Figure 24.6 Thin-section image (ppl) of an aragonite coral fragment (C) whose corallites are filled with HMC micropeloids (red) that have accumulated as sediment from the water column, Holocene Belize shallow reef margin. Image width 1.5 mm.

cement from calcareous mud sediment that was simply deposited between the grains. A common fabric in many limestones is the association of this cement with micropeloids (Figure 24.6). The peloids, 20–100 μm in size, are surrounded by micrite cement. This is the 'structure grumeleuse' of early workers, which is formed by numerous clots or vaguely defined small peloids (aggregates of Mg-calcite rhombs) that are surrounded by micrite and microspar. The origin of these peloids, which is a matter of debate, has been attributed to: (1) spontaneous nucleation and precipitation of Mg-calcite in the water column that then settles out of suspension as tiny particles; or (2) bacterially influenced precipitation, given that bacterial rods have been found in the nucleus and are organic-rich (as demonstrated by epifluorescence). These precipitates commonly merge with automicrite (Figure 3.12), which has similar characteristics.

- *Fibrous crystals*: This cement generally occurs as rinds up to 200 μm thick. The elongate crystals, with a length: width ratio of >6: 1, are arranged in a picket-fence style on the substrate (Figure 24.4) or grow as small spherulites that, with growth, merge laterally with each other. The spherulites are commonly arranged in tiers.
- *Blocky or equant crystals*: These small crystals, 20–50 μm in size, are rare in shallow marine systems but occur in hardgrounds and reefs.

Aragonite cements. Aragonite is either microcrystalline (Figure 24.5) or is present as needles up to 300 μm

long. The needles have several habits. They can occur as rinds, especially in sands, and as epitaxial extensions of skeletal crystals in corals, green algae, and gastropods. The most spectacular growths are botryoids (Figures 24.3, 24.5) that range in diameter from tens of micrometers to centimeters. These cements are locally developed in carbonate sands but are most spectacular in reef cavities.

Alteration

Mineral-driven diagenesis

Neritic environments. In most cases, aragonite and Mg-calcite do not seem to be affected while they are on the seafloor. There are, however, some reports of neomorphism taking place via mineral-driven diagenesis, including: (1) Mg-calcite foraminifers and coralline algae that are altered to aragonite; (2) aragonite cements that have replaced aragonite shells; and (3) parts of Mg-calcite shells altering to LMC. Although rare, these changes highlight the subtle nature of the factors that govern precipitation of various carbonate minerals in seawater and emphasize that possibilities for marine neomorphism do exist in the modern oceans.

Deep-water environments. When neritic sediments composed of metastable minerals are transported into deep water via sediment gravity flows, they are unstable in these colder waters under high hydrostatic pressure and so can neomorphose to dLMC.

Microbial- and other biotic-driven diagenesis

Even though neomorphism is a rare process in shallow-water carbonates, particles and rocks can undergo profound alteration to micrite or mudstone via a combination of biological and chemical processes. This alteration involves infestation of the carbonate host by a variety of boring organisms (endoliths) and, upon their death, the filling of their drill holes by sediment and cement (Figures 24.7, 24.8). On a microscale this process involves endolithic algae or fungi boring into sand grains. Cyanobacteria are most effective in the photic zone, whereas fungi can extend to 500 mwd and heterotrophic bacteria range into abyssal depths. In shallow tropical settings this commonly leads to the development of a *micrite envelope* that is produced by repeated cycles of boring, vacation of the boring once the microbe dies, and filling of the boring with microcrystalline aragonite or Mg-calcite

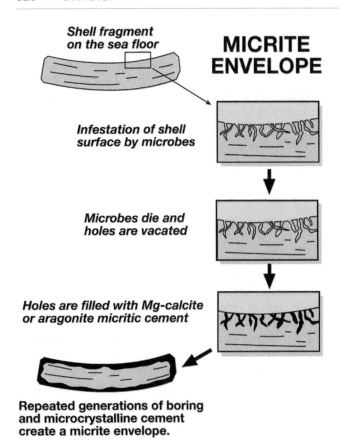

Shell fragment on the sea floor

MICRITE ENVELOPE

Infestation of shell surface by microbes

Microbes die and holes are vacated

Holes are filled with Mg-calcite or aragonite micritic cement

Repeated generations of boring and microcrystalline cement create a micrite envelope.

Figure 24.7 Sketch illustrating the formation of micrite envelopes, in this case on an aragonitic bivalve fragment.

cement (Figures 24.7, 24.8). If infestation is intense and prolonged, then the entire grain can be altered to micrite and becomes a diagenetic peloid. Although boring endoliths are active in cool-temperate environments, micrite envelopes do not form because the borings are generally not filled with micrite precipitates. The presence of this envelope has profound consequences for later diagenesis. The micrite is quickly altered to dLMC, thereby, preserving the shape of the particle which itself was dissolved.

Macroscale borings produced by endolithic sponges, bivalves, and polychaete worms are common in many settings. In tropical seas, numerous corals are extensively bored by sponges (especially *Cliona*) and boring bivalves commonly attack dead corals while they are on the seafloor. Once vacated, these borings are typically filled with various types of sediment (internal sediment) that can be cemented. With repeated cycles of boring and sediment filling of vacated borings and cementation, traces of the original skeletons are progressively lost and can, in extreme cases, be completely destroyed.

Synsedimentary limestone

Synsedimentary cementation results in the formation of limestone on the seafloor, either as hardgrounds or as lithified reefs. Although not ubiquitous, this "rock-making" cementation can exert an important influence over the development of some successions.

Hardgrounds

A hardground is defined as a lithified carbonate sediment that formed on the seafloor (Figure 24.9). A firmground is an intermediate stage between the original soft sediment and the hardground. Although a firmground has some inherent strength, a few burrowers, including decapod crustaceans, can still dig into it. By contrast, only boring organisms (e.g., boring bivalves, sponges, microbes) can penetrate solid hardgrounds.

The processes of hardground formation (Figure 24.10) are tricky because two diametrically opposed conditions must be operative, namely: (1) there must be water flow through the sediment pores in order to provide the materials needed for continued cement precipitation; and (2) the grains cannot be moved. The simplest explanation would seem to be that the cementation takes place at some depth below the seafloor (a few decimeters at most) where the sediment is not moving. Later, high-energy conditions (e.g., storms, tidal currents) strip away the overlying sediment and expose the lithified layer. Continued current activity keeps the surface clear of sediment and mediates further cementation and induration of the seafloor.

The formation of hardgrounds has profound environmental consequences because, once on the seafloor, the soft sediments are cemented to form a hard substrate. This has a radical impact on the seafloor biota, as a benthic community specifically adapted to life on a hard substrate replaces the biota that inhabited the soft sediment. Encrusting organisms that can attach themselves to the hard substrate (e.g., crinoids, corals) and boring organisms dominate hardground biotas. Many reefs are, for example, rooted on hardgrounds. An additional consequence is that these largely impermeable horizons in the rock record impede vertical fluid flow.

Hardgrounds, despite their diversity (Figure 24.11), can be recognized in the field through the integration of the following criteria (Figure 24.12): (1) a highly irregular surface that commonly includes overhanging ledges that could not develop in soft sediments; (2) the presence of

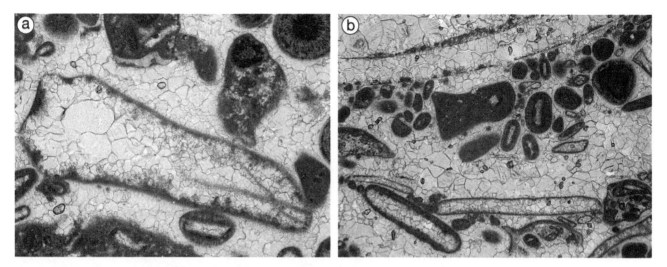

Figure 24.8 Micrite envelopes, Middle Ordovician Limestone, Kingston, Canada. (a) Bivalve mold with a micrite envelope filled with and surrounded by LMC cement. Image width 3 mm. (b) Radial ooids and several bivalve molds with micrite envelopes encased by and filled with LMC cement. Image width 3.5 mm.

Figure 24.9 Image of an excavated modern hardground in the Bahamas illustrating the protruding cemented layers, underlying soft sediment, and numerous benthic organisms attached to the hardground surface. Image width ~6 m. Photograph by E. A. Shinn. Reproduced with permission.

borings; (3) surfaces colonized by encrusting organisms; (4) surface impregnation by iron, manganese, or phosphate; (5) clasts in the overlying beds that were derived from the hardground; (6) local expansion ridges or teepees; (7) decrease in cementation downwards; and (8) evidence of an eroded surface. Microscale evidence in thin-section includes the presence of: (1) truncated grains; (2) fibrous or micritic cements (typically isopachous); (3) internal sediments; and (4) borings that cut

HARDGROUND FORMATION

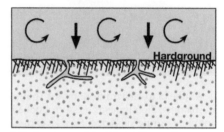

Marine erosion of sediment - Lowering of seafloor - Hardground exposure

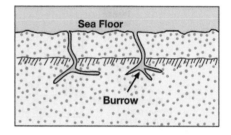

Shallow subsurface cementation

Figure 24.10 Sketch of how hardgrounds likely form, cementation below the seafloor, followed by exhumation and further cementation. The burrow was produced while the sediment was still soft, and a part was preserved in the cemented hardground.

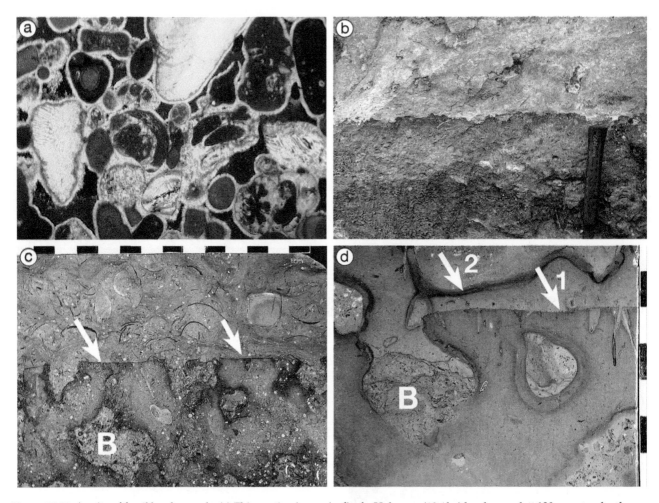

Figure 24.11 A suite of fossil hardgrounds. (a) Thin-section image (ppl) of a Holocene (18.4 ka) hardground at 130 m water depth off southern Australia. Biofragments are surrounded by isopachous IMC and the subsequent voids are filled with muddy planktic sediment infiltrated during subsequent sea-level rise. Image width 2.5 mm. (b) Hardground impregnated with Mn and PO_4, Oligocene, North Carolina, USA. Hammer handle 25 cm. (c) Burrowed (B) and beveled (arrows) hardground, in core Upper Devonian Western Canadian Sedimentary Basin. Centimeter scale. Photograph by W. Martindale. Reproduced with permission. (d) Multiple hardgrounds (arrows) in core Upper Devonian Western Canadian Sedimentary Basin. 1, burrowed (B), bored, and impregnated with Mn-PO_4; 2, burrowed and impregnated. Centimeter scale. Photograph by W. Martindale. Reproduced with permission.

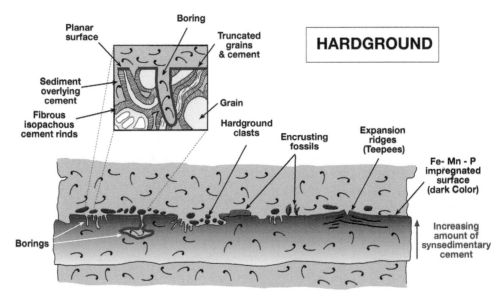

Figure 24.12 Sketch illustrating the main macroscale and petrographic (box) attributes of hardgrounds. Source: Adapted from James and Choquette (1990). Reproduced with permission of the Geological Association of Canada.

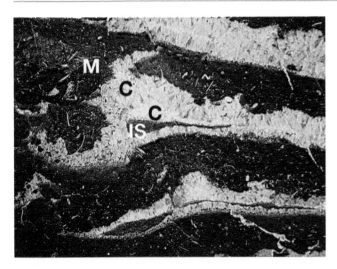

Figure 24.13 *Stromatactis* synsedimentary fibrous calcite (C) and internal sediment (IS) filling growth cavities in a Mississippian biogenic mud mound (M), Belgium. Image width 3 cm. Photograph by B. Pratt. Reproduced with permission.

across the cement crystals and the hardground surface. It is important to use as many of these criteria as possible to establish the presence of a hardground because they can, in some circumstances, be confused with subaerial exposure surfaces.

Reefs and mounds

Synsedimentary cements commonly develop inside reefs and mounds where conditions are ideal for cementation because: (1) the rigidity of the reef framework means that cavities are permanent; (2) seawater can flow unimpeded through the porous structures; and (3) many reefs grow in high-energy settings and therefore have large volumes of water pumped through them on a daily basis. Isopachous fibrous cements, micritic cements, and internal sediments are characteristic of skeletal reefs. Such cements are almost universal in reef mounds of all ages.

Stromatactis (Figure 24.13) is one of the most common but puzzling cement fabrics in the rock record (see Chapter 15). It comprises irregular masses of fibrous calcite that typically have a flat base. Originally thought to be a recrystallized skeleton (hence the italicized name), it is now realized that *Stromatactis* are sediment-floored cavities that have been filled with fibrous, calcite cement. The problem is therefore now one of explaining the origin of the pores, with suggestions ranging from dissolved sponges, to calcimicrobial cemented voids, to

synsedimentary cemented cavities. Although *Stromatactis* is common in Paleozoic mud mounds, they disappeared after the Cretaceous.

Spatial distribution of early lithification

Synsedimentary cementation is most common in platform margin settings, especially on the windward sides, where saturated waters are commonly upwelling into warm waters (thus becoming supersaturated), reefs are most prolific, and oolitic sand shoals are common (Figure 24.14). This situation is also true for those periods in the past when platform margins were dominated by microbial deposition and cementation (see Chapter 15). By contrast, lagoonal sediments and the reefs that grow there are not usually cemented, probably because of sluggish water movement, continuous bioturbation that moves the grains around, subsurface reducing conditions, and poor subsurface water circulation.

On ramps, most cementation generally takes place on the inner ramp rather than on the mid-ramp where storm-dominated deposition is the norm. This is particularly the case on the shallow seafloor of the Persian Gulf off Abu Dhabi, where human artifacts and pearl oysters are cemented into modern hardgrounds. Synsedimentary cements are also common in outer-ramp reef mounds.

Slope sediments, especially those adjacent to large platforms, undergo variable degrees of cementation. The upper slope on modern platforms that have reefs and sand shoals along the rim can exhibit extensive cementation in some areas. As cementation decreases with increasing water depth, hardgrounds give way to nodules and then to unlithified sediment. These partially to completely lithified sediments are commonly subject to failure and the lithified nodules then become clasts in sediment gravity-flow deposits.

Stable isotopes of modern marine cements

Synsedimentary cements precipitated in modern sediments and reefs at low latitudes in warm shallow-shelf and platform settings (Figure 24.15) have high $\delta^{18}O$ values (from -1 to $+2‰$) and their ^{13}C values are higher than would be expected (from $+2.5$ to $+5.5‰$) if they had been precipitated in isotopic equilibrium with seawater ($\delta^{13}C$ of $+2.0$ to $+2.5‰$ at 25 °C). Specifically, aragonite cements are enriched by $1.0–1.4‰$ $\delta^{13}C$ and $1.5–2.0‰$ $\delta^{18}O$, whereas Mg-calcites (average 16 mole % $MgCO_3$)

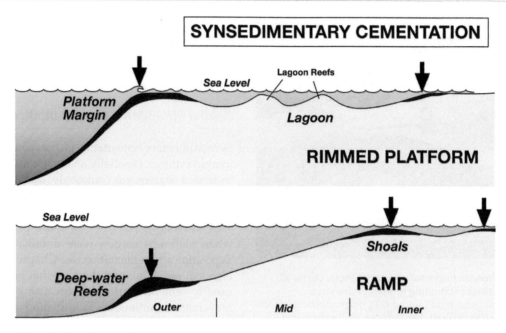

Figure 24.14 Spatial distribution of synsedimentary marine cements (arrows) on rimmed platforms and ramps.

STABLE ISOTOPIC COMPOSITION OF MODERN SYNSEDIMENTARY CARBONATE CEMENTS

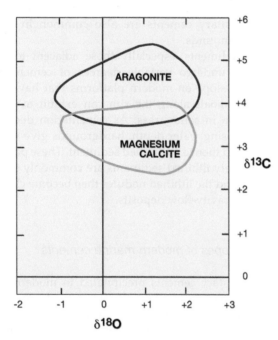

Figure 24.15 Isotopic composition of aragonite and Mg-calcite in modern reefs from Bermuda, Enewetak, Bikini and Belize. Source: Adapted from James and Choquette (1990). Reproduced with permission of the Geological Association of Canada.

are enriched by 0.3–0.4‰ $\delta^{13}C$ and 1.8–1.9‰ $\delta^{18}O$ (Figure 24.15). These are not equilibrium precipitates! Several workers have proposed that microbes are somehow involved in such precipitation, but this implication has yet to be proven.

Strandline diagenesis

The strandline is bathed largely in marine waters, and the processes that take place there are typically part of the synsedimentary marine diagenetic environment, albeit marginal marine with some meteoric influences. Beaches and muddy tidal flats are typical areas where this type of diagenesis occurs.

Beaches

Beachrock (Frontispiece) is a recurring feature of many tropical beaches. It is typically hard, well-cemented layers of beach grainstone consisting of the same grains that are found in the beach sand. Beachrock layers dip seaward at the same angle as the beach sediments (Frontispiece). Viewed from above, the limestone sheets are commonly arranged in an arcuate pattern.

Beachrocks generally extend from the intertidal zone, which is constantly awash with seawater from onshore waves, to the shallow subtidal zone where they usually grade seaward into unconsolidated sediment. The rock is well-laminated in cross-section. The deposit ranges from scattered nodules, to isolated slab-like layers, to thick successions of limestone with layers stacked on top of one another. There is, however, little pattern to the distribution of beachrock. Some tropical intertidal zones are all beachrock, others have only scattered layers, and many do not have any beachrock at all. On some beaches, loose beach sands pass laterally into areas covered with beachrock, even though conditions appear constant throughout.

Coins, bottles, skeletons, and other artifacts cemented into the beachrock demonstrate that the layers are a product of rapid cementation. On some of the remote islands in the Pacific Ocean, beachrock is harvested on a regular basis as a natural source of building stone. On the other hand, beachrock is the scourge of beachfront hotel owners who want nice sandy beaches!

The surface of exposed beachrock (Frontispiece) is typically pitted, displays dissolution basins or potholes, and is commonly bored. The rock is characteristically jointed into blocks and textured by cracks and channels (Figure 24.16). Erosion of rock slabs and reworking during intense storms leads to undermining of the slabs and the formation of conglomerates locally as ramparts of beachrock clasts.

Induration ranges from loosely cemented sand to well-lithified rock that can only be broken with a hammer. Cements (Figure 24.17) can be aragonite and Mg-calcite but microcrystalline aragonite is particularly common. The presence of meniscus or microstalactitic cements in some pores indicates precipitation in air-filled voids. Shoreward parts of the beachrock can have meteoric LMC cements. Much of the cementation has been attributed to precipitation that is triggered by the evaporation and degassing that takes place during low tide. Other alternatives are degassing of CO_2-rich carbonate-saturated groundwaters or biologically induced precipitation.

The origin of beachrock is open to debate. Most exposed beachrock shows evidence of erosion that seems to have post-dated the lithification of the beachrock. It is therefore entirely possible that the beds of beachrock developed in the shallow subsurface and only became exposed as storms removed the overlying sediment. Once exposed, erosive processes that gradually lead to the destruction of the beachrock replace the processes of cementation.

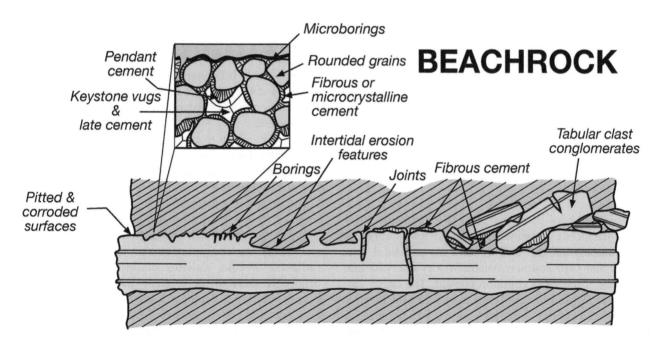

Figure 24.16 Sketch illustrating the main macroscale and petrographic (box) attributes of beachrock. Source: Adapted from James and Choquette (1990). Reproduced with permission of the Geological Association of Canada.

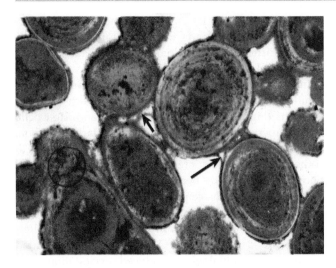

Figure 24.17 Thin-section image of beachrock cement, Bahamas, where the aragonite cement (arrows) has a meniscus fabric. Image width 3 mm.

Figure 24.18 Polygonal teepees on tidal flats, Fisherman Bay, South Australia, ground view of polygon teepee margin. Hammer 30 cm long (circled).

The presence of beachrock on tiny tropical islands and along extensive coastlines probably points to a multiplicity of origins. This is particularly true where the limestone is developed on cool-temperate beaches.

Aragonite crusts

Lustrous aragonite crusts (pelagosite) up to 20 cm thick (coniatolites) can coat beachrock and carbonate grains in the intertidal zone. Looking like cream-colored enamel paint, they resemble tufa or travertine. They are best developed on impermeable surfaces and are thought to form from evaporated sea spray. They can easily be mistaken for marine cements in the rock record.

Muddy tidal flats

Cementation here is in the form of cemented crusts which are locally prominent on humid tidal flats (e.g., the Bahamas) but can be extensive on arid tidal flats (e.g., Persian Gulf, southern Australia). Cryptocrystalline aragonite is the most common cement. The crusts can pass seaward into marine hardgrounds. Rapid precipitation produces large expansion fractures and arcuate to polygonal ridges and teepees (Figure 24.18). Broken crusts can be reworked into pavements of tabular clasts by storms.

The rock record

Cements

Synsedimentary marine cements can be recognized in rocks of all ages and, regardless of their original mineralogy, are now all LMC. The major problem in deciphering their origin is ascertaining if they are altered Mg-calcite or aragonite or if they are original LMC and unaltered. This latter case is especially pertinent because, as seawater composition has changed through time (see Chapter 20), there were probably periods when there were calcite cements with little magnesium. This is compounded by that fact that some of these cements are not identical to their modern counterparts because they might be neomorphic.

It is imperative to remember the period of geologic time when the seafloor cements formed. First, because the Mg:Ca ratio of seawater has changed through time, it would seem likely that low-Mg calcite should have been precipitating as a marine cement in greenhouse times. If this was the case, then many of the somewhat enigmatic cement textures that are rare or not seen today, such as inclusion-rich epitaxial cement and radiaxial fibrous calcite, could have been LMC originally. Second, it is clear that the amounts of synsedimentary cement have decreased through geologic time; they were prolific in the Precambrian, abundant in the Paleozoic and early Mesozoic, but relatively minor in the post-calcareous plankton age. The reasons for this are unclear but it could be because these prolific microfossils, by extracting so much $CaCO_3$ from seawater, have lowered carbonate saturation levels.

Fibrous calcite cements. These cements (Figure 24.19) are divided into two distinct types: (1) fascicular optic; and (2) radiaxial fibrous. Fascicular optic calcite (FOC) comprises cones of fibrous calcite with divergent optic axes (Figure 24.20), whereas radiaxial fibrous calcite (RAX) has curved to wavy boundaries and convergent optic axes (Figures 24.20, 24.21). These two types can grade laterally into one another along a single cement fringe.

Radiaxial calcite has long been enigmatic because it is common in ancient reefs and reef mounds but almost unknown from modern settings. The problem with RAX is deciding if it is original or altered high-Mg calcite. The consensus is that both RAX and FOC were probably precipitated as HMC and LMC. The RAX is thought to form by a process of asymmetric growth within calcite crystals that undergo split crystal growth. It would appear that those that formed from waters with a low Mg:Ca ratio are likely original, whereas those that formed with a high Mg:Ca ratio are altered Mg-calcites.

Epitaxial calcite cement. There is abundant evidence from the rock record that epitaxial (also called syntaxial) overgrowths of marine Mg-calcite cement on echinoderm particles were also common in the past (Figure 21.3). In some localities, marine epitaxial cements are also common on brachiopods, corals, stromatoporoids, calcispheres, and ooids where crystals are (or were) oriented normal to the particle surface.

Spherulitic calcite cements. There are numerous examples of spherulitic calcite that mimic the geometry of Holocene botryoidal aragonite (Figure 24.19). The fabric is preserved either as a mosaic of tiny crystals or a paramorphic (similar to calcitized) replacement. Actual aragonite crystallites preserved in the arrays have been discovered in rocks as old as Pennsylvanian.

Micrite cements. These are the most difficult to interpret as cements. Regardless, there have been relict aragonite crystallites with pitted surfaces found in Paleozoic muds that seem to be cemnets. In contrast, those with uniform textures, no relicts, and no pitted surfaces have been interpreted as original calcite cements.

ANCIENT SHALLOW MARINE CEMENTS

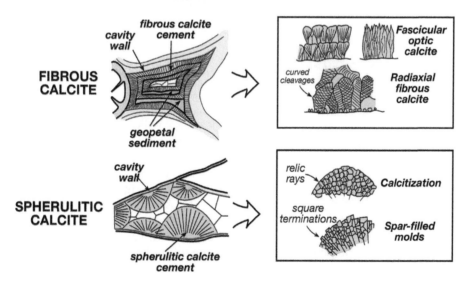

Figure 24.19 Fabrics and morphotypes of coarse fossil synsedimentary marine cements. Fibrous calcite is generally interpreted as derived from LMC or HMC and spherulitic calcite from aragonite. Source: James and Choquette (1983). Reproduced with permission of the Geological Association of Canada.

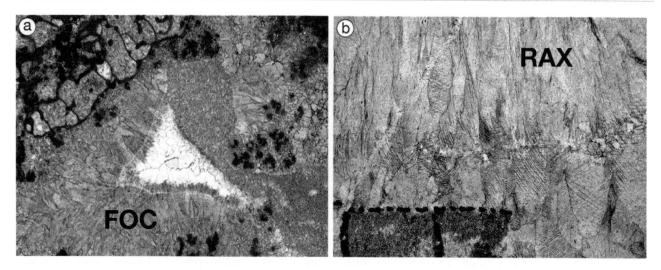

Figure 24.20 Thin-section images of ancient fibrous marine cements. (a) Fascicular optic calcite (FOC), lower Cambrian, Labrador, Canada. Image width 3 mm. (b) Radiaxial fibrous calcite (RAX), lower Cambrian Flinders Ranges, South Australia. Image width 2 mm.

RADIAXIAL FIBROUS CALCITE CEMENT

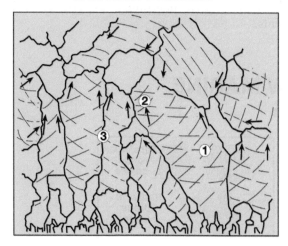

**(1) curved twins (2) convergent optic axes (arrows)
3) concertal boundaries**

Figure 24.21 Sketch depicting the three key features of radiaxial fibrous calcite (RAX): (1) curved twin lamellae; (2) convergent optic axes; and (3) concertal (non-planar) crystal boundaries.

Isotopic composition of ancient marine cements

Numerous careful studies of ancient calcite cements indicate that many such precipitates have by and large preserved much their original geochemical signature via microscale closed-system diagenesis. The technique is to

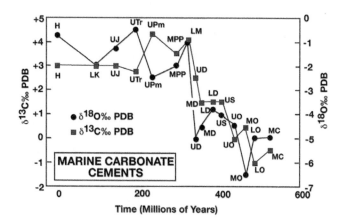

Figure 24.22 A plot of the $\delta^{13}C$ and $\delta^{18}O$ values for isotopically highest ("heaviest") values of marine cements through Phanerozoic time and believed to approximate the original marine compositions of the cements. Source: Adapted from Lohmann & Walker 1989. Reproduced with permission of John Wiley & Sons.

take the highest isotope values and assume that they are closest to original composition. When compared to coeval fossil LMC biofragments (brachiopods, bryozoans, trilobites), the marine cements have roughly the same $\delta^{13}C$ values but $\delta^{18}O$ values are more variable. Using this technique, the fibrous and botryoidal precipitates display systematic variations through Phanerozoic time that transcend regional differences in diagenetic history. Such variations seem to somewhat parallel the compositions of whole rocks and brachiopods (Figure 24.22). When compared to aragonite sea and calcite sea times, it further appears that carbon and oxygen isotope values in cements and skeletons from carbonates in aragonite seas are higher than those in calcite seas.

So what does all this mean? It seems that we cannot use modern isotope cement values and simply take them directly into the geologic record because the isotopic composition of seawater has changed through time. Instead, we must look at each geologic period separately when we want to work out diagenesis. The best way is to take the values for carbonates during each time period, as illustrated in Figure 24.22, and use these values as a starting point (e.g., Silurian: +1.5 δ^{13}C and −4 δ^{18}O) rather than Holocene values (+4.0 δ^{13}C and −1.0 δ^{18}O).

Further reading

Bathurst, R.G.C. (1982) Genesis of *stromatactis* cavities between submarine crusts in Paleozoic carbonate mud buildups. *Journal of the Geological Society of London*, 139, 165–181.

Beier, J.A. (1985) Diagenesis of Quaternary Bahamian beachrock: petrographic and isotopic evidence. *Journal of Sedimentary Petrology*, 55, 755–761.

Davies, G.R. (1977) Former magnesian calcite and aragonite submarine cements in Upper Paleozoic reefs of the Canadian Arctic: a summary. *Geology*, 5, 11–15.

De Choudens-Sánchez, V. and González, L.A. (2009) Calcite and aragonite precipitation under controlled instantaneous supersaturation: Elucidating the role of CaCO$_3$ saturation state and Mg/Ca Ratio on calcium carbonate polymorphism. *Journal of Sedimentary Research*, 79, 363–376.

Ferguson, J., Chambers, L.A., Donnelly, T.H., and Burne, R.V. (1988) Carbon and oxygen isotope composition of a recent megapolygon-spelean limestone, Fisherman Bay, South Australia. *Chemical Geology*, 72, 63–76.

Ginsburg, R.N. and Schroeder, J.H. (1973) Growth and submarine fossilization of algal cup reefs, Bermuda. *Sedimentology*, 20, 575–614.

Gischler, E. (2008) Beachrock and intertidal precipitates. In: Nash D.J. and McLaren S.J. (eds) *Geochemical Sediments and Landscapes*. Blackwell Publishing Ltd, pp. 365–390.

Given, R.K. and Lohmann, K.C. (1985) Derivation of the original isotopic composition of Permian marine cements. *Journal of Sedimentary Petrology*, 55, 430–439.

González, L.A. and Lohmann, K.C. (1985) Carbon and oxygen isotopic composition of Holocene reefal carbonates. *Geology*, 13, 811–814.

James, N.P. and Ginsburg, R.N. (1979) *The Seaward Margin of Belize Barrier and Atoll Reefs*. Oxford: International Association of Sedimentologists, Special Publication no. 3.

James, N.P. and Choquette, P.W. (1990) Limestone: The sea floor diagenetic environment. In: McIlreath I. and Morrow D. (eds) *Diagenesis*. St John's, ND, Canada: Geological Association of Canada, Reprint Series 4, pp. 13–34.

James, N.P., Bone, Y., and Kyser, T.K. (2005) Where has all the aragonite gone? Mineralogy of Holocene neritic cool-water carbonates, Southern Australia. *Journal of Sedimentary Research*, 75, 454–463.

Kendall, A.C. (1977) Fascicular-optic calcite: a replacement of bundled acicular carbonate cements. *Journal of Sedimentary Petrology*, 47, 1056–1062.

Lohmann, K.C. and Walker, J.C.G. (1989) The δ^{18}O record of Phanerozoic abiotic marine calcite cements. *Geophysical Research Letters*, 16, 319–322.

Schlager, W. and James, N.P. (1978) Low-magnesian calcite limestones forming at the deep-sea floor, Tongue of the Ocean, Bahamas. *Sedimentology*, 25, 675–702.

Schroeder, J.H. and Purser, B.H. (eds) (1986) *Reef Diagenesis*. New York: Springer-Verlag.

Shinn, E.A. (1969) Submarine lithification of Holocene carbonate sediments in the Persian Gulf. *Sedimentology*, 12, 109–144.

Walker, J.C.G. and Lohmann, K.C. (1989) Why the oxygen isotopic composition of sea water changes with time. *Geophysical Research Letters*, 16, 323–326.

Wright, V.P. and Cherns, L. (2008) The subtle thief: selective dissolution of aragonite during shallow burial and the implications for carbonate sedimentology. In: Lukasik J. and Simo J.A.T. (eds) *Controls on Carbonate Platform and Reef Development*. Society for Sedimentary Geology, Special Publication no. 89, pp. 47–54.

CHAPTER 25
METEORIC DIAGENESIS OF YOUNG LIMESTONES

Frontispiece Young Pleistocene limestones lithified in the meteoric zone along the shore of Bimini, Bahamas facing the Gulf Stream.

Origin of Carbonate Sedimentary Rocks, First Edition. Noel P. James and Brian Jones.
© 2016 Noel P. James and Brian Jones. Published 2016 by John Wiley & Sons, Ltd.
Companion website: www.wiley.com/go/james/carbonaterocks

Introduction

The interactions between carbonate sediments or limestones and fresh meteoric waters result in profound diagenetic change. Two alteration processes are active here: mineral-driven diagenesis and water-driven diagenesis. Mineral-driven diagenesis takes place when carbonate sediments are exposed to meteoric waters for the first time and all metastable components alter to diagenetic LMC (dLMC) and then stop. Water-driven diagenesis takes place at the same time but continues as long as the carbonates stay in the meteoric realm, sometimes for millions of years. This chapter focuses on mineral-driven diagenesis, the maturing of young carbonate sediments into limestones.

Mineral-driven diagenesis leads to dissolution, cementation, and porosity formation as the metastable minerals are transformed to dLMC (Figure 25.1). The changes to the limestones are extensive. Soft sediments are lithified and transformed to limestone, often quickly, in mere thousands of years. Marine particles and cements of all types are prone to significant alteration, resulting in their variable preservation. The Ca and CO_3 ions liberated by these processes go into the groundwater and may then be precipitated elsewhere as cement or transported out of the system and lost forever. The original porosity and permeability of the sediments are typically altered as intergranular pores are filled with cement and moldic and intercrystalline pores develop. Such changes are typically

METEORIC DIAGENESIS

DIAGENESIS OF YOUNG LIMESTONES

MINERAL SPECIFIC

MINERALOGY	ARAGONITE ⟶ LMC & HMC	
PROCESS	FABRIC SPECIFIC • DISSOLUTION • CEMENTATION • NEOMORPHISM	
CONTROLS	ORIGINAL MINERALOGY WATER FLOW RATE SEDIMENT TEXTURE CLIMATE	
POROSITY & PERMEABILITY	ENHANCED BY DISSOLUTION REDUCED BY CEMENTATION	

Figure 25.1 Processes and products of mineral driven meteoric diagenesis of young limestones.

fabric-specific because the reactions are keyed to the composition of the grains in the sediments (Figure 25.2). Textures are therefore commonly preserved, even though fabric is lost. Although the processes are mainly mineral-driven, water and microbes can play important roles in these processes. All of these modifications are important because once the sediment/rock has been altered to dLMC, it can remain largely unchanged for the rest of geologic time.

The products of such diagenesis depend on: (1) intrinsic factors, including original mineralogy and original grain size of the sediment; and (2) extrinsic factors such as time and local climate (Figure 25.3). Nevertheless, the underlying processes that govern the way in which aragonite and Mg-calcite are altered and the nature of dLMC cements that are precipitated remain the same. The fact that fabrics produced by diagenesis in the vadose, phreatic, and mixed water zones are different means that they can be recognized throughout the geological record by numerous well-established criteria.

Processes

When exposed to percolating fresh waters, young metastable carbonate components alter to calcite in the following ways (Figure 25.4):

1. Soon after water begins to percolate through the sediment, small elongate or equidimensional calcite crystals begin to precipitate on grain surfaces, growing equally on Mg-calcite or aragonite particles.
2. As more cement is added, the Mg-calcite grains begin to alter to low Mg-calcite (dLMC) via dissolution–reprecipitation processes that operate on a microscale.
3. Once most of the grains are coated with a complete rind of calcite cement, the aragonite components start to dissolve or undergo calcitization. The former (which is the most common) leads to porosity development, whereas the latter may lead to the preservation of the original grains despite the change in crystallography.
4. Dissolution of the aragonite releases large amounts of $CaCO_3$ into the groundwater. Much of that Ca and CO_3 is then reprecipitated as dLMC cement between particles or inside molds that were created by the dissolution of the aragonite grains (Figure 25.5).
5. Once all the aragonite has gone, the remaining sediment is transformed into a hard calcite limestone.
6. The original interparticle porosity is transformed from interparticle to mostly moldic in nature.

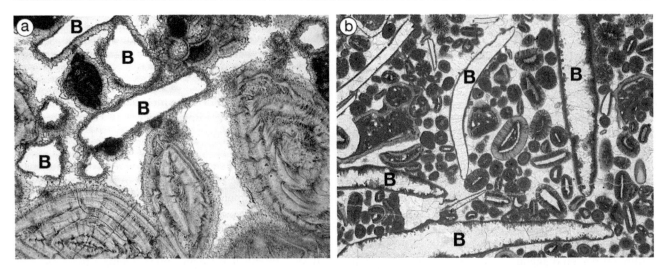

Figure 25.2 Fabric specific diagenesis, thin-section (PPL) images. (a) Well-preserved Pliocene benthic foraminifers (*Amphistegina* sp.), Barbados, that were originally HMC and are now dLMC. Note the molds of bivalves (B) that were originally aragonite (with micrite envelopes) and the thin rind of cement which is dLMC. Image width 3 mm. (b) Well-preserved Middle Ordovician radial ooids that were originally HMC and are now dLMC, Kingston, Ontario, Canada. Bivalves (B) were originally aragonite. Micrite envelopes dissolved and are now filled with late burial cement. Image width 4 mm.

METEORIC DIAGENESIS

CONTROLS

INTRINSIC

MINERALOGY

GRAIN SIZE

POROSITY & PERMEABILITY

EXTRINSIC

CLIMATE

VEGETATION

TIME

SETTINGS

VADOSE ZONE

WATER TABLE

PHREATIC ZONE

MIXING ZONE

Figure 25.3 The intrinsic and extrinsic factors that most affect meteoric diagenesis. Source: Adapted from James and Choquette (1984). Reproduced with permission of the Geological Association of Canada.

Alteration of Mg-calcite components

Mg-calcite particles and cements formed in the marine environment are either microcrystalline or composed of fibrous calcite. Alteration to dLMC in the meteoric diagenetic realm is generally via dissolution–reprecipitation on a microscale or nanoscale. Given that the particles were originally calcite, albeit Mg-rich, there is no associated crystallographic change. As a result, the original microtexture is preserved in the dLMC (Figure 25.2) and does not appear different from the original Mg-calcite particle, even under the petrographic microscope; there is complete textural preservation. Only under SEM is it clear that there has been some crystal enlargement.

Alteration of aragonite components

Aragonite particles and cements precipitated in the marine environment range from microcrystalline to platy to fibrous. Since aragonite has an orthorhombic rather than the hexagonal crystal habit of calcite, alteration also involves crystallographic change via complete dissolution of aragonite and precipitation of dLMC. There are two ways in which this takes place (Figure 25.6):

1. *Macroscale dissolution*: in this case the whole component, fossil, grain, or cement is dissolved to leave a pore (Chapter 21 Frontispiece, Figure 25.7). That pore may then be filled with cement in the meteoric zone or can remain open for long periods of time. The preservation of such molds is enhanced by the presence of a micrite envelope. Micrite envelopes are critical to preservation in grainstones. The microcrystalline nature of the envelope means that it will be preserved as dLMC to frame the original aragonite particle, even though it is dissolved. This is not necessarily the case with cool-water

STAGES OF METEORIC DIAGENESIS

MINERAL CONTROLLED, FABRIC SPECIFIC METEORIC DIAGENESIS

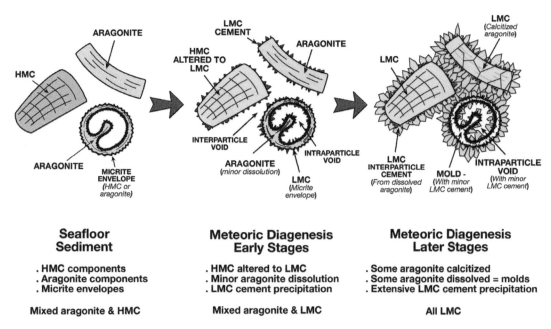

Figure 25.4 Sketch illustrating the mineral controlled diagenesis of young limestones in the meteoric zone.

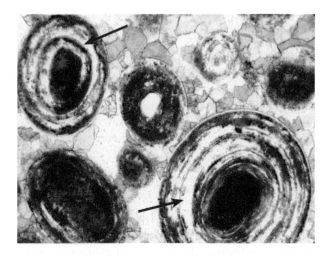

Figure 25.5 Meteoric diagenesis caught in the act, Pleistocene Miami Limestone, Florida, USA. Partially dissolved aragonite ooids (white areas; arrows). The dissolved carbonate has been precipitated as intergranular dLMC cement. Image width 1.5 mm.

carbonates where there are few micrite envelopes and so any record of the original aragonite grain is lost.

2. *Calcitization*: this process involves a micrometer- to nanometer-thick water front that moves across the grain with aragonite being dissolved on one side and calcite being precipitated on the other side (Figures 25.8–25.10). This results in a mosaic of calcite crystals that cross-cut the original fabric. Nevertheless, relics of organic matter, other insoluble material (Figure 25.10), or some of the original aragonite crystals can be trapped in the replacive calcite and attest to an original aragonite composition.

Mg-calcite molds

Although aragonite particles most often dissolve to form molds and Mg-calcite grains are well-preserved during the change to dLMC, this is not always the case. Just as aragonite components can be calcitized if water rates are

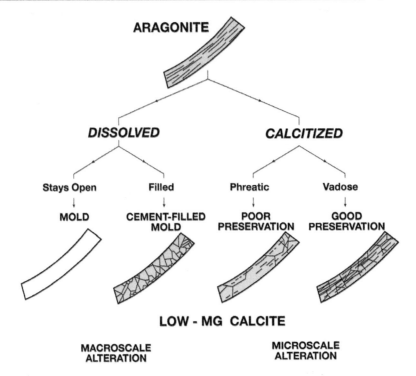

Figure 25.6 Sketch illustrating the four possible ways that an originally aragonite bivalve shell can be altered in the meteoric diagenetic environment. Source: Adapted from James and Choquette (1984). Reproduced with permission of the Geological Association of Canada.

Figure 25.7 Large vugs of former aragonite corals generated by dissolution (Miocene, Gulf of Suez, Egypt). Scale divisions 2 cm.

slow, Mg-calcite particles can be dissolved if flow rates are high and water are rich in CO_2. Although this is not common, it should be anticipated. Thus, not all molds signify formerly aragonite components, especially near subaerial unconformities.

Carbonate muds

Those muddy limestones that were originally aragonite-rich can contain crystal relics or crystal molds in the microcrystalline dLMC and are usually well cemented with little porosity. By contrast, those muddy limestones that were originally calcite sediment are micrite with no micromolds or relics.

The zone in which such alteration takes place appears to be important. Fabrics altered in the vadose zone have fabric-selective mosaics with relatively small crystal sizes, show excellent replacement, and contain some organic residue. By contrast, although calcitized those in the phreatic zone do not have such good relic preservation.

Cements and cementation

Cements that form during meteoric diagenesis are generally clear (limpid) crystals and contain few inclusions. This is because the water itself contains comparatively few cations, other than Ca and perhaps Mg. There are generally three types of crystal fabrics: (1) drusy; (2) epitaxial (Figure 25.11); and (3) blocky.

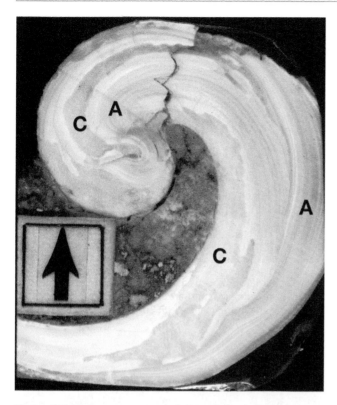

Figure 25.8 Pleistocene gastropod (*Strombus* sp.), Barbados, that is in the process of being calcitized. Most of the skeleton is still original aragonite (A), but some has been altered to dLMC (C) with partial preservation of microfabric (calcitized). Sides of box around arrow are 1 cm.

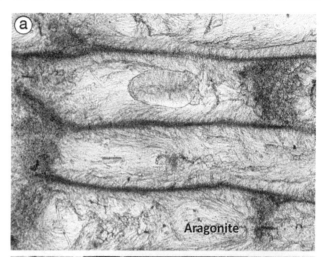

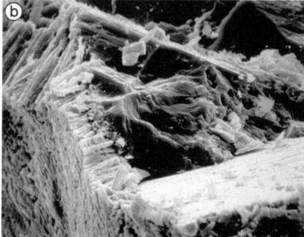

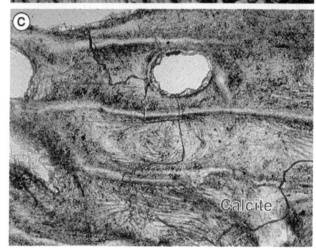

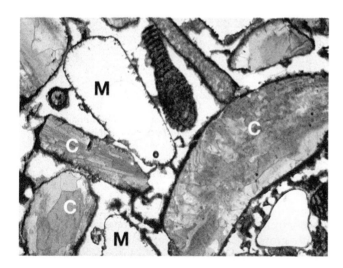

Figure 25.9 Bivalve alteration, thin-section partially polarized light. Bivalves in the Pleistocene Anastasia Formation of northern Florida, USA where some of the bivalves have been completely dissolved to form molds (M, with micrite envelopes), whereas others have been calcitized to dLMC (C). Image width 2 mm.

Figure 25.10 Calcitization of the aragonite coral *Acropora palmata*, Pleistocene Barbados. (a) Thin-section (PPL) image of a living aragonite specimen. Image width 5 mm. (b) SEM image of a partially altered specimen with fibrous aragonite at left grading into replacing dLMC at right. Image width 80 μm. (c) Thin-section (PPL) image of completely calcitized dLMC skeleton illustrating fairly good preservation of original fabric in the new calcite crystals. Image width 5 mm.

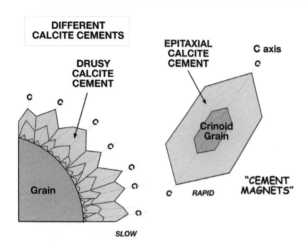

Figure 25.11 Diagram of the main attributes of drusy and epitaxial dLMC cement fabrics. Because the Crinoids are already single crystals, dLMC cement precipitates more easily and therefore more quickly there than on other grains that have drusy fabrics; they are cement magnets.

Figure 25.12 dLMC cements. Drusy calcite cement growing upward from carbonate substrate at bottom (Pleistocene, Barbados, Caribbean). Image width 2 mm.

Drusy calcite

Drusy fabrics (Chapter 23; Figures 25.11, 25.12) are characterized by an increase in crystal size but a decrease in the numbers of crystals into the pore space. Most junctions between crystals are triple and many are enfacial (Figure 21.6); it is a common and easily recognizable fabric. If the crystals grow into a water-filled (phreatic) void, they usually have nice pyramidal terminations; if they grow against a water–air interface (vadose), they will have smooth terminations reflecting crystal growth impeded by air.

Blocky calcite

Blocky crystal fabrics result from a low number of nucleation sites along the substrate surface such that the widely separated crystals grow until they interfere with one another in the void space (Figure 25.12a).

Syntaxial (epitaxial) calcite

These cements are most common on biofragments that are formed of one large crystal (e.g., echinoid plates, crinoid ossicles). In modern settings, such cements are also found on the surface of foraminifers (Figure 25.13) and bivalves that have a prismatic skeletal crystal structure. Echinoids, in particular, precipitate their skeletons as optically single crystals of HMC; the fragment is

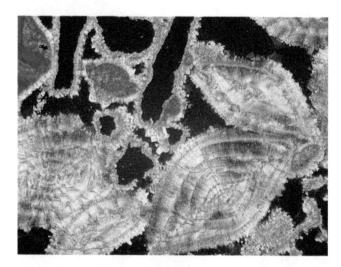

Figure 25.13 Epitaxial dLMC cement thin-section image XPL. Benthic foraminifers (*Amphistegina sp.*) in the Pliocene of Barbados, with fibrous HMC skeletons that have altered to dLMC where the dLMC cement has precipitated in optical continuity (epitaxial) with the skeletal fibers. Image width 2 mm.

essentially a single crystal of calcite. The epitaxial cements simply grow as extensions in optical continuity with the original skeletal crystal. Given that little thermodynamic energy is needed, these overgrowths can form easily and rapidly. In rocks that contain numerous echinoid fragments or crinoid ossicles, epitaxial cement is commonly the dominant cement.

Vadose cements

Cements (Figure 25.14) that precipitate into vadose voids that are at different times filled with water, water and air, or just air, are either drusy or blocky. Such cements are quite irregular in their distribution. In many Pleistocene limestones, for example, it is common to see well-cemented areas just millimeters away from porous zones that lack cements. Cements that develop in the vadose zone are variable because the chemical microenvironments in the pores change frequently through time. Following rainfall, air-filled pores may become filled with water that can lead to the development of isopachous bands of cement that cover the pore walls. As that water drains from the pores, conditions change so that water is restricted to menisci between neighboring grains or may drip from the bottom of grains into the open pores. The former gives rise to meniscus cements, whereas the latter leads to the formation of pendant (also known as microstalactitic) cements. Although typically formed of dLMC, these cements may be characterized by compositional banding that reflects changes in the composition of the waters that flowed through the pores. These fabrics are good evidence of vadose precipitation, but are often lost as cementation continues in later diagenetic environments.

Phreatic cements

Pores in the phreatic zone are always filled with water and crystals can therefore grow unimpeded except for intercrystalline competition. Crystal rinds are generally drusy or blocky and isopachous around the pore walls (Figure 25.14). Cements are generally larger than in the vadose zone. Be aware, however, that such crystals can also develop in burial environments and so are really only typical of water-filled pore spaces.

Syntaxial cements in meteoric environments

These cements do not obey the rules and are similar in both vadose and phreatic diagenetic environments. Syntaxial cements therefore yield no information as to which environment they were precipitated in, except for their CL signature.

Diagenesis of calcite sediments

Carbonate sediments that were originally Mg-calcite or calcite with little associated aragonite follow a different

METEORIC CEMENTS

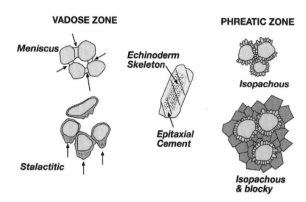

Figure 25.14 Sketch of the different drusy and blocky dLMC cement morphologies that precipitate in meteoric vadose as opposed to meteoric phreatic environments; epitaxial cements on echinoderms can form in either environment. Source: Adapted from James and Choquette (1984). Reproduced with permission of the Geological Association of Canada.

alteration pathway in the meteoric environment than those containing aragonite grains. Such changes are especially evident in chalks (calcite components) or temperate carbonates (that have lost their aragonite due to seafloor diagenesis). Many neritic cool-water carbonates, for example, remain largely unlithified even though they may have been in the meteoric realm for millions of years (Figure 25.15). The lack of Ca and CO_3 from which the cements can precipitate is attributed largely to the paucity of aragonite in the original sediments. This observation is especially important when considering the geological record of components that were formed of calcite originally (e.g., the middle Paleozoic and middle Mesozoic). In the absence of karst features (see Chapter 26), many of these rocks will display little or no record of diagenesis because they would have passed virtually unlithified into the burial diagenetic environment. In short, no aragonite means little meteoric cementation.

Importance of grain size

In general, the smaller the grain size of the original sediment the more quickly the deposit is altered to dLMC. Muds, therefore, alter more quickly than sands, which change more quickly than large particles such as whole corals. This relationship holds because the sediments with the smaller grains collectively offer a greater surface area for reactions. In Pleistocene

Figure 25.15 Diagenesis of originally calcitic sediments. (a) Outcrop of poorly cemented Miocene grainstone–rudstone in southern Australia, in which all of the components are well preserved because they were originally calcitic (HMC, IMC and LMC). Centimeter scale. (b) Close-up of friable bryozoan Oligocene limestone, Chatham Islands, New Zealand that has virtually no dLMC cement. Finger 1 cm wide. (c) B. Jones illustrating the poorly cemented nature of the originally calcitic sediment illustrated in (b). (d) Thin-section image (PPL) of originally aragonite-free Oligocene bryozoan grainstone, southern Australia, without any obvious cement even though it has been flushed with meteoric waters for ~15 Myr. Image width 6 mm.

limestones, for example, large corals commonly retain their aragonitic skeleton even though the surrounding matrix has evolved into a well-cemented dLMC mud. The fact that the finer-grained layers are cemented first means that porosity and permeability heterogeneities develop within a succession that may then control the flow of groundwater. The preferential flow of water through uncemented coarser-grained layers, for example, may have a profound effect on subsurface karst development.

Diagenesis in different meteoric settings

Vadose versus phreatic diagenesis

There are several generalities possible when comparing diagenesis in the freshwater vadose versus phreatic zones.

Vadose zone. Diagenesis of sediments is slower because the pores are filled and drained episodically and are not always full of water. Sediments in the phreatic zone can all be dLMC and well cemented, whereas those in the vadose zone are still metastable and poorly cemented.

Phreatic zone. Waters are more varied in composition than in the vadose zone because they can come from far afield as opposed to simple downward seepage. Marine sediments alter to dLMC more quickly in the phreatic zone because they are continuously surrounded by flowing fresh waters.

Freshwater–saltwater mixing zone diagenesis

The most important process here is dissolution (see Chapter 26). Such dissolution should be most intensive in the upper dilute part of the zone, whereas precipitation, if it takes place, should occur in the lower saline part of the zone where seawater is oversaturated with respect to $CaCO_3$. It has been suggested that calcite crystals with Mg-inclusions, including microdolomite, should be present in such mixing zones.

Importance of climate

Climate is critically important in meteoric diagenesis because it controls: (1) the amount of water that flows through the vadose zone; and (2) the air temperature, which may influence the rate of diagenetic changes mediated by the rainwater (Figure 25.16).

Hot and arid (desert-like)

Since there is little rainfall, diagenesis in the vadose zone is slow and there is little or no cementation. By contrast, effects in the phreatic zone could be extensive because the fluids can originate far afield and not be a product of the local climate. Regardless, water flow is likely to be slow because of low recharge.

Warm and semi-arid (tropical and temperate)

Alteration in the vadose zone should be rapid, with: (1) fabric-selective preservation; (2) relatively good neomorphism of former aragonite components; and (3) extensive calcite cements. Diagenesis in the phreatic zone should be minimal and caves relatively small due to limited recharge and slow groundwater flow.

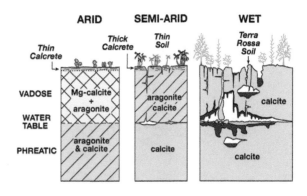

Figure 25.16 Sketch illustrating the relative rates of meteoric diagenesis of similar sediment in arid, semi-arid, and humid climates. Source: Adapted from James and Choquette (1984). Reproduced with permission of the Geological Association of Canada.

Warm and rainy (subtropical and tropical)

Dissolution should prevail in these perpetually wet conditions. Precipitation should be minimal and non-fabric-selective diagenesis should dominate; caves and enlarged fractures should be extensive.

How long does it take?

Climate is critically important in meteoric diagenesis because it controls: (1) the amount of freshwater that flows through the vadose zone; and (2) the air temperature, which may influence the rate of diagenetic changes mediated by the rainwater. Numerous studies have tried to evaluate these controls by examining limestone diagenesis in different climatic situations, all with the view of providing ideas on the rapidity of the diagenesis. Sadly, it seems that rates change and vary greatly from place to place!

For example, oolitic carbonate sediments on cays in the Bahamas, a humid but not year-round rainy zone, composed of aragonite grains originally, have been relatively well cemented in ~1000 years (Frontispiece, Figure 25.17). In most similar situations where there is a mixture of aragonite and Mg-calcite particles, complete cementation and stabilization to dLMC is accomplished in about 120,000 years. By contrast, Pliocene carbonates in a semi-arid to arid climate in Australia are still soft and totally composed of aragonite and Mg-calcite. In fact, tiny amounts of relict aragonite entombed in calcite occur in

Figure 25.17 The shoreline of Josie Cay in the Joulters Cay complex, western Bahamas. The limestone on the right is ~1000 years old and partially cemented by meniscus dLMC (see Figure 21.5a). Metal box in background (circled) is 30 cm long.

limestones as old as Mississippian. Even though these situations highlight the importance of climate, especially rainfall, these periods are but a blink of an eye in geologic time; the changes take place quickly.

The ultimate product

Original particle composition and grain size is fundamental in determining the character of any final limestone. Two end-members are illustrated in Figure 25.18. A suite of particles of mixed mineralogy in a muddy matrix (float-stone or wackestone texture) alters in a predictable way: (1) HMC alters to LMC with textural change; (2) aragonite is dissolved to form a mold; and (3) the released $CaCO_3$ precipitates in the mud micropores and the deposit is lithified. The process is entirely mineral-controlled. Once the deposit is all LMC, there is no longer any mineralogical drive for change, but the ongoing percolation of meteoric waters undersaturated with respect to LMC (for whatever reasons) circulate through the earlier-formed

EXAMPLES OF METEORIC DIAGENESIS

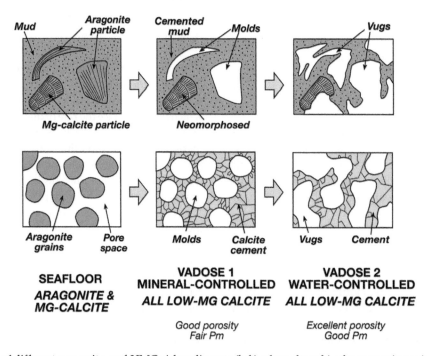

Figure 25.18 The fate of different aragonite- and HMC-rich sediments (left) when altered in the meteoric environment (center), and their fate if left in that environment once all components have changed to dLMC and alteration is water controlled (right). Source: Adapted from James and Choquette (1990). Reproduced with permission of the Geological Association of Canada.

molds and continue to dissolve the carbonate, forming vugs. This is now a water-controlled system.

By contrast, a sand (grainstone) composed only of aragonite grains (think ooids or peloids) also follows the same alteration rules: (1) the aragonite particles gradually dissolve forming a series of molds; and (2) the dissolved carbonate precipitates as LMC cement on and between the dissolving particles. There has been a complete inversion of fabric; what were once grains are now holes, what were once interparticle pores are now cement. Similar to the muddy sediment considered above, once everything is LMC it becomes a water-controlled system with dissolution being the dominant process.

Geochemistry

The geochemical changes that accompany meteoric diagenesis have been well established from numerous studies of Pleistocene limestones that were deposited in the last few hundred thousand years and subsequently exposed to freshwater reactions. These changes are best understood when placed against a stratigraphic column (Figure 25.19) or as an isotopic cross-plot (Figure 25.20).

Trace elements

Dissolution of aragonite releases Sr^{2+} into the pore waters whereas dissolution of Mg-calcite releases Mg^{2+} into pore waters. The partition coefficients for these elements relative to dLMC result in a relative decrease in both of these elements in meteoric dLMC precipitates compared to the original sediments (Figure 25.19).

Above the water table, where conditions are generally oxidizing, the bivalent cations (Fe^{2+} and Mn^{2+}) cannot be incorporated into the dLMC crystals. Below the water table, where conditions are generally reducing, Fe^{2+} and Mn^{2+} are commonly incorporated into the precipitating dLMC. As a result, vadose cements are usually black and non-luminescent under CL, whereas phreatic cements are often brightly luminescent.

Meteoric isotope geochemistry

As described above, the components involved are the original sediment, carbon dioxide, and meteoric water. The reaction involves dissolution of metastable minerals (aragonite, HMC, and minor LMC) in meteoric water with variable amounts of dissolved carbon dioxide and their reprecipitation as diagenetic LMC (Figure 25.20).

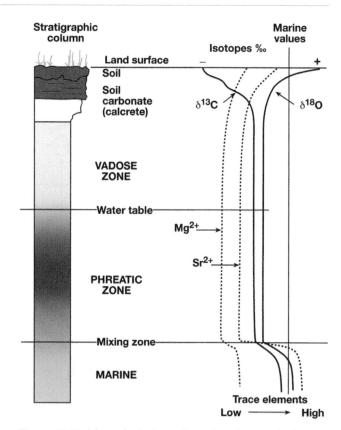

Figure 25.19 A hypothetical core through carbonates in the meteoric zone under a semi-arid climate. Chemical curves show the expected variation in trace elements and stable isotopes of bulk samples in the various zones. Source: Adapted from James and Choquette (1990). Reproduced with permission of the Geological Association of Canada.

Carbon is the most variable reactant because it can come from the original sediment ($CaCO_3$), from the atmosphere (CO_2), or from the soil as the rainwater trickles through it. The important thing is that if there is a soil, the microbes there tend to fractionate the carbon by extracting the ^{13}C, thereby leaving the soil gas enriched in ^{12}C (lower, typically negative $\delta^{13}C$ values). This gas is easily dissolved in the meteoric water.

The important factors controlling stable isotopic compositions of the replaced components (grains and marine cements) that are now dLMC are: (1) the origin of the oxygen; and (2) the origin of the carbon. The oxygen isotopic compositions of the precipitates are inherited largely from the meteoric water and most new precipitates will have the values of fresh water (the meteoric water line on Figure 25.20). Although such values are generally between –4 and –6‰ in the tropics and subtropics, they can be as low as –10‰ in

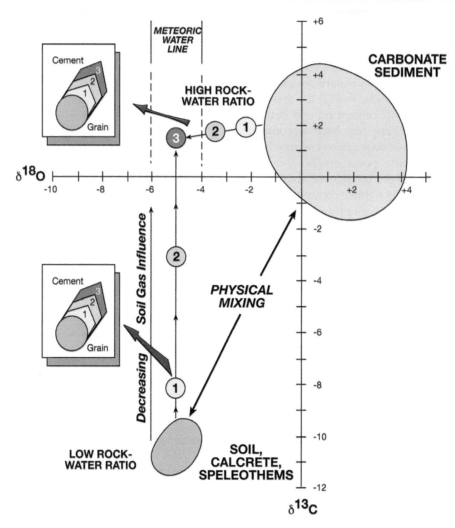

Figure 25.20 A δ¹⁸O versus δ¹³C stable isotope cross-plot illustrating the values expected as carbonate sediment (upper right) changes to dLMC under conditions of a high water–rock ratio (lower left) and low rock–water ratio (upper left). The cements (in boxes) will have progressively different values deeper below the exposure surface, as illustrated by the numbers in the zones.

higher latitudes. The carbon isotopes, by contrast, come from either the soil gas or the original sediment or generally some mixture of both. There are two end-member situations: (1) one with well-developed soils; and (2) another with little or no soil.

Little or no soil. Such conditions prevail in arid conditions and at high latitudes in arctic and sub-arctic climates where soils are poorly developed. The carbon comes almost entirely from the metastable sediments, whereas the oxygen comes from the meteoric water. The isotopic composition of the new dLMC will reflect the carbon isotopic composition of the sediment and the oxygen values will reflect those of the waters. As the waters descend through the meteoric zone, the δ¹³C values will

not change much but the δ¹⁸O values will move to reflect the surrounding waters.

Soil-rich horizon. This situation is likely under a humid climate where clays are abundant and the decay of leaf litter and grasses dominate. Semi-arid climates are less important here because of the development of soil carbonates, especially calcrete (see Chapter 26). The oxygen in this situation, like the preceding example, is largely meteoric and so similar to the above. When the water emerges from the base of the soil horizon charged with soil CO_2, it reacts with the immediately underlying sediments by dissolving various components before reprecipitating dLMC with a strong soil gas signature (−10‰ in Figure 25.20). The first cements therefore have a very negative $\delta^{13}C$ signature (Figure 25.20, Stage 1). As the water continues to move down through the sediment, there will be a series of dissolution–reprecipitation reactions (Figure 25.20, Stages 2, 3). During each reaction, the original sediment donates carbon but there is no replenishment of the soil gas carbon and so the effect of the depleted ^{13}C is reduced until eventually all the carbon comes from the sediment; the signature then resembles that of the original material on the order of 0–3‰ $\delta^{13}C$.

Rock–water ratio.   An important and additional factor is whether or not the diagenetic system is "open" or "closed" (see Chapter 22), that is, if the system has a low or a high rock–water ratio, respectively. This relationship is best understood in Figure 25.20. When there is little water involved and the rock–water ratio is high, most of the isotopes in dLMC will be contributed by the sediments that are being altered and produce mostly positive $\delta^{13}C$ values and slightly negative $\delta^{18}O$ values. By contrast, if the system is flushed by actively flowing fresh waters, the rock–water ratio will be low (dominated by the water) and so the values will reflect the waters, whereas the sediment contribution will be relatively insignificant until into the lower vadose and phreatic zones (see above).

An important consequence of this is that in systems where the soil is preserved we can, with care, extract information about ancient atmospheres from the $\delta^{13}C$ signatures of the soil carbonates. Finally, the application of clumped isotopes has the potential to determine the specific environments of precipitation more accurately.

Implications for the rock record

1. Diagenetic LMC of replaced components and of new cements will usually have relatively low trace element concentrations compared to the original materials except if the sediments were originally calcite or the replacement takes place in a very closed system with a continuing high rock–water ratio. Such systems are rare in the meteoric environment.
2. Mn^{2+}, the element that promotes luminescence in calcites, is only abundant in the phreatic zone.
3. Limestones immediately adjacent to an exposure surface with a soil cover (that could have been eroded during subsequent transgression) can have negative $\delta^{13}C$ values compared to the rest of the unit. The isotopes signal exposure and thus meteoric diagenesis of the underlying carbonates.

4. Both $\delta^{13}C$ and $\delta^{18}O$ remain largely unchanged once set in the meteoric environment because the dLMC does not change.

Further reading

Allan, J.R. and Matthews, R.K. (1982) Isotope signatures associated with early meteoric diagenesis. *Sedimentology*, 29, 797–817.

Budd, D.A. (1992) Dissolution of high-Mg calcite fossils and the formation of biomolds during mineralogical stabilization. *Carbonates and Evaporites*, 7, 74–81.

Budd, D.A. and Land, L.S. (1990) Geochemical imprint of meteoric diagenesis in Holocene ooid sands, Schooner Cays, Bahamas: correlation of calcite cement geochemistry with extant groundwaters. *Journal of Sedimentary Petrology*, 60, 361–378.

Dodd, J.R. and Nelson, C.S. (1998) Diagenetic comparisons between non-tropical Cenozoic limestones of New Zealand and tropical Mississippian limestones from Indiana, USA: Is the non-tropical model better than the tropical model? *Sedimentary Geology*, 121, 1–21.

Evans, C.C. and Ginsburg, R.N. (1987) Fabric-selective diagenesis in the late Pleistocene Miami Limestone. *Journal of Sedimentary Petrology*, 57, 311–328.

Frank, T.D. and Lohmann, K.C. (1995) Early cementation during marine-meteoric fluid mixing; Mississippian Lake Valley Formation, New Mexico. *Journal of Sedimentary Research, Section A: Sedimentary Petrology and Processes*, 65, 263–273.

Goldstein, R.H. (1988) Cement stratigraphy of Pennsylvanian Holder Formation, Sacramento Mountains, New Mexico. *American Association of Petroleum Geologists Bulletin*, 72, 425–438.

Halley, R.B. and Harris, P.M. (1979) Fresh water cementation of a 1,000 year-old oolite. *Journal of Sedimentary Petrology*, 49, 969–988.

Huntington, K.W., Budd, D.A., Wernicke, B.P., and Eiler, J.M. (2011) Use of clumped-isotope thermometry to constrain the crystallization temperature of diagenetic calcite. *Journal of Sedimentary Research*, 81, 656–669.

James, N.P. (1974) Diagenesis of scleractinian corals in the subaerial vadose environment. *Journal of Paleontology*, 48, 785–799.

James, N.P. and Choquette, P.W. (1984) Diagenesis 9. Limestones: the meteoric diagenetic environment. *Geoscience Canada*, 11, 161–194.

James, N.P. and Bone, Y. (1989) Petrogenesis of Cenozoic, temperate water calcarenites, South Australia: a model for meteoric/shallow burial diagenesis of shallow water calcite sediments. *Journal of Sedimentary Petrology*, 59, 191–203.

James, N.P. and Choquette, P.W. (1990) Limestone: The meteoric diagenetic environment. In: McIlreath I. and Morrow D. (eds) *Sediment Diagenesis*. St John's, ND, Canada: Geological Association of Canada, Reprint Series 4, pp. 35–74.

Land, L.S. (1970) Phreatic versus vadose meteoric diagenesis of limestones: evidence from a fossil water table. *Sedimentology*, 14, 175–185.

Lasemi, Z. and Sandberg, P.A. (1984) Transformation of aragonite-dominated lime muds to microcrystalline limestones. *Geology*, 12, 420–423.

Melim, L.A., Swart, P.K., and Eberli, G.P. (2004) Mixing-zone diagenesis in the subsurface of Florida and the Bahamas. *Journal of Sedimentary Research*, 74, 904–913.

Meyers, W.J. and Lohmann, K.C. (1978) Microdolomite-rich syntaxial cements: proposed meteoric-marine mixing zone phreatic cements from Mississippian limestones, New Mexico. *Journal of Sedimentary Petrology*, 48, 475–488.

Niemann, J.C. and Read, J.F. (1988) Regional cementation from unconformity-recharged aquifer and burial fluids, Mississippian Newman Limestone, Kentucky. *Journal of Sedimentary Petrology*, 58, 688–705.

Quinn, T.M. (1991) Meteoric diagenesis of Plio-Pleistocene limestones at Enewetak Atoll. *Journal of Sedimentary Petrology*, 61, 681–703.

Vollbrecht, R. and Meischner, D. (1996) Diagenesis in coastal carbonates related to Pleistocene sea level, Bermuda Platform. *Journal of Sedimentology Research*, 66, 243–258.

CHAPTER 26
KARST AND WATER-CONTROLLED DIAGENESIS

Frontispiece Flowstone along a cave wall, Bermuda. Image width ~4 m.

Origin of Carbonate Sedimentary Rocks, First Edition. Noel P. James and Brian Jones.
© 2016 Noel P. James and Brian Jones. Published 2016 by John Wiley & Sons, Ltd.
Companion website: www.wiley.com/go/james/carbonaterocks

Introduction

Limestones, because they are so soluble in fresh meteoric waters, have produced some of the most breathtaking natural landforms on the planet. The tower karst and the Stone Forest in China, the impenetrable rain forest topography in New Guinea, and the cockpit country of Jamaica are all the result of surface dissolution. The underground caverns, galleries, shafts, tunnels, and cathedral-scale chambers of Lascaux in France, Waitomo in New Zealand, Carlsbad in New Mexico, Slovenia, and Mexico to name only a few, are products of subsurface corrosion. Such caves are festooned with spectacular precipitates (speleothems) in the form of stalactites, stalagmites, columns, cave pearls, and moonmilk that attract tourists everywhere. At the same time, precipitation of soil carbonates has turned arable landscapes into grasslands suitable only for grazing animals.

These processes, which have been active since the Archean, are also of economic importance in terms of natural resources. Lead and zinc sulfide ores worldwide are localized in karst cavities that typically formed before the minerals were emplaced. Spectacular oil fields in the Permian of Texas and the Cretaceous of Mexico are, for example, localized in subsurface paleokarst pore systems.

This chapter concentrates on the processes and products of this water-driven diagenesis (Figure 26.1). The discussion begins with an investigation of surface karst and calcrete, and concludes with an analysis of subsurface karst.

METEORIC DIAGENESIS

UNIVERSAL ASPECTS

WATER CONTROLLED

MINERALOGY	*NOT RELEVANT*

PROCESS	*DISSOLUTION*
	    • *SURFACE*
	    • *SUBSURFACE*
	PRECIPITATION - SPELEOTHEMS
	CALCRETE FORMATION

CONTROLS	*WATER FLOW RATE* *MIXING OF DIFFERENT WATERS* *CLIMATE* *TIME* *VEGETATION*
POROSITY & PERMEABILITY	*ENHANCED* *REDUCED BY CALCRETE*

Figure 26.1 The main processes and products of water controlled meteoric diagenesis of carbonate rocks.

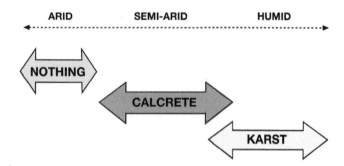

Figure 26.2 The relationship between climate and near-surface diagenesis in carbonate terranes.

Surficial processes and products

Climate is the overriding control in this environment (Figure 26.2). Extensive surface karst terrains develop in humid climates where thick organic-rich soils and dense vegetation are generally the norm. In contrast, areas under semi-arid climates, especially those with winter rainfall but long dry summers, are typified by calcareous soils (calcretes). Limestones in truly arid climates, with essentially no rain year-round except during unpredictable storms, are largely unaffected because there is no water to mediate diagenetic change. There are therefore two end-member situations: (1) the surface karst facies; and (2) the calcrete facies. These diagenetic facies are, however, not mutually exclusive; karst and calcrete can coexist at any one time and overlap in any given area.

Surface karst facies

Surface karst (*exokarst* of geomorphologists) encompasses those dissolution features that form at the carbonate rock surface that is (was) either exposed to the air or covered by soil (Figure 26.3). The nomenclature of karst features is voluminous, with several names for similar structures in a variety of languages. Water is the critical component: in deserts there is minor karst; in semi-arid areas both karst and calcrete form in slightly different situations; and in humid, especially tropical regions, karst is everywhere. Surface karst formation in high latitudes is slow, whereas in temperate Mediterranean-type climates, karst and calcrete are common but development is seasonal. In areas of high rainfall, dissolution is profound with karst forming a fantastic array of towers, jagged ridges, hogbacks, canyons, and sinkholes that, together with extensive vegetation, make such areas almost impenetrable to surface travel.

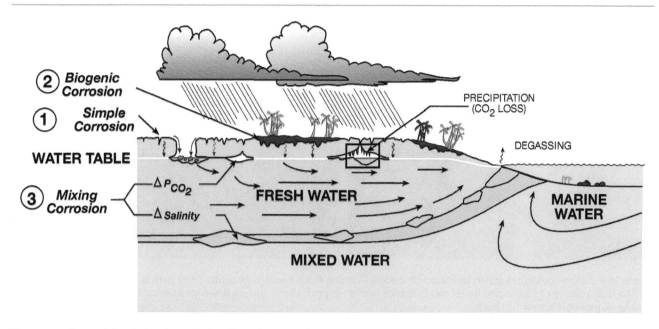

Figure 26.3 Areas of dissolution (corrosion) and precipitation of carbonate during the exposure of calcite limestone to percolating meteoric waters. Source: Adapted from James and Choquette (1990). Reproduced with permission of the Geological Association of Canada.

Syngenetic karst

Syngenetic karst forms while young sediments are being altered to dLMC. Solution sculpture is variable but solution pipes are widespread, especially where associated with tree roots. Subsurface caves are well developed, with roof collapse especially common. Sediment at the surface in these exposure situations quickly becomes lithified as thin millimeter-thick grey calcite cement crusts are precipitated. Many quarry operators take advantage of these natural hardening processes by cutting soft limestones into construction-sized blocks and then leaving them exposed to the atmosphere to weather for a few years until they harden and become suitable building stone. Cement crusts are infested with algae and fungi that can colonize the substrates within months of them being exposed. White sediment exposed in new road cuts, for example, becomes discolored within months of being exposed. The grey color of the crusts is generally attributed to organic matter produced by the resident microbes.

Processes

Whereas surface corrosion via acid rain can be impressive, dissolution is especially active beneath a soil cover and most intensive in the tropics because of high temperatures and increased volumes of vegetative litter and soil (Figure 26.4a). Although biogenic dissolution via carbonic acid is the most important process beneath tropical soils, corrosion due to other acids (e.g., fulvic, crenic, sulfuric, and nitric acids), mostly generated by microbes is also important but, unlike the reactions associated with carbonic acid, many of these reactions are irreversible. The style of development also depends a great deal on the porosity and permeability of the limestone or sediment, with surface features best developed on well-lithified carbonates.

Simple corrosion. The amount of CO_2 dissolved in water open to the air depends upon the partial pressure of CO_2 (PCO_2) in the air (~0.00035 atmospheres) and at the air–water interface. Rainfall is in equilibrium with atmospheric CO_2 and is, therefore, an extremely weak acid. Simple corrosion (Figure 26.4a) is brought about by rainfall on barren rock surfaces. This is readily apparent in tombstones and buildings made of limestone that are subject to acid rain.

Biogenic corrosion. Air in the soil has a significantly higher PCO_2 than the atmosphere because of plant respiration, the decay of organic matter, and microbial activity. CO_2 is added to the water as rainwater percolates through the soil and PCO_2 is between 0.003 and 0.2 atmospheres. Water that emerges from the base of the soil is therefore very aggressive and can dissolve considerable amounts of carbonate (Figure 26.4b), much more than ordinary rainwater.

Figure 26.4 Surface karst. (a) Karsted Devonian limestone, Canning Basin Western Australia. Note person for scale (circled). (b) Roadside outcrop of Pleistocene limestone, Barbados with an irregular karst surface developed beneath the thick soil of a banana plantation. Outcrop 2 m high.

Products

From a practical geological viewpoint, surface karst features can be divided into small-scale meter-sized solution sculptures that can be easily recognized on the outcrop or drill-core scale and large-scale karst features that are meters to kilometers in size, better thought of as karst landforms. Regardless of size, all of these sculptured features are referred to as *karren*.

Solution sculpture. Small karren in temperate climates are generally smooth surfaces. In tropical areas, particularly where rainfall is high and mist is common, the rock surface is characterized by small, slightly elongate cup-like pits <3 cm in size (cockling) that intersect one another in knife-sharp edges (Figure 26.5) and give the rock a crinkly or cindery appearance. In other tropical regions, a similar texture of black-coated jagged pinnacles of many sizes and marked by delicate lacy dissection is termed *phytokarst* (Figure 26.6) because of the intensive activity of endolithic microflora and microbes in its formation.

Medium-size karren usually take the form of razor-sharp, finely chiseled runnels or rounded runnels called *rillenkarren* and *rinnekarren* respectively, and solution pans. The solution pans (*klaminitza* or *kamenica*) are small flat-bottomed basins that typically have overhanging edges. They can form on horizontal surfaces where rainwater collects in small depressions beneath humus patches or in the intertidal zone through the combined action of mixing corrosion and bioerosion.

Figure 26.5 Surface karst: sharp spitzkarren (also known as a manketa surface) on Pleistocene limestone, supratidal zone, Bermuda. Hammer 20 cm long (circled).

Large karren are generally termed *rundkarren* because the presence of soil over the surface promotes diffuse water flow and a tendency to smooth out some of the sharp features. The forms that seem to develop best under these conditions are clints and grikes, that is, flat top blocks (clints) surrounded on all sides by solution-widened joints (grikes) (Figure 26.7). Vegetation plays an important role here, with plant and tree roots extending into the subsurface and quickly transporting acidic waters deep into the sediment or rock.

Figure 26.6 Black phytokarst developed on white Miocene limestone, Grand Cayman, Caribbean: (a) phytokarst at Hell (image width ~15m); and (b) B. Jones standing in a phytokarst hole.

Figure 26.7 Surface karst. Joint controlled clints (blocks) and grikes (solution enlarged fractures) developed on Carboniferous limestone, Burren National Park, Ireland. Foreground width ~4m. Photograph by T. Frank. Reproduced with permission.

Figure 26.8 Water-filled doline in Oligocene limestone, southern Australia.

Karst landforms. The most impressive of these features are sinkholes or dolines (Figure 26.8) and tower karst (Figure 26.9). Sinkholes have also been called cenotes, mogotes, uvalas, karst valleys, and poljes. The sinkholes/ dolines are usually funnel–bowl–dish-shaped depressions meters to kilometers in diameter and up to 100m deep. They form via dissolution in the subsurface that works its way up to the surface, resulting in either gradual subsidence or collapse. In actively developing karst terrains, sinkholes are commonly filled with water. The lower parts of the sinkholes are chocked with breccias that

are chaotically arranged lithoclasts of variable size and lithology mixed with soils. Speleothems may be found on the walls of some sinkholes. Beds with a V-shaped structure are due to subsurface dissolution and partial roof collapse.

Tower karst, with steep-sided cylindrical pillars in the form of small mountains that are separated by flat-sediment floored valleys, result form prolonged dissolution of the carbonate surface. The most impressive examples are in China with ancient counterparts in the Mississippian Madison Limestone in North America. In both cases dissolution seems to be focused along joints.

Figure 26.9 (a) Small-scale tower karst, Miocene Limestone, New Zealand Pillars ~3m high. (b) Tower karst developed in Mississippian limestones, Guilin, China. Photograph by N. Chow. Reproduced with permission.

Precipitation

Near-surface diagenetic calcite precipitation is rare. Nevertheless, this is a zone of evaporation or evapo-transpiration, especially in the tropics. Waters percolate downward only to be drawn up again via evaporation and precipitate carbonate as they evaporate. During these processes, the loss of CO_2 from the water due to heating, plant activity, or microbial activity can lead to precipitation.

Calcrete facies

Calcrete (or caliche or duricrust) is the carbonate-lithified part of the soil profile that is commonly, but not exclusively, developed on carbonate sediments and rocks (Figure 26.10). As such, they are excellent indicators of subaerial exposure in the rock record. Calcretes also develop through accretionary processes and are therefore also useful in paleoclimate studies.

Calcrete zones, which can be up to 7m thick, are characterized by diverse arrays of diagenetic fabrics. They are the integrated result of temporal variations in limestone alteration and repeated phases of calcite precipitation. These processes are modified or partially controlled by microbes and plants of various types. As such, this facies is an example of both water-driven and microbe-driven diagenesis. Distinguishing attributes include laminated crusts, rhizoconcretions, diagenetic peloids, ooids and pisoids, breccias, clotted micrite and chalky carbonate, vuggy porosity, microborings, and iron oxides.

Lithologies

A typical calcrete succession grades vertically upward from bedrock into: (1) massive chalky carbonate; (2) nodular and crumbly carbonate; (3) irregular crusts and hardpans; and (4) a compact crust or hardpan. Calcretes are notoriously variable and many different arrangements of these facies are evident (Figure 26.10).

Massive chalky carbonate. This zone is variably altered host substrate and generally formed of original carbonate particles or rock fragments that display various stages of dissolution or neomorphism, microspar, and scattered nodules.

Nodules and crumbly carbonate. There are few recognizable original sediment grains in this part of the profile, that contains many peloids and coated grains in the form of ooids and pisoids (Figure 26.11). These dLMC glaebules, of soil terminology, are silt- to pebble-sized, vary from spherical to irregular or cylindrical, and can be isolated to coalesced in arrangement. The matrix between the particles is typically calcite microspar.

Crusts and hardpans. Crusts or surficial hardpans (Figure 26.12), which can be millimeters to centimeters thick, are composed of microcrystalline to cryptocrystalline low Mg-calcite. They range from structureless to horizontally laminated, and superficially resemble laminar stromatolites. Most crusts below the surficial hardpan are thin, commonly friable, horizontal to subhorizontal plates or sheets that are separated from one another but join and bifurcate laterally.

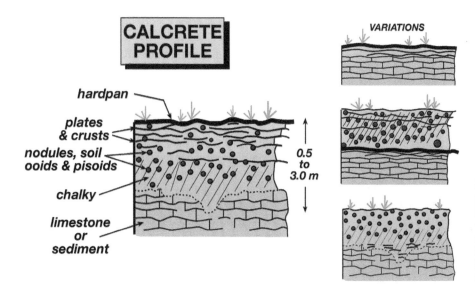

Figure 26.10 Sketch of a calcrete profile showing the major components (left) and some of the observed variations from different modern examples (right). Source: Adapted from James and Choquette (1990). Reproduced with permission of the Geological Association of Canada.

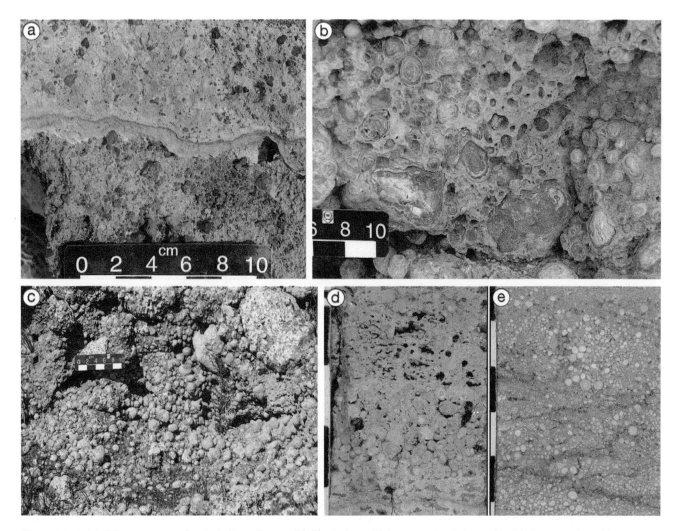

Figure 26.11 (a) Calcrete crust and soft chalky calcrete with black clasts, Pleistocene, South Australia. (b) Cross-section of large calcrete pisoids, Pleistocene, South Australia. (c) Surface exposure of abundant calcrete pisoids, Pleistocene, South Australia. (d, e) Core images of calcrete pisoids, Mississippian Saskatchewan, Canada. Centimeter scale in all panels. Photographs (d, e) by W. Martindale. Reproduced with permission.

Figure 26.12 (a) Numerous subhorizontal calcrete crusts developed in Pleistocene lagoonal facies. Pleistocene, Barbados. Shovel 1 m long (circled). (b) Laminated crusts (arrows) in paleocalcrete profile, Mississippian limestone, Kentucky, USA. Finger 1 cm wide.

Rhizoconcretions. Roots can be pervasive throughout and are typically calcified or entombed in calcite to form *rhizoconcretions*. Roots also lead to the formation of crusts and can penetrate and brecciate mature calcrete, resulting in teepee or pseudo-anticlinal structures. The holes made by the roots can be confused with the burrows of marine invertebrates (*Skolithos*, *Thalassinoides*, etc.) but

their downward bifurcation, inclusion of original grains, and downward thinning help distinguish them.

Petrography

The predominant fabric (Figure 26.13) in calcrete is clotted peloidal micrite and microspar with spar-filled channels and cracks. Grains and fragments of limestone are fractured and separated from the surrounding matrix by circumgranular cracks that form through expansion and shrinkage. Such grains are typically coated by micrite laminae that can in turn be connected by thin laminations to neighboring grains. Voids can be alveolar (full of tiny holes), vermicular (wormlike), or dense networks of micrite tubules and rods in a micrite matrix yielding a spaghetti-like texture. Needle-fibers of calcite are typical as a crystal mesh. *Microcodium*, elongate petal-shaped calcite crystals, prisms, ellipsoids, or bell-shaped clusters are common in late Mesozoic and Cenozoic examples. They form through morphological transformation of root tissue by growth of calcite within the cortical cells, which distorts the cell shape.

Variations

The different styles of calcrete (Figure 26.10) reflect length of exposure and local climate. Some researchers have also differentiated between vadose or rhizoconcretionary calcrete (as described above) and groundwater calcrete (massive, mottled, and laminar) in the zone of capillary rise just above the water table.

Subsurface karst facies

Water flow

Unconfined flow. The most common aquifers are unconfined or open to the atmosphere. The two types of flow found in these aquifers are: (1) *diffuse flow* where there is a well-defined water table and water movement obeys or nearly obeys Darcy's Law; or (2) *free flow* where the water table is poorly defined and water moves through integrated conduits like a series of underground rivers. Diffuse flow characterizes newly formed limestones but with time they evolve into conduit flow.

Confined flow. Flow in confined aquifers occurs when they are bound by aquitards (impermeable layers) and there is no vertical connection to the atmosphere. These aquifers are important because they can transport fresh waters deep into basins or beneath the continental shelf.

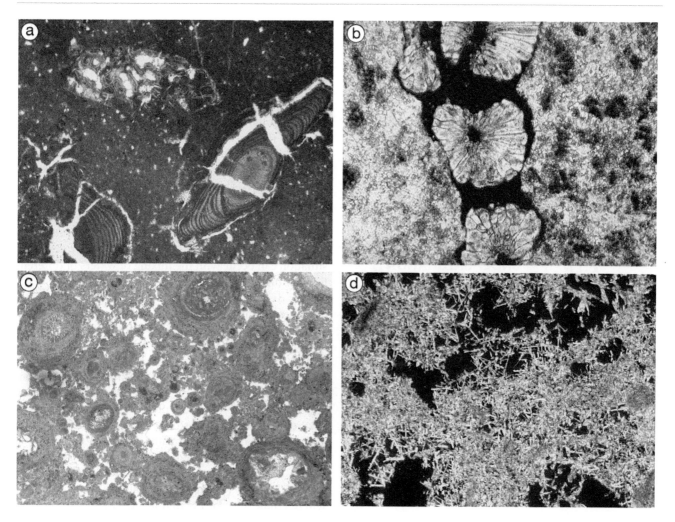

Figure 26.13 Thin-section images of calcrete fabrics, Pleistocene, Barbados. (a) Circumgranular cracking (PPL). Image width 1 cm. (b) *Microcodium* (XPL). Image width 500 μm. (c) Clotted to peloidal to oolitic texture (PPL). Image width 400 μm. (d) Needle-fibre calcite (XPL). Image width 500 μm.

The vadose zone

Processes of intensive physicochemical and biological dissolution dominate the zone of infiltration. Vertical caves are produced and collapse breccias can be common. Precipitates are usually finely crystalline.

When the CO_2-rich waters emerge from beneath the soil, they dissolve carbonate, quickly become saturated, and lose their chemical aggressiveness a few decimeters below the soil zone. Once the waters are saturated with respect to calcite, no further dissolution occurs. Under conditions of diffuse flow, there is little reaction, but if flow is free, then rapidly flowing waters can be aggressive to depths of 100 m or more. Dissolution can also occur when vadose seepage and vadose flow are mixed.

Many caves in the vadose zone can be due to perched water tables or are relict water-table caves (see below) that were abandoned when the water table dropped.

The water table

This is an area of intense chemical activity. The region includes the base of the vadose zone, the water-table surface, and the upper part of the phreatic zone.

Dissolution. When the system is closed to outside air, the zone in which vadose and phreatic waters mix is a zone of intense corrosion (Figure 26.3). Contrary to common sense, it appears that the mixing of two waters, both of which are at equilibrium and saturated with

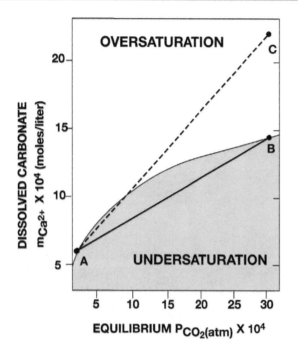

Figure 26.14 The non-linear relationship between equilibrium PCO_2 and dissolved $CaCO_3$; the points A, B, and C are waters of different composition. The physical mixture of waters A and B will fall somewhere along the solid line and will always be undersaturated (even though the waters at A and B are at equilibrium and saturated with the $CaCO_3$), resulting in dissolution. The mixture of waters A and C will only be undersaturated when a large proportion of water A is in the mixture. Source: Adapted from James and Choquette (1990). Reproduced with permission of the Geological Association of Canada.

respect to carbonate, can lead to water that is supersaturated or undersaturated, depending upon the character of the solutions and the minerals in question. This situation can result in considerable subsurface limestone dissolution. The simplest type of mixing occurs when two bodies of meteoric water, isolated from contact with the air and each at equilibrium but with different CO_2 content, come together (Figure 26.14). The resulting mixture will lie along a straight line joining the fluid concentrations and depending upon the proportions of each (e.g., A and B in Figure 26.14). The composition will move up from and to the left of this line until the curved saturation line is reached, with dissolution occurring in the process. This situation commonly takes place when vadose water with high PCO_2 mixes with low PCO_2 phreatic water. A similar situation can occur with the mixing of waters of different temperatures.

Such dissolution results in caves (Figures 26.15, 26.16), passages, channels, and shafts that are related to water flow and carbonate lithology. Walls and ceilings can be smooth, pockmarked with corrosion pockets, dimples, and pits, or textured with rills and grooves. Scallops or crescentic shell-shaped dissolution features are common. Water movement is largely vertical in the vadose zone and mostly horizontal in the phreatic zone; large solution cavities are therefore mostly vertical in the vadose zone and subhorizontal in the phreatic zone.

The water table is not static and will rise during the wet season and fall during the dry season. Furthermore, in marginal marine systems and on islands, the water table will fluctuate in accord with tidal activity.

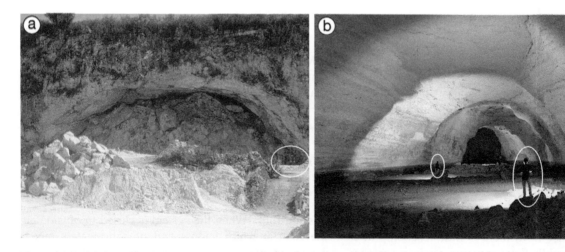

Figure 26.15 (a) A small cave developed in Pleistocene limestone that has been breached during quarrying and is filled with collapsed breccia, Bermuda. Students for scale (circled). (b) Cave at Nullarbor Plain, Australia. People for scale (circled). Image courtesy of S. Milner and the Cave Exploration Group of South Australia.

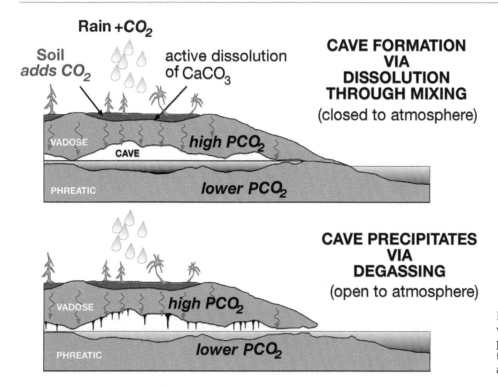

Figure 26.16 The formation of water table caves (top) and the precipitation of speleothems once the system is open to atmospheric air (bottom).

Precipitation. Precipitates that form in caves are collectively known as speleothems (Frontispiece, Figures 26.17, 26.18) and include stalactites (pendant with water flow downward through a central canal), stalagmites (columnar or mound-like from the cone flow but with no canal), draperies (furled sheets formed by dripping and trickling water), and flowstone (layered deposits formed by flowing sheets of water). Crystals that grow from the substrate terminate at the air–water interface. As a result, the outer surface of the precipitate is smooth and there are never scalenohedral crystal terminations.

Most carbonate precipitation takes place in the upper parts of caves and cavities above or just below the water table. If the system is open to the atmospheric air, high PCO_2 vadose waters emerging at the top of caves equilibrate with low $_pCO_2$ atmospheric air, degas, and precipitate carbonates (Figure 26.16).

Precipitation begins as a water droplet emerges from the cave ceiling in the form of a ring of calcite around the drip. With time and the progressive addition of ring after ring, a tubular structure known as a soda straw develops (Figure 26.18b). Water can then flow down the outside of the soda straw or through the soda straw (providing it does not become blocked by precipitates). Precipitates from the water that flows over the outer surfaces of the soda straw add layer upon layer to the initial tube to form the familiar laminated stalactites. Water that drips from

the cave ceiling or from the tips of stalactites commonly leads to the formation of stalagmites. Like stalactites, these structures are internally laminated. Unlike stalactites, stalagmites do not have a soda straw at their centre. If growth continues for a long period of time, a stalactite and stalagmite may join to form a column. Maximum growth takes place during the wet season, whereas growth may cease during the dry season. Stalactites are commonly used in paleoclimate studies because the concentric laminae record seasonal growth and the calcite in the different layers can be dated using U/Th techniques, whereas the stable isotopes reflect variations in climate conditions.

Flowing waters in the caves are also at equilibrium with atmospheric air and so precipitation only occurs because of variations in water temperature and turbulence. Cave floors are covered with flowstone (a general term applied to calcite and aragonite or both) that is precipitated from waters that flow over the cave floor. Complex architectural structures commonly develop, with rimstone pools being common features. Rimstone dams typically develop where small obstructions on the cave floor cause increased water agitation that, in turn, promotes CO_2 degassing. Spectacular terraces formed of a series of rimstone pools develop on sloping floors. Such pools range from small (<10 cm long), shallow (<5 cm deep), to large pools that are meters in diameter and depth. Such pools are, in effect, phreatic environments

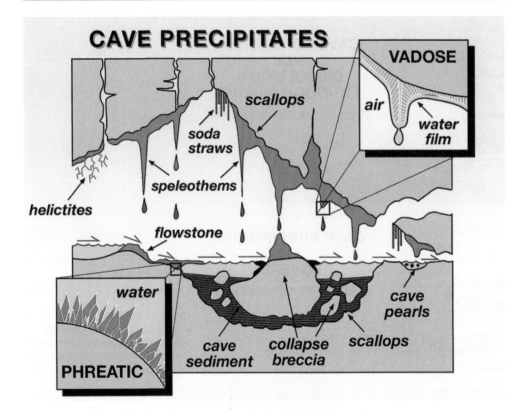

Figure 26.17 The styles of precipitates formed above and below the water table in a cave open to the atmosphere. Source: Adapted from James and Choquette (1990). Reproduced with permission of the Geological Association of Canada.

Figure 26.18 (a) Speleothems in an active cave system with numerous soda straws, Bermuda. Individual large stalagmites ~4 m tall. (b) Soda straws, Bermuda. Image width 1 m.

where spectacular speleothemic precipitates develop. Common features include cave pearls (coated grains), calcite rafts, and ledges that form around the edge of the pool.

Erratic forms of precipitates are mostly moonmilk and helictites. Moonmilk resembles cottage cheese when wet and is composed of microscopic carbonate crystallites or random crystals with 35–75% water when precipitated; when dry however, the structure turns into powder. The crystals are enmeshed in a net of bacterial filaments, actinomycetes, microbes, and algae because plants are critical in removing CO_2 from the water. Helictites are stalactites that grow in a variety of directions generally controlled by rapid growth rate.

The phreatic zone

Dissolution. The *lenticular zone* of phreatic waters that float on underlying marine waters is principally a region of dissolution and little, if any, precipitation. In particular, there is a brackish zone of mixing at the base of the freshwater lens that is a zone of corrosion and porosity formation. Dissolution is most intense when fresh and marine waters mix, and is particularly well-developed along or adjacent to shorelines of relatively young carbonates. In other situations, the mixing of waters from the upper phreatic and more static mixed fresh waters and saline waters in the lower phreatic zone can mingle and also cause dissolution. Caves tend to be larger and more numerous toward the coast where flow rates are highest. In short, corrosion would seem to be the dominant process with little or no precipitation.

Although the precise curves for any given situation are different, the results follow a similar trend (Figure 26.19). As seawater is mixed with carbonate-bearing fresh groundwater, the fluid initially becomes progressively more undersaturated with respect to a mineral, in this case calcite. With continuing addition of seawater, undersaturation decreases so that mixtures containing large amounts of seawater are oversaturated. Once in the field of oversaturation, precipitation should occur but is generally prevented by kinetic factors (especially high Mg^{2+}, PO_4^{3-}, SO_4^{2-}). Dissolution by this process is particularly

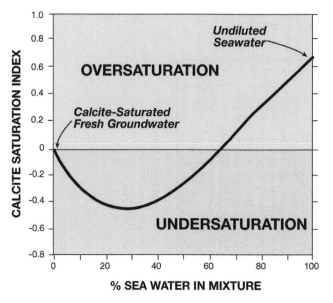

Figure 26.19 The effect on calcite saturation in groundwaters by mixing waters of different ionic strength, in this case seawater and meteoric groundwater. Source: Adapted from James and Choquette (1990). Reproduced with permission of the Geological Association of Canada.

evident near the shore where fresh and marine waters actively mix (Figure 26.20).

Karst breccias. Whereas much subsurface corrosion results in small to large to spectacular open caves and cave systems, just as often the cavities are choked with limestone breccia (Figures 26.15a, 26.21). Although dissolution excavates the subsurface void, the space is architecturally unsound and so the roof collapses inward until a stable ceiling arch or dome is formed. The resultant breccia can fill the cave because of irregular clast packing. A similar phenomenon can develop if the water table rises with accompanying upward stoping to eventually form a breccia-floored doline when the process breaks surface. If an unexpected carbonate breccia is encountered in limestone stratigraphy, it just may be of karst origin.

Precipitation. PCO_2 in the phreatic zone should remain constant because the waters are out of contact with the atmosphere and surrounded by carbonate. Under natural conditions, groundwaters are typically supersaturated, but precipitation is inhibited by crystal growth kinetics and ion inhibition (the presence of high numbers of other bivalent cations in solution). It is well known, for example, that more energy is needed to grow a crystal than to dissolve it, and the rate at which CO_2 dissolves in water is very fast but the rate at which it precipitates out of water is comparatively slow. Overall, it appears that the phreatic zone, if the limestones are LMC, is mostly one of dissolution and not precipitation.

If phreatic crystallization does take place, because all precipitation is in water, crystals grow in all directions and have scalenohedral terminations (Figure 26.17). Crystals can form at the water surface or grow laterally as rafts and then sink to the water floor. Cave pearls (concentrically laminate peloids) form in splash pools or pools of moving water. The black sooty coating on cave walls and pebbles is a variety of manganese minerals precipitated by specialized microbes, especially bacteria. The red color of many cave sediments is caused by chemo-autotrophic iron bacteria that obtain N_2 from the atmosphere and carbon from iron carbonates, liberating ferrous iron in the process.

Paleokarst

Karst is present in carbonates of all ages from Proterozoic to Holocene (Figure 26.22). There are several different modes: (1) depositional paleokarst, formed at the top of meter-scale successions; (2) local paleokarst, when small areas of a platform are exposed by tectonism or small drops

Figure 26.20 Dissolution via coastal mixing in Pleistocene limestone, Yucatan Peninsula, Mexico. (a) Aerial image of dissolution along fractures where freshwater and seawater mix. Small road at upper left for scale. (b) Dissolution along elongate fracture. Note persons for scale. (c) Close-up of numerous dissolution vugs. Image width 3 m.

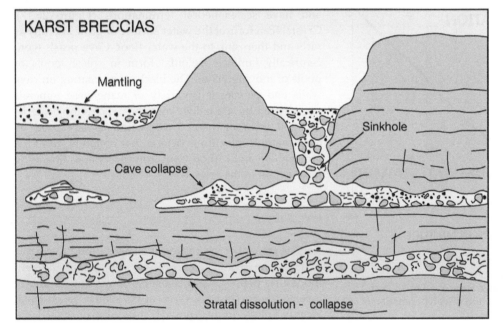

Figure 26.21 Various kinds of subsurface breccias associated with karst. Source: Choquette and James (1988). Reproduced with permission of Springer Science+Business Media.

Figure 26.22 (a) Cross-section of a scalloped karst surface, Paleoproterozoic, Bathurst Inlet, Arctic Canada. Sledgehammer 15 cm long. (b) Paleokarst bedding surface at the top of Lower Ordovician Sauk Sequence, Mingan Islands, Gulf of St Lawrence, Canada.

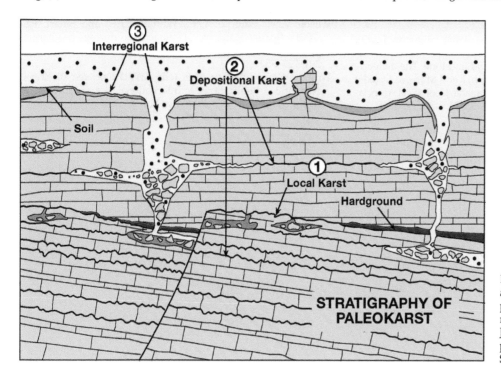

Figure 26.23 The various scales and temporal disposition of paleokarst in the geological record. Source: Choquette and James (1988). Reproduced with permission of Springer Science+Business Media.

in sea level; and (3) interregional paleokarst where large areas are exposed for long periods of time (Figure 26.23).

Surface and subsurface carbonate geochemistry

Most dripstone and flowstone deposits are colored shades of yellow, orange, or brown. There are two main sources of the colors: (1) oranges and browns are from organic compounds, probably fulvic and humic acids, brought from the surface; and (2) deep browns and blacks are from hydroxides of iron and manganese or iron-pigmented clays on the calcite crystal surface. There are relatively few trace elements, except if the phreatic waters are reducing.

$\delta^{13}C$ values of dLMC precipitated as calcretes, soil carbonates, or speleothems are typically negative (–7 to –13‰) because of the involvement of soil gas. $\delta^{18}O$ values reflect the composition of the meteoric water line, that is, the rainwater with low oxygen isotope values.

Further reading

Alonso-Zarza, A.M. (2003) Palaeoenvironmental significance of palustrine carbonates and calcretes in the geological record. *Earth-Science Reviews*, 60, 261–298.

Back, W., Hanshaw, B.B., Herman, J.S., and Van Driel, J.N. (1986) Differential dissolution of a Pleistocene reef in the groundwater mixing zone of coastal Yucatan, Mexico. *Geology*, 14, 137–140.

Choquette, P.W. and James, N.P. (1988) Introduction. In James, N.P. and Choquette, P.W. (eds) *Paleokarst*. Berlin: Springer-Verlag, pp. 1–24.

Colson, J. and Cojan, I. (1996) Groundwater dolocretes in a lake-marginal environment: an alternative model for dolocrete formation in continental settings (Danian of the Provence Basin, France). *Sedimentology*, 43, 175–188.

Ford, D.C. and Williams, P.W. (1989) *Karst Geomorphology and Hydrology*. Boston, MA: Unwin Hyman Academic.

Goldstein, R.H. (1991) Stable isotope signatures associated with paleosols, Pennsylvanian Holder Formation, New Mexico. *Sedimentology*, 38, 67–78.

James, N.P. (1972) Holocene and Pleistocene calcareous crust (caliche) profiles: criteria for subaerial exposure. *Journal of Sedimentary Petrology*, 42, 817–836.

James, N.P. and Choquette, P.W. (eds) (1988) *Paleokarst*. Berlin: Springer-Verlag.

James, N.P. and Choquette, P.W. (1990) Limestone: The meteoric diagenetic environment. In: McIlreath, I. and Morrow, D. (eds) *Sediment Diagenesis*. St John's, ND, Canada: Geological Association of Canada, Reprint Series 4, pp. 35–74.

Melim, L.A. and Spilde, M.N. (2011) Rapid growth and recrystallization of cave pearls in an underground limestone mine. *Journal of Sedimentary Research*, 81, 775–786.

Miller, C.R., James, N.P., and Bone, Y. (2012) Prolonged carbonate diagenesis under an evolving late Cenozoic climate; Nullarbor Plain, southern Australia. *Sedimentary Geology*, 261–262, 33–49.

Semeniuk, V. and Searle, D.J. (1985) Distribution of calcrete in Holocene coastal sands in relationship to climate, southwestern Australia. *Journal of Sedimentary Petrology*, 55, 86–95.

Spötl, C. and Wright, V.P. (1992) Groundwater dolocretes from the Upper Triassic of the Paris Basin, France: a case study of an arid, continental diagenetic facies. *Sedimentology*, 39, 1119–1136.

Stoessell, R.K., Ward, W.C., Ford, B.H., and Schuffert, J.D. (1989) Water chemistry and $CaCO_3$ dissolution in the saline part of an open-flow mixing zone, coastal Yucatan Peninsula, Mexico. *Geological Society of America Bulletin*, 101, 159–169.

Weidlich, O. (2010) Meteoric diagenesis in carbonates below karst unconformities: heterogeneity and control factors. In: van Buchem F.S.P., Geredes K.D., and Esteban M. (eds) *Mesozoic and Cenozoic Carbonate Systems of the Mediterranean and the Middle East: Stratigraphic and Diagenetic Reference Models*. London: Geological Society, Special Publications no. 329, pp. 291–315.

Wilson, A.M., Sanford, W.E., Whitaker, F.F., and Smart, P.L. (2001) Spatial patterns of diagenesis during geothermal circulation in carbonate platforms. *American Journal of Science*, 301, 727–752.

Wright, V.P. and Tucker, M.E. (eds) (1991) *Calcretes*. Blackwell Scientific Publications, International Association of Sedimentologists, Reprint Series Vol. 2.

CHAPTER 27
BURIAL DIAGENESIS OF LIMESTONE

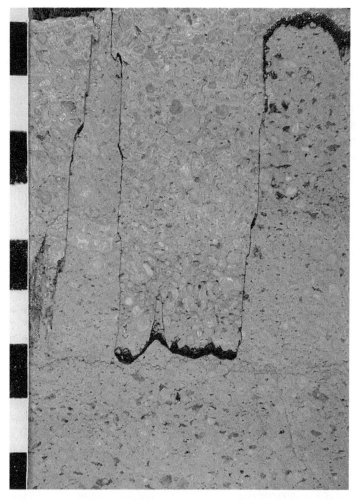

Frontispiece Slabbed core of Mississippian limestone, subsurface Western Canadian Sedimentary Basin, with a stylolite representing extensive burial dissolution. Centimeter scale. Photograph by W. Martindale. Reproduced with permission.

Origin of Carbonate Sedimentary Rocks, First Edition. Noel P. James and Brian Jones.
© 2016 Noel P. James and Brian Jones. Published 2016 by John Wiley & Sons, Ltd.
Companion website: www.wiley.com/go/james/carbonaterocks

Introduction

Limestones spend most of their lives in the burial realm. Those carbonate sediments that were altered in the synsedimentary marine diagenetic environment, as well as those that were modified by a plethora of processes in the meteoric realm, are further changed in this deep, somewhat mysterious, environment. Even those basinal carbonates that escaped synsedimentary and meteoric diagenesis are inevitably subjected to burial diagenesis. Remember, all limestones that you investigate (except perhaps some Cenozoic examples) have spent time in the burial diagenetic environment. More important, from an economic standpoint, is that all conventional hydrocarbon reservoir rocks have undergone burial diagenesis.

The setting

Burial diagenesis (Figures 27.1, 27.2) can be defined as "any change or series of changes that take place below the zone of near-surface diagenesis and above the realm of low-grade metamorphism." Near-surface means the marine or meteoric diagenetic realms. The base of the meteoric realm and the top of the burial realm are a bit fuzzy, and will vary from basin to basin.

In contrast to these shallow realms where alteration is geologically instantaneous, burial diagenesis is usually prolonged and involves a more complex array of chemical and physical change. The most important processes are: (1) physical (mechanical) compaction and dewatering; (2) chemical compaction (pressure-solution); (3) cementation; (4) dissolution; (5) dolomitization; (6) alteration of hydrous to anhydrous minerals; (7) thermally driven mineral stabilization; and (8) alteration and maturation of organic matter. The overall long-term trend is reduction of porosity and permeability at increasing temperatures and pressures over time spans of 10^6–10^8 years.

BURIAL DIAGENESIS

- The collection of changes that takes place below the meteoric and seafloor diagenetic environments and above the realm of low grade metamorphism

- Temp.   = 40-200°C (commonly 50-150 °C)
  Depth   = a few meters to several kilometers
  Pressure = slightly > 7000 kPa to ~200,000 kPa

- 1000s to 1,000,000s of years (time!)

- Where limestones spend most of their lives

Figure 27.1 The main processes, controls, and products of burial diagenesis.

BURIAL DIAGENESIS

MINERALOGY	Aragonite & Mg-Calcite $\rightarrow$ Calcite
	No Change - if already LMC
PROCESSES	Physical (mechanical) compaction
	Chemical compaction
	Cementation
	Dissolution
CONTROLS	

Intrinsic

    Purity - clay content
    Grain Size
    Previous Diagenesis
    Porosity & Permeability
    Mineralogy

Extrinsic

    Pressure
    Temperature
    Pore Fluid Chemistry
    Burial time
    Interaction with organics

POROSITY &   *Reduced by compaction*
PERMEABILITY

Figure 27.2 The principal controls operative during burial diagenesis.

Controlling factors

Intrinsic

Mineralogy is perhaps the most important control (Figure 27.2). If sediment has not been completely altered to dLMC, then it will have more diagenetic potential than LMC limestones in the burial realm. On the other hand, if the sediment has spent a long period of time in the marine or meteoric realm and is well cemented, then it will largely resist the mechanical effects of burial. The most critical intrinsic attributes are as follows (Figure 27.2):

- *Purity*: Limestones that contain a few percent clay minerals or unstable organic matter are more susceptible to the effects of compaction than pure limestones. It seems that ~10% silt or clay content determines which chemical solution fabrics will be generated, but it is not certain why.
- *Grain size*: Fine sediments compact more readily than coarse sediments. Lime muds, for example, can undergo so much physical compaction that they become diagenetic wackestones or packstones.
- *Previous diagenesis*: Early precipitated cement or dolomitization will increase the bearing strength of sediment so that it compacts less readily.

Porosity and permeability: Coarser-grained sediments will transmit fluids in larger volumes and at higher flow rates than their finer-grained equivalents, and therefore result in more rapid burial diagenesis.

Extrinsic

The most important extrinsic factors that are inherent in the burial system are as follows:

- *Pressure*: Pressures that act on carbonate rocks in the subsurface include: (1) lithostatic (generally vertical, transmitted by gravity through the solid framework); (2) hydrostatic (generally vertical, transmitted by gravity through the pore-system); and (3) linear (tectonic directed pressure). Abnormally high pore-fluid pressure (where the water cannot escape) can inhibit burial diagenesis by propping open pore spaces and can shut off export and import of pore waters. Such over-pressuring can be caused by rapid sedimentation or confinement below aquicludes such as evaporites, shales, or hardgrounds.
- *Temperature*: With most basins having temperature gradients on the order of 15–35°C km^{-1}, increased temperatures during burial can lead to: (1) calcite precipitation; (2) release of water from hydrated minerals (e.g., gypsum, certain clays, iron hydroxides such as limonite); or (3) sedimentary organic matter being changed from unstable to more stable forms and, in the process, releasing carboxylic and other organic acids that are capable of dissolving carbonates.
- *Pore-fluid chemistry*: The composition of pore fluids, commonly modified by the ions generated by dissolution, will dictate the types of minerals that can be precipitated during burial diagenesis. The introduction of liquid hydrocarbons during burial effectively prevents cementation.
- *Burial time*: The longer sediment remains in this environment, the more drastic the diagenetic changes will be.

Processes and products

The plethora of controls on burial diagenesis can be intimidating when trying to resolve the burial diagenetic history of any given limestone. Fortunately, the sum of all these processes usually results in only a few major changes, namely: (1) physical (mechanical) compaction; (2) chemical compaction; (3) LMC cement precipitation; and (4) water-controlled dissolution.

Physical compaction

As sediments compact under load, they dewater. They lose porosity and decrease in thickness as sedimentary particles and sedimentary structures are modified or rearranged (Figure 27.3).

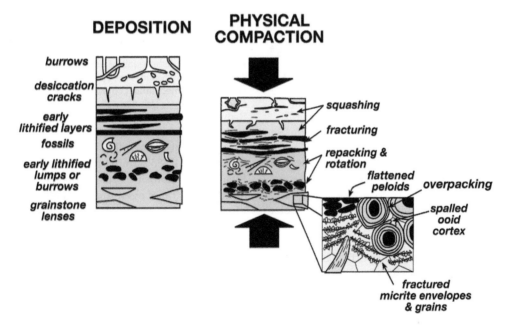

Figure 27.3 The principal products of physical (mechanical) compaction during burial diagenesis. Source: Choquette and James (1987). Reproduced with permission of the Geological Association of Canada.

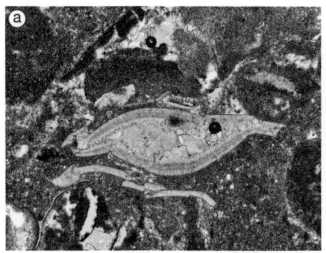

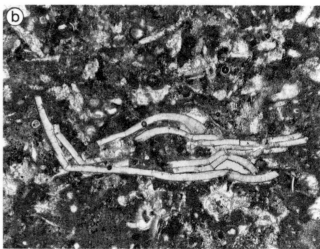

Figure 27.4 Mechanical compaction. Thin-sections (PPL impregnated with blue epoxy) of Mississippian packstones from the subsurface of the Williston Basin, Saskatchewan. Ostracod skeletons have been broken by overburden pressure and part of the remaining pore space filled by burial cement. Image widths (a) 2 mm; and (b) 3 mm. Photographs by W. Martindale. Reproduced with permission.

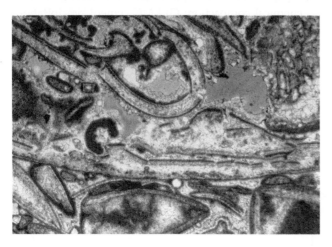

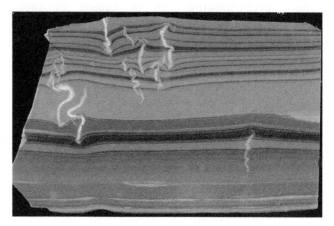

Figure 27.5 Mechanical compaction. Thin-section image (PPL impregnated with blue epoxy) of Miocene dolomitized biofragmental grainstone in the Gulf of Suez, Egypt. Micrite envelopes have been crushed and broken during mechanical compaction. Image width 4 mm.

Figure 27.6 Polished slab of Lower Ordovician argillaceous carbonate mudstone where spar-filled tension gashes have been crumpled by mechanical compaction, Western Newfoundland, Canada. Image width 8 cm. Photograph by M. Coniglio. Reproduced with permission.

Processes. Physical compaction can be visualized as taking place in three stages:

1. particle settling and repacking during the first meter or so of burial;
2. rearrangement of particles, dewatering, and compaction of muddy sediment until a self-supporting framework is achieved with porosities of ~40%; and
3. increasing overburden stress at grain contacts, resulting in particle deformation as ductile squeezing or brittle fracture and breakage takes place.

Experiments show that modern shallow-water sediments with 47–83% initial porosity compact to 30–65% of their original porosity. Thus, carbonates can compact under as little as 100 m of overburden to half their thickness, with losses of 50–60% of original pore volume.

Products. Physical compaction in mud-supported carbonates can lead to thinning of laminae, fracturing of early lithified layers, squashing of burrows, crumpled fenestrae, rotation of shells into the horizontal, flattened pellets, crushed grains, and the obliteration of desiccation cracks and fenestral pores (Figures 27.3–27.6). Physical

DEPOSITIONAL WACKESTONE

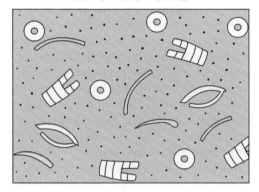

DIAGENETIC PACKSTONE

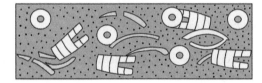

Figure 27.7 The effect of mechanical compaction on what was a depositional wackestone (top) that has been transformed into a diagenetic packstone by physical (mechanical) compaction. Source: Adapted from Choquette and James (1987). Reproduced with permission of the Geological Association of Canada.

compaction in grain-supported carbonates can lead to plastic deformation of grains, fractured and crushed grains (e.g., spalled ooids), and fractured micrite envelopes (Figure 27.5). As noted above, these processes can change the nature of the rock from a depositional wackestone to a diagenetic packstone (Figure 27.7), so be careful when interpreting such rocks as sediments! One of the most important overall results is stratigraphic thinning of up to 30% of the original thickness.

Chemical compaction

This process is of great importance and can result in an additional reduction in bed thickness of 20–35% following physical compaction. The limestone undergoing compaction serves as both a donor and receptor of carbonate cement (Figure 27.8).

Processes. The processes involve pressure due to load or tectonic stress. The stress is transmitted to and concentrated at contact points or surfaces between grains, crystals, or larger entities where the increased solubility of the stressed component may cause dissolution (Figure 27.8). This creates a microscopic solution film, a few water molecules thick, at the contact. Such a pressure-induced solubility increase sets up a chemical potential gradient that decreases away from the high-stress locus to areas of lower stress. Ions move down this gradient by solution transfer or diffusion and then are precipitated as new calcite cement at nearby sites where stresses and solubility are lower, or they are taken into the pore waters and transported to more distant sites. An analogy is the layer of water that forms beneath ice skates due to the localized pressure, which refreezes when the skater has passed by and the pressure is released. The difference in the subsurface is that the pressure is maintained.

Products. The most common products (Figures 27.9–27.11) are *stylolites* or *sutured seams*, which are jagged surfaces coated with insoluble clays, organic materials, or other minerals and made up of interlocking pillars, sockets, and variously shaped "teeth." Conversely, *non-sutured seams* are wispy surfaces that are formed in the same way as sutured seams but develop in sediments that have significant amounts (>10%) of clay, silt-sized particles, or organic matter. In some cases, they are so abundant as to impart a diagenetic bedding that is strikingly like depositional bedding.

Stylolites (Figure 27.11) typically have an insoluble residue at the top and bottom of each pillar. The amplitudes of a stylolite, which can vary from a few millimeters to 5–6 cm, can provide a measure of the minimum amount of the host rock that has been lost to dissolution. Only isolated stylolites occur in some carbonates, whereas in others there will be swarms of stylolites. Careful examination of the rock commonly reveals allochems that have been truncated by dissolution associated with the development of the stylolite. Finally, do not forget that each stylolite seam is formed by fluid migration and can continue to act as a conduit for horizontal fluid movement. Transmission of Mg-rich fluids can promote dolomitization along the seams. Conversely, the concentration of clays and other insolubles along the stylolite can act as a barrier to vertical fluid movement.

Non-sutured seams (Figure 27.12) can be pervasive or localized along specific argillaceous horizons. Diagenetic nodules are particularly common in sections of alternating limestone and shale. Dissolution of these limestone beds can leave only remnants of the original carbonate (Figure 27.12c, d).

Influencing factors.   The important factors are: (1) burial depth; (2) pore water composition; (3) metastable mineralogy; (4) clay minerals; (5) early dolomitization; and (6) liquid hydrocarbons. Pressure solution begins at a few tens of meters depth in the form of particle to particle contact dissolution, but it is generally accepted that significant pressure solution begins at depths of ~300 m. Whenever aragonite and Mg-calcite persist into the burial realm, they will be susceptible to pressure solution. By contrast, LMC limestones will have comparatively lower diagenetic potential. Clay minerals appear to foster pressure solution as discussed above.

Early dolomitization can inhibit the formation of stylolites because of the crystal-supported framework. Liquid hydrocarbons introduced into the system relatively early also tend to shut off the process.

Burial cementation

Shallow burial (marine) cements

These precipitates include displacive fibrous calcite (colloquially referred to as "beef" and "cone in cone" structures) that is developed particularly well in argillaceous mudstones during shallow burial just below the seafloor.

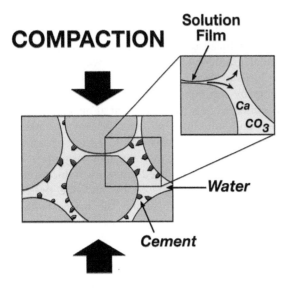

Figure 27.8 The effect of chemical compaction at grain contacts where, because of stress, carbonate is dissolved and put into solution, only to be precipitated locally at grain surfaces where pressure is less. Source: Adapted from Choquette and James (1987). Reproduced with permission of the Geological Association of Canada.

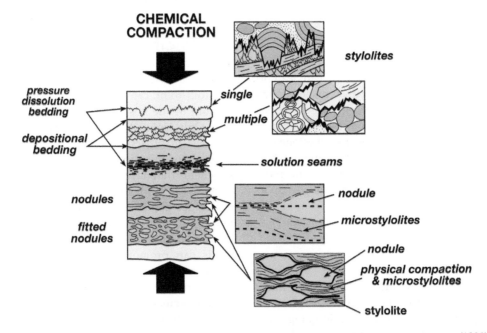

Figure 27.9 The processes and products of chemical compaction. Source: Adapted from Choquette and James (1990). Reproduced with permission of the Geological Association of Canada.

STYLES OF PRESSURE-SOLUTION

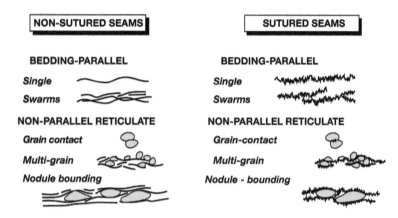

Figure 27.10 The different styles of pressure solution highlighting the difference between sutured seams (stylolites) and non-sutured solution seams. Source: Adapted from Wanless (1979) and Choquette and James (1990).

Deeper burial cements

Such cements are generally clear, coarse, bladed-prismatic (Figure 27.13) or coarse mosaic spar commonly described as "blocky." Cements are generally precipitated from reducing fluids and are therefore ferroan (>500 ppm Fe) and enriched in Mn^{2+} but impoverished in Sr^{2+} compared to earlier cements. CL is usually dull because even though the Mn^{2+} values can be high, the accompanying high Fe^{2+} in the crystals tends to quench the luminescence (Figure 22.4). Crystals can show compositional zonation along with aqueous fluid and hydrocarbon inclusions that reflect the composition of pore fluids, which can change as the basin matures.

Coarse mosaic calcite. Plane-sided equant (blocky) crystals from tens to hundreds of millimeters in size that can be ferroan and are typically zoned (Figure 27.14).

Bladed-prismatic calcite. Elongate scalenohedral crystals are generally found growing either directly on grains or on earlier cements.

Poikilotopic calcite. These large crystals enclose other grains and have a dull CL because of the high Fe:Mn ratio which quenches luminescence.

Cement isotope geochemistry

Calcite $\delta^{13}C$ values do not change much (they decline only slightly) with burial because the system is rock-dominated but the $\delta^{18}O$ values decrease sharply by as

much as 10–15‰. Burial cements are almost universally isotopically low in $\delta^{18}O$ values compared to earlier marine and meteoric cements (Figure 27.15). The decrease in $\delta^{18}O$ is primarily a function of increasing water temperature; the values can, for example, be as much as 8‰ different for cements precipitated about 70°C apart. Clumped isotopes are increasingly being used in an attempt to more precisely constrain the temperatures of cement precipitation.

Criteria for recognition of burial cements

The best criteria for determining if a cement is of burial origin are when the cement:

- succeeds obvious synsedimentary marine or meteoric cements;
- heals micrite envelopes broken by physical compaction;
- surrounds particles fractured by physical compaction;
- overgrows stylolites produced by chemical compaction;
- surrounds fractures filled with earlier cement; and
- encloses or post-dates hydrocarbon alteration products such as asphalt or pyrobitumen.

Origin of burial cements

Factors that promote burial cementation include: (1) most (but not all) subsurface pore waters are supersaturated with respect to calcite; (2) increasing temperature (but remember that increasing pressure counteracts this effect somewhat); and (3) CO_2 outgassing where

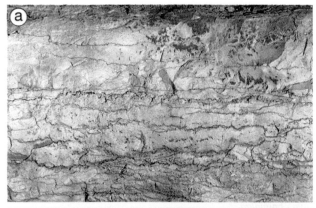

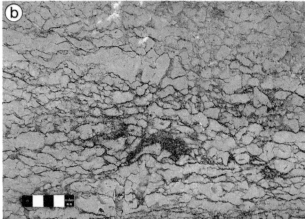

Figure 27.11 Stylolites (sutured seams). (a) Numerous stylolitic seams in lagoonal mudstones in the Upper Devonian, Canning Basin, Western Australia. Image width 0.5 m. (b) Innumerable stylolites in Middle Ordovician wackestone, western Newfoundland, Canada. Scale divisions 1 cm. (c) Stylolites choked with dolomite, Middle Ordovician wackestone, western Newfoundland. Image width 25 cm.

subsurface fractures connect the deep subsurface with the shallow near-surface and so abruptly reduce pore pressure while dropping the PCO_2 and causing carbonates to precipitate. Factors that inhibit precipitation are mostly the entry of hydrocarbons into pore systems.

Autocementation

There is abundant evidence that pressure solution in partially closed systems along solution seams might be the main agent for cementation in limestones undergoing burial diagenesis. In this case, carbonate dissolved along the seam is reprecipitated locally, making these zones sites of low porosity.

Water-controlled dissolution

Most burial dissolution is ascribed to formation waters with anomalously high concentrations of CO_2, supplied by the thermal decarboxilization of organic matter (kerogen) or hydrous pyrolysis reactions between organic carbon and oxygen in water to yield CO_2 or organic acids.

Another source of solution porosity can be attributed to H_2S or CO_2 generated during burial and dissolving in pore waters to produce sulfuric and carbonic acids. The dissolution can be fabric-selective or non-fabric-selective, with the former being related to earlier diagenetic events. More often, however, dissolution is non-fabric selective, (Figure 27.16). Because of differences in crystal size, composition, microporosity, and inclusions of organic matter, the solubilities of different components will be slightly different; biogenic calcite > calcitized aragonite or Mg-calcite > dLMC calcite cement. This is also demonstrated where biogenic LMC is typically replaced by silica but calcitized skeletons are not. Much microporosity (pores <0.03 mm in diameter) is developed in the burial realm.

In summary, it is clear that: (1) solution porosity can be generated in the deep subsurface; and (2) organic compounds are implicated.

Changes in porosity with depth

Most information comes from Mesozoic and Cenozoic chalks where there is an exponential decrease in porosity with depth (Figure 27.17). Chalks were used because of their uniform grain-size but also because they were monomineralic (LMC) at the time of deposition and so had very low diagenetic potential. Lithification was therefore accomplished almost entirely by burial processes. Calculations suggest that: (1) a reduction in volume from 80 to 75% in the top ~1 m of the sediment column is

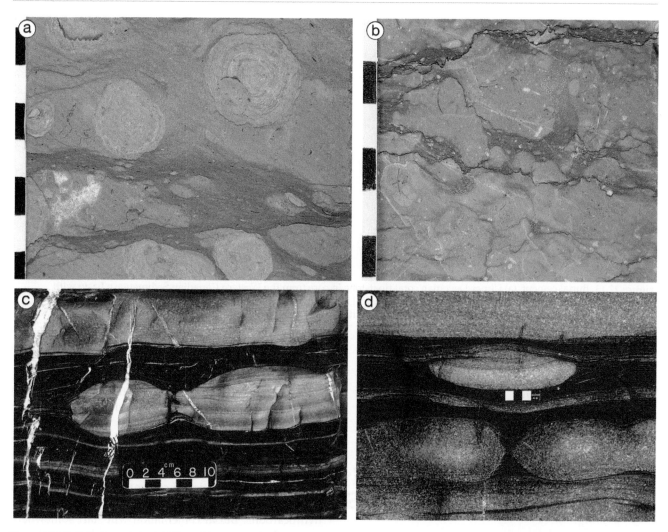

Figure 27.12 Solution seams. (a) Upper Devonian oncolite floatstone in core from the Western Canadian Sedimentary Basin with a central zone of non-sutured solution seams. Centimeter scale. Photograph by W. Martindale. Reproduced with permission. (b) Upper Devonian mudstone and floatstone in core from the Western Canadian Sedimentary Basin with both sutured (stylolites) and non-sutured solution seams. Centimeter scale. Photograph by W. Martindale. Reproduced with permission. (c) Lower Ordovician limestone and shale from Western Newfoundland, Canada, where the original cross-laminated packstone (center) has been dissolved away and only nodules remain. Centimeter scale. (d) Upper Cambrian limestone and shale from Western Newfoundland, Canada, where the limestone has been dissolved by chemical compaction. Centimeter scale.

accomplished in ~50,000 years; (2) a further reduction from 75 to 60% volume by physical compaction needs about 200–500 m of overburden and takes roughly 10 my; and (3) reduction from 60 to 40%, occurring to depths of ~1000 m, takes about 120 my.

By applying these concepts as well as previous diagenetic history, a series of situations can be predicted (Figure 27.17): (1) a normal curve; (2) a curve for carbonate sediment that was affected very early by meteoric or marine diagenesis; (3) a curve expected if solution porosity was created after burial and continued to considerable depth; and (4) a curve caused by early buildup of high fluid pressure.

Burial diagenetic models

Basinal burial domain

These sediments pass directly into the burial realm (Figure 27.18), carrying marine pore waters into the subsurface. These waters are then modified by reactions with burial fluids, organic matter, and inorganic sediment constituents. The system is partially closed hydrochemically because pore waters are being expelled upward and toward the basin margins.

Most sediments will remain unlithified (except for hardgrounds) and are either calcite (in the case of chalks)

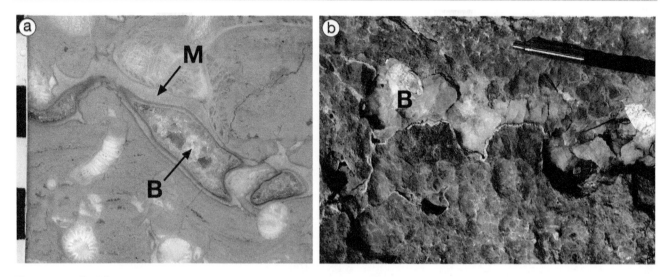

Figure 27.13 Burial cement. (a) Upper Devonian coral floatstone in core from the Western Canadian Sedimentary Basin with synsedimentary marine fibrous cement (M) and clear euhedral burial calcite cement (B). Centimeter scale. Photograph by W. Martindale. Reproduced with permission. (b) Void filled with clear burial calcite cement (B) in the Permian reef complex of New Mexico, USA. Pen is 10 cm long.

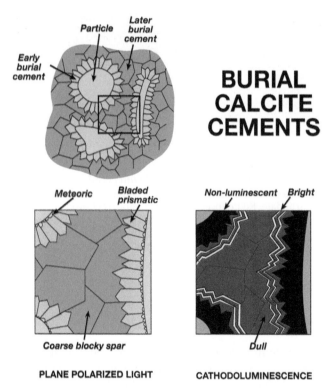

Figure 27.14 The two major types of burial cements: bladed prismatic and coarse blocky spar. Source: Adapted from Choquette and James (1987). Reproduced with permission of the Geological Association of Canada.

or neritic Mg-calcite and aragonite, the former with little diagenetic potential and the latter with great diagenetic potential.

The main trends, after expulsion of pore waters, are physical compaction initially and later pressure solution compaction, long-term reduction in thickness and pore volume, production of cement via pressure solution, and thermochemical conversion of organic matter to hydrocarbons. Burial cementation will start somewhere below 200–250 m and be dominant below 1000 m. Pressure-solution bedding will be common.

Shelf-platform burial domain

These neritic deposits either pass directly into the burial realm after undergoing synsedimentary marine diagenesis, or go into the meteoric realm and then downward into the burial realm (Figure 27.18). Regardless, they have been substantially altered before they are buried. Consequently, sediments are far more varied in composition and also include marginal marine terrigenous clastics and evaporites of all sorts.

Pores are typically filled with meteoric groundwater before burial, which may be replaced by other fluids of varying composition during burial. The dominant trends will be the evolution of pore waters from a varied marine–freshwater suite to mainly saline or hypersaline brines.

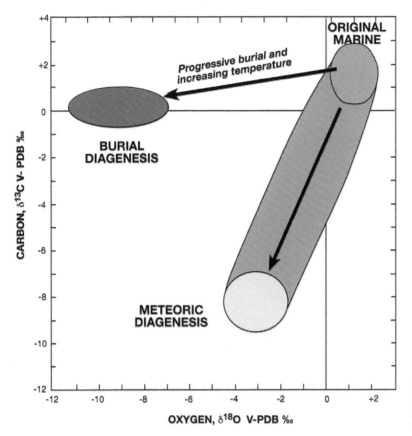

Figure 27.15 Simplified stable isotope cross-plot of $\delta^{18}O$ versus $\delta^{13}C$, showing the different trends in values that take place during meteoric and burial diagenesis.

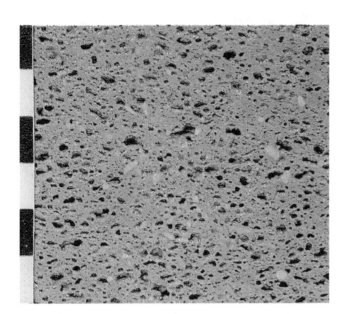

Figure 27.16 Mississippian paleocalcrete grainstone in core from the Williston Basin Saskatchewan, depicting extensive burial dissolution and generation of vuggy porosity. Centimeter scale. Photograph by W. Martindale. Reproduced with permission.

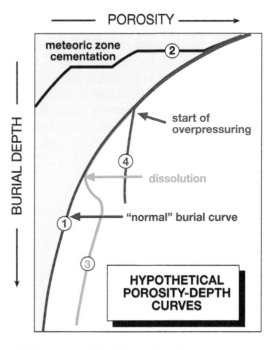

Figure 27.17 A synthetic burial curve based on numerous subsurface studies illustrating the general decrease in porosity with burial depth and how this trend can be altered by early meteoric diagenesis, onset of over-pressuring, and late-stage dissolution. Source: Adapted from Choquette and James (1987). Reproduced with permission of the Geological Association of Canada.

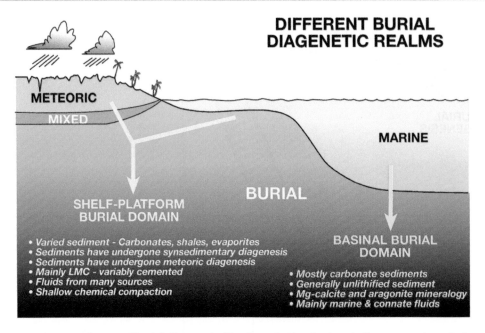

DIFFERENT BURIAL DIAGENETIC REALMS

METEORIC

MIXED

MARINE

SHELF-PLATFORM BURIAL DOMAIN

BURIAL

BASINAL BURIAL DOMAIN

- *Varied sediment - Carbonates, shales, evaporites*
- *Sediments have undergone synsedimentary diagenesis*
- *Sediments have undergone meteoric diagenesis*
- *Mainly LMC - variably cemented*
- *Fluids from many sources*
- *Shallow chemical compaction*

- *Mostly carbonate sediments*
- *Generally unlithified sediment*
- *Mg-calcite and aragonite mineralogy*
- *Mainly marine & connate fluids*

Figure 27.18 The two different pathways of burial diagenesis: (1) where shelf and other shallow-water deposits have been subject to synsedimentary marine or meteoric diagenesis, or both, before entering the burial realm; and (2) where basinal sediments have simply been buried with little early diagenesis.

Sediments entering the burial realm will consist of a variety of mineralogies. Resistance to physical compaction will be highly varied because some deposits are lithified and others are not. Processes should involve physical compaction, variable cementation from pressure solution and externally-derived fluids, dehydration of minerals such as gypsum and smectite, and in most cases, a reduction in porosity. Burial diagenetic features will be determined largely by the previous diagenetic history and will be highly variable. Pressure solution features seem to develop shallower than in basinal carbonates.

Paragenesis via cement stratigraphy

Since grainy sediments are cemented by different calcite cements in succeeding diagenetic environments, it is possible that the diagenetic history of a limestone could be recorded in the sequence of cementation events (Figure 27.19). This is called "cement stratigraphy" and has been used with great success, not only to work out alteration history but also to deduce paleohydrology. This is most successful when all petrographic techniques, such as staining and CL, are utilized. The example in Figure 27.20 is a lower Cambrian reef limestone where the inside of an individual archaeocyath (calcareous sponge shaped like an ice-cream cone) has been encrusted by the calcimicrobe *Renalcis*, then encrusted with synsedimentary fibrous

CEMENT STRATIGRAPHY

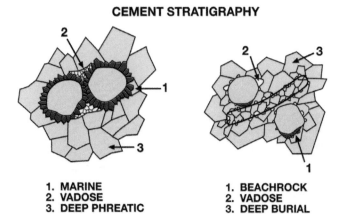

1. MARINE
2. VADOSE
3. DEEP PHREATIC

1. BEACHROCK
2. VADOSE
3. DEEP BURIAL

Figure 27.19 Cement stratigraphy: how the sequence of different cements can be used to unravel the diagenetic history of a limestone that is either now in the burial realm, or has been exposed after residing in the burial realm for a long period of time. Source: Choquette and James (1990). Reproduced with permission of the Geological Association of Canada.

cement on the seafloor, followed by clear meteoric cement, and, when buried, the remaining pore space was filled by the precipitation of burial dLMC. We can therefore read the history of this limestone from the seafloor to the barren primitive Cambrian landscape to the deep burial realm, all in a thin-section! Now it is exposed along the coast of Labrador will that exposure leave a signature? Time will tell.

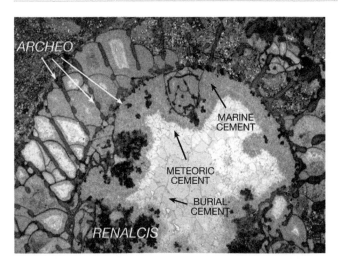

Figure 27.20 Cement stratigraphy. Thin-section (PPL, stained blue with potassium ferricyanide to emphasize Fe-rich calcite) lower Cambrian reef limestone, Labrador, Canada. The inside of an individual archaeocyath (calcareous sponge shaped like an ice-cream cone) that has locally been encrusted by the calcimicrobe *Renalcis*, then covered with synsedimentary fibrous calcite cement on the seafloor, followed by clear meteoric cement. When buried, the remaining pore space was filled by the precipitation of burial LMC. Image width 9 mm.

Further reading

Bathurst, R.G.C. (1987) Diagenetically enhanced bedding in argillaceous platform limestones: stratified cementation and selective compaction. *Sedimentology*, 34, 749–778.

Bathurst, R.G.C. (1995) Burial diagenesis of limestones under simple overburden; stylolites, cementation and feedback. *Bulletin de la Societe Geologique de France*, 166, 181–192.

Choquette, P.W. and James, N.P. (1987) Limestone diagenesis: the deep-burial environment. Geoscience Canada, v. 14, p. 3–35.

Choquette, P.W. and James, N.P. (1990) Limestone: The burial diagenetic environment. In: McIlreath I. and Morrow D. (eds) *Diagenesis*. St John's, ND, Canada: Geological Association of Canada, Reprint Series 4, pp. 75–111.

Dickson, J.A.D., Montanez, I.P., and Saller, A.H. (2001) Hypersaline burial diagenesis delineated by component isotopic analysis, late Paleozoic limestones, West Texas. *Journal of Sedimentary Research*, 71, 372–379.

Grover, G., Jr. and Read, J.F. (1983) Paleoaquifer and deep burial related cements defined by regional cathodoluminescent patterns, Middle Ordovician carbonates, Virginia. *American Association of Petroleum Geologists Bulletin*, 67, 1275–1303.

Jones, G.D. and Xiao, Y. (2006) Geothermal convection in the Tengiz carbonate platform, Kazakhstan: Reactive transport models of diagenesis and reservoir quality. *AAPG Bulletin*, 90, 1251–1272.

Machel, H.G., Cavell, P.A., and Patey, K.S. (1996) Isotopic evidence for carbonate cementation and recrystallization, and for tectonic expulsion of fluids into the Western Canada Sedimentary Basin. *Geological Society of America Bulletin*, 108, 1108–1119.

Meyers, W.J. (1978) Carbonate cements: their regional distribution and interpretation in Mississippian limestones of southwestern New Mexico. *Sedimentology*, 25, 371–400.

Meyers, W.J. and Hill, B.E. (1983) Quantitative studies of compaction in Mississippian skeletal limestones, New Mexico. *Journal of Sedimentary Petrology*, 53, 231–242.

Railsback, L.B. (1993) Intergranular pressure dissolution in a Plio-Pleistocene grainstone buried no more than 30 meters: Shoofly oolite, Southwestern Idaho. *Carbonates and Evaporites*, 8, 163–169.

Ricken, W. (1987) The carbonate compaction law: a new tool. *Sedimentology*, 34, 571–584.

Scholle, P.A. (1977) Chalk diagenesis and its relation to petroleum exploration: oil from chalks, a modern miracle? *American Association of Petroleum Geologists Bulletin*, 61, 982–1009.

Shinn, E.A. and Robbin, D.M. (1983) Mechanical and chemical compaction in fine-grained shallow-water limestones. *Journal of Sedimentary Petrology*, 53, 595–618.

Wanless, H.R. (1979) Limestone response to stress: pressure solution and dolomitization. *Journal of Sedimentary Petrology*, 49, 437–462.

CHAPTER 28
DOLOMITE AND DOLOMITIZATION

Frontispiece The Sella Platform, a completely dolomitized carbonate edifice where Count Dolomieu first described the mineral dolomite in 1791. Note the road and chateaux for scale.

Origin of Carbonate Sedimentary Rocks, First Edition. Noel P. James and Brian Jones.
© 2016 Noel P. James and Brian Jones. Published 2016 by John Wiley & Sons, Ltd.
Companion website: www.wiley.com/go/james/carbonaterocks

Introduction

In 1791, Count Déodat du Dolomieu described rocks he found in northeastern Italy (Frontispiece) that looked like limestone but did not effervesce in weak acid. In the following year, the mineral was formally named *dolomie*, or *dolomite* in English. Although one of the oldest mineral names in geology, dolomite still remains somewhat enigmatic and poorly understood. Even usage of the name can be problematic because dolomite is sometimes used for both the mineral and the rock that is formed largely of the mineral dolomite. In this chapter, dolomite refers to the mineral and dolostone refers to a rock formed of more than 75% dolomite. Dolostones are as widespread as limestones in the geological record and, from an economic perspective, probably even more important. Globally, there are many large hydrocarbon reservoirs (oil and gas) and mineral deposits (mainly lead–zinc) that are hosted in porous and permeable dolostones of all ages.

The "dolomite problem" is one of the oldest conundrums in geology. Specifically, how can we explain the origin of thick, pervasively dolomitized successions of rocks that cover thousands of square kilometers? At the core of this problem are the facts that: (1) dolomite has never been synthesized in the laboratory under low-temperature low-pressure abiogenic conditions; and (2) dolomite is rarely precipitated from modern seawater even though ocean water is supersaturated with respect to the mineral. By necessity, therefore, explanations of dolomite formation rely largely on interpretations of the dolostones themselves. At present there is a consensus that dolomite can form early in synsedimentary (authigenic) situations, during early diagenesis in the relatively shallow subsurface, or during late diagenesis under deep burial conditions.

Scientific approach

The way in which we investigate limestones and dolostones is not the same. This is because the birthplace of limestone and the processes by which it changes into a rock, as illustrated in the previous chapters, can mostly be visited and investigated with ease. By contrast, the cradle of dolomite genesis is shrouded in mystery; dolomite should be forming everywhere today but it is not, even though the chemistry seems simple. Consequently, we cannot use the same methods as we did to solve the origin of limestones and must rely on different techniques, mostly inference and deduction, instead of direct observation.

As a result of these problems, this chapter and Chapters 29 and 30 are organized in a somewhat different way than the previous chapters discussing limestone. This chapter outlines the attributes of dolomite the mineral and dolostone the rock. Recognizing that there is not one universal origin of dolomite, Chapter 29 describes current thinking about the different ways that these rocks might form and the proposed models that seem to explain most deposits. Chapter 30 then brings together all of these attributes with our current understanding of dolomite geochemistry to constrain the ways in which different dolomites in geologic history could have formed. Unlike limestone, whose geohistory can sometimes be resolved in a single thin-section, unraveling dolostone genesis involves integration of all possible attributes. Understanding dolomite genesis is a formidable challenge!

Dolomite: the mineral

Dolomite is $CaMg(CO_3)_2$ where equal amounts of Ca and Mg ions bond with CO_3 radicals to form a mineral with a hexagonal crystal structure (Figure 2.2). Although crystallographically similar to calcite, dolomite has 50% of the bivalent cation sites occupied by Mg^{2+} instead of Ca^{2+} (Figure 28.1). Furthermore, dolomite has a tripartite

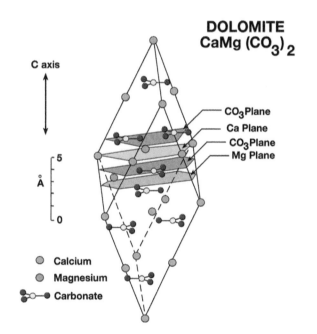

Figure 28.1 Diagram of four rhombohedral unit cells of dolomite showing its layered superstructure perpendicular to the c-axis direction, composed of regularly alternating sheets of calcium, magnesium, and carbonate (also see Figure 28.2). Source: Adapted from Morrow (1990). Reproduced with permission of the Geological Association of Canada.

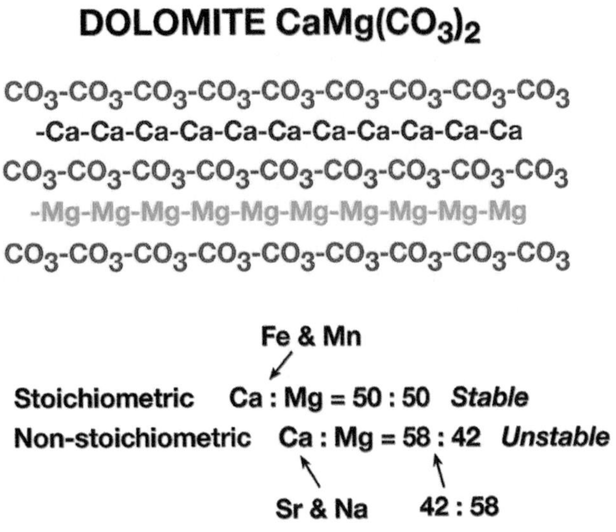

DOLOMITE CaMg(CO₃)₂

CO₃-CO₃-CO₃-CO₃-CO₃-CO₃-CO₃-CO₃-CO₃
-Ca-Ca-Ca-Ca-Ca-Ca-Ca-Ca-Ca-Ca-Ca
CO₃-CO₃-CO₃-CO₃-CO₃-CO₃-CO₃-CO₃-CO₃
-Mg-Mg-Mg-Mg-Mg-Mg-Mg-Mg-Mg-Mg
CO₃-CO₃-CO₃-CO₃-CO₃-CO₃-CO₃-CO₃-CO₃

Fe & Mn

Stoichiometric   Ca : Mg = 50 : 50   *Stable*
Non-stoichiometric Ca : Mg = 58 : 42   *Unstable*

Sr & Na   42 : 58

Figure 28.2 The layered chemical crystal structure of dolomite and chemical variability of the mineral. Fe and Mn can substitute for Ca in stoichiometric dolomite whereas Sr and Mg can substitute for Ca in non-stoichiometric dolomite with ratios ranging from 58:42 to 42:58, respectively.

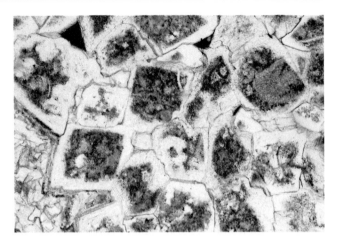

Figure 28.3 Euhedral dolomite crystals with cloudy cores and clear rims. Note ghosts of fossil fragments in some of the cloudy cores. Brac Formation (Oligocene), Cayman Brac. Image width 0.25 mm.

ordered structure with one layer of Mg ions overlain by a layer of CO_3 that is, in turn, overlain by a layer of Ca ions (Figures 28.1, 28.2). These layers are arranged normal to the c-axis. Stoichiometric dolomite is characterized by perfect ordering of the layers in addition to a uniform chemical composition, and so the formula can therefore be written as $Ca_{50}Mg_{50}(CO_3)_2$.

Not all dolomites are stoichiometric and some crystals can contain up to 52% Mg^{2+} and 62% Ca^{2+}. Most synsedimentary dolomites are Ca-rich and are generally referred to as "calcian dolomite." Although the precise manner in which the dolomite crystal accommodates this excess Ca^{2+} is unclear, it is probably achieved by having the excess Ca^{2+} replace some of the Mg^{2+} in the Mg-layers or by adding extra layers of Ca^{2+} ions into the crystal structure. Regardless, calcian dolomite is disordered and comparatively unstable relative to stoichiometric counterparts. The amount of excess Ca^{2+} is critical because the relationship between the Ca^{2+} level and the relative solubility or rate of dissolution is not linear, and even small amounts of excess Ca^{2+} can cause large increases in solubility. The percentage of Ca in the crystal structure therefore plays a critical role in the early diagenetic behavior of dolomite.

The term *protodolomite* was originally used for a poorly ordered dolomite generated in the laboratory and it was assumed that, with time, it would change to stoichiometric dolomite. Some non-stoichiometric synsedimentary dolomites have been called protodolomite, but use of the term is not encouraged and it is better to state the chemical composition of the mineral or call it calcian dolomite.

Non-stoichiometric dolomite is common in Cenozoic dolostone successions. Such dolomites can be divided into low-calcium calcian dolomite (LCD) with 51–55 %Ca and high-calcium calcian dolomite (HCD) with 56–62 %Ca. Individual crystals in these dolostones can be entirely LCD, HCD, or alternating zones of LCD and HCD. HCD is characterized by numerous growth defects that increase the solubility of these crystals. Determining the %Ca (mole % $CaCO_3$) in dolomite cannot be done by standard thin-section petrography. Instead, the %Ca must be determined by X-ray diffraction analysis of powdered material or electron microprobe analysis of polished samples.

Zoned dolomite crystals, which are common in dolostones of all ages, can be due to variations in: (1) the trace element (e.g., Fe, Mn) content of the crystal; or (2) the %Ca content of the dolomite (Figure 28.2). Such zoning may be: (1) oscillatory with repeated alternations of thin zones; or (2) simple with a large core that is encased by a thin cortical zone (Figure 28.3). Viewed in thin-section, the cores typically have a "dirty" appearance whereas the cortical zone is "clean." Zones defined by variations in the trace elements are generally easy to recognize through the use of CL microscopy. Zoning based on the % Ca content is only evident in SEM images of etched surfaces, on backscattered electron images, or from mapping of variations in the % Ca.

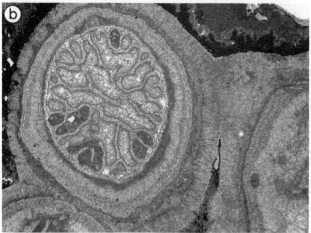

Figure 28.4 Mimetic dolomite, Miocene, Gulf of Suez, Egypt (see Figure 15.18). (a) Well-preserved but dolomitized scleractinian coral (*Montastraea* sp.), centimeter scale. (b) Thin-section (PPL) of mimetically replaced corals. Image width 5 mm. (c) Thin-section (PPL) of perfectly replaced coralline algae where the cells are clearly visible. Image width 3 mm.

Dolostone: the rock

The classification of dolostones is difficult because they are collectively characterized by a perplexing array of textures and fabrics. As a result, the terms applied to dolostones are: (1) general; (2) specific to certain types of dolostones; or (3) embodied in different classifications.

General terms

Dolostones, formed by replacement of precursor limestones, can be divided into *fabric-retentive* and *fabric-destructive* types. In fabric-retentive dolostones, the fabrics of the original limestones are completely or partially inherited with the allochems and cements still being evident. They range from mimetic (excellent retention of original fabric) to preferential replacement of specific components such as grains, cements, or burrow fills (Figure 28.3). The fabrics of the precursor limestones in fabric-destructive dolostones are lost because they were destroyed as the dolomite replaced the original calcitic components.

Specific terms

Some types of dolomite have been given specific names because they are widespread, highly distinctive, or generally indicative of specific modes of formation. Among these are *mimetic* dolomite, *limpid* dolomite, and *saddle* dolomite.

Mimetic dolomite. The original limestone fabrics are completely preserved (fabric-retentive) because of mimetic replacement whereby the original components were replaced by dolomite nanocrystals that cannot be resolved with a petrographic microscope (Figure 28.4). Preservation of the original fabrics is so perfect that the rocks still seem to be limestones and their true composition is only revealed with acid testing, staining, or X-ray diffraction. Such dolostones can be classified according to either Folk's or Dunham's limestone classification schemes (see Chapter 4).

Limpid dolomite. Limpid dolomite is characterized by clear crystals that look like ice cubes and are usually produced by clear overgrowths on inclusion-rich cores

(Figures 28.3, 28.5). Clear rims often interlock with neighboring crystals, forming a porous but rigid framework and producing a rock with a sucrosic (sugary) texture.

Saddle dolomite. Saddle dolomite (also known as *baroque* dolomite; Figure 28.6) is so named because warping of the lattice produces saddle-like curved crystals. The crystals are characterized by: (1) curved crystal faces; (2) undulatory extinction patterns in thin-section under cross-polarized light; and (3) crystal faces that are typically stepped (Figure 28.5d). Saddle dolomite, which can contain as much as 15 mole % $FeCO_3$, is commonly

associated with metallic sulfide ores, barite, fluorite, or hydrocarbons.

General classification schemes

Two main schemes have been proposed for the classification of dolostones: one based on crystal size and the other based on fabric attributes. The textural classification focused on crystal size is elegant because it is simple, easy to apply, and carries no genetic connotations. Under this scheme, dolostones are divided into: very finely crystalline (<100 μm long); finely crystalline (100–250 μm long);

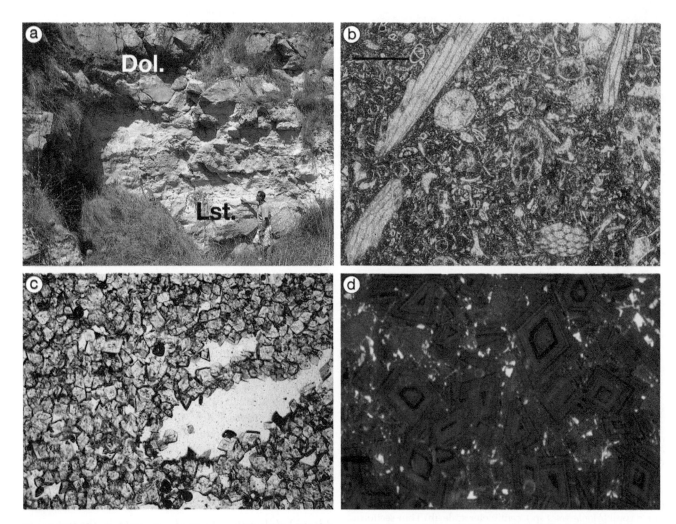

Figure 28.5 Fabric-destructive medium crystalline dolomite, Oligocene, South Australia. (a) Outcrop image illustrating lower limestone (Lst) and upper dolomite (Dol). (b) Limestone, bryozoan floatstone to wackestone. Scale bar: 1 mm. (c) Dolomite has replaced limestone in (b), where the rock is composed of numerous euhedral planar crystals with no vestige of the original limestone. Image width 4.4 mm. (d) CL image illustrating strong crystal zonation. Image width 4.4 mm.

Figure 28.6 Saddle dolomite. (a) 'Zebra' saddle dolomite resulting from hydrofracturing and white sparry saddle dolomite in the pores, Upper Cambrian, Alberta, Canada. Centimetre scale. (b) Close-up of coarse crystals with conspicuous curved faces, Devonian subsurface core. Image width 3 cm. (c) Thin-section (PPL) impregnated with blue epoxy showing the curved crystal faces, Devonian subsurface core. Image width 2 cm. (d) Same thin-section (XPL) illustrating sweeping extinction of the saddle dolomite crystals. Photographs by W. Martindale. Reproduced with permission.

medium crystalline (250–1000 µm long); coarsely crystalline (500–1000 µm long); and very coarsely crystalline (>1000 µm long) without any consideration of crystal type or morphology. There is a general relationship between crystal size and the water from which it precipitates (Figure 28.6). Crystal size is largely related to the density of nucleation sites and the fluid saturation state (Table 28.1).

An alternative classification is one that reflects the shape of the crystals and their location in the dolostone (Figure 28.7). Medium and coarsely crystalline dolostones are composed of: (1) idiotopic dolomite that is

Table 28.1 The general relationship between dolomite crystal size versus number of nucleation sites and water saturation.

	Dolomite crystal size	
	Small crystal size	Large crystal size
Nucleation sites	High density	Low density
Water saturation	High	Low
	Results in large number of crystal per unit volume	Small number crystals per unit volume

PRACTICAL CLASSIFICATION OF DOLOMITE TEXTURES

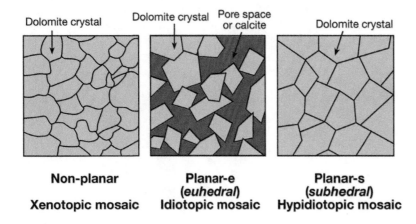

Figure 28.7 Dolomite classification. Source: Adapted from Sibley and Gregg (1987).

characterized by rhombic-shaped euhedral to subhedral crystals; and 2) xenotopic dolomite that is typified by non-rhombic, usually anhedral, crystals. The divisions in each group determine if the crystals are euhedral (E) or subhedral (S), if they formed cements (C), or if the rock displays a porphyrotopic texture (P) with dolomite crystals floating in a limestone matrix.

The limestone to dolostone transition

There is a continuous spectrum between limestones formed of 100% calcite and dolostones formed of 100% dolomite. Potentially, the fabrics evident in the transitional dolomitic limestones and calcareous dolostones can provide insights into the way replacive dolomite developed. As outlined earlier, the replacive dolomite can be considered fabric-retentive or fabric-destructive relative to its precursor limestone. Fabric-retentive dolomite can be either: (1) totally fabric-retentive (mimetic); (2) fabric-specific whereby certain fabrics are altered and others are not and remain limestone; or (3) multigeneration when the limestone is altered by two or more phases of dolomitization.

Mimetic dolomitization is described above and illustrated in Figure 28.4. Other fabric-specific dolostones, in the simplest sense, can be divided into particle-selective and matrix-selective. In the former, only the grains (e.g., fossils) are dolomitized and the matrix remains calcitic (Figure 28.8a, b). In the latter, the matrix is replaced by dolomite whereas the allochems remain calcitic. In some cases, the fabric-specific dolomite may target specific fossils (e.g., coralline algae), probably because their high-Mg skeletons facilitated dolomite development. Multigeneration fabric-specific

dolomitization usually occurs when a limestone undergoes an early phase of fabric-specific dolomitization and parts of the rock remain limestone. A second later phase of dolomite, commonly with a different crystal size, then alters the remaining limestone such that the resulting dolostone is characterized by two different dolomite crystal types (Figure 28.8d).

Fabric-independent destructive dolomitization develops in situations where dolomitization indiscriminately replaces the particles and the matrix. The resultant rock is a suite of dolomite crystals, commonly with good porosity, which is called *sucrosic* dolomite because of its similarity to crystalline sugar (Figure 28.5c).

Early diagenetic alteration of dolomite

For a long time, it was tacitly assumed that dolomite, once formed, remained immune from further diagenetic change. The nature of the crystals, the contained trace elements, their stable isotopic signatures, and radiogenic isotopic signatures were therefore usually considered indicative of the fluids from which the dolomite formed. Recent research, however, suggests that dolomite is also prone to early diagenetic modifications that are similar, in many respects, to those that affect calcite and aragonite (Figure 28.9). Such early diagenetic changes include: (1) modification of crystal shape; and (2) dissolution of the crystal core and its replacement by either calcite or new dolomite. Detrital dolomite should also be added to this list even though it is not, strictly speaking, a product of diagenesis. Finally, the trace element stable isotope and radiogenic isotope signature will also change.

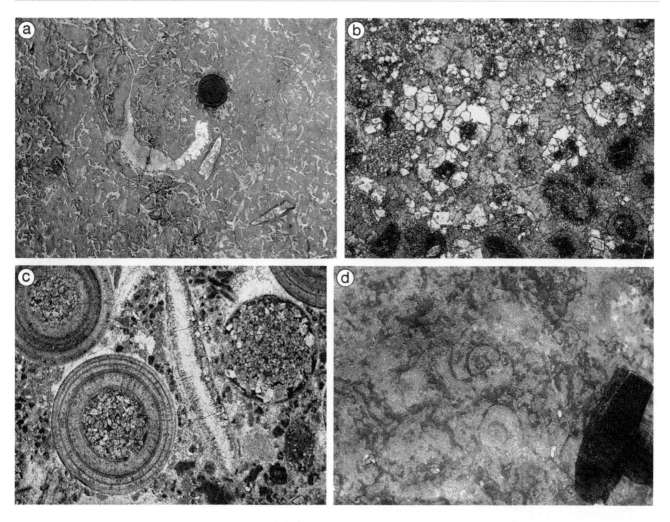

Figure 28.8 Fabric-retentive medium crystalline dolomite. (a) Bedding plane view of Lower Ordovician limestone where burrows, gastropods, and sponges are preferentially replaced by dolomite. Lens cap 50 mm. (b) Thin-section (PPL) calcite stained red where ooids were replaced by dolomite but all the other components are calcite, Upper Cambrian, Utah, USA. Image width 6 mm. (c) Thin-section (PPL) in which the peloids and peloid cores to ooids are dolomite (turquoise) but all the other components are calcite, lower Cambrian, Labrador, Canada. Image width 4.4 mm. (d) Outcrop surface of same unit as (a) but where the former limestone matrix is replaced by a later-stage dolomite (a true multistage dolostone). Sledgehammer head 10 cm long.

Modification of crystal shape

The possibility of dolomite crystal shape modification during early diagenesis has been largely overlooked because such changes are generally not apparent during standard petrographic analyses. Imaging that shows zoning by virtue of the % Ca in the dolomite can illustrate that some dolomite crystals are two-phased. The first phase is an euhedral crystal and the second phase is an irregular anhedral crystal overgrowth. The second phase produces a syntaxial dolomite cement overgrowth that typically leads to the development of an irregularly shaped crystal (Figure 28.10). As with the growth of syntaxial calcite cement overgrowths in limestones, the shape of the overgrowth, and hence the crystal, is usually a function of competition for space with neighboring crystals.

Some studies suggest that crystal shape is indicative of particular styles of dolomitization. This assertion must, however, be treated with caution because the crystal form may reflect diagenetic modifications rather than original crystal morphology.

Crystal-specific dissolution

The solubility of dolomite is directly related to the % Ca in the crystal. This is particularly common in "dirty-core–clear-rimmed" crystals whose cores are formed of

DOLOMITE STABILIZATION

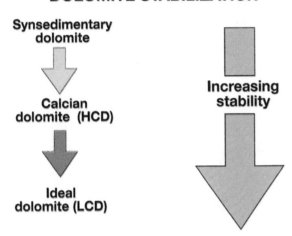

Figure 28.9 The progression of synsedimentary 'protodolomite' to sedimentary non-stoichiometric dolomite to stoichiometric dolomite with time.

Figure 28.11 SEM image of hollow dolomite crystal showing etched crystal interior, Cayman Formation (Miocene), Grand Cayman. Image width 70 μm.

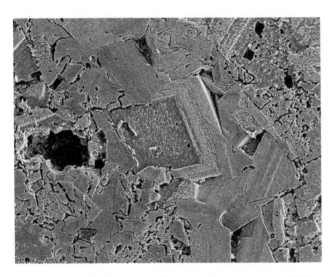

Figure 28.10 SEM image of polished surface etched with HCl, showing dolomite crystal with euhedral core that developed into an anhedral crystal as a result of preferential overgrowth on right side. Zoning in crystal reflects variations in the mole % CaCO₃ content of the dolomite. Cayman Formation (Miocene), Grand Cayman. Sample from 3 m below ground level. Image width 50 μm.

HCD and rims are formed of LCD. The HCD with its high concentration of growth defects and, in many cases, minute inclusions of calcite, aragonite, apatite, or all three, is far more susceptible to dissolution than the surrounding LCD that contains less growth defects and few, if any, non-dolomite inclusions. Minor breaches in the cortical zone provide aggressive fluids access to the core that can then dissolve rapidly while the outer cortical zone remains largely intact. Such

dissolution leads to the development of "hollow crystals" (Figure 28.11). Dolostones that are formed entirely of empty crystals are soft, commonly appear chalky, and can often be crushed between the fingers. The leached dolomite crystal cores can remain open, be filled with calcite cement, or be filled with younger dolomite.

Some dolomite crystals are characterized by oscillatory zoning with alternating zones of LCD and HCD (Figure 28.10). Selective dissolution of the HCD zones in these crystals can produce crystals that have intact LCD zones that alternate with hollow zones generated by dissolution of the HCD.

Core replacement by calcite (dedolomite). Calcite cement can partially to completely fill the dissolved crystal core, and the resultant crystal is commonly referred to as *dedolomite*. Some have argued that dedolomite is a bad term because it refers to the precursor mineral rather than the final mineral and should, instead, be referred to as calcitization.

Most calcitized dolomites result from core dissolution (Figure 28.11), formation of a hole, and later cement precipitation in that cavity. It has also been argued, however, that some dedolomite might develop without an intermediate cavity phase. In this case, fluids entering the core of a dolomite crystal are thought to dissolve the dolomite and immediately precipitate calcite, probably across a dissolution–precipitation front akin to that commonly invoked to explain the transformation of aragonite to calcite. Irrespective of the process involved, the resultant fabric is a rhomb-shaped crystal that has a thin outer zone

 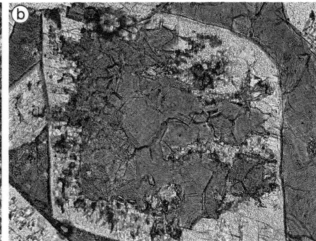

Figure 28.12 Thin-section images (PPL and stained with Alizarin red S) of Oligocene Limestone, Southern Australia. (a) Numerous dolomite crystals that have been dissolved and their hollow cores are the site of later calcite precipitation (red). Image width 1.4 mm. (b) Close-up of a single crystal, image width 200 μm.

of limpid dolomite (the cortical zones of the original crystal) surrounding a calcite core (Figure 28.12). In some cases, however, even the dolomite rim is lost and all trace of the dolomite crystals vanishes. Calcite cement can also be precipitated between the crystals. Limestone may therefore be altered to a dolostone and then changed back to a limestone. Although difficult to prove, it is entirely possible that some limestones can develop through this process without retaining any trace of the dolomite crystals.

Core replacement by dolomite (inside-out dolomite). Instead of calcite precipitation in the void following core dissolution, it is also possible that new dolomite could precipitate there. The dolomite that is precipitated inside the hollow dolomite crystals will therefore be younger than the dolomite that formed the rim of the original dolomite crystal (Figure 28.13). Crystals that develop in this manner are known as *inside-out dolomite*.

Recognizing inside-out dolomite crystals is difficult because it is only possible if: (1) there is a contrast in the compositions of the two phases of dolomite; and (2) there is evidence of a suture line that denotes the inner boundary of the original cortical zone (Figure 28.13). If the composition of the dolomite in the two zones is close to being identical, then recognition of this dolomite becomes nearly impossible.

Detrital dolomite

The concept that some dolomite crystals could be detrital in origin is commonly ignored even though it can have a profound impact on the interpretation of the dolomite. Numerous examples have been documented where fine

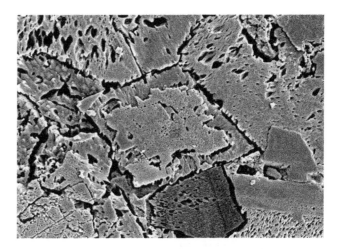

Figure 28.13 SEM image of polished surface etched with HCl showing inside-out dolomite (center of image) showing irregular core surrounded by rim. Zoning is evident because of differential etching due to varying mole % $CaCO_3$ content of the dolomite. Cayman Formation (Miocene), Grand Cayman, image width 14 μm.

dolomite crystals that formed on muddy tidal flats were deflated by arid winds and swept into the adjacent ocean. Alternatively, poorly lithified dolostones can be eroded and the crystals transported as particles by rivers to be eventually deposited in the shallow marine realm. This dolomite is not easy to differentiate from replacement crystals, but characteristics that can help include: (1) identification of abraded rhombs with truncated zones, especially using CL; (2) rare isolated rhombs or rhomb clusters; and (3) rhombs that are distinctively different from one another in the same sediment (Figure 28.14).

Figure 28.14 Detrital dolomite. (a) Thin-section image of a compound Pleistocene dolostone clast incorporated in Holocene sediment, Lacepede Shelf, southern Australia. Image width 0.7 mm. (b) CL image of individual Pleistocene dolomite crystals and clusters from Holocene sediment, Lacepede Shelf, southern Australia. Image width 0.8 mm. Photographs by Y. Bone. Reproduced with permission.

It is estimated that up to 10% of the dolomite in some neritic and basinal settings is detrital in origin. The critical point is that these crystals can easily act as "seed crystals" for later dolomite overgrowth; once the nucleus is provided, precipitation of a crystal overgrowth requires much less energy.

Detrital dolomite can be varied depending on the nature of the original source material, but is often inclusion- and Ca-rich and poorly ordered. Because the detrital cores are less stable relative to stoichiometric dolomite, it often undergoes recrystallization and replacement, making it difficult to identify in ancient examples.

Dolomite geochemistry

Introduction

There are traditionally three geochemical techniques that have proven useful in determining the origin of any given dolomite: (1) trace element concentrations, including imaging by CL; (2) stable isotope analyses; and (3) radiogenic isotopes. Clumped isotopes are becoming more useful but are not yet in the mainstream of analysis.

Geochemical data can be difficult to interpret and, given the variables involved, does not always provide conclusive evidence of dolomite genesis. The first difficulty in interpreting geochemical data results from the fact that dolomite has *not* been synthesized in the laboratory at low temperatures except in the presence of certain microbes. This makes determination of trace element partition coefficients difficult to determine. Second, calcian

TRACE ELEMENTS IN DOLOMITE

ELEMENT	IONIC RADIUS	PARTITION CO-EFFICIENT (substitution for Ca in dolomite)	MOVEMENT IN DIAGENESIS
Sr	1.13 A°	0.13 – 0.25	decrease
Ca	1.00 A°	1.00	–
Mn	0.83 A°	5 – 30	increase
Fe	0.78 A°	1 – 20	increase

Figure 28.15 Trace elements in dolomite.

dolomites (HCD) formed early are prone to alteration and recrystallization, and original geochemical signatures are lost during diagenesis. Third, in many cases we can only analyze the whole crystal or groups of crystals, and not individual precipitation zones.

Trace elements

Fe and Mn concentrations provide the most useful information regarding interpretations of dolomite genesis. Sr concentrations are also used, but with lesser success. This contrast can be attributed to the fact that Fe and Mn are gained, whereas Sr is lost during diagenesis (Figure 28.15).

Recall that the abundance of a trace element in a crystal is reflected by its partition coefficient and its abundance in the fluid (see Chapter 23). The partition coefficient will approach 1 as the ionic radius of the trace element approaches the ionic radius of the element it replaces (Figure 28.15). Depending on the element and the composition of the porewater, the coefficient will increase with rising water temperature. When precipitation is slow then values approach expected equilibrium, but if the precipitation rate is fast then the composition approaches that of the fluid: the "kinetic effect."

Strontium. Sr is relatively high in modern sediments, especially in organic and inorganic aragonitic precipitates (see Chapter 23). It is also comparatively high in synsedimentary dolomites. The partition coefficient for dolomite values range from 0.13 to 0.25. Essentially Sr, which is larger than Ca, remains behind in the water during dolomitization. Sr is not added because concentrations, although high in seawater, are low in most other diagenetic fluids.

Iron and manganese. In the reduced state, both Fe^{+2} and Mn^{+2} fit easily into the dolomite structure, producing partition coefficients greater than unity (Figure 28.15). Both ions are smaller than Mg (ionic radius 0.72 Å) and substitute easily for both Mg and Ca in the crystal lattice. This can result in relatively high concentrations in dolomite, even when their concentrations are relatively low in the surrounding water. They generally increase in concentration during diagenesis, but the degree to which they increase is strongly dependent on the oxidation state of the porewater.

Remember that Mn is the source of CL activation whereas Fe quenches or retards luminescence. The strength of luminescence depends on the Mn:Fe ratio in the dolomite, so the effects of Mn and Fe tend to counteract one another. These relationships can result in spectacular zonations (Figure 28.5d). The variations, however, depend on: (1) the oxidation state of the pore water; and (2) the H_2S content of the pore water (that governs the availability of the reduced Fe in the pore water for precipitation). There are other factors that are important, such as crystal growth rate and bulk fluid equilibrium, but are beyond the requirements for general interpretation.

Stable isotopes

Carbon and oxygen stable isotopes in carbonates were described in general terms in Chapter 23. The focus here is on their presence in dolomite and dolostone.

Oxygen. Most workers agree that the $\delta^{18}O$ value of dolomite precipitated from seawater at 25°C is around 3‰ VPDB. Calcite and dolomite precipitated from the same

fluid therefore have $\delta^{18}O$ values of 0‰ and +3‰ VPDB, respectively. The most significant factors governing oxygen isotope composition are the temperature of precipitation and the composition of the dolomitizing fluid. Also important is whether the system was closed or open, that is, rock-dominated or water-dominated. Although most oxygen comes from the fluid, settings with high rock–water ratios will have a significant contribution from the precursor $CaCO_3$ or dolomite. By contrast, those with a low rock–water ratio will derive most oxygen from the water. Finally, it must be remembered that if the isotopic composition of seawater has changed through time and all minerals are subject to isotopic exchange with pore water, the isotopic signature of dolomite must be interpreted with caution.

Carbon. The carbon isotopic composition of dolomite is strongly influenced by the nature of the precursor $CaCO_3$ that is being replaced because that is the source of most of the carbon that goes into dolomite. Calcite formed from seawater usually has a $\delta^{13}C$ value of between 0 and +4‰, but carbon can also be derived from organic matter. If the values are extreme then they are possibly caused by addition of carbon by sulfate reduction, methane oxidation (–22‰ to –30‰), or fermentation via methanogenesis (up to +15‰).

Clumped stable isotopes. Clumped stable isotopes have recently been used to constrain the temperature of dolomite formation. This is especially important when the dolostone is multi-staged.

Radiogenic isotopes

Strontium If dolomite formed in seawater, then the $^{87}Sr{:}^{86}Sr$ of the dolostone can be used to interpret the timing of dolomite formation. Seawater at any given time has a relatively distinctive $^{87}Sr{:}^{86}Sr$ ratio (Figure 23.14). The $^{87}Sr{:}^{86}Sr$ value of dolomite can therefore be compared to published seawater curves and may provide clues as to when it formed. This approach has proven particularly useful for the Cenozoic because $^{87}Sr{:}^{86}Sr$ values have gradually increased over the last 40 million years, which has the potential to impart a time-dependent unique value (Figure 23.14).

Strontium isotope ratios are also useful when interpreting the origin of late-stage dolomites. If the dolomitizing fluids have previously come into contact with continental basement or siliciclastic aquifers, then they will be enriched in ^{87}Sr. This is especially true for fault-associated saddle dolomites.

Further reading

The published scientific literature on dolomite and dolomitization is prodigious. This is partly because of the dolomite problem and the fact that dolostone is such a common hydrocarbon reservoir rock, as well as a host lithology for mineral deposits. The interested reader is first referred to books in the Introduction to Diagenesis. Lists in this chapter and the following are not complete, but cover many different aspects of the subject in both recent and older contributions.

Banner, J.L., Hanson, G.N., and Meyers, W.J. (1988) Water-rock interaction history of regionally extensive dolomites of the Burlington-Keokuk Formation (Mississippian): isotopic evidence. In: Shukla V. and Baker P.A. (eds) *Sedimentology and Geochemistry of Dolostones*. Tulsa, OK: SEPM, Special Publication no. 43, pp. 97–113.

Bone, Y., James, N.P., and Kyser, T.K. (1992) Synsedimentary detrital dolomite in Quaternary cool-water carbonate sediments, Lacepede shelf, South Australia. *Geology*, 20, 109–112.

Choquette, P.W. and Hiatt, E.E. (2008) Shallow-burial dolomite cement: a major component of many ancient sucrosic dolomites. *Sedimentology*, 55, 423–460.

Coniglio, M., James, N.P., and Aissaoui, D.M. (1988) Dolomitization of Miocene carbonates, Gulf of Suez, Egypt. *Journal of Sedimentary Petrology*, 58, 100–119.

Dawans, J.M. and Swart, P.K. (1988) Textural and geochemical alternations in Late Cenozoic Bahamian dolomites. *Sedimentology*, 35, 385–403.

Ferry, J.M., Passey, B.H., Vasconcelos, C., and Eiler, J.M. (2011) Formation of dolomite at 40-80°C in the Latemar carbonate buildup, Dolomites, Italy, from clumped isotope thermometry. *Geology*, 39, 571–574.

Given, R.K. and Wilkinson, B.H. (1987) Dolomite abundance and stratigraphic age: constraints on rates and mechanisms of Phanerozoic dolostone formation. *Journal of Sedimentary Petrology*, 57, 1068–1078.

Gregg, J.M. and Sibley, D.F. (1984) Epigenetic dolomitization and the origin of xenotopic dolomite texture. *Journal of Sedimentary Petrology*, 54, 908–931.

James, N.P., Bone, Y., and Kyser, T.K. (1993) Shallow burial dolomitization and dedolomitization of mid-Cenozoic, cool-water, calcitic, deep-shelf limestones, southern Australia. *Journal of Sedimentary Petrology*, 63, 528–538.

Jones, B. (1989) Syntaxial overgrowths on dolomite crystals in the Bluff Formation, Grand Cayman, British West Indies. *Journal of Sedimentary Petrology*, 59, 839–847.

Jones, B. (2004) Petrography and significance of zoned dolomite cements from the Cayman Formation (Miocene) of Cayman Brac, British West Indies. *Journal of Sedimentary Research*, 74, 95–109.

Jones, B. (2005) Dolomite crystal architecture: Genetic implications for the origin of the Tertiary dolostones of the Cayman Islands. *Journal of Sedimentary Research*, 75, 177–189.

Jones, B. (2007) Inside-out dolomite. *Journal of Sedimentary Research*, 77, 539–551.

King, R.J. (1995) Minerals explained 19: The dolomite group. *Geology Today*, 11, 105–107.

Land, L.S. (1982) *Dolomitization*. Tulsa, OK: American Association of Petroleum Geologists, Short Course Note Series no. 24.

Morrow, D.W. (1990) Dolomite – Part 1: The chemistry of dolomitization and dolomite precipitation. In: McIlreath I. and Morrow D. (eds) *Diagenesis*. St John's, ND, Canada: Geological Association of Canada, Reprint Series 4, pp. 113–123.

Murray, R.C. and Lucia, F.J. (1967) Cause and control of dolomite distribution by rock selectivity. *Geological Society of America Bulletin*, 78, 21–36.

Reeder, R.J. and Prosky, J.L. (1986) Compositional sector zoning in dolomite. *Journal of Sedimentary Petrology*, 56, 237–247.

Shukla, V. and Baker, P.A. (eds) (1988) *Sedimentology and Geochemistry of a Regional Dolostone*. Tulsa, OK: Society for Sedimentary Geology, Special Publication no. 43.

Sibley, D.F. and Gregg, J.M. (1987) Classification of dolomite rock textures. *Journal of Sedimentary Petrology*, 57, 967–975.

CHAPTER 29
DOLOMITIZATION PROCESSES AND SYNSEDIMENTARY DOLOMITE

Frontispiece A shallow marginal marine saline lake south of the Coorong Lagoon, South Australia, where synsedimentary dolomite was discovered for the first time in the desiccated carbonate muds almost a century ago. Note people for scale.

Origin of Carbonate Sedimentary Rocks, First Edition. Noel P. James and Brian Jones.
© 2016 Noel P. James and Brian Jones. Published 2016 by John Wiley & Sons, Ltd.
Companion website: www.wiley.com/go/james/carbonaterocks

Introduction

This chapter builds on the mineralogical and geochemical attributes of dolomite outlined in the previous chapter, focusing on: (1) the reasons why dolomite does not form easily from modern seawater, and how those constraints on dolomite precipitation can be overcome; and (2) the occurrence of dolomite in modern carbonate depositional environments.

Dolomite precipitation from waters of varying composition can be expressed as either of the following two equations:

$$Ca^{2+} + Mg^{2+} + 2CO_3^{2-} \leftrightharpoons CaMg(CO_3)_2 \qquad (28.1)$$

$$Mg^{2+} + Ca^{2+} + 4HCO_3^- \leftrightharpoons CaMg(CO_3)_2 + 2CO_2 + 2H_2O \qquad (28.2)$$

Dolomite replacement of calcite is generally expressed as:

$$2CaCO_3 + Mg^{2+} \leftrightharpoons CaMg(CO_3)_2 + Ca^{2+} \qquad (28.3)$$

What limits dolomite formation?

Modern seawater, with a Mg:Ca ratio of ~5:1, is supersaturated with respect to dolomite by about two orders of magnitude. The fact that dolomite is not being precipitated from modern seawater indicates that crystallization is inhibited by a variety of factors, such as mineral kinetics, crystallization rates, Mg hydration, CO_2 activity, and sulfate levels (Table 29.1). The abundance of dolostones throughout the geological record implies that these constraints could have been suppressed at various times in the past (see Chapter 18 on the Precambrian).

Mineral kinetics

Surface seawater is oversaturated with respect to various carbonate minerals including dolomite, calcite, and aragonite. Dolomite, however, is a highly ordered mineral and therefore requires more thermodynamic energy to form than is necessary for calcite or aragonite precipitation. Calcite and aragonite therefore precipitate more readily than dolomite in modern seawater.

This barrier to dolomite formation can be overcome if the Mg:Ca ratio of the fluid is raised to about 10:1 so that more Mg ions become available. This can, for example, be achieved through evaporation that leads to the preferential removal of Ca as LMC, HMC, aragonite, gypsum, or anhydrite, are precipitated. Conversely, dilution of seawater with fresh water will also overcome this kinetic problem. As seawater is diluted, decreased ion–ion interaction increases their availability for crystallization. Furthermore, dilution causes a reduction in the sulfate concentration, thus reducing the effect of an important inhibitor to dolomite precipitation.

Crystallization rates

Rapid precipitation from supersaturated solutions favors the formation of calcium-rich dolomite (HCD) because there is insufficient time to allow segregation of the

Table 29.1 Conditions governing low-temperature dolomite formation.

Conditions inhibiting low-temperature dolomite formation	Conditions favoring low-temperature dolomite formation
Low temperatures (most seawater is <10°C)	Elevated temperature (tidal flat sediments can reach >40°C during summer months)
Time! Slow crystallization rate from supersaturated waters	Time! Given 10^3–10^6 years, dolomite will precipitate
Relatively low Mg:Ca (~3:1) in seawater	High Mg:Ca (~10:1) • Precipitate $CaSO_4$ or $CaCO_3$ and so remove Ca from the water.
Strong Mg hydration (Mg–OH)	Break Mg–OH bond • Microbes or increase temperature
Relatively low CO_3^{2-} • Most carbonate in seawater is HCO_3^-	Raise pH above 9 • Most carbonate in seawater is now CO_3^{2-} • Sulfate reduction increases alkalinity (produces more HCO_3^-)
Presence of SO_4	Remove SO_4 by • Sulfate mineral precipitation • Sulfate-reducing bacteria • Fresh water dilution • Microbe reduction of organic-rich seafloor sediment removes SO_4 and increases HCO_3^-

Ca and Mg ions into distinct layers. Dilution of the fluid by fresh water decreases the precipitation rate and generally leads to the growth of more stoichiometric dolomite (LCD).

Hydration

Mg has a smaller cation radius than Ca and the electrostatic attraction of Mg to other ions is therefore stronger. The strength of the electrostatic attraction of Mg to the OH^- radical is about 20% greater than that for Ca and much higher than that for the CO_3 ion. A cloud of loosely attached OH ions therefore surrounds each Mg ion. This hydration sheath must be broken for the Ca or Mg to bond in the crystal lattice. Critically, it takes less energy to break the Ca–OH bond than it does to break the Mg–OH bond. Once again, $CaCO_3$ forms more readily than dolomite. The difficulty of severing the OH–Mg bond is evident by the large number of naturally occurring hydrous Mg carbonates (e.g., hydromagnesite, nesquehonite). From a theoretical perspective, dolomitization should be easiest in hot, saline fluids because high water temperature and high salinity can help overcome the hydration barrier.

CO_3 activity

Mg and Ca cations bond most easily with CO_3. Normal seawater, however, has a pH of 7–8 with most carbonate in the form of HCO_3 and therefore not readily available for bonding (Figure 29.1). In fluids with a pH ~9 or higher, most of the carbonate will be in the form of CO_3

and more readily available for bonding. Naturally occurring alkaline fluids include: (1) continental groundwaters that have been involved with the weathering of siliceous rocks; (2) waters containing products derived by the dissolution of continental alkaline minerals; or (3) continental waters that have undergone anaerobic bacterial sulfate reduction. As a consequence, dolomite can form from a variety of terrestrial surface (lacustrine) and subsurface waters.

Sulfate (SO_4^{2-})

For whatever reason, sulfate seems to prevent dolomite formation. The abundant sulfate in seawater therefore acts as an inhibitor to dolomite precipitation. The sulfate in solution can be reduced by: (1) the action of sulfate-reducing bacteria; (2) the precipitation of sulfate minerals such as gypsum and anhydrite; (3) dilution with fresh water in coastal aquifers; or (4) microbial reduction of organic-rich seafloor sediments.

How to form extensive dolomite

Given enough time, dolomite should form as long as some of the above constraints can be overcome (Table 29.1). But that is not enough; there must also be: (1) a source of abundant Mg; (2) hydrologic conditions that allow for the delivery of the Mg to the zone of dolomitization; and (3) physicochemical conditions that are suitable for the formation of dolomite.

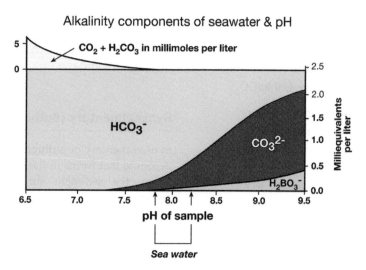

Figure 29.1 Variation of the alkalinity components in seawater with pH. Source: Adapted from Bathurst (1975). Reproduced with permission of Elsevier.

The source of magnesium

Modern seawater (Figure 29.2) contains ~1250 ppm Mg^{2+} and ~411 ppm Ca^{2+} with a Mg:Ca ratio of ~5:1, whereas river water has ~4 ppm Mg^{2+} and ~15 ppm Ca^{2+} and a Mg:Ca of 0.25:1. Formation waters are highly variable, but generally have Mg:Ca ratios from 1.8:1 to 0.4:1. In short, *seawater must be involved in any processes that create large volumes of dolomite.*

As temperature rises, the amount of Mg needed for dolomitization falls. At 25°C, ~650 m^3 of normal seawater are needed to dolomitize 1 m^3 of limestone with a porosity of 40%. By contrast, at 50°C only ~450 m^3 of fluid are needed to dolomitize 1 m^3. Dolomitization of a limestone with 6% $MgCO_3$ and 40% porosity requires 807 pore volumes of seawater. If diluted 10 times by fresh water, then approximately 8100 pore volumes are needed, whereas in a halite-saturated brine only 44 pore volumes are needed. If the fluid is rich in K and Mg salts (bitterns), then 10–12 pore volumes are necessary.

Fluid flow

Dolomitization, especially when involved in the formation of large rock bodies, requires that the reactants must be transported to and away from the sites of mineral formation. This is critical for both synsedimentary and post-depositional dolomites. Dolomitization is therefore, in many respects, an issue of hydrodynamics. Fluid flow can be driven by: (1) storms that drive ocean waters onto nearly tidal flats; (2) elevation in the form of a hydraulic head of meteoric or marine waters; (3) gradients in fluid density that are brought about by differences in water temperature or salinity; and (4) pressure due to burial or tectonic compaction. It is important to recognize, however,

DOLOMITE CaMg(CO$_3$)$_2$

REPLACEMENT

$$CaCO_3 + Mg^{2+} + CO_3^{2-} \rightleftharpoons CaMg(CO_3)_2$$

1 m^3 limestone at 25°C requires 640 m^3 of seawater

1 m^3 limestone at 50°C requires 450 m^3 of seawater

SOURCE OF Mg^{2+}

* Seawater = 1250 ppm Mg^{2+} : 411 ppm Ca^{2+} 3 : 1

River water = 04 ppm Mg^{2+} : 15 ppm Ca^{2+} 0.25 : 1

Formation water = 1.8 : 1 ⟶ 0.04 : 1

Figure 29.2 The amounts of Mg needed per unit time and the sources of that Mg to form dolomite.

DOLOSTONE TYPES AND THEIR ORIGINS

SYNSEDIMENTARY (Authigenic)
Finely crystalline; Fabric retentive; Near surface
Tidal flats (Sabkha), microbial, lacustrine,
organic-rich marine sediments

EARLY DIAGENETIC
< 300 m subsurface; replaces all types of carbonates
Fabric retentive to fabric destructive
Mixed water, brine reflux, thermal convection

LATE DIAGENETIC
> 300 m subsurface
Fabric destructive; saddle dolomite & base metals common
Warm-hot saline brines, larger regional aquifers

Figure 29.3 The different modes of dolomite formation.

that these water movements are usually a combination of several different drivers acting simultaneously. This situation arises because the balance between drives will change through time as a function of variations in relative sea level, climate, platform geometry, and paleoceanography. Finally, the host rocks themselves determine the character of low fluid flow with dolomitization focused in the most permeable beds and those with high reactive surface areas (fine grain sizes).

The different types of dolomite and dolostone

At the broadest scale, there are really three sorts of dolomites (Figure 29.3): (1) synsedimentary (authigenic) dolomites that form in the depositional environment soon after sediment accumulation; (2) early diagenetic dolomites that form in the shallow subsurface by replacing sediment, limestone, or dolostone; and (3) late diagenetic dolomites that replace limestone or dolostone in the deep subsurface or form cement overgrowths on pre-existing crystals. The rest of this chapter is an overview of synsedimentary dolomite.

Synsedimentary (authigenic) dolomite

Synsedimentary or authigenic dolomite is defined here as dolomite that forms in the environment of deposition just below the sediment surface during or shortly after sedimentation. It is most common as a precipitate from saline waters in processes that usually involve microbes or penecontemporaneous replacement of carbonate sediment. The most frequent settings are muddy tidal flats, lakes, and organic-rich marine sediments beneath sites of active ocean upwelling. Modern dolomites formed in these settings are usually microcrystalline euhedral rhombs, 1–3 μm long, with most being precipitated directly from the water with or without the influence of microbes.

Dolostones or partially dolomitized limestones of this type are generally stratiform and range from a few meters to ~10 m in thickness. They are usually lenticular and not horizontally extensive because they are facies-specific and do not cross-cut facies. Such dolomites do not generate large rock bodies but, when similar units or cycles are stacked on top of one another, considerable thicknesses of dolostone can result. Associated sediments range from terrestrial to peritidal to neritic to deep marine. Sedimentary structures in the dolostones can be numerous, diagnostic, and generally well preserved.

Overall, the dolomite crystals are Ca- to locally Mg-rich, and generally non-stoichiometric. They have high Sr contents but low amounts of Fe and Mn. $\delta^{18}O$ values are typically positive, usually reflecting normal to highly evaporated seawater. $\delta^{13}C$ values record the relative importance of different carbon sources. Whereas most have slightly positive $\delta^{13}C$ values because of an ubiquitous microbial or organogenic carbon component, those related to methanogenesis can have values >+10‰ VPDB.

Muddy tidal flat dolomite

Today dolomite forms in the muddy sediments on tidal flats in both humid (e.g., the Bahamas) and arid environments (e.g., southwest coast of the Persian Gulf) (Figure 29.4, Chapter 12).

Humid peritidal settings. The amount of dolomite in peritidal environments such as those in the Bahamas (see Chapter 12) and south Florida is generally less than 30% of the sediment. Most of the crystals are precipitated as microcrystalline cement. Dolomitization is promoted by the precipitation of aragonite from the interstitial seawater that causes an increase in the Mg:Ca ratio of the remaining fluid. This microcrystalline dolomite is generally localized to upper parts of a peritidal cycle that is composed of normal to slightly saline marine facies. The dolomitic strata tend to be lenticular on the kilometer-scale but can extend across many hundreds of square kilometers.

The lack of gypsum or anhydrite precipitation under humid climates means that there is not as rapid an increase in the Mg:Ca ratio. Consequently, dolomite in such peritidal sediments are less abundant than in arid environments.

Arid peritidal settings. The fundamental processes in arid settings, such as those in the sabkhas of the Persian Gulf, involve: (1) oceanic waters that are driven onto the sabkha during storms; (2) intense evaporation of the seawater under the hot, arid climate of the area that causes the density of the water to increase; (3) sinking of the dense fluids into the sediment where precipitation of gypsum and anhydrite removes Ca, and thereby passively increases the Mg:Ca ratio of the pore waters; (4) removal of sulfate via the precipitation of gypsum and anhydrite; and (5) dolomitization of the calcitic and aragonitic sediments that takes place when the Mg:Ca ratio exceeds 10:1.

As the seawater evaporates, however, the hydraulic head is lowered and this situation generates inflow of marine groundwaters from the ocean or of continental

Figure 29.4 (a) A pit dug into the supratidal sabkha on the coast of the United Arab Emirates along the southern margin of the Persian Gulf. The white material is anhydrite and the brown sediment is dolomite. Centimeter scale. (b) Permian peritidal, fenestral dolomite, Wyoming, USA. Image width 1 m. Photographs by P. Choquette. Reproduced with permission.

groundwaters from land. This process is called *evaporative pumping* and can modify the chemistry of the dolomitizing fluids, most importantly by providing a source of additional Mg from inflowing seawater.

Dolomite can be as extensive (>65%) as it is on modern sabkhas (tidal flats) along the southern shore of the Persian Gulf. Prograding cycles can be hundreds to thousands of square kilometers in areal extent. Later evaporite dissolution might result in intrastratal solution breccias that testify to the former presence of anhydrite. The tops of cycles can be veneered with quartzose aeolian sand. Whole cycles are typically dolomitized and, when stacked, can form dolomite bodies tens to hundreds of meters in thickness.

Tidal-flat dolomite geochemistry.   Dolomites from tidal flat settings (e.g., Abu Dhabi) have $\delta^{18}O$ values of +1.0 to +4.0‰ VPDB (due largely to evaporation) and $\delta^{13}C$ values of –2 to +4‰ VPDB, reflecting precipitation from seawater (Figure 29.5).

Microbial dolomite

Some coastal lakes, lagoons, and bays filled with mixed marine–continental groundwaters are subject to seasonal flooding (wet and oxic) and desiccation (dry and anoxic). The marginal marine Coorong Lakes in Australia (Frontispiece, Figure 29.6) and others along the Brazilian

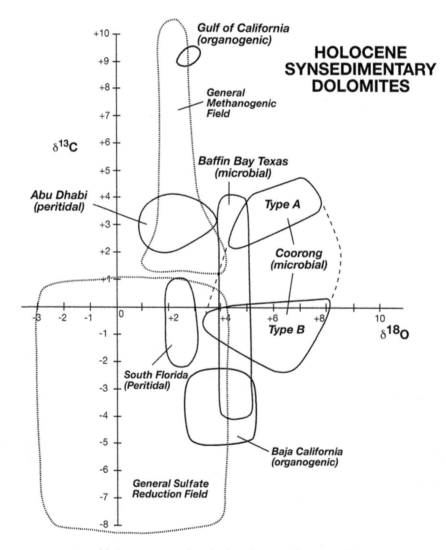

Figure 29.5 A carbon versus oxygen stable isotope cross-plot of values from modern synsedimentary dolomites in deep and shallow organic-rich marine, evaporitic lacustrine, and peritidal environments. Data from Tucker and Wright (1990) and Mazzolo (2000).

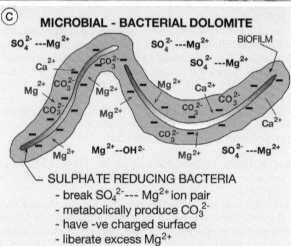

Figure 29.6 Synsedimentary microcrystalline dolomite in a saline lake at the southern end of the Coorong Lagoon, southern Australia. (a) A group of geologists standing on the light-colored dolomite-rich carbonate muds. The Southern Ocean is on the other side of the trees. (b) The yogurt-like dolomite-rich mud that is dark, reducing, and contains abundant sulfate-reducing bacteria below the surface. Note feet for scale. (c) Sketch illustrating the various chemical attributes of microbes and the surrounding biofilm that could influence dolomite precipitation.

coast are modern examples of this type of setting. Dolomite precipitation in these situations is tied to the activities of aerobic and anaerobic microbes (Figure 29.7c). Such processes have only recently been duplicated in the laboratory at surface temperatures and pressures, and it is now clear that specific microbes are instrumental in dolomite precipitation.

The role of microbes. Heterotrophic aerobic bacteria (those that use organic carbon-containing compounds as a source of energy and carbon) are prokaryotes that derive their energy and carbon from the organic compounds found in the upper 5 cm or so of the lake or lagoon sediment. The bacterial metabolism and Ca–Mg precipitation in this case leads to changes in pH (increased alkalinity) and ion concentration. The extracellular polysaccharide (EPS) that covers the surfaces of the bacteria absorbs Ca and Mg ions and provides microenvironments where dolomite crystals can grow.

Sulfate-reducing bacteria are prokaryotes that obtain energy by oxidizing organic compounds or molecular hydrogen (H_2) while reducing sulfate (SO_4) to hydrogen sulfide (H_2S). These bacteria can break the strong bond between SO_4 and Mg and utilize the SO_4 and the Mg (vital for some physiological functions) in their metabolic processes. During bacterial metabolism, excess Mg along with other byproducts of sulfate reduction such as bicarbonate and H_2S, are expelled into the EPS envelope. This creates the microenvironment necessary for dolomite precipitation. A continuous source of SO_4 is needed for this process to continue. Once the dolomite nucleates, this crystal template allows for continued crystal growth by inorganic processes changing from Ca-rich HCD to more stoichiometric crystals of LCD.

The dolomite crystals are micro-sized, and can be spherical, spindle-shaped, and dumbbell-shaped (both natural and laboratory generated). In most cases the bacterial dolomite recrystallizes with depth (a few tens of meters), becoming more stoichiometric.

Microbial dolomite geochemistry. Dolomites precipitated in association with halophylic bacteria under laboratory conditions can have Sr values as high as ~2900 and 6300 ppm. Natural crystals containing high amounts of Sr have also been found in modern hypersaline waters (e.g., the Coorong saline lakes, 3000–7000 ppm). Evaporative Coorong dolomites have very positive $\delta^{18}O$ values (+3 to +8‰) and widely varying $\delta^{13}C$ values (−2 to +4‰) because of the changing organic influence (Figure 29.5).

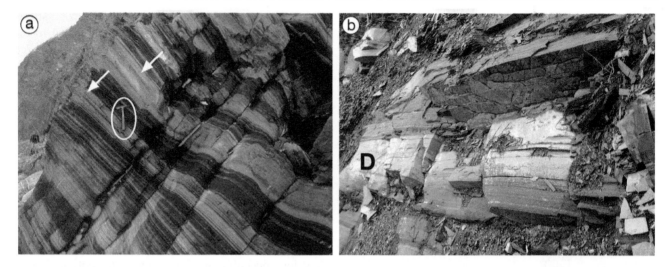

Figure 29.7 (a) Interbedded lower Cambrian organic-rich dolomites (arrows), shales, and limestones, western Newfoundland, Canada. Hammer length 15 cm. (b) Close-up of Lower Ordovician laminated organic dolomite (D) in bed 0.5 m thick enclosed by organic-rich dark shale, Lévis, Québec, Canada.

Organogenic marine dolomite

Dolomite also precipitates in a variety of modern marine environments as a result of not only bacterial sulfate reduction as outlined above, but also methanogenesis (Figure 29.7). The sediments are uniformly anoxic and organic-rich. The dolomite crystals are mostly <10 μm in size, Ca-rich, well-ordered and mostly in the form of cement. Again, the most favorable conditions are where microbial respiration of sedimentary organic matter increases alkalinity and pH of pore waters, lowers the sulfate concentration, and simultaneously alters Mg–Ca solution and crystal surface chemistries in order to promote dolomite precipitation.

Environments of formation range from peritidal to shallow marine to open ocean settings. They are common in shallow marine mud banks and in peritidal muds. Today, some of the most extensive are found in areas of upwelling beneath the O_2-minimum zone in the uppermost tens to hundreds of meters of open ocean slope sediments off Peru and southwest Africa. Dolomite occurs there as indurated layers interbedded with Holocene and Pleistocene silts, clays, and phosphates.

Organogenic marine dolomite geochemistry. These dolomites contain ~300–700 ppm Sr. Since a significant portion of the carbon is derived from microbially degraded organic matter, crystals are characterized by distinctive $\delta^{13}C$ values. Values are wide ranging, from high $\delta^{13}C$ values due to methanogenesis (up to +16‰ VPDB) to low values (down to –8‰ VPDB) in situations

dominated by bacterial sulfate reduction (Figure 29.5). The crystals have $\delta^{18}O$ values of 0 to +4‰ VPDB.

Further reading

Baker, P.A. and Burns, S.J. (1985) Occurrence and formation of dolomite in organic-rich continental margin sediments. *American Association of Petroleum Geologists Bulletin*, 69, 1917–1930.

Bathurst, R.G.C. (1975) *Carbonate Sediments and their Diagenesis*. Amsterdam: Elsevier Science, Developments in Sedimentology.

Bontognali, T.R.R., Vasconcelos, C., Warthmann, R.J., *et al.* (2010) Dolomite formation within microbial mats in the coastal sabkha of Abu Dhabi (United Arab Emirates). *Sedimentology*, 57, 824–844.

Bontognali, T.R.R., Vasconcelos, C., Warthmann, R.J., Lundberg, R., and McKenzie, J.A. (2012) Dolomite-mediating bacterium isolated from the sabkha of Abu Dhabi (UAE). *Terra Nova*, 24, 248–254.

Dix, G.R. (1997) Stratigraphic patterns of deep-water dolomite, Northeast Australia. *Journal of Sedimentary Research*, 67, 1083–1096.

Krause, S., Liebetrau, V., Gorb, S., Sánchez-Román, M., McKenzie, J.A., and Treude, T. (2012) Microbial nucleation of Mg-rich dolomite in exopolymeric substances under anoxic modern seawater salinity: new insight into an old enigma. *Geology*, 40, 587–590.

Lasemi, Z., Boardman, M.R., and Sandberg, P.A. (1989) Cement origin of supratidal dolomite, Andros Island, Bahamas. *Journal of Sedimentary Petrology*, 59, 249–257.

Mazzullo, S.J. (2000) Organogenic dolomitization in peritidal to deep-sea sediments. *Journal of Sedimentary Research*, 70, 10–23.

Meister, P., Reyes, C., Beaumont, W., *et al.* (2011) Calcium and magnesium-limited dolomite precipitation at Deep Springs Lake, California. *Sedimentology*, 58, 1810–1830.

Tucker, M.E. and Wright, V.P. (1990) *Carbonate Sedimentology*. Blackwell Scientific Publications: Oxford.

Vasconcelos, C. and McKenzie, J.A. (1997) Microbial mediation of modern dolomite precipitation and diagenesis under anoxic conditions (Lagoa Vermelha, Rio de Janeiro, Brazil). *Journal of Sedimentary Research*, A67, 378–390.

Warren, J.K. (1990) Sedimentology and mineralogy of dolomitic Coorong lakes, South Australia. *Journal of Sedimentary Petrology*, 60, 843–858.

Wright, D.T. and Wacey, D. (2005) Precipitation of dolomite using sulphate-reducing bacteria from the Coorong region, South Australia; significance and implications. *Sedimentology*, 52, 987–1008.

CHAPTER 30
SUBSURFACE DOLOMITIZATION AND DOLOSTONE PARAGENESIS

Frontispiece An open pit lead–zinc mine where dark beds are Lower Ordovician limestones that have been twice dolomitized and light beds are massive replacement saddle dolomite with sphalerite and galena, Daniels Harbour, Newfoundland, Canada.

Origin of Carbonate Sedimentary Rocks, First Edition. Noel P. James and Brian Jones.
© 2016 Noel P. James and Brian Jones. Published 2016 by John Wiley & Sons, Ltd.
Companion website: www.wiley.com/go/james/carbonaterocks

Introduction

What was originally perceived as "the dolomite problem" (see Chapter 28) has been largely resolved over the last several decades through a combination of laboratory experimentation, analysis of innumerable ancient dolomites and dolostones, and the application of hydrological principles. It is now clear that most dolostone in the rock record is largely a subsurface diagenetic phenomenon, and many dolostones are multigenerational. Paradoxically, however, large amounts of seawater are always needed if extensive dolomitization is to take place.

This chapter examines the geochemical and hydrological models that have been proposed for the origin of such dolomites. After examining the shallow subsurface diagenetic models, the focus is then placed on the deeper subsurface models that mostly involve hot fluids. This chapter also provides a practical synthesis of dolomitization and concludes with a narrative on dolomite paragenesis.

Shallow-burial early-diagenetic dolomites

Dolomites formed in the shallow subsurface are more diverse than their synsedimentary cousins and generally form from meteoric or marine waters or combinations of both that have temperatures of <60°C. The environments range from: (1) sediments below lagoons whose pores are filled with dense *refluxed evaporated brines*; to (2) aquifers filled with *mixed seawater* and *freshwater*; to (3) platform margins where cold seawater is drawn into the structure by *thermal convection* to mix with circulating groundwater. Dolomites formed in all of these settings range from

mimetic and fabric-retentive to fabric-selective to completely fabric-destructive. The original mineralogy of the sediments/rocks is important because it is easier to dolomitize aragonite or HMC than it is LMC. This is because both aragonite and HMC are more soluble than LMC and therefore more susceptible to alteration.

Resultant dolostone bodies cross-cut all depositional facies. The dolomite replaces both sediment and limestone with a wide variety of crystal types, and dolomite cement commonly lines existing pore spaces. Rhombs range from submicron to as long as 100 μm and locally to a few millimeters in size. Those associated with reflux and thermal convection are generally petrographically uniform planar-e or planar-s rhombs that can be strikingly zoned when viewed under CL. Crystals formed in the freshwater–seawater mixing zone are usually limpid with cloudy cores and clear rims.

Dolomites that replace carbonate, especially those that do so under high rock–water conditions, inherit much of the Sr from their precursor carbonate. For example, if the precursor was aragonite, then the dolomite has 500–600 ppm. If, however, the original sediment or rock was calcite, then values are much lower at 200–300 ppm. Low rock–water conditions dilute the Sr and so values can be even lower. Mn and Fe subsurface dolomites are also relatively low; generally Fe is <300 ppm and Mn is <35 ppm.

Seepage reflux

Reflux circulation occurs when platform-top shallow-marine waters concentrated by evaporation flow downward through the underlying sediment and displace the less-dense underlying groundwaters (Figure 30.1).

REFLUX DOLOMITIZATION

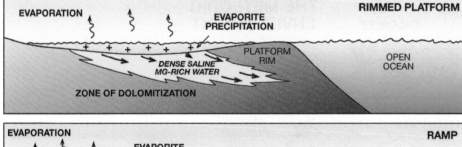

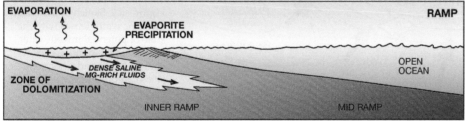

Figure 30.1 Two possible situations where brine reflux could potentially dolomitize underlying carbonate sediments.

Evaporation of seawater lagoons with variable connection to the open ocean will increase the salinity and density of the remaining seawater. This seawater can be mesohaline, with salinities of 40–100‰, or hypersaline with salinities >100‰ (gypsum saturation = 180‰, halite saturation = 249‰). Dolomitization by mesohaline waters is called *latent reflux*, whereas the process associated with hypersaline waters is termed *brine reflux*. In the latter case, gypsum and anhydrite have already been precipitated and the resulting relatively dense, Mg-rich, commonly sulfate-reduced fluids then seep down through the porous carbonates beneath the lagoon floor and mix with marine-derived connate fluids (Figure 30.1). Mixing of these two fluids triggers the precipitation of dolomite, which replaces the sediment and fills the pores. Although no convincing modern analogs are known, recurring stratigraphic relationships show that it must have been operative in the past.

Hydrochemical simulations suggest that hypersaline reflux is capable of extensive (tens of kilometers in lateral extent and hundreds of meters in depth) pervasive dolomitization at hundred-of-thousands years timescales. Modeling has also generated dolomite bodies that were thickest close to the brine source and thinned away from it (similar to relationships found in the rock record). Dolostone rock bodies are therefore localized to the platform interior but can range from tens to hundreds of meters in thickness. Sr contents and $\delta^{18}O$ values generally decrease downward away from the lagoon floor, reflecting the decreasing effects of seawater seepage.

Replacement dolomitization can increase porosity up to 8%, but post-replacement reflux could decrease porosity because of dolomite cement precipitation.

$\delta^{18}O$ values are generally quite positive, reflecting the evaporated nature of the refluxing waters. The absence of any covariant $\delta^{18}O$–$\delta^{13}C$ trends are thought to separate these dolomites from mixed water types.

Evaporative drawdown

A variation on this theme is the falling sea level or evaporative drawdown model. In this version, a TST leads to the growth of a shelf-marginal reef and patch reefs on the shelf. If sea level subsequently drops below the top of the shelf-margin reef, then waters can be trapped on the shelf where they will be subject to intense evaporation. As the Mg:Ca ratio increases, dolomitization of the marginal reef, the patch reefs, or rocks below the shelf floor can take place.

Mixed meteoric–marine waters

Solution chemistry predicts that dolomite can form in the freshwater–seawater mixing zone, whether in a confined or unconfined aquifer (Figure 30.2). Dolomite should form because: (1) Mg is being supplied by the marine waters; (2) mixing results in undersaturation with respect to calcite, aragonite, and Mg-calcite; (3) sulfate content is lowered; and (4) dolomite precipitates more easily from a dilute solution than from seawater (Figure 30.3).

This model has, however, fallen into disfavor because: (1) dolomite precipitation is minimal in most modern mixing zones; and (2) it has been argued that the production of thick dolostone successions would be impossible via such a process. Despite this, interpretations of many ancient dolostone successions, especially their chemistry, point to a mixing-zone origin. It has also been postulated that the flow of mixed waters seaward could lead to compensatory movement of replacement seawater into

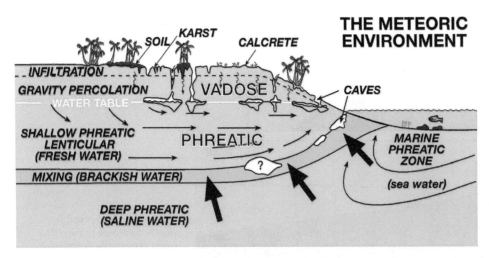

Figure 30.2 Sketch of the major meteoric water lenses with the zone of potential mixed-water dolomite precipitation (arrows). Source: James and Choquette (1984). Reproduced with permission of the Geological Association of Canada.

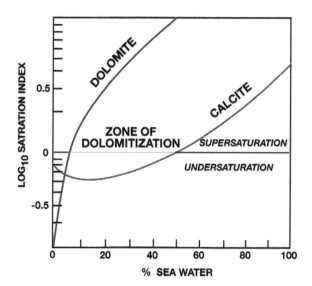

Figure 30.3 Relationship between the mixing of seawater and the relative saturation of dolomite and calcite. There is a certain mixture (in this case between ~10 and ~50% seawater) where the fluid is oversaturated with respect to dolomite and undersaturated with respect to calcite, thus promoting calcite dissolution and contemporaneous dolomite precipitation. Source: Hardie (1987). Reproduced with permission of the SEPM Society for Sedimentary Geology.

the platform, resulting in dolomitization. This latter theory seems to be an acceptable alternative.

Given that modern sea level has only been at its present position for the last 6000 years, the time for dolomite formation in modern mixing zones may simply have been insufficient. Furthermore, dolomitization of relatively thick limestone successions is possible because the mixing-zone is tied to sea level and will therefore move up and down through the strata in concert with changes in sea level. Finally, because it is a near-surface process, fluctuations in sea level resulting in subaerial exposure, alternating with seawater reflux, could lead to alternating periods of brine reflux and mixed meteoric and seawater dolomitization of the same units.

Such dolomites can be associated with subaerial exposure, karst surfaces, or sequence boundaries in the rock record. The rock bodies are not extensive, possibly tens of meters in thickness and lenticular in character (except if related to slowly rising or falling sea level over an extended period of time when they can be more widespread). The crystals are variable but limpid rhombs with cloudy cores, and clear rims are characteristic. Recent research suggests that these clear rims also extend into intercrystalline pores as dolomite cement.

Dolomites are highly variable but consistently have 100–250 ppm Sr. Plio-Pleistocene interpreted mixed-water dolomites have a wide range of $\delta^{18}O$ values (+4 to –2‰

VPDB), with $\delta^{13}C$ values (–1 to +3.5‰ VPDB) generally reflecting the role of soil gas or sulfate reduction or both. The common covariant relationship (Figure 30.4) supports a mixed-water origin with the two end-members being seawater and fresh water.

Thermal convection

Limestones in the seaward parts of carbonate platforms that lie adjacent to deep marine waters can be invaded by cold, Mg-rich seawater (Figure 30.5). This situation is promoted by the temperature contrast between seawater and platform interior groundwaters warmed by geothermal heat flux. The result is an upward convective flow in the form of a half-cell with waters drawn in at depth and expelled back into the ocean at shallow zones. Dolomitization takes place during the mixing of seawater with fresh or evaporated platform interior waters. The temperatures of dolomitization are predicted to range from 20 to 30°C. Extensive dissolution and dolomitization takes place during fluid mixing. This process can supply endless amounts of Mg^{2+} to enable dolomitization and generate excellent porosity and permeability at the same time.

Recent modeling suggests that parts of an isolated platform could be completely dolomitized by such thermal

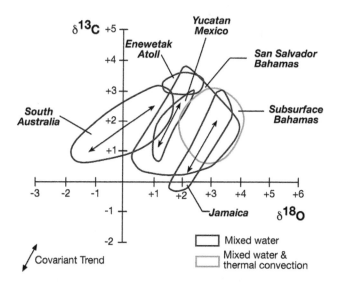

Figure 30.4 A carbon versus oxygen stable isotope cross-plot of values from late Cenozoic dolomites formed in interpreted thermal convection and mixed freshwater–seawater shallow subsurface environments (arrows indicate covariant isotope trends). Data from many different sources.

THERMAL CONVECTION

Half Cell Convection of Mixed Seawater and Freshwater at the Shelf Margin

(Kohout Convection)

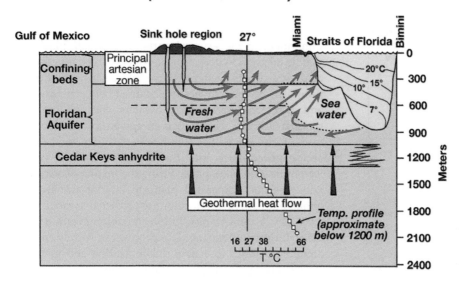

Figure 30.5 Sketch, based on data from the Florida subsurface, illustrating the half-convection cell setup due to the interaction between cold seawater and warm freshwaters that could potentially dolomitize carbonate sediments. Source: Kohout (1967). Reproduced with permission of the American Association of Petroleum Geologists.

convection within 10–15 million years. The dolomite appears to form most rapidly in about the upper 50 m of the platform in a zone up to 10 km inboard from the platform margin. The resultant dolomite body in an isolated platform would predictably be thickest along the margin and thin towards the platform interior. The dolomitization process releases calcium into solution, resulting in the precipitation of anhydrite cement downstream away from the zone of dolomitization where temperatures are >50°C.

The geochemistry is also variable, but most crystals have low Sr contents with positive $\delta^{18}O$ and $\delta^{13}C$ values. Most such dolomites do not have a covariant $\delta^{18}O$–$\delta^{13}C$ trend (e.g., Bahamas; Figure 30.4).

Sea-level-driven flow

Much smaller-scale potential dolomitization by seawater is also postulated to occur when seawater is driven through the platform by up-and-down variations in the sea surface. This can be achieved by the interaction between platform topography, winds, waves, and especially tides. For example, the sea surface on one side of an isolated platform could be temporarily higher than on the other, thus driving water through the platform from the area of high sea surface to the zone of low sea level, even if the difference is only a few meters. Typically, such dolomites would be near-surface and not as extensive as those generated via thermal convection.

Integration

Whereas mixed-water dolomitization can create local lenticular dolostone bodies, brine reflux and thermal convection can operate on a regional scale and are likely responsible for many of the massive and extensive dolostone bodies in the rock record. There seems to be a general relationship between thermal convection and seawater reflux. Geothermal circulation is most active at platform margins, but becomes restricted and eventually shut off by reflux downward from the platform interior. The persistence of geothermal circulation is therefore dependent upon the rate of platform-top saline water reflux. Low permeability evaporites can severely restrict reflux, whereas high permeability units in hydraulic connection can enhance brine transport.

Deep-burial late-diagenetic dolomites

These dolomites are the result of precipitation from hot, generally saline, fluids and are typically associated with base metal deposits and very porous hydrocarbon reservoirs. They are universally fabric-destructive such that, in many cases, there is no vestige of the original limestone or dolostone remaining. The dolomites are either hydrothermal or the result of burial and long-distance tectonically driven fluid flow. The latter long-distance transport

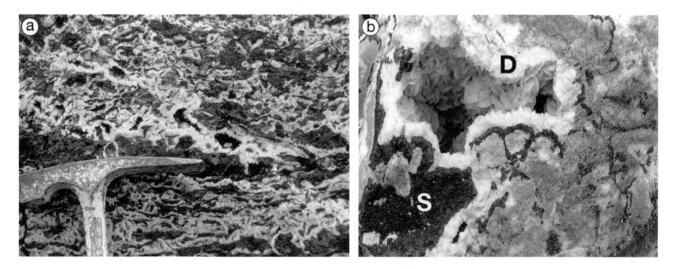

Figure 30.6 Saddle dolomites. (a) Saddle dolomite and extensive porosity, Lower Ordovician, western Newfoundland. Hammer head 15 cm. (b) Original dolomite, saddle dolomite (D) and sphalerite (S), Middle Devonian, Pine Point Mine, Northwest Territories, Canada. Image width 20 cm. Photograph by W. Martindale. Reproduced with permission.

model is either: (1) related to uplift of the regional terrane and meteoric recharge with much magnesium derived from rock–water interaction before dolomitization; or (2) related to tectonically driven flow of metamorphic fluids expelled from structurally loaded orogenic belts.

Most ancient dolomites that have been buried for significant lengths of time and have undergone prolonged diagenesis possess Sr values of ~20 and 70 ppm, but have high Fe and Mn values. Ferroan dolomite, with >2 mole % $FeCO_3$, is found in burial environments where the reduced Fe is picked up from clays, especially during the smectite to illite transition. Likewise, saddle dolomite can be ferroan. Be aware, however, that burial dolomites can be low in Fe where the pore fluids are sour. This is because of high H_2S concentrations that aid in forming pyrite (FeS_2), thus removing Fe from the fluids. $\delta^{13}C$ values are generally slightly positive (+1‰ to +4‰ VPDB), reflecting the values of replaced carbonate. $\delta^{18}O$ values are usually strongly negative (>–4 to –12 VPDB) because of the very hot waters (50°C to >120°C) from which the dolomite precipitated.

Hydrothermal

Dolomitization mediated by hydrothermal fluids that actively move upward along faults or fractures can produce dolostones that are thousands of cubic meters in volume. The dolomite is typically coarsely crystalline. Saddle dolomite, which is a common product of this style of dolomitization, is both replacive and void filling (Figure 30.6). The curved crystals are due to high rates of

SADDLE DOLOMITE: global characteristics

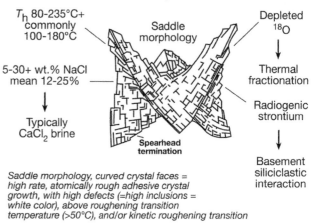

T_h 80-235°C+ commonly 100-180°C

5-30+ wt.% NaCl mean 12-25%

Typically $CaCl_2$ brine

Saddle morphology

Spearhead termination

Depleted ^{18}O

Thermal fractionation

Radiogenic strontium

Basement siliciclastic interaction

Saddle morphology, curved crystal faces = high rate, atomically rough adhesive crystal growth, with high defects (=high inclusions = white color), above roughening transition temperature (>50°C), and/or kinetic roughening transition

Figure 30.7 Crystallographic and geochemical attributes of saddle dolomite crystals. Source: Davies and Smith (2006). Reproduced with permission of the American Association of Petroleum Geologists.

crystal growth with many defects, while the white color is due to many inclusions. Fluid inclusions in the dolomite crystals confirm their formation from highly saline $CaCl_2$ brines under high temperatures, or as a product of thermal sulfate reduction (Figure 30.7) at temperatures that commonly ranged from 100 to 180°C.

Given that such dolostones are linked to a fault or fracture, the dolostone bodies are either vertically columnar or restricted to certain stratigraphic levels (Figure 30.8). Associated fabrics include extensive hydrofracturing and brittle deformation, geopetal dolomitic internal

Fault controlled fluid flow and diagenetic model for hydrothermal dolomite

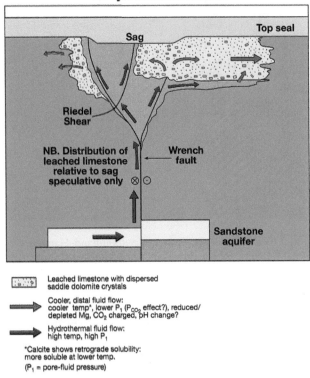

Leached limestone with dispersed saddle dolomite crystals

Cooler, distal fluid flow:
cooler temp*, lower P_1 (P_{CO_2} effect?), reduced/depleted Mg, CO_2 charged, pH change?

Hydrothermal fluid flow:
high temp, high P_1

*Calcite shows retrograde solubility:
more soluble at lower temp.
(P_1 = pore-fluid pressure)

Figure 30.8 Sketch illustrating the dynamics of saddle dolomite precipitation and replacement associated with wrench faulting. Source: Davies and Smith (2006). Reproduced with permission of the American Association of Petroleum Geologists.

sediments, sphalerite or galena, and hydrocarbons (typically bitumen). Saddle dolomites can be part of a spectrum of deposits that range from basinal sedimentary-exhalative-type mineral deposits (SEDEX), to Mississippi-Valley-type mineral deposits (MVT), to hydrocarbon accumulations (Figure 30.9). The lateral extensions out from the main fault system can be stratiform in layers ~1–6 m thick.

The crystals are low in Sr but can have high Fe values. Mn is elevated, but less so than Fe, with the result that crystals have bright CL. $\delta^{13}C$ values are slightly positive whereas $\delta^{18}O$ values are very negative as the result of thermal fractionation. Radiogenic strontium values are high, reflecting fluid interaction with crystalline basement sources or contact with siliciclastic aquifers.

Burial dolomitization: long-distance-transport deep-burial dolomites. The drive for dolomitization at depth within depositional basins is accelerated by the high temperatures found in those environments. Although it has been suggested that the Mg needed for dolomitization comes from dewatering of shales, volumetric calculations indicate that shales cannot supply the vast quantities of Mg that are needed. Therefore, it seems more probable that the Mg is brought in by topographically driven or compression-driven tectonic fluid flow (Figure 30.10). This is a difficult process because the compaction associated with these depths leads to reduced rock permeability that impedes all fluid flow.

The "Hydrothermal" Mineral Association

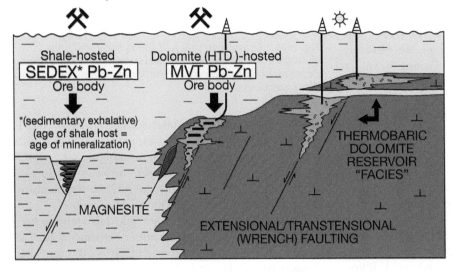

Figure 30.9 Sketch of the spatial association between mineral deposits and hydrocarbon accumulations associated with hydrothermal dolomitization. Source: Davies and Smith (2006). Reproduced with permission of the American Association of Petroleum Geologists.

Gravity or hydrodynamic head drive fluids through the rock

- Dolomitizing fluids include modified seawater and Mg-rich residual evaporite brines

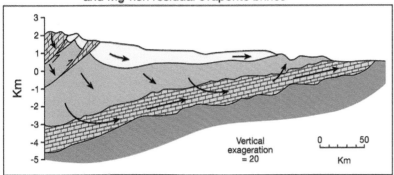

Figure 30.10 Sketch of gravity-driven fluid flow where surface meteoric waters flow downward from a mountain belt, are chemically modified and warmed at depth, and then flow upward potentially altering the surrounding carbonates. Developed from the Western Canadian Sedimentary Basin. Source: Adapted from Garven and Freeze (1984). Reproduced with permission of American Journal of Science.

Figure 30.11 Outcrop of Upper Devonian dolomite reef carbonate with numerous partial and whole vugs derived from leached stromatoporoids, Rocky Mountain Front Ranges, Alberta, Canada. Pencil 12 cm long.

These dolostones typically cross-cut pre-existing fabrics with no relationship to platform or ramp geometry. Crystals are generally non-planar-a with associated extensive vuggy porosity. Most dolomites are fabric-destructive, except that large fossils such as stromatoporoids can be partially preserved as coarse dolomite but with most of their skeletons preserved in the form of crystal-lined vugs (Figure 30.11).

Sr contents are typically low, whereas Fe and Mn can have elevated values. Stable isotopes reflect the warm fluid temperatures in the form of very negative $\delta^{18}O$ values. The $\delta^{13}C$ values are generally slightly positive because of carbon derived from buried organic material.

Synthesis

The main attributes of synsedimentary early shallow burial and late deep burial dolomites discussed in this chapter and the previous chapter are outlined in Tables 30.1 and 30.2. These tables are a general summary, and it is important to remember that any particular dolostone may not exhibit all of the outlined characteristics. It must be emphasized that many dolostones evolved through multiple generations of dolomitization and are typically composed of early dolomites that have been overprinted and altered by later shallow and deep burial processes.

Dolomite paragenesis

To unravel the origin or origins of multigeneration dolostones, it is critical to identify each different dolomite phase and then pinpoint the petrographic and geochemical attributes of each stage. As stressed above, these stages can be a succession of dolomite and dolomitizing events from synsedimentary to shallow diagenetic to late diagenetic. Remember, however, that not all dolostones will have all of these phases. For example, the only dolomitizing event could be shallow burial in origin, or this dolomite could be followed by precipitation of burial dolomite. Each dolostone has a unique geohistory!

Table 30.1 Characteristics of different synsedimentary dolostone bodies.

	Lacustrine	Organic marine	Normal marine	Humid peritidal	Arid peritidal
Crystal size	Fine < 1 μm	Micritic–microspar	Micritic	Fine <10 μm	Fine <10 μm
Stratigraphy	Stratiform and facies specific	Stratiform and facies specific	Stratiform and facies specific	Stratiform and facies specific	Stratiform and facies specific
Primary sedimentary structures	Laminations, teepees, intraclasts, mud cracks, stromatolites, dolocretes	Laminated	Laminated to burrowed	Normal peritidal mudflat sediment structures	Normal peritidal mudflat sediment structures and nodular and displacive evaporites
Associated facies	Terrestrial or marine	Deep-water marine	Shallow neritic	Adjacent normal marine–lagoonal	Adjacent normal marine or evaporative lagoon
Size	Lenticular, km-scale, 1–5 m thick	Lenticular beds 1–2 m thick, not much of the section	Not well constrained	1–10 m thick, lenticular and local to stratiform, 10^3 km²	1–10 m thick, lenticular and local to stratiform, 10^3 km²
Petrography	Metastable Ca-rich	Metastable Ca-rich	Metastable Ca-rich	Metastable Ca-rich	Metastable Ca-rich
Trace elements	High Sr	High Sr	High Sr	High Sr	High Sr
Isotopes	Positive $\delta^{18}O$, negative and positive $\delta^{13}C$	Positive $\delta^{18}O$, if methano-genesis can be high >10	Positive $\delta^{18}O$, slightly negative $\delta^{13}C$	Positive $\delta^{18}O$, slightly positive $\delta^{13}C$	Positive $\delta^{18}O$, positive $\delta^{13}C$

Table 30.2 Characteristics of different early and late diagenetic dolostone bodies.

	Early diagenetic			Late diagenetic	
	Brine reflux	Mixed water	Thermal convection	Deep aquifer	Hydrothermal
Crystal size	Medium crystal 10–100 μm	Medium crystal 10–100 μm	Medium crystal 10–100 μm	Coarse to very coarse, sparry, non planar–planar	Coarse to very coarse, saddle
Stratigraphy	Beneath widespread subaqueous evaporites	Cross-cutting depositional facies	Cross-cutting facies boundaries, near platform margins and interiors	Cross-cutting, fault associated	Halos around high-angle faults
Primary sedimentary structures	Normal marine–variably preserved	Normal marine to evaporitic–variably preserved		Usually destroyed, except large fossils (implies porosity)	Usually destroyed, hydro-fracturing, brecciation
Associated facies	Evaporites, solution breccias	Below karst surfaces, sequence boundaries	Subsurface karst	None	None
Size	10–100 m thick over large areas	10s–100s m thick	10–100s m thick over very large areas	Regionally extensive	Vertically deep but not laterally extensive
Petrography	Fabric-retentive–fabric-destructive Ca-rich metastable to stoichiometric	Fabric-retentive–fabric-destructive mimetic, limpid crystals stoichiometric	Fabric-retentive–fabric-destructive	Fabric destructive, vug filling cements stoichio-metric	Completely fabric destructive, Ca-rich
Trace elements	Sr decreasing downward	Low Sr	Low Sr	Low Sr, implies elevated Fe, Mn	Low Sr (but can have high $^{87}Sr/^{86}Sr$ values); high Fe and Mn
Isotopes	^{18}O values decrease away from dolo evap contact	Highly variable, positive and negative $\delta^{18}O$, $\delta^{13}C$, covariant trends	Negative $\delta^{18}O$ positive $\delta^{13}C$	Highly negative $\delta^{18}O$, slightly positive $\delta^{13}C$	Very negative $\delta^{18}O$ slightly positive $\delta^{13}C$
Miscellaneous	Dolomitization decreases downward from evaporites	Best developed around basin margins			Associated with base metals

In assessing the origin of a particular body of dolostone, the following questions should be asked:

1. What are the interpreted depositional environments of the enclosing facies?
2. Is there a difference in crystal size related to fabric (e.g., matrix is microcrystalline, fossils are coarser crystalline) that can signal two stages of dolomitization?
3. Is the dolomite fabric-retentive or fabric-destructive?
4. Did the dolomite predate or post-date stylolites?
5. Are there any obvious cross-cutting relationships between different dolomite phases?

6. Are the crystals iron-rich or iron-poor?
7. Is the Sr content low or high?
8. If saddle dolomite is present, what was the nature of the preexisting dolomite?

Some of these questions can be answered via simple petrographic observation. In Figure 30.12, for example, the dolomite crystals that are cut by a stylolite must have formed before the rock was buried deep enough for chemical compaction to occur. CL often provides the clearest delineation of distinct dolomite phases (Figure 22.4b).

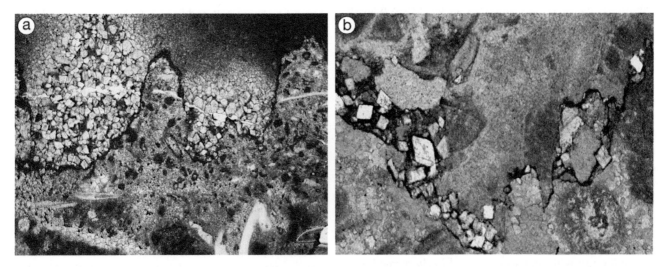

Figure 30.12 Dolomites and burial. (a) A stylolite that cuts across euhedral dolomite crystals confirming their formation prior to chemical compaction, Lower Ordovician, Western Newfoundland, Canada. Image width 3 mm. (b) Dolomite crystals precipitated after stylolite formation, indicating development during burial and after pressure solution, Middle Ordovician, Western Newfoundland, Canada. Image width 5 mm.

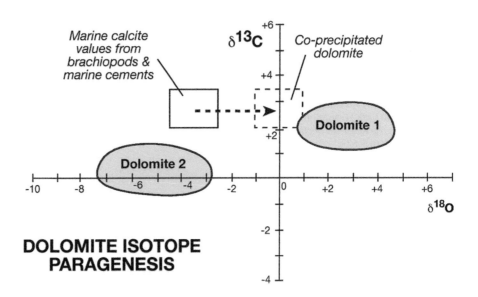

Figure 30.13 A theoretical example of using a carbon versus oxygen stable isotope cross-plot to interpret two different dolomites.

Stable isotope data can be particularly useful in interpreting dolomite genesis. The oxygen and carbon isotopic values of dolomite reflect the sources of oxygen and carbon that go into dolomite. This depends on if it was precipitated from water or came from the parent rock in the case of replacive dolomite. When well-preserved low-magnesium-calcite skeletal grains such as brachiopods and pelagic foraminifers are present, seawater isotopic values can be estimated. This can provide a baseline to interpret synsedimentary dolomites and provide a basis for precursor rock isotopic values. Early dolomites should have values close to the calcite values (remembering the fractionation factors between water and calcite and water and dolomite are different), whereas burial dolomites should be ^{18}O low relative to such values. Ideally, each dolomite phase should be analyzed, but this might not be possible given the difficulty in sampling individual phases.

A theoretical example that demonstrates the relationship between calcite and dolomite phases is shown in Figure 30.13. There are two unknown dolomites: dolomite 1 is microcrystalline and dolomite 2 is medium crystalline. Stable isotope values for calcites precipitated from seawater at this time are outlined in the solid square. A dolomite precipitated from the same seawater would have a value as illustrated in the dashed square ($\delta^{18}O$ partition = +3‰ VPDB). Dolomite 1 is close to the value of this dolomite and so is likely synsedimentary. The very positive values suggest that it might even be evaporitic. Dolomite 2 has much lower $\delta^{18}O$ values and so is clearly both out of equilibrium with that seawater and likely formed at a much higher temperature, probably during burial.

Further reading

Al-Helal, A.B., Whitaker, F.F., and Xiao, Y. (2012) Reactive transport modeling of brine reflux: dolomitization, anhydrite precipitation, and porosity evolution. *Journal of Sedimentary Research*, 82, 196–215.

Budd, D.A. (1997) Cenozoic dolomites of carbonate islands: their attributes and origin. *Earth-Science Reviews*, 42, 1–47.

Cander, H.S. (1994) An example of mixing-zone dolomite, middle Eocene Avon Park Formation, Floridan Aquifer system. *Journal of Sedimentary Research, Section A: Sedimentary Petrology and Processes*, 64, 615–629.

Davies, G.R. and Smith, L.B. (2006) Structurally controlled hydrothermal dolomite reservoir facies: An overview. *AAPG Bulletin*, 90, 1641–1690.

Ferry, J.M., Passey, B.H., Vasconcelos, C., and Eiler, J.M. (2011) Formation of dolomite at 40-80°C in the Latemar carbonate buildup, Dolomites, Italy, from clumped isotope thermometry. *Geology*, 39, 571–574.

Garven, G. and Freeze, R.A. (1984) Theoretical analysis of the role of groundwater flow in the genesis of stratabound ore deposits. 2. Quantitative results. *American Journal of Science*, 284, 1125–1174.

Hardie, L.A. (1987) Dolomitization; a critical view of some current views. *Journal of Sedimentary Research*, 57, 166–183.

Kohout, F.A. (1967) Ground-water flow and the geothermal regime of the Floridian Plateau. *Transactions of the Gulf Coast Association of Geological Societies*, 17, 339–354.

Kyser, T.K., James, N.P., and Bone, Y. (2002) Shallow burial dolomitization and dedolomitization of Cenozoic cool-water limestones, southern Australia; geochemistry and origin. *Journal of Sedimentary Research*, 72, 146–157.

Machel, H.G. (2004) Concepts and models of dolomitization: a critical reappraisal. In: Braithwaite C.J.R., Rizzi G., and Drake G. (eds) *The Geometry and Petrogenesis of Dolomite Hydrocarbon Reservoirs*. London: Geological Society, Special Publication no. 235, pp. 7–63.

Machel, H.G. and Buschkuehle, B.E. (2008) Diagenesis of the Devonian Southesk-Cairn Carbonate Complex, Alberta, Canada: Marine cementation, burial dolomitization, thermochemical sulfate reduction, anhydritization, and squeegee fluid flow. *Journal of Sedimentary Research*, 78, 366–389.

Qing, H., Bosence, D.W., and Rose, E.P.F. (2001) Dolomitization by penesaline sea water in Early Jurassic peritidal platform carbonates, Gibraltar, western Mediterranean. *Sedimentology*, 48, 153–163.

Rivers, J.M., Kyser, T.K., and James, N.P. (2012) Salinity reflux and dolomitization of southern Australian slope sediments: the importance of low carbonate saturation levels. *Sedimentology*, 59, 445–465.

Swart, P.K. and Melim, L.A. (2000) The origin of dolomites in Tertiary sediments from the margin of Great Bahama Bank. *Journal of Sedimentary Research*, 70, 738–748.

Ward, W.C. and Halley, R.B. (1985) Dolomitization in a mixing zone of near-seawater composition, Late Pleistocene, northeastern Yucatan Peninsula. *Journal of Sedimentary Petrology*, 55, 407–420.

Whitaker, F.F. and Xiao, Y. (2010) Reactive transport modeling of early burial dolomitization of carbonate platforms by geothermal convection. *AAPG Bulletin*, 94, 889–917.

Zhao, H. and Jones, B. (2012) Genesis of fabric-destructive dolostones: A case study of the Brac Formation (Oligocene), Cayman Brac, British West Indies. *Sedimentary Geology*, 267–268, 36–54.

CHAPTER 31
DIAGENESIS AND GEOHISTORY

Frontispiece Mountainside of neritic Jurassic carbonates in Oman that are equivalent to the prolific hydrocarbon reservoirs in Saudi Arabia. Houses at base of cliff for scale.

Origin of Carbonate Sedimentary Rocks, First Edition. Noel P. James and Brian Jones.
© 2016 Noel P. James and Brian Jones. Published 2016 by John Wiley & Sons, Ltd.
Companion website: www.wiley.com/go/james/carbonaterocks

Introduction

When we encounter limestones or dolostones in the deep subsurface or in outcrop we are, in most cases, seeing the combined effects of rock– or sediment–water reactions that have taken place in a variety of diagenetic environments over long periods of time. The older the rocks, of course, the more diagenesis they have seen.

The preceding chapters have detailed the diagenesis of carbonate sediments, limestones, and dolostones in the major different diagenetic realms. This chapter takes these processes and the results of these changes and places them within a geohistorical framework: specifically, what are the results of these processes acting separately and together through time. Using this perspective, the zones of diagenesis have usefully been grouped in a slightly different way (Figure 31.1). The *eogenetic zone* comprises all changes that take place early in the synsedimentary and meteoric diagenetic realms. The *mesogenetic zone* encompasses all changes that the carbonates undergo during shallow and deep burial. The *telogenetic zone* is where rocks, once buried, are re-exposed via tectonics or prolonged denudation, and typically karsted. Some successions may go through this cycle many times. It is also important to realize that our knowledge of the ongoing processes in the eogenetic zone is far better than those operative in the mesogenetic zone. This is simply because we can observe and monitor the shallow diagenetic zones, whereas information from greater depths is more difficult to obtain.

ZONES OF DIAGENESIS

EOGENETIC
- SYNSEDIMENTARY REALM - LIMESTONE
- METEORIC REALM - LIMESTONE
- SYNSEDIMENTARY REALM - DOLOMITE
- EARLY DIAGENETIC REALM -DOLOMITE

MESOGENETIC
- BURIAL REALM - LIMESTONE
- LATE DIAGENETIC REALM - DOLOMITE

TELOGENETIC ⟶ KARST - POST MESOGENETIC

Figure 31.1 Relationship between limestone and dolomite diagenetic realms and diagenetic zones. Source: Choquette and Pray (1970). Reproduced with permission of the American Association of Petroleum Geologists.

The greatest variability of change takes place in the eogenetic zone where, although the time spent there is comparatively short, there is a wide spectrum of controls such as diagenetic setting, local climate, and sequence stratigraphy. By contrast, subsequent diagenetic history in the mesogenetic zone is prolonged, gradual, and variably intense, but characterized by fewer variables.

Eogenetic diagenesis

This section concentrates on how shallow surface and subsurface diagenesis changes as the environment evolves. The most important attributes in this context are rock body geometry, specifically platforms versus ramps, the local climate, mostly humid versus arid, fluid dynamics or hydrogeology, and sequence stratigraphy.

Rock body geometry

At the largest scale, the principal rock bodies are either platforms or ramps (Figure 31.2). What is also critical is if these structures were isolated or connected to a hinterland. Carbonate platforms can be shelves built on continental margins, on tectonic terranes, or around volcanoes (connected platforms; see Chapter 1). Isolated carbonate platforms can also develop on continental relicts or on volcanic edifices. The principal difference from the perspective of diagenesis is that connected platforms can be flushed by continental groundwaters, whereas isolated platforms have no such source of waters and all fluids are local. Ramps are always connected to land and are therefore much like connect platforms.

Local climate

Regardless of all other factors, water is needed for diagenesis to occur. As emphasized in Chapters 25 and 26, the rapidity of near-surface diagenesis and the alteration products are highly dependent on local rainfall. Tropical rainforests are areas of massive karst; areas with semi-arid Mediterranean climates are characterized by calcretes and relatively slow diagenesis; and deserts are regions of profoundly slow diagenesis or, in some cases, no diagenesis at all. Whereas fresh water is the control in these situations, arid climates are areas of intense evaporation and thus the formation of dense brines from seawater.

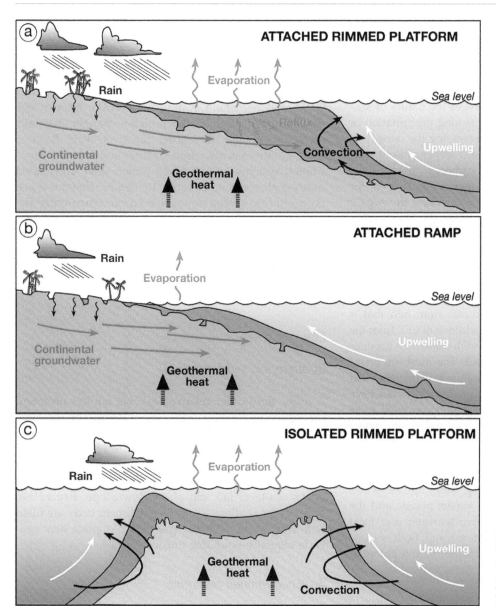

Figure 31.2 The attributes of rock bodies, the different dynamics of diagenetic fluids, and the climatic effects on: (a) attached rimmed platforms; (b) attached ramps; and (c) isolated platforms.

Sequence stratigraphy

The discussion from Chapter 19, which details the principles of sequence stratigraphy, should serve as a background as you read this chapter. Ever-changing sea levels make this a dynamic system. The most important concept in terms of early diagenesis is the relationship between different orders of sea-level fluctuation (Figures 19.3, 19.4). Third-order relatively long-term fluctuations control the major systems tracts, that is, the TST, HST, and LST. Higher fourth- and fifth-order fluctuations usually do not change sea level much, but can have a dramatic effect on diagenesis, especially for shallow-water rimmed platforms.

Fluid dynamics

The principal fluids involved with early diagenesis are seawater and freshwater (Figure 31.2). These waters can be, and usually are, profoundly changed within different environments.

Seawater. Open ocean seawater can be diluted by fresh water, concentrated by evaporation, altered by mineral precipitation, and chemically changed by upwelling. Surface dilution is most noticeable in humid regions where rain can lower the salinity by several parts per thousand. Although not greatly important in the open ocean, it can be important in wide shallow epeiric-like seas and semi-isolated lagoons. Evaporation changes the nature of seawater by concentrating the ions, making precipitation of some minerals easier as well as increasing the density of the fluid, thus promoting downward seepage into underlying sediment. Precipitation of carbonate changes the nature of the fluids from which they were derived; for example, precipitation of aragonite increases the Mg:Ca ratio. Growth of anhydrite also increases the Mg:Ca ratio, but also removes SO_4^- from the seawater. Once brought into shallow warm environments, upwelling cold ocean waters with abundant dissolved carbonate commonly precipitate carbonate as synsedimentary cements and ooids.

Freshwater. Locally, freshwater comes from rain that is generally quickly modified by the addition of CO_2 from the air and soil cover and mixed with with subsurface groundwaters that may be distally sourced. The most important source of near-surface CO_2 is the community of microbes that infest the soil profile. Far-travelled groundwaters can have high or low carbonate contents, depending on what rocks they have flowed through, and their trace and major element contents can be highly modified. For example, increased Mg^{2+} can be derived from subsurface dolomite host rocks.

Mixed waters. The major mixing zones are the water table where vadose and phreatic waters mingle and the base of the freshwater zone where seawater and local phreatic or continental groundwaters blend. Both of these zones are regions of carbonate dissolution and potential dolomitization.

Fluid flow. There are two major drivers for fluid flow in the shallow subsurface: gravity and thermal convection. Gravity-driven flow is manifest as the downward percolation of vadose waters and the sublateral flow of continental groundwaters towards base level which, in most cases, is the ocean. Such flow is also down a gentle hydraulic gradient. Convective heating is most effective when two water bodies of different temperatures come in contact. This process is well illustrated along platform margins where waters inside the bank are relatively warm because of geothermal heating, whereas surrounding ocean waters are comparatively cold. This sets up a thermal convective half-cell where cold ocean waters are pulled into the platform interior and mixed with generally fresh platform interior waters. These mixed waters are then expelled into the ocean at shallower depths.

Approach

The following discussion is framed in the context of sequence stratigraphy from the perspective of a lowstand system tract (LST), transgressive system tract (TST), and highstand system tract (HST), respectively. Analysis will focus on the possible changes that take place during one third-order sea-level cycle. Each of these situations is discussed for three end-member depositional systems: an attached rimmed shelf, an isolated rimmed platform, and a ramp. Remember that this is just the prelude to similar deposition and accompanying diagenesis that takes place during the next depositional cycle (followed by the next cycle, and so on). The complexity lies in the fact that during each succeeding cycle of diagenesis also affects, to a greater or lesser degree, all of the underlying deposits.

Lowstand systems tract

Stratigraphic sedimentology

A major third-order relative sea-level fall has profound consequences for all depositional systems. The sediment factory on rimmed platforms, attached or unattached, is shut down except perhaps for a narrow region on the slope, and deposition essentially ceases. The situation is different for a ramp because the sediment factory merely tracks sea-level fall and shifts downslope. Regardless, pores in the shallow subsurface sediment body are filled with fresh water. The changes that take place here generate sequence-bounding unconformities.

Attached rimmed platforms

There is a gentle seaward hydraulic gradient whose fresh waters come from local rainfall and an updip recharge area (Figure 31.3a). The most profound changes are widespread surface and subsurface karst and rapid alteration of metastable carbonates to diagenetic LMC. Surface karst is especially profound under humid climates with rich soils, whereas subsurface karst is best developed at the water table and in the lenticular zone, especially adjacent to the ocean (Figure 26.3). The recharge area updip is a zone of mostly metastable grain dissolution with cementation being more prevalent downdip towards the ocean. The mixing of seawater and fresh water can lead to dolomitization.

Reduction in freshwater input under semi-arid and arid climates considerably reduces the amount and

LOWSTAND SYSTEMS TRACT

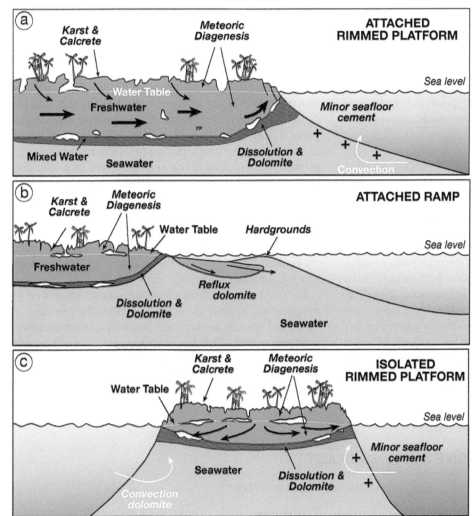

Figure 31.3 The nature of subsurface waters and diagenesis during a LST on: (a) an exposed attached rimmed platform; (b) a ramp; and (c) an isolated rimmed platform.

rate of meteoric diagenesis. On the other hand, soil-related calcretes typically develop on the exposed surface, particularly after the evolution of extensive land vegetation and complex soils (especially in the Devonian).

Isolated rimmed platforms

Exposure of these banks (Figure 31.3b) has the same consequences as attached rimmed platforms: no sedimentation and profound meteoric alteration. The difference is that there is no external source of fresh water and the amount of meteoric water is completely dependent upon local climate. Platform subsidence is generally uniform and so there is a platform-wide subsurface meteoric lens that floats on marine water. The size of the lens depends

on the rate of recharge via rain. Flow in this lens is directed towards the margins.

Diagenesis is similar to that on an attached platform, except that it is somewhat slower because of less subsurface fresh water. Alteration is concentrated along the water table which, do not forget, can migrate up and down depending on rainwater flux and sea-level change.

Ramps

The downslope shift in deposition is accompanied by landward exposure of recently deposited ramp sediments in an enlarged recharge area (Figure 31.3b). The meteoric diagenesis is similar to that in attached rimmed platforms, where continental groundwaters extend beneath the now-exposed older ramp sediments.

Sequence boundaries

The fall in sea level and extensive diagenesis associated with third-order sequences is one of the major causes of sequence boundaries in carbonates. The diagenesis associated with LST and meteoric alteration, especially in the form of karst surfaces and calcrete horizons, creates horizons that are recognizable throughout the geological record, allowing separation of one major depositional sequence from another.

Transgressive systems tract

Stratigraphic sedimentology

The relatively rapid sea-level rise results in mostly aggradation with an elevated rim (reefs or sand shoals), relatively steep slope, deep lagoon, and muddy tidal flats or beaches along the landward moving shoreline. Accommodation is generally not filled. Inundation fills holes in the underlying succession with seawater, displacing earlier freshwater aquifers and lenses. Such filling is relatively rapid in terms of rimmed platforms, but gradual in the case of ramps.

Attached rimmed platforms

The seaward margin at the top of the accreting rim can be the site of synsedimentary cementation, but it is likely not extensive because of rapid reef growth (Figure 31.4a). Seawater convection at the platform margin can promote diagenesis and possibly dolomitization. Lagoonal sedimentation is not as rapid as that at the rim, and the generally muddy sediments do not undergo much diagenesis. Lagoonal sediments can, however, fall below the zone of active productivity, resulting in hardgrounds or minor dissolution if the sediments are in a reducing environment. This new marine sediment can also filter down into underlying karst cavities and depressions. There is not much diagenesis on the tidal flats because they are continuously moving landward and being progressively submerged by rising sea level. This landward movement also results in the gradual movement of the mixing zone updip.

Elevated carbonate seawater saturation brought on by evaporation in arid or semi-arid climates can lead to increased synsedimentary cementation along the shallow platform margin. This rise in salinity will simultaneously result in a decreased carbonate-producing biota in the lagoon and therefore a reduction in sedimentation rates.

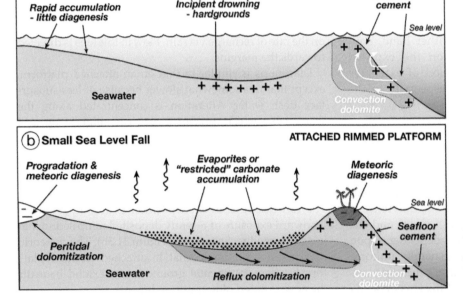

Figure 31.4 The nature of early diagenesis on an attached rimmed platform: (a) during major third-order TST; and (b) during a fourth- or fifth-order sea-level fall.

A small temporary fourth- or fifth-order fall in sea level (Figure 31.4b) would result in local exposure of the margin (expressed as a series of islands), restriction of ocean water circulation into the lagoon, shallowing of the lagoon via sedimentation, and slight progradation of the muddy tidal flats.

There are several important consequences of this situation. Isolated freshwater lenses would develop beneath the small marginal islands, resulting in local meteoric diagenesis. The lagoon could become restricted, resulting in a profound reduction in the biota if in the tropics, or subtidal evaporites if in arid climates. Dense brines left over from such evaporite precipitation could seep down through the lagoon floor via brine reflux, promoting dolomitization. This process will eventually be shut off as the evaporites seal the lagoon floor. Muddy tidal flats would prograde locally with attendant dolomitization or meteoric alteration in the freshwater lens below.

Isolated rimmed platforms

These rimmed structures have the same general accretionary style as their attached rimmed platform cousins (Figure 31.5a). The stratigraphy is one of a keep-up or catch-up motif where even though sedimentation is generally shallow-water in nature, it rarely reaches sea level. The structures can be asymmetric with extensive reef growth on the windward margins, but somewhat suppressed growth on the leeward margin with possible development of carbonate shoals.

Overall, the same diagenetic processes as those present in attached rimmed platforms dominate. The major difference is that there is no large recharge area or source of continental groundwater such that meteoric diagenesis associated with small fourth- and fifth-order drops in sea level are restricted to marginal facies (Figure 31.5b). The leeward grainstone shoals have localized development of hardgrounds. There will likely be thermal convection and

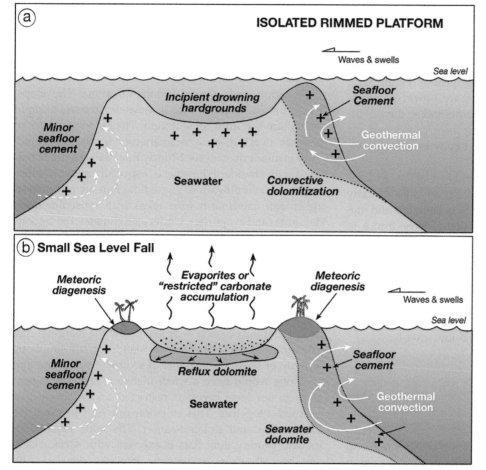

Figure 31.5 The nature of early diagenesis on an isolated rimmed platform: (a) during major third-order TST; and (b) during a fourth- or fifth-order sea-level fall.

TRANSGRESSIVE & HIGHSTANDS SYSTEMS TRACTS

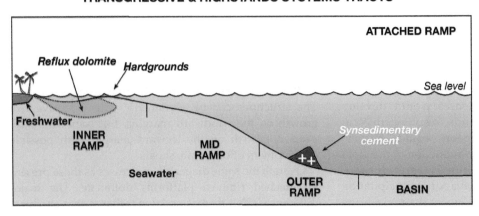

Figure 31.6 The nature of early diagenesis on a ramp during major third-order TST and highstand.

possible dolomitization if the platform is large enough. Reflux tends to alter subsurface sediments downward from the lagoon, whereas thermal convective fluid flow is most intensive along the margin and decreases inward toward the bank interior.

Ramps

Sedimentation just keeps pace with the rate of sea-level rise during the TST, and all facies migrate landward. The inner ramp is close to sea level and therefore susceptible to both synsedimentary and meteoric diagenesis (Figure 31.6). The mid- and outer-ramp sediments are, however, continuously perturbed by waves and storms and so there is little chance for any marine diagenesis there, except perhaps for episodic hardground formation between disturbance events.

The narrow lagoons adjacent to muddy tidal flats and carbonate sand shoals of the inner ramp are sites of early marine diagenesis. The lagoons can be evaporative and precipitate sulfates, allowing the remaining dense brines to seep down into sediments and promote dolomitization. The tidal flats can be subject to the same diagenetic processes described for rimmed shelves. The sand shoals (especially those that are oolitic), however, typically contain centimeter-thick hardgrounds or cemented subsurface layers. The other remaining site of synsedimentary cementation is in downslope biogenic or reef mounds, well below wave base.

Highstand systems tract

Stratigraphic sedimentology

The slowed rate of sea-level rise allows sedimentation to fill all accommodation and so there is slow accretion and progradation on rimmed platforms and pronounced progradation in ramp systems. There is little basinal sedimentation at maximum flooding in ramp systems and biogenic or reef mounds tend to develop in the outer ramp until buried by later progradation.

Attached rimmed platforms

The filling of accommodation and pronounced progradation generates a shelf that comes close to resembling a LST situation with much subaerial exposure (Figure 31.7a). The shallow seaward margin is a zone of synsedimentary cementation and potential dolomitization. The slope, which progrades basinward is, however, the site of rapid deposition but little cementation. The lagoon accretes to near sea level but, since sedimentation is relatively rapid, there is minor diagenesis. Muddy tidal flats prograde seaward with attendant diagenesis, especially meteoric alteration and dolomitization. These tidal flats are also sites of ephemeral marginal marine lakes with synsedimentary dolomites and evaporites. Under arid climates, the lagoons can become centers for evaporites with subsurface dolomitization. The tidal flats will be sabkhas with attendant dolomitization and synsedimentary evaporites.

A small drop in sea level (Figure 31.7b) generally results in almost total exposure of the platform, shutdown of the factory, and meteoric diagenesis similar to the LST situation except that the ocean is at or just below the marginal rim. The rim itself is a series of small islands formed simply by buildup or sweeping together of sediment by storms, and each one would have a small meteoric lens. The lagoon close to sea level becomes restricted and evaporitic, or exposed altogether. This situation can lead to extensive brine reflux and dolomitization. The wide expanse of inboard muddy tidal flats is exposed with surface karst and beneath them, a meteoric lens can develop with attendant alteration. The margin is also exposed as a wide

EARLY HIGHSTAND SYSTEMS TRACT

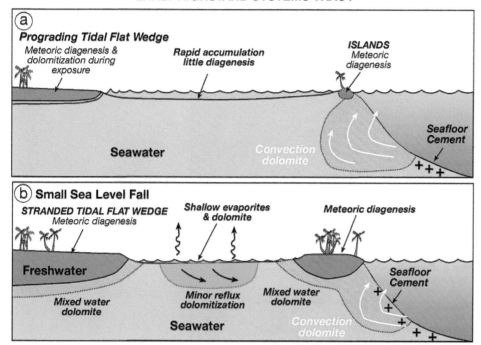

Figure 31.7 The nature of early diagenesis on an attached rimmed platform: (a) during major third-order HST; and (b) during a fourth- or fifth-order sea-level fall.

tract beneath which a freshwater lens develops with likelihood of both meteoric diagenesis and dolomitization.

Isolated rimmed platforms

With most of the platform at or close to sea level because of overall filling of accommodation, there is little diagenesis except along the steep margins, where synsedimentary cementation or dolomitization via thermal convection can be important (Figure 31.8a).

If the seafloor is close to the ocean surface, even small fourth- or fifth-order falls in sea level can lead to extensive early diagenesis (Figure 31.8b), specifically: localized meteoric diagenesis beneath large islands along the margin; mixed water diagenesis in the underlying lens; reflux dolomite beneath restricted and possibly evaporitic lagoons; and dolomitization or cementation along the ocean facing margins. Tidal flats can form around lagoonal islands and windward margins prograding into the lagoon.

Ramps

Ramps tend to prograde during this phase in sea-level rise and so diagenesis is much the same as during TST, except that meteoric diagenesis is somewhat more widespread. Progradation can lead to complexes of overlapping meteoric water lenses. If progradation is slow, then there is time

for extensive diagenesis, particularly centered at the water table. Calcrete is extensive on nearshore aeolianites under semi-arid climates, and corridors between the ridges can be sites of synsedimentary lacustrine dolomite.

Post-eogenetic diagenesis

The diagenetic history of a hypothetical succession of carbonates is outlined in Figures 31.9 and 31.10. This sequence of events is typical of many successions in the rock record and is particularly useful as a background to interpret any series of diagenetic changes. In almost all cases, the rocks pass from the eogenetic zone to the mesogenetic zone with the possibility of re-exposure in the telogenetic zone along the way.

Time–process diagram

Once the sediments and rocks are buried below a few hundred meters of sediment, below the eogenetic zone, then our best tools are thin-sections, petrography, and CL. There, the diagenetic history is mostly revealed by cross-cutting relationships, timing of stylolitization, dissolution, and dolomitization.

Perhaps the most utilized technique integrating these observations is to prepare a time–process diagram (Figure 31.9). This is not only a stringent intellectual

HIGHSTAND SYSTEMS TRACT

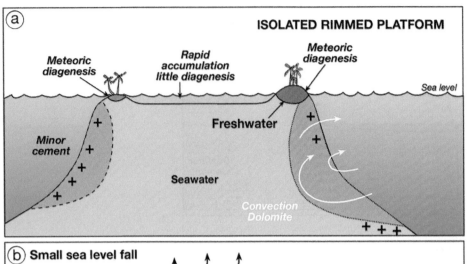

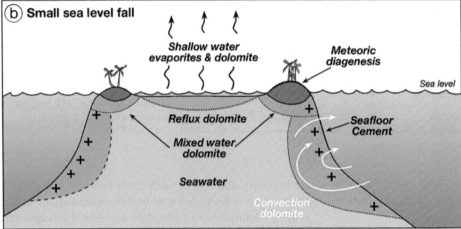

Figure 31.8 The nature of early diagenesis on an isolated rimmed platform: (a) during major third-order HST; and (b) during a fourth- or fifth-order sea-level fall.

CARBONATE PARAGENESIS

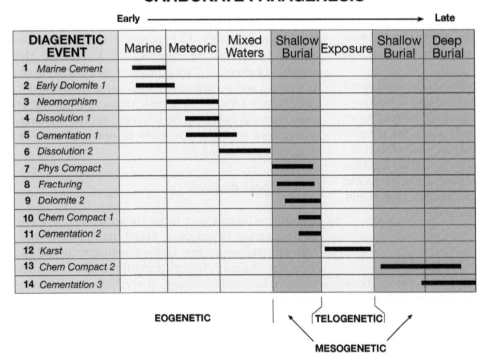

DIAGENETIC EVENT	Marine	Meteoric	Mixed Waters	Shallow Burial	Exposure	Shallow Burial	Deep Burial
1 Marine Cement	▬▬						
2 Early Dolomite 1	▬▬						
3 Neomorphism		▬▬					
4 Dissolution 1		▬▬					
5 Cementation 1		▬▬▬					
6 Dissolution 2			▬▬▬				
7 Phys Compact				▬▬			
8 Fracturing				▬▬			
9 Dolomite 2				▬▬			
10 Chem Compact 1				▬▬			
11 Cementation 2				▬▬			
12 Karst					▬▬		
13 Chem Compact 2						▬▬▬	
14 Cementation 3							▬▬

EOGENETIC

TELOGENETIC

MESOGENETIC

Figure 31.9 Hypothetical plot of diagenetic events versus their relative timing in different diagenetic realms. Dark bars: timing.

DIAGENESIS - BURIAL - TIME PLOT

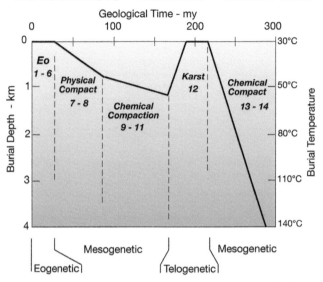

Figure 31.10 A simplified burial plot- time curve illustrating the various diagenetic events plotted against burial depth with an intervening period of exposure; numbers correspond to those in Figure 31.9.

exercise for the investigator but also an extremely useful tool for illustrating the history of the rocks to others.

Eogenetic diagenesis, as outlined earlier in this chapter, begins on the seafloor (Figure 31.9), with local cementation in neritic environments and potential dolomite formation on muddy tidal flats. With a fall in relative sea level and flushing by fresh waters in the meteoric realm, the entire spectrum of changes including neomorphism, dissolution, and cementation can take place. Massive dissolution with the possibility of some additional calcite cementation is predicted as the carbonates pass into the zone of mixed seawater and fresh water.

Physical or mechanical compaction dominates processes in shallow parts of the mesogenetic zone with the potential for dolomitization via a variety of shallow subsurface processes. Rocks here can also be subject to fracturing via compaction or even tectonics, but not always. Chemical compaction becomes more important deeper in the mesogenetic zone with ever-increasing burial, and is manifest as stylolites, solution seams, major pressure-related dissolution, and cementation.

These rocks can continue to be buried or exposed via a variety of processes and, once again, flushed by fresh waters. Now the rocks are all calcite, relatively well cemented, and with comparatively little porosity; the

main processes in the telogenetic zone during re-exposure is karst. This can be of variable duration, but generally results in increased porosity. On the other hand, if burial continues or resumes after karstification, then burial diagenesis continues until almost all pore space is lost. As the petroleum geologist would say: "it is tombstone."

Diagenesis–geohistory diagram

Another useful and particularly helpful way of illustrating progressive diagenesis is by a geohistory diagram (Figure 31.10). Such a figure is generated by plotting burial depth against time and burial temperature if known. For this approach to work, it is critical to know the exact age of the units in question, the geological and tectonic history, and the relative timing of diagenetic events. It quickly becomes clear how little time is spent in the eogenetic zone and how much in the mesogenetic zone.

This approach also allows the investigator to place the events related to hydrocarbon or mineral formation in a temporal and paragenetic framework. Perhaps more importantly, it frames the generation or obliteration of primary and secondary porosity in an understandable framework.

Further reading

Braithwaite, C.J.R. and Montaggioni, L.F. (2009) The Great Barrier Reef: a 700 000 year diagenetic history. *Sedimentology*, 56, 1591–1622.

Braithwaite, C.J.R. and Camoin, G.F. (2011) Diagenesis and sea-level change: lessons from Mururoa, French Polynesia. *Sedimentology*, 58, 259–284.

Caron, V., Nelson, C.S., and Kamp, P.J.J. (2005) Sequence stratigraphic context of syndepositional diagenesis in cool-water shelf carbonates: Pliocene limestones, New Zealand. *Journal of Sedimentary Petrology*, 75, 231–250.

Choquette, P.W. and Pray, L.C. (1970) Geologic nomenclature and classification of porosity in sedimentary carbonates. *American Association of Petroleum Geologists Bulletin*, 54, 207–250.

Moore, C.H. (2001) *Carbonate Reservoirs; Porosity Evolution and Diagenesis in a Sequence Stratigraphic Framework*. Netherlands: Elsevier, Developments in Sedimentology.

Morad, S., Ketzer, J.M., and De Ros, L.F. (2013) *Linking Diagenesis to Sequence Stratigraphy*. Wiley-Blackwell, International Association of Sedimentologists, Special Publication no. 45.

Tucker, M.E. (1993) Carbonate Diagenesis and Sequence Stratigraphy. In: Wright V.P. (ed.) *Sedimentology Review*. Blackwell Publishing Ltd., Vol 1, pp 51–72.

CHAPTER 32
CARBONATE POROSITY

Frontispiece Molds of the stick-like coral *Acropora cervicornis*, Pleistocene, Barbados. Hammer 13 cm long.

Origin of Carbonate Sedimentary Rocks, First Edition. Noel P. James and Brian Jones.
© 2016 Noel P. James and Brian Jones. Published 2016 by John Wiley & Sons, Ltd.
Companion website: www.wiley.com/go/james/carbonaterocks

Introduction

Limestones and dolostones are hosts to some of the world's most prolific hydrocarbon reservoirs and richest metallic ore deposits. The ability of these rocks to house and transmit these resources is a function of their porosity and permeability patterns. The world's largest conventional oil reservoir, the Ghawar field in Saudi Arabia, exists because of the intergranular and intercrystalline pores that developed in Jurassic carbonates. Most Pb–Zn mines on the planet are in middle Paleozoic dolomites (Mississippi Valley type), where sulfides were precipitated in intercrystalline pores and dissolution voids.

In terms of their porosity and permeability, carbonate rocks are significantly different to siliciclastic rocks. Porosity in sandstones, for example, is largely the product of grain size and sorting and therefore fundamentally related to the hydrodynamic regimen that controlled deposition of the sands. Diagenesis generally has little impact on sandstone porosity, apart from those situations where quartz overgrowths develop or authigenic clays are precipitated in the pores. Irrespective, porosity in sandstones is generally homogeneous at all scales and hence predictable. Porosity levels derived from the analysis of small samples can generally be upscaled with little trouble, a fact that is especially important when modeling reservoirs.

Porosity

Porosity in limestones and dolostones (Frontispiece) is typically heterogeneous at all scales because a myriad of diagenetic processes can be superimposed on the original depositional porosity. This is well illustrated by the fact that carbonate sediment with a depositional porosity of 40% may evolve into a lithified limestone with 40% porosity; the porosity in the limestone, however, can be completely different in style from that found in the original sediment. As limestones and dolostones evolve through different diagenetic regimes, porosity will be destroyed and created. Such changes typically follow a spatially variable pattern that is controlled by the manner in which fluids pass through the rocks at the microscale. Irrespective, the end result is generally a carbonate rock with a highly variable pattern of porosity and permeability. As a result, porosity in carbonates is notoriously difficult to predict and upscaling from small samples to the reservoir scale can be problematic.

Porosity measurement

Porosity, which is expressed as the percentage of spaces in a sample, can be measured and characterized using many different techniques. Even with the most sophisticated techniques available, however, the determination of porosity in carbonate rocks can be a challenge. Methods include whole-core analysis, thin-section analysis, computed tomography (CT), micro-CT, and mercury injection porosimetry. It should be noted, however, that the same sample analyzed using different methods can produce different results! This reflects the variability of carbonate porosity, the limitations associated with each method, and user-selected parameters that are usually required with these methods. Some methods simply provide the porosity whereas other methods will also provide information regarding the size and shape of the pores.

Whole-core porosity

Cores for this analysis, which must be at least twice as long as their diameter, need to be free of fractures or holes that cut across the core. In this method, the core is completely encased by a rubber sleeve and a gas (usually helium) is injected into the core through one end. Boyle's Law is then used to calculate the total porosity of the sample. As such, the porosity is the average for the entire sample.

Thin-section analysis

Thin-sections impregnated with blue epoxy are used to determine porosity by visual estimates, point counting, or computer-assisted analysis. All of these methods assess the area of pores (highlighted by the blue epoxy) relative to the area of the rock evident in thin-section. Visual estimates depend on the skill of the analyst, whereas point counting yields a more rigorously defined number. Computerized image analysis systems, based on scans of thin-sections, have also been developed for such calculations. Thin-sections cut from cores that have porosity determined by whole-core porosity analysis will commonly yield significantly lower porosity values, irrespective of the method used.

Computed tomography

Borrowed from the medical profession, this system uses computer-processed X-rays to produce three-dimensional views of the porosity in a rock. As a core is fed through the system, a large number of two-dimensional X-rays are acquired with each one being a transverse slice across the core. This system works because the pores are air filled and hence have a different density than the rock. The computer then assembles those images into a three-dimensional picture that shows the internal porosity. Such images can also be manipulated so that they can be viewed from any direction and artificial cross-sections through the rock can be generated. Other analytical techniques can then be used to calculate the porosity of the rock and measure the size of the pores.

CT analyses can be performed at two different scales. Normal CT scans can deal with rocks of virtually any size with lengths of large-diameter core being easily accommodated. If details of the microporosity are needed, then micro-CT can be used. This system, which usually uses rock cubes with sides that are about 1.5 cm long, can delineate pores that are as small as 8–10 μm. This operates on the same principle as the larger CT scanners, but is designed to deal with the microporosity that is common and so important in many carbonates (e.g., chalk).

Mercury injection porosimetry

This method is based on mercury being injected into the pores within a rock. With ever-increasing pressure, mercury is forced into increasingly smaller pores. This relationship allows calculation of the size of the pores in the rock with injection pressures up to 15,000 psi, enabling detection of pores as small as 0.01 μm. Such analyses not only yield total porosity but also produce statistical data regarding the size ranges of the pores in a rock sample. Given the highly toxic nature of mercury, samples must be destroyed following this type of analysis.

Potential problems

As with any analytical technique, care must be taken in using the different methods of assessing porosity in a sample because different analyses of samples from the same parent rock can produce different results for the following reasons:

- *Scale*: Some analyses are based on very small samples (e.g., micro-CT with samples no bigger than 1.5 cm), whereas other analyses are based on samples that are orders of magnitude larger (e.g., whole-core analysis).
- *User-defined thresholds*: Many analytical systems function on user-defined parameters (e.g., color thresholds defined for image analysis of thin-sections impregnated with blue epoxy). Different results can be obtained simply by changing the threshold levels for a particular parameter.

Permeability

Although the focus of this chapter is porosity, permeability is a critical aspect of the discussion. Permeability (K) is a measure of the ability of gases or fluids to flow through a rock. In effect, it is a measure of the connection between the pores in a rock. It is important to note, however, that high porosity does not necessarily mean high permeability, because individual pores may be isolated from each other (think Swiss cheese). Similarly, fractured rocks will have a high permeability even though the rock itself may have low porosity. It is virtually impossible to visually estimate the permeability of a carbonate rock. Rough assessments can be made based on how readily and quickly a rock absorbs water that is poured onto its surface.

Permeability is usually measured in the laboratory using whole-core samples or small plugs that have been taken from a sample. The permeability, which is typically derived by passing air or nitrogen through the rock, is generally expressed in terms of a Darcy (D) or, more commonly, as milliDarcy (mD). Tight limestones and dolostones will have permeabilities of <1 mD, porous carbonates typical of oil reservoirs have permeabilities of 100–10,000 mD, and fractured rocks may have permeabilities as high as 10^8 mD. Given that most limestones and dolostones are heterogeneous, each core sample is usually characterized by three permeability values: (1) K_{max}, which is the maximum permeability measured at 90° to the core length; (2) K_{90}, which is the permeability measured at 90° to K_{max}; and (3) K_{vert}, which is permeability measured in a direction that is at 90° to both K_{max} and K_{90}.

Types of porosity

The simplest, non-genetic scheme for classifying porosity in a limestone or dolostone is based on the notion that porosity may be fabric-selective or non-fabric-selective. Fabric-selective porosity refers to any openings that can be directly related to the original components and fabric of the rock (e.g., particles, fossils, crystals). Non-fabric-selective porosity includes all those pores and cavities that cut across the grains and depositional fabrics of the original rock. As with many schemes, there are those attributes that can be either, such as breccias, borings, burrows, and desiccation cracks.

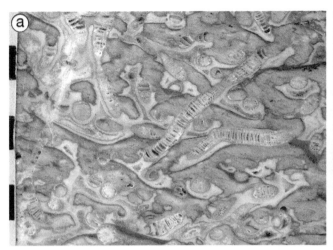
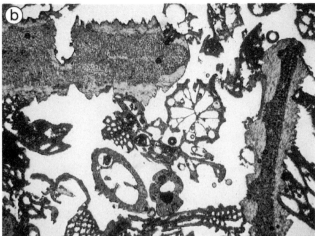

Figure 32.1 Primary depositional porosity. (a) Intraparticle porosity in the colonial rugose coral *Phacellophyllum* sp., subsurface Upper Devonian, Western Canadian Sedimentary Basin. Centimeter scale. Photograph by W. Martindale. Reproduced with permission. (b) Interparticle porosity between echinoids and bryozoans in Oligocene limestone, southern Australia, thin-section (PPL). Image width 2 cm.

Types of fabric-selective porosity

Intraparticle pores. This type of porosity, which is inside the grains, reflects the fact that biofragments are common elements of many limestones. The size, shape, and volume of such porosity is directly related to the skeletal architecture of the biofragments. Different genera of modern corals, for example, have vastly different porosities because of their different skeletal architectures. Biofragments with high internal porosity include corals of all types (Figure 32.1a), benthic and pelagic foraminifers, gastropods, and rudist bivalves.

Interparticle pores. Referring to pore space between particles (Figure 32.1b), this porosity element depends on the size and packing of the grains that form the sediment. In most modern grainy sediments, porosity is between 40 and 50% because of the irregular nature of the grains and loose packing. Sand shoals, especially those formed of ooids, commonly have very high interparticle porosity, whereas bioclastic sands tend to have slightly lower porosities. Surprisingly, modern muddy sediments can have primary porosities of between 44 and 75%.

Intercrystalline and intracrystalline porosity. Common in dolostones, intercrystalline porosity refers to the pores between the constituent crystals (Figure 32.2), whereas intracrystalline porosity refers to those pores that are inside the constituent crystals. The size and shape of the intercrystalline pores depend on the size, shape, and packing of the constituent crystals. The size and shape of the intracrystalline pores depends on the size of the

Figure 32.2 SEM image of finely crystalline dolomite rhombs with excellent intercrystalline porosity, Cayman Formation, Miocene. Grand Cayman.

crystals and the degree to which their interiors have been dissolved. In some dolostones, the amount of porosity associated with the development of hollow rhombs can be as important as the intercrystalline porosity.

Moldic porosity. Moldic porosity (Figure 32.3) develops through the preferential dissolution of various components of a limestone or dolostone. In many limestones, for example, the aragonitic skeletons of corals, bivalves, or gastropods may be completely lost through dissolution. In other cases, selective dissolution of ooids can produce

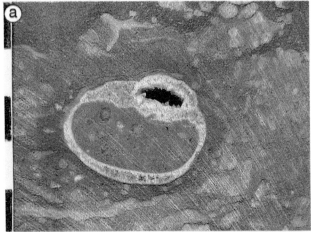

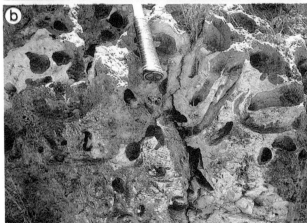

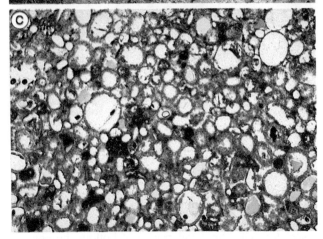

Figure 32.3 Secondary diagenetic porosity. (a) A gastropod mold partially filled with burial cement, Upper Devonian subsurface, Western Canadian Sedimentary Basin. Centimeter scale. (b) Molds of the aragonitic portion of caprinid rudists, Mexico. Hammer handle 3 cm wide. (c) Molds of aragonite ooids partially surrounded by LMC cement, Pleistocene, South Florida. Image width 4 mm.

oomoldic porosity. The size and shape of the pores will reflect the morphology of the original components. The loss of gastropod shells will typically produce small pores, whereas the loss of large domal corals can produce voids that are up to 3 m in diameter. The total of this style of porosity obviously depends on the density of the particles in the original limestone. Moldic porosity in some limestones and dolostones can be as high as 40–50%.

Fenestral porosity.   Also known as "bird's eye porosity," fenestral pores (Figure 12.8b) are typically less than 5 mm long and elongate parallel to sediment laminations. Generally associated with calcareous mud that forms on tidal flats, this type of porosity is attributed to gases that develop with the decay of organic matter (from cyanobacterial mats). Porosities of up to 45% are known.

Shelter porosity.   Shelter porosity refers to the spaces beneath shells that were deposited in a convex-up position (Figure 32.4). In general, this forms only a small part of the overall porosity of a rock.

Growth framework porosity.   This type of porosity develops as colonial organisms, such as corals and stromatoporoids, produce reefs and related structures. Voids of this type can be of any size and shape because it depends solely on the growth patterns of the formative organisms. The total porosity produced in this manner can be high.

Types of non-fabric selective porosity

Fractures.   Fractures, which reflect the stress regimes that have affected a rock, can have variable apertures (width of fracture) and vertical extent. Because of their potential

Figure 32.4 Brachiopods with shelter porosity beneath the shells, Permian, West Texas. Finger width 1 cm.

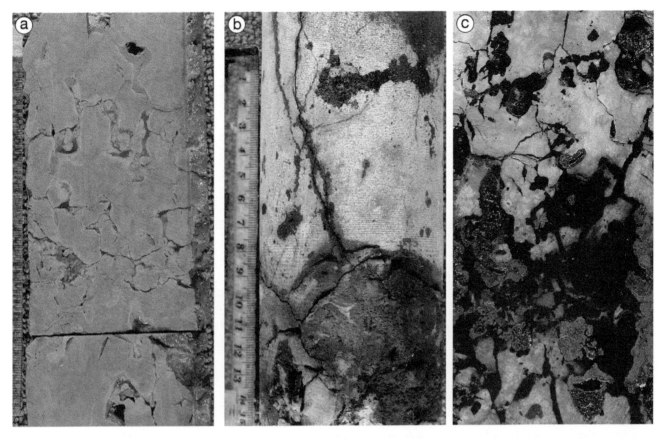

Figure 32.5 Several examples of fracture porosity in subsurface cores of finely crystalline dolostones from the Upper Devonian Grosmont Formation, Western Canadian Sedimentary Basin. (a) Fractures connecting vugs; and (b, c) Subvertical to vertical fractures highlighted by bitumen (black) that connect vugs. All cores 7.5 cm wide.

interaction and ramifying nature, fractures can create excellent porosity but are perhaps more important as permeability conduits. This is especially true when the fractures connect pores in relatively impermeable limestones and dolostones (Figure 32.5).

Fractures can develop from either extensional or compressional stress. Most fractures, however, are the result of faulting, folding, solution collapse, salt dome movement, and overpressuring. Fractures commonly become wider as aggressive waters flow along them.

Channels, vugs, and caverns. These types of porosity (Figure 32.6), which develop through bedrock dissolution, are usually associated with subsurface karst development. Distinguishing between channels, vugs, and caverns is difficult because there are no official definitions for each type. In general, however, caves and caverns have been defined as cavities in which a person can stand. In most cases, the development of these features is not related to the original fabric of the limestones and dolostones.

The size, shape, and distribution of these features are controlled by length of time and the acidity of the fluids that passed through the rocks. Long networks that include vast caves connected by channels that are many kilometers long characterize some areas.

Fabric-selective or not

Breccias. High porosity is commonly associated with breccias that developed as a result of solution collapse, karst, and hydrothermal processes.

Solution-collapse breccias form where groundwaters dissolve evaporites that lie beneath a limestone–dolostone sequence. As the evaporites are dissolved, support for the overlying rocks is lost and collapse occurs. Commonly sudden and catastrophic, this leads to brecciation of the rocks, with porosity developing between the chaotically arranged lithoclasts. The size and shape of the pores is a function of the size and packing of the breccia clasts.

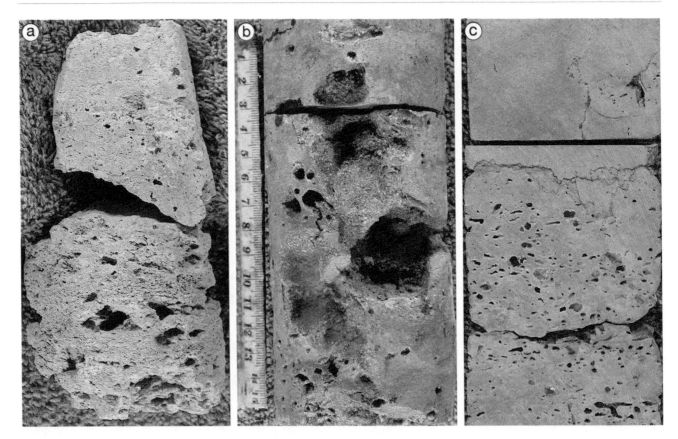

Figure 32.6 Vuggy porosity in finely crystalline dolostones (after bitumen removed) of the Upper Devonian Grosmont Formation, subsurface of Western Canadian Sedimentary Basin. (a) Mixture of small, irregular-shaped vugs of unknown origin and pinpoint porosity. (b) Large irregular-shaped vugs (possibly formed after burrows) with small vugs of unknown origin. (c) Core with little visible porosity (upper part) overlying core with well-developed small vuggy porosity. Boundary between the different parts of the core is a low-amplitude stylolite. All cores 7.5 cm wide.

Karst solution-collapse breccias are a common component of subsurface karst systems. Subsurface dissolution of the limestone can lead to a structurally unstable situation where the underground chamber will collapse inward. The breccias that result from this process have irregular particle packing and high porosity (Figure 32.7). These breccias are typically associated with large-scale unconformities.

Hydrothermal breccia porosity (see Chapter 29) is caused by hydrofracturing (Figure 32.8) as hot, pressurized corrosive fluids ascend from depth and dissolve large amounts of the host carbonate. Although these fractures are commonly filled with saddle dolomite, base metals (galena and sphalerite), or hydrocarbons, some will remain open and hence contribute to the overall porosity of the rock body.

Borings and burrows.   Borings (hard substrates) and burrows (soft substrates), irrespective of their origin, produce tubular structures that may or may not be related to the original fabric of the rock. Such structures range from the submicron borings produced by microbes to large-diameter (up

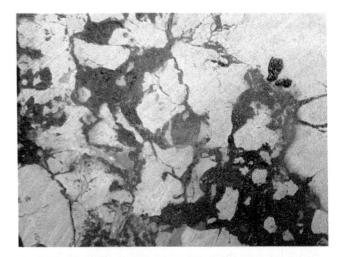

Figure 32.7 Breccia porosity in finely crystalline dolostone from the Upper Devonian Grosmont Formation, Western Canadian Sedimentary Basin. Bitumen (black) impregnation is confined to intercrystalline matrix porosity between the dolostone fragments. Image width 3 cm.

Figure 32.8 Hydrofractures filled with saddle dolomite, Cambrian, front ranges of the Rocky Mountains, Alberta. Scale 2 cm increments.

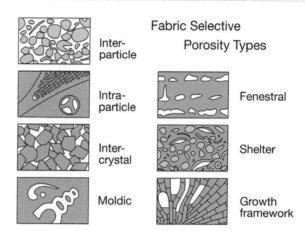

Figure 32.10 Choquette and Pray (1970) porosity classification: sketches of fabric-selective pore types. Source: Choquette and Pray (1970). Reproduced with permission of the American Association of Petroleum Geologists.

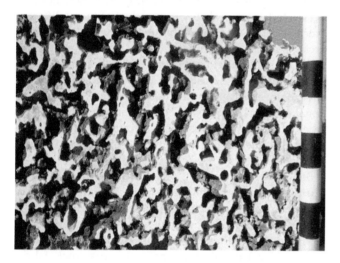

Figure 32.9 Slab of Pleistocene Miami Limestone where the oolite grainstone to packstone was intensively burrowed by shrimp (*Callianassa* sp.). These burrows have remained open because the rock was never buried. Scale divisions 10 cm.

to 5 cm) long burrow systems produced by various burrowing organisms, such as worms and shrimps (Figure 32.9). While open, these structures can form a significant part of the permeability and porosity in a limestone or dolostone.

Shrinkage. Shrinkage, typically evident as desiccation cracks, generally forms only a minor part of the porosity in a limestone or dolostone.

Porosity classification

The classification of porosity in carbonate rocks is fraught with problems because it is heterogeneous and commonly reflects numerous processes that have developed through many diagenetic cycles. The classification systems proposed by Choquette and Pray (1970) and Lucia (1999) approach the issue from two different directions. Choquette and Pray (1970) based their classification on the relationship between the pores and the fabric of the rock, so it is therefore largely descriptive in nature. In contrast, the classification proposed by Lucia (1999) is genetically based with the notion that the pore-size distribution controls permeability and fluid saturation with recognition that pore size distribution is related to rock fabric.

Choquette and Pray classification

This classification system is based on the spatial relationship between the pores and the original fabric of the rock (Figures 32.10, 32.11). Given that different types of porosity characterize many limestones and dolostones, it is important to recognize that full characterization of the porosity may require the use of multiple terms.

This classification scheme, at its basic level, employs descriptive terms to reflect the nature of the porosity. As such, it avoids many of the difficulties and problems that are associated with genetically based systems. Nevertheless, a genetic facet can be introduced by using modifiers that reflect some of the processes that have led to the development of the pores. For example, a solution-enlarged fracture reflects the fact that an original fracture was widened as a result of dissolution of the fracture walls.

Lucia classification

This more recent scheme is focused towards the petroleum industry. The classification (Figures 32.12, 32.13) is based on the concept that pore-size distribution controls permeability and the pore-size distribution is related to rock fabric. To establish this relationship, it is necessary to determine whether the pore space belongs to one of three major pore-type classes: *interparticle*; *separate-vug*; or *touching-vug*. It is equally as important to determine the volume of pore space in these various classes. The classification is based on the same component relationships as the rock classification of Dunham (1962; see Chapter 4) except that total rock fabric, including both depositional and diagenetic attributes, is utilized. Fabrics are therefore divided into *grain-dominated* and *mud-dominated*.

Intergranular pore space (Figure 32.12) is present in grainstones and some packstones where the pores are only partially filled with mud. Separate-vug pores (Figure 32.13) are defined as pore space that is interconnected through the interparticle porosity. By contrast, touching-vug pores (Figure 32.13) are defined as pore space that forms an interconnected pore system independent of the interparticle porosity.

Porosity evolution through time

Complex fabrics characterize many limestones and dolostones. The porosity and permeability characteristics that we see today are probably the product of many different processes that developed through numerous diagenetic cycles that are superimposed upon the original depositional fabric. Central to this issue is the fact that porosity in limestones and dolostones can be created or destroyed at virtually any stage during their evolution. Post-depositional creation of porosity comes about by dissolution, whereas the loss of porosity can be caused by the precipitation of cements, the deposition of internal sediments, and compaction.

Subaerial exposure of carbonate successions, be it during eogenesis or telogenesis, is particularly important because subsurface karst with aggressive waters is one of the most potent porosity-creating processes as it leads to the creation of vugs, channels, and caves (Frontispiece) of all sizes that cut across rock fabric. Some of the largest hydrocarbon reservoirs in the world

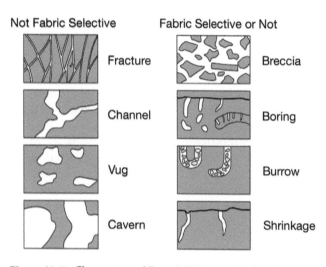

Figure 32.11 Choquette and Pray (1970) porosity classification: sketches of non-fabric-selective and other pore types. Source: Choquette and Pray (1970). Reproduced with permission of the American Association of Petroleum Geologists.

INTERPARTICLE PORE SPACE
Particle size ≠ sorting
(matrix interconnection)

GRAIN DOMINATED FABRIC   **MUD DOMINATED FABRIC**   **CRYSTALLINE**

GRAINSTONE/PACKSTONE   WACKESTONE/MUDSTONE

PERCENT INTERPARTICLE POROSITY

LIMESTONE or DOLOSTONE

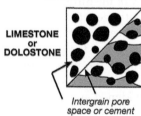

Intergrain pore space or cement

Intercrystal pore space

Figure 32.12 Lucia (1999) porosity classification: a simplified version of the classification illustrating the different types of pores present in grain-dominated, mud-dominated, and crystal-dominated carbonate rocks. Source: Lucia (1999). Reproduced with permission of Springer Science+Business Media.

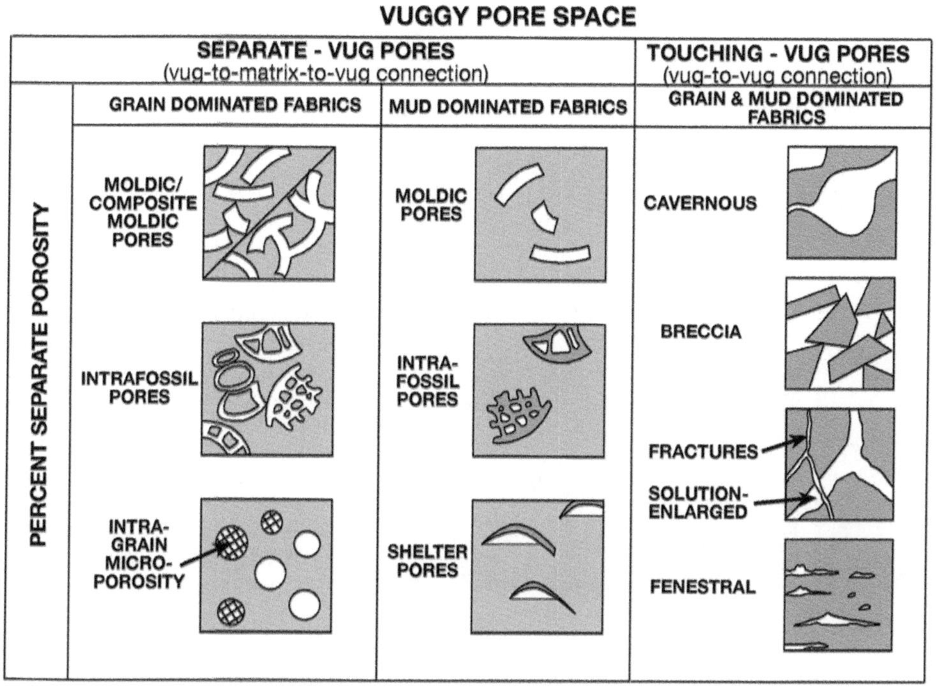

Figure 32.13 Lucia (1999) porosity classification: a simplified version of the classification based on the characteristics of vug interconnection. Source: Lucia (1999). Reproduced with permission of Springer Science+Business Media.

are found in successions that are characterized by karst-generated non-fabric-selective cavities. It must also be remembered however that subsurface karst systems can also be characterized by the precipitation of cements or the deposition of internal sediments, processes that will lead to porosity reduction.

As limestones and dolostones are buried, the dominant diagenetic change is compaction due to overburden pressure. Compaction involving both chemical and mechanical processes generally leads to porosity reduction. Nevertheless, there is dissolution in the subsurface. The most important processes seem to be related to hydrocarbon maturation, shale dewatering, dolomitization, and deep penetration of meteoric waters related to orogenesis. These reactions, combined with increased temperatures related to the geothermal gradient and high lithostatic and hydrostatic pressures, can result in aggressive fluids that lead to water-driven dissolution that typically produces non-fabric-selective porosity.

Porosity and dolomitization

Many dolostones are characterized by high porosities, with some of the most productive hydrocarbon reservoirs and freshwater aquifers being housed in these rocks. Such dolostones typically have high intercrystalline porosity (Figure 32.6) and many also have high intracrystalline

porosity as hollow dolomite crystals develop. With fabric-retentive dolostones, fossil-moldic porosity may also be common (Figure 32.14). Irrespective of the style of dolomitization, intercrystalline porosity can only develop if there has been significant carbonate dissolution that caused selective dissolution of the sediments that once existed between the dolomite crystals. The density difference between calcite and dolomite means that when calcite is dolomitized, there is an increase of 7% porosity. Although replicated in the laboratory, this seems to have little relevance in the real world.

Hydrothermal dolostones (Figure 32.15), which can have the most impressive porosity of any carbonates, are typified by hydrofracturing that forces the rock apart to create voids of various morphologies.

The evolution of porosity

One of the most useful ways to understand and convey the evolution of porosity in a formation is to plot a time–porosity diagram. Two examples are taken from Middle and Upper Devonian reef hydrocarbon reservoirs in the Western Canadian Sedimentary Basin (Figure 32.16). At the time of writing, these limestones and dolomites have reserves of roughly 15 billion barrels of oil and 35 trillion cubic feet of gas. Each carbonate reservoir is a reef-rimmed carbonate platform or bank dominated by

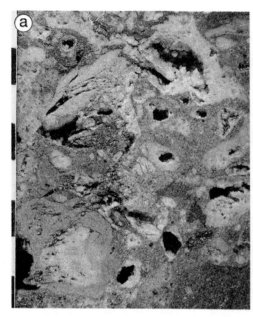

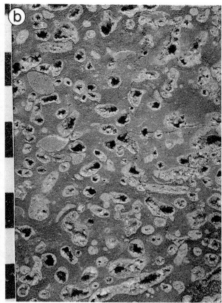

Figure 32.14 Burial dolomite in core middle Upper Devonian Swan Hills Formation, subsurface west-central Alberta, Canada. (a) Vuggy porosity developed in dolomitized stromatoporoids; and (b) moldic-vuggy porosity developed in dolomitized *Amphipora* sp. sticks. Centimeter scale in both panels. Photographs by W. Martindale. Reproduced with permission.

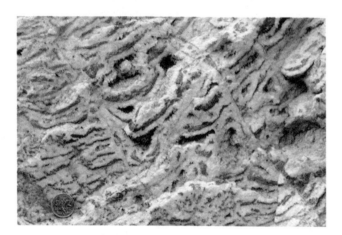

Figure 32.15 Hydrothermal saddle dolomite with excellent porosity, Middle Cambrian, Western Alberta. Coin scale 1 cm. Photograph by W. Martindale. Reproduced with permission.

stromatoporoids that has undergone (depending on the reef) diagenesis in marine, meteoric, and burial environments, as well as dolomitization in different realms. The depth–porosity plots for each example are for the bank margin and bank interior.

Upper Devonian Leduc Formation limestone reef

Bank margin. Synsedimentary marine cement almost filled all depositional porosity and reduced an initial 50% depositional porosity to <20%. Subsequent meteoric diagenesis lowered porosity by only a few percent. This was probably because the climate was semi-arid, periods of exposure of the reef margin were short-lived, and the sediments are largely calcitic to begin with and so cement precipitation was minor. Burial cementation decreased final porosity to levels of ~7%.

Bank interior. The bank interior had less than half of the synsedimentary cement compared to the margin and so porosity was initially better. Meteoric diagenesis was variable and reduced porosity by ~10%. The limestones contain much burial cement due to extensive chemical compaction of the comparatively poorly cemented sediments, reducing the porosity by a further 7–10% to final values of ~14%.

Middle Devonian Slave Point Formation dolomite reefs

Reef margin. These reefs are similar to the Leduc example, but had less depositional relief above the seafloor with only low-relief margins. As a result, there is only about one-half the synsedimentary cement that reduced initial porosity from ~50% to ~35%. Reefs show little evidence of subaerial exposure, so meteoric diagenesis was largely ineffective. Shallow-burial diagenetic dolomitization was profound and increased porosity by

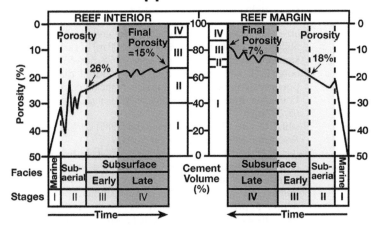

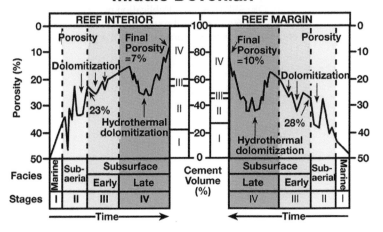

Figure 32.16 Diagenesis and reservoir history for (a) Leduc Formation reefs limestone in the Golden Spike field; and (b) Slave Formation reefs. Reef interiors and reef margins are compared. Porosity evolution through time and cement volume in % are estimated. Source: Adapted from Walls and Burrowes (1985). Reproduced with permission of the SEPM Society for Sedimentary Geology.

as much as 10%. Hydrothermal dolomitization initially increased porosity to values as high as 35%, only to have them reduced to final values of ~10% by saddle dolomite precipitation.

Reef interior. The trend is similar in the interior of the above reefs with only ~15% porosity loss via marine cements. Diagenetic dolomite decreased porosity via precipitation and then increased porosity by dissolution with a net porosity loss of ~10%. Burial and hydrothermal dolomitization, however, lead to: (1) porosity loss because of dolomitization; (2) porosity increase because of dissolution; and (3) porosity loss via saddle dolomite precipitation. The final porosity in these rocks is about 7%.

Integration

These two actual examples highlight several critical aspects of the history of carbonate porosity development through time. The major implications include the following:

- porosity can be increased or decreased as the sediments and rocks pass from one to another diagenetic realm to another;
- porosity is generally lost in the marine diagenetic realm;
- porosity can be gained or lost in the meteoric realm;
- overall, if deeply buried, original depositional porosity is reduced;
- dolomitization can increase porosity but in the final analysis, it does not always increase pore space; and

- carbonate units in a sedimentary basin can have widely divergent diagenetic histories and therefore different porosities.

Further reading

Braithwaite, C.J.R., Rizzi, G., and Darke, G. (eds) (2004) *The Geometry and Petrogenesis of Dolomite Hydrocarbon Reservoirs*. London: Geological Society, Special Publication no. 235.

Choquette, P.W. and Pray, L.C. (1970) Geologic nomenclature and classification of porosity in sedimentary carbonates. *American Association of Petroleum Geologists Bulletin*, 54, 207–250.

Choquette, P.W., Cox, A., and Meyers, W.J. (1992) Characteristics, distribution and origin of porosity in shelf dolostones: Burlington-Keokuk Formation (Mississippian), US mid-continent. *Journal of Sedimentary Petrology*, 62, 167–189.

Dunham, R.J. (1962) Classification of carbonate rocks according to their depositional texture. In: Ham W.E. (ed.) *Classification of Carbonate Rocks*. Tulsa, OK: American Association of Petroleum Geologists, Memoir no. 1, pp. 108–121.

Lucia, F.L. (1999) *Carbonate Reservoir Characterization*. Berlin, Heidelberg, New York: Springer-Verlag.

Melim, L.A., Anselmetti, F.S., and Eberli, G.P. (2001) The importance of pore type on permeability of Neogene carbonates, Great Bahama Bank. In: Ginsburg R.N. (ed.) *Subsurface Geology of a Prograding Carbonate Platform Margin, Great Bahama Bank; Results of the Bahamas Drilling Project*. Tulsa, OK: SEPM (Society for Sedimentary Geology), Special Publication no. 70, pp. 217–238.

Melzer, S.E. and Budd, D.A. (2008) Retention of high permeability during shallow burial (300 to 500 m) of carbonate grainstones. *Journal of Sedimentary Research*, 78, 548–561.

Moore, C.H. and Druckman, Y. (1981) Burial diagenesis and porosity evolution, Upper Jurassic Smackover, Arkansas and Louisiana. *American Association of Petroleum Geologists Bulletin*, 65, 597–628.

Roehl, P.O. and Choquette, P.W. (eds) (1985) *Carbonate Petroleum Reservoirs*. New York: Springer-Verlag.

Saller, A.H. and Henderson, N. (1998) Distribution of porosity and permeability in platform dolomites: insight from the Permian of West Texas. *AAPG Bulletin*, 82, 1528–1550.

Schmoker, J.W. and Halley, R.B. (1982) Carbonate porosity *versus* depth: a predictable relation for south Florida. *American Association of Petroleum Geologists Bulletin*, 66, 2561–2570.

Walls, R.A. and Burrowes, G. (1985) The role of cementation in the diagenetic history of Devonian reefs, western Canada. In: Schneidermann N. and Harris P.M. (eds) *Carbonate Cements*. Tulsa, OK: SEPM, Special Publication no. 36, pp. 185–220.

Weyl, P.K. (1960) Porosity through dolomitization: conservation-of-mass requirements. *Journal of Sedimentary Petrology*, 30, 85–90.

GLOSSARY

Abiogenic: Formed by inorganic processes. No involvement of animals or plants.

Acicular: A fibrous or needle-like growth form of calcite in which crystals have a length to width ratio greater than 6:1 and are less than 10 μm wide.

Accommodation: Space available for sediments to fill, between the seafloor and sea level.

Accretionary slope: Slope where there is active sediment accumulation.

Aeolianite: Carbonate sediments that accumulate as aeolian sand dunes in marginal marine environments.

Aggregate grains: Particles formed of various types of grains that are bound together by biofilms, encrusting organisms, or synsedimentary cement.

Ahermatypic corals: Non-reef building corals that lack photosymbionts such as zooxanthellae.

Allochem: One or several varieties of discrete and organized carbonate aggregates (skeletal fragments, pellets, peloids, intraclasts, ooids, and others).

Allochthonous: Materials that formed somewhere other than their present location.

Allogenic: Sediment components that form elsewhere and are transported to the place of deposition.

Allomicrite: Micrite formed by (a) precipitation in water column; (b) disintegration of algae and invertebrate skeletons; (c) byproduct of bioerosion; (d) secretions by fish; or (e) accumulation of calcareous plankton.

Anhedral: A single crystal that does not show well-defined or typical crystallographic forms (i.e., crystal faces are absent).

Anoxia: Complete lack of oxygen in water.

Aphotic: Zone in water column below the base of light penetration, no light.

Aquitard: Impermeable layer of rock.

Aragonite: Anhydrous crystalline form of $CaCO_3$, orthorhombic system.

Atoll: Bank with reef around the edge surrounding a central lagoon, commonly associated with a subsiding volcano.

Authigenic: Minerals that grow in place with a rock or sediment, rather than having been transported and deposited.

Autochthonous: Formed or produced in the place where now found.

Automicrite: Micrite precipitated as a cement on seafloor or in sediment.

Autotroph: Organism that makes its own food by photosynthesis or chemosynthesis.

Back reef: Area on landward or lagoonal side of barrier reef.

Balance-filled lake: Intermediate between underfilled and overfilled lakes.

Bank: Isolated platform surrounded by deep water.

Beachrock: Beach sands that are cemented to produce hard sheets of limestone that generally dip seaward.

Benthos: Organisms that live on the seafloor.

Bioerosion: Breakdown of hard substrates by actions of various organisms.

Biogenic: Formed through direct or indirect processes mediated by animals or plants.

Biogenic ooze: Very-fine-grained pelagic sediment with at least 30% skeletal remains.

Origin of Carbonate Sedimentary Rocks, First Edition. Noel P. James and Brian Jones.
© 2016 Noel P. James and Brian Jones. Published 2016 by John Wiley & Sons, Ltd.
Companion website: www.wiley.com/go/james/carbonaterocks

Bioherm: Lens-shaped bioaccumulation with all the attributes of a reef.

Biostrome: Tabular-shaped bioaccumulation with all the attributes of a reef.

Boring: Hole produced in solid substrate by an organism (e.g., microbe, boring bivalve).

Breccia: A rock formed of large angular fragments (lithoclasts) held in a fine-grained matrix.

Brine: Water containing more dissolved inorganic salt than typical seawater.

Burrows: Holes produced in soft sediment by various organisms (e.g., worms, shrimps).

Bypass slope: Also known as an "escarpment" slope, sediment bypasses the slope and accumulates on adjacent toe-of-slope or on the basin floor.

Calcarenite: A limestone composed largely (more than 50%) of sand-sized calcium carbonate grains (a lithified carbonate sand).

Calcian dolomite: Dolomite with excess Ca (>50%) and disordered crystal chemistry.

Calcimicrobe: Calcification of a microbe cell wall before the organism dies, thus preserving the cell.

Calcite raft: Thin floating mass of precipitated $CaCO_3$ crystals that forms at surface of a pool, usually due to degassing.

Calcite: Anhydrous form of $CaCO_3$, hexagonal system.

Calcitization: Alteration of mineral to calcite (commonly applied to change of aragonite to calcite) with partial preservation of fabric.

Calcrete (caliche): Sand, gravel, or cobble-sized sediment that is cemented by calcium carbonate in semi-arid climates as a result of evaporative concentration of $CaCO_3$ by microbes in soils or protosoils.

Cap carbonate: Limestone and dolostone, deposited directly on top of glacigene sediments (mostly Neoproterozoic).

Carbonate buildup: Non-genetic term applied to any carbonate mass that rises above surrounding seafloor.

Carbonate compensation depth: The water depth at which a carbonate becomes completely dissolved in the ocean.

Cathodoluminescence: Light emitted by a substance undergoing bombardment by cathode rays.

Cave pearl: Coated grain that forms in a cave, commonly with a shiny surface.

Cement: Passively precipitated mineral between grains that serves to bind them together.

Cementstone: Carbonate rock in which cements form a significant volume of the rock (may exceed skeletal grains).

Chalk: A microcrystalline, pelagic, post-Jurassic limestone composed of calcareous plankton, especially planktic foraminifers and coccolithophorids.

Chemocline: Thin layer in a water body where water chemistry changes more rapidly than in the layers above and below.

Cnidarian: Invertebrate marine animals characterized by a radially symmetrical body with a sac-like internal cavity that includes jellyfish, hydroids, sea anemones, and corals.

Column: Formed when a stalactite and stalagmite merge together.

Compromise boundary: Boundary developed and shared by two neighboring cement crystals as a result of competition for growth space.

Condensed succession: Thin successions of carbonates that formed over a long period of time.

Conoform stromatolite: Stromatolite that appears to be formed of stacked, inverted cones.

Constratal growth: Applied to reefs where vertical growth is about the same as the rate of sediment accumulation on surrounding seafloor, hence tops of organisms are only just above sediment level.

Contour current: Ocean current that follows bathymetric contours.

Contourite: Deposit formed from a contour current.

Cool spring: Spring with water temperature <20°C at vent.

Coral bleaching: Corals that turn white because of expulsion of the photosymbionts (zooxanthellae), usually due to environmental stress.

Corrosion: A geomorphic term equivalent to dissolution of carbonates.

Cross-bedding: Sediment layers that are internally composed of inclined laminations.

Debrite: Deposit from a debris flow.

Dedolomite: Dolomite that is partly to completely replaced by calcite.

Dendrite crystal: Crystal with branches similar to a tree.

Detritus: Particles of rock derived from the mechanical breakdown of preexisting rocks by weathering and erosion.

Dewatering: Loss of water from a sediment, usually during burial.

Diachronous: A sedimentary layer that is time transgressive (i.e., different parts of same bed are different in age).

Diagenesis: All processes involved in transformation of sediment to hard sedimentary rock above the zone of metamorphism.

Diamictite: Sedimentary rock formed of a wide range of lithified, non-sorted to sorted, detrital sediment (commonly glacigenic).

Diastasis cracks: Cracks in ooid-muddy facies produced by combined action of waves and currents.

Discharge apron: Area around a spring vent that waters discharged from spring flow over.

Dolomite: Mineral formed of CaMg(CO$_3$)$_2$, hexagonal system.

Draperies: Curtain-like structure hanging from cave ceiling formed of precipitated CaCO$_3$ that drips from a fissure in cave ceiling.

Drusy fabric: Crust formed of fine crystals (druse) lining rock cavity, generally with crystal sizes increasing from the edges to the centers of the pores.

Dysoxic: Water with very low oxygen content.

Ecological bioherm: Reef with significant constructional relief above surrounding seafloor.

Ekman spiral: Wind blowing across ocean surface causes water movement to right in Northern Hemisphere and to left in Southern Hemisphere.

Endolith: Organism that lives in a hard substrate.

Enfacial junction: Crystal triple junction where one angle is 180°.

Eogenetic Zone: Shallow near-surface diagenesis in the time interval between final deposition and burial to below the zone near-surface alteration processes.

Epeiric basins: Local, vast shallow depressions on platforms covered by epeiric seas.

Epeiric platform: Flooded craton, many thousands of kilometers in size, shallow seawater throughout.

Epeiric ramp: Low-gradient seafloors that rimmed intracratonic depressions.

Epeiric sea: Vast, shallow water sea that covered large areas of cratons.

Epifauna: Animal that lives on sediment.

Epilimnion layer: Topmost layer in a thermally stratified lake, overlying the hypolimnion layer.

Epilith: Organism that lives on a hard surface.

Epiphyte: Organism (e.g., diatom, foraminifer) that lives on plants (e.g., seagrass blade).

Epitaxial (also syntaxial): Cement overgrowth that is in optical continuity with the parent grain.

Eukaryotic organism: Organism with cell that contains a nucleus and other organelles enclosed in a membrane.

Euphotic zone: High light conditions in water column, ideal for photosynthetic organisms.

Eutrophic: High supply of nutrients resulting in oxygen-depleted water.

Exokarst: Surface karst.

Fabric: Describes orientation in space of particles, matrix, and cement.

Fabric-specific: Process or feature that is related to specific feature or component of a limestone or dolostone.

Fabric-destructive: Original fabric of rock lost due to recrystallization or replacement of original minerals.

Fabric-retentive: Original fabric of rock retained despite replacement and/or recrystallization of original minerals.

Facies: Area of deposition characterized by a particular type of sediment or biota.

Fair-weather wave base: Maximum depth at which the passage of a water wave under normal conditions causes significant water motion.

Fascicular-optic fibrous calcite: Cavity-filling calcite mosaic formed of fibrous crystals radiating away from the initial growth surface and showing characteristic divergent optic-axis pattern within crystals.

Fecal pellet: Oval to spherical grains, typically formed of micrite, generated by detritus feeding animals.

Fenestral porosity: Small pores, typically less than 5 mm long, probably a result of gas bubbles. Also known as "bird's eye porosity" when filled with cement.

Firmground: Intermediate state between a hardground and soft sediment. Sediment is firm enough to prevent burrowing by some organisms.

Flaser: Alternating sand and mud layers that form as a result of intermittent water flow, common in tidal settings.

Flowstone: CaCO$_3$ precipitated from flowing water.

Fore reef: Area on oceanward side of reef.

Fractionation: Process whereby a certain amount of a chemical mixture is divided into smaller quantities in which the composition varies according to a gradient.

Geopetal sediment: Sediment with a horizontal surface on the floor of cavity, commonly overlain by spar calcite cement.

Geyser: Periodic eruptions of forcefully ejected columns of water and steam.

Glendonite: Calcite pseudomorph of ikaite.

Grapestone: Cluster of ooids or peloids held together by filamentous microbes or cement in a grape-like cluster.

Growth-framework porosity: Pores that form as a result of growth of irregular-shaped colonial organisms in reefs.

Halohaline: Global seawater circulation driven by variations in density brought about by differences in salinity.

Hardground: Hard surface on seafloor created by early marine cementation of sediment.

Hermatypic corals: Corals that have photosymbionts (zooxanthellae) and are reef-building.

Herringbone cross-bedding: Cross-bedding with laminae going in opposite directions.

Heterotrophs: Organisms that derive their nourishment from consuming other organisms.

Heterozoan factory: Carbonate factory in which most calcareous organisms are heterotrophs.

Hypercalcifier: Organisms, including corals, that can rapidly produce large calcareous skeletons.

Hypertrophic: Extremely high supply of nutrients and resulting low oxygen levels in the water.

Ikaite: Hydrous CaCO$_3$ monoclinic system that forms in near-freezing waters (<7°C) of high alkalinity.

Infauna: Animals that live in the sediment.

Inside-out dolomite: Dolomite crystal in which core has been leached out and then subsequently filled with younger dolomite cement.

Intercrystalline porosity: Porosity between crystals.

Intergranular porosity: Pore spaces between grains.

Internal sediment: Sediment that is deposited in a cavity in a lithified rock or reef.

Intertidal zone: Zone between low water mark and high water mark.

Intraclast: Clasts (fragments) formed in the environment of deposition.

Intracrystalline porosity: Pores inside crystals.

Intragranular porosity: Pores inside grains.

Isopachous: Band of cement of equal thickness coating a particle or cavity wall.

Kelp: Large seaweeds belonging to the brown algae (Phaeophyceae) in the order Laminariales.

Keystone vug: Small- to medium-sized cavities inside a grainy limestone that form when air moves upward as water moves downward in sediment during wave surge up a beach, defining a hole somewhat larger than a simple intergranular pore.

Lag time: Time interval between initial submergence of an area and a fully functioning carbonate factory.

Leiolites: Microbialites with a microcrystalline microfabric.

Limpid dolomite: Dolomite that is clear and looks like an ice cube.

Littoral: Shallow water around lake or ocean margin.

Lysocline: Depth at which a carbonate mineral starts to dissolve in the ocean.

Macrophyte: A large aquatic alga that is emergent, submergent, or floating.

Macrotidal: Tidal range of >4 m.

Marine bar sand bodies: More massive than tidal bar sand bodies, composed of parabolic bars or spillover lobes with numerous superimposed elongate bars.

Marl: A mixture of clay and fine-grained calcium carbonate in varying proportions between 35 and 65% each; formed under marine or freshwater conditions.

Matrix: Synsedimentary mechanically deposited material between grains (as distinct from precipitated cement).

Megabreccia: Breccia with large numbers of very large (>1 m long) lithoclasts.

Mesogenetic zone: Time period when sediments are buried below the zone where near-surface processes control diagenesis.

Mesotidal: Tidal range of 2–4 m.

Mesotrophic: Nutrients are in good supply.

Metalimnion layer: Intermediate layer between the epilimnion and hypolimnion layers in a water body.

Micrite: Abbreviation for microcrystalline calcite, with grains < 4 µm in size.

Micrite envelope: Envelope of micrite around a grain, formed by repeated cycles of microbe boring followed by micrite filling.

Microbes: General term used for all micro-organisms such as bacteria, fungi, and cyanobacteria.

Micropeloids: Peloids, formed of micrite, 20–100 µm in size.

Microspar: Spar calcite crystals up to 50 µm in size, but typically 5–6 µm in size.

Microsparstones: As for microspar, but with crystals <10 µm long.

Microtidal: Tidal range of 0–2 m.

Mimetic dolomite: Replacement of original limestone by very small dolomite crystals so that original fabric is preserved.

Minimicrite: Like micrite but with grains <1 µm in size.

Mixotrophs: Heterotrophs containing photosymbionts.

Molar tooth structure: Structure formed of vertical to ptygmatically crumpled sheets of finely crystalline calcite spar.

Moldic porosity: Pore that retains the shape of original grain that was dissolved (e.g., coral).

Moonmilk: Soft mass of calcite fibers, high water content, commonly associated with microbes in caves.

Mud drapes: Commonly associated with flaser bedding, mud drapes develop during quiet periods and cover sand ripples that formed when currents were flowing.

Neomorphism: Applied to recrystallized limestones, commonly used when process of recrystallization is unknown.

Neptunian dike: A synsedimentary fissure that cross-cuts marine-cemented carbonates and can be filled with marine cement, encrusting organisms, or sediment.

Neritic: Marine zone between the shoreline and a depth of 200 m.

Non-fabric-specific: Process or feature that is not related to any specific limestone or dolostone feature or component.

Non-sutured seams: Wispy surfaces that form in sediments with high (>10%) clay content due to overburden pressure, commonly impart a bedded appearance to rock.

Obligate calcifier: Organisms such as corals that produce a calcareous skeleton as a result of their metabolism.

Ocean gyre: Large-scale circular movement of water at the ocean surface.

Oligophotic zone: Low light conditions in water column.

Oligotrophic: Nutrient deficient.

Olistolith: Large (>100 m long) blocks found in megabreccias, typically associated with slope deposits.

Oncolite (oncoid, oncolith): Spherical, concentrically laminated stromatolite that is not attached to the substrate.

Ooid (oolite, oolith): Spherical sediment grain <2 mm in diameter, formed of concentric laminae around a nucleus.

Ooze: Fine-grained sediments on the ocean floor, containing at least 30% biogenous pelagic material.

Open platform: A flat-topped carbonate platform but open to the ocean with no structures along platform edge to impede wave or swells from sweeping into shallow water.

Overfilled lake: Water and sediment supplies exceed the low rate of accommodation creation and watershed remains open.

Palustrine: Area around a lake margin, commonly marsh.

Paragenesis: The sequential order of mineral formation or transformation.

Partition coefficient: Ratio of concentrations of a compound in two immiscible phases at equilibrium.

Pedogenic: Relating to soil development.

Pelagic zone: The shallow part of the open ocean, usually <200 m water depth.

Pelagosite (coniatolite): Generally formed of aragonite as thin, white, grey, or brownish crust precipitated from saltwater spray along the shoreline, possibly through action of microbes.

Peloid: Like a fecal pellet but formed by (a) diagenetic alteration of other grains; or (b) spontaneous carbonate precipitation.

Pendant cement: Cement in pore that hangs from roof of cavity (like a stalactite); also known as microstalactitic cement.

Peri-platform ooze: Fine-grained carbonate sediment derived from shallow-water neritic environments but deposited in deep water adjacent to carbonate platforms.

Photic zone: Zone in water column that is illuminated by sunlight.

Photosymbionts: Symbiotic organisms that are phototrophs.

Phototroph: Organism that obtains energy from photosynthesis.

Photozoan factory: Carbonate factory in which most organisms are phototrophs, autotrophs, or mixotrophs.

Phreatic zone: Zone below water table (water-filled pores).

Phytoclasts: Produced by breakup of calcareous crust formed on and around plants.

Pisoid (pisolite, pisolith): Similar to an ooid but >2 mm in diameter.

Planktic/plantkton: Organisms that live in the water column.

Profundal: Deep, aphotic zone of lake.

Progradation: Lateral accretion of sediments.

Prokaryote: Organism with a cell that lacks a nucleus.

Protodolomite: Term used for poorly ordered dolomite.

Pycnocline: Depth in ocean water column where there is a sudden increase in water density.

Radial-fibrous calcite: A cement with crystal fibers arranged with their long axes radiating from a center (opposite to tangential) or perpendicular to a substrate.

Radiaxial-fibrous calcite: A cavity-filling calcite mosaic consisting of fibrous calcite radiating away from the initial growth surface and allied optic axes that converge away from the wall. RFC is characterized by curved cleavage, undulose extinction, and irregular intergranular boundaries that distinguish this fabric from simple radial-fibrous calcite.

Ramp: Marginal marine depositional system attached to land, gradual slope offshore, eventually merging with basin floor.

Ravinement: A time transgressive or diachronous subaqueous erosional surface resulting from nearshore marine and shoreline erosion associated with a sea-level rise.

Reactivation surface: Discontinuity cutting across a foreset, generated by erosion or changing flow strength before resumption of foreset migration.

Reef crest: Highest part of reef, typically in shallow water, separates open ocean from lagoon.

Reef flat: Area on landward or lagoonward side of reef crest, commonly shallow water with hard, cemented substrate.

Reef front: Zone with water 10–100 m deep, oceanward side of reef, generally characterized by diverse reef-building organisms.

Regression: Oceanward movement of shoreline and associated facies.

Relict grains: Grains formed during highstand then exposed and altered before being incorporated in sediment that forms during the next highstand.

Rhizoconcretions: Focused around plant roots, concretions formed of calcium carbonate precipitates that cement grains together.

Rhodolith (rhodoid, rhodolite): An irregular laminated calcareous nodule composed largely of encrusting coralline algae (and possibly encrusting foraminifers) arranged in more or less concentric layers about a core; generally cream to pink, and spheroidal but with a somewhat knobby surface.

Rhythmites: Sediment couplets of alternating composition that form over an unknown time frame.

Rimmed platform: Carbonate platform, either attached to land or unattached as a bank with a distinct break in slope at platform edge, rimmed with islands, reefs, or shoals that impede waves and swells from sweeping inboard.

Sabkha: A supratidal environment under arid to semiarid conditions on restricted coastal plains just above normal high-tide level.

Saddle dolomite: Dolomite with warped crystal lattice, curved and stepped crystal faces, and undulatory extinction patterns.

Seiche: Large-scale long-period wave caused by piling-up of water at one end of lake as result of wind-shear and low atmospheric pressure. Once wind stops, water then rebounds to other side of lake and may continue to oscillate back and forth for many hours or days.

Sequence boundary: Subaerial unconformities and correlative conformities in deeper water.

Shelter porosity: Pore beneath a shell that is in a convex-up position.

SMOW: Standard mean ocean water.

Soda straw: Hollow tube-like structure in center of stalactite.

Sparry calcite: A mosaic of calcite crystals, formed either as cement or by neomorphic processes, sufficiently coarsely crystalline to appear fairly transparent in thin-section; commonly termed "spar."

Sparstone: Limestone altered to spar so that original components are no longer recognizable.

Speleothem: General term for all precipitates found in caves.

Spring: Where underground water emerges at the Earth's surface.

Stalactite: Structure formed of laminated $CaCO_3$ precipitates hanging from ceiling of cave, commonly with central soda straw.

Stalagmite: Similar to stalactite but grows upward from floor of cave, lacks soda straw.

Stoichiometric dolomite: Dolomite with Ca:Mg ratio of 50 50 and perfectly ordered.

Stranded grains: Grains deposited in shallow water during a lowstand, but then marooned in deeper water as water rises to next highstand.

Stratigraphic bioherm: Reef with minimal original relief above seafloor, but builds up vertically as reef and surrounding sediments amass.

Stromaclasts: Intraclasts that display stromatolitic laminations and appear to be derived from a stromatolite.

Stromatactis: Irregular-shaped cavity (of unknown origin) with a smooth floor and digitate roof; geopetal sediment on floor and synsedimentary sparry calcite filling most or all of cavity.

Stromatolite: A laminated organosedimentary structure produced by sediment trapping, binding, or precipitation as a result of microbial organism (principally cyanophytes) growth and metabolic activity.

Structure grumeleuse: Clotted texture that appears as numerous small, globular clots of very finely crystalline calcite that contrast with surrounding groundmass of dark grey granular calcite.

Stylolite: Jagged surfaces coated with insoluble materials formed as result of pressure solution; also known as sutured seams.

Sublittoral: Transition zone between littoral and open marine in the ocean and profundal zone in a lake.

Subtidal: Zone below low water mark.

Sulfate-reducing bacteria: Bacteria and Archaea that obtain energy by oxidizing organic compounds or molecular hydrogen (H_2) while reducing sulfate (SO_{2-4}) to hydrogen sulfide (H_2S). These organisms "breathe" sulfate rather than oxygen, in a form of anaerobic respiration.

Suprastratal growth: Applied to reefs where skeletons project tens of centimeters above the seafloor, creating positive relief.

Supratidal: Zone above high water mark typically inundated by storms.

Symbionts: Two or more organisms living together to the mutual benefit of both.

Synaeresis cracks: Sedimentary structures formed without desiccation as a result of expulsion of fluids from the sediment, also known as subaqueous shrinkage cracks.

Syngenetic karst: Karst that forms at an early diagenetic stage, typically while young sediments are being altered.

Synoptic relief: Height between seafloor and top of buildup or stromatolite.

Synsedimentary cementation: Cementation of sediments while on seafloor shortly after deposition.

Syntaxial: Synonym of epitaxial.

Teepee structure: Neighboring plates of cemented carbonates thrust upwards (due to expansion), forming an inverted V-shaped structure that resembles a teepee.

Telogenetic zone: Time period when long-buried rocks are brought back to the surface and come under the influence of near-surface processes once again.

Tempestite: Deposit formed by storm activity.

Texture: Collectively refers to size, shape, and arrangement of constituent elements in a carbonate rock.

Thermal spring: Spring with water temperature >20°C at vent.

Thermocline: Thin layer in water body where water temperature changes more rapidly than in the layers above and below.

Thrombolite: Like a stromatolite but with a clotted fabric as opposed to a laminated fabric.

Tidal bar sand bodies: Long linear sand bodies oriented roughly perpendicular to shelf edge.

Tidal delta sand bodies: Complex lobate features formed of flood and ebb tidal deltas, commonly found at ends of tidal inlets between islands.

Tidal prism: Volume of water that passes a point in each tidal half-cycle.

Transgression: Landward movement of shoreline and associated facies.

Travertine: Dense low-porosity crystalline spring deposits, generally associated with warm to hot spring water.

Triple junction: Meeting place of three intercrystalline boundaries.

Tufa: Porous spring deposit that shows clear evidence of forming in and around plants, generally associated with cooler water.

Underfilled lake: Lake in which water and sediment supplies are low compared to available accommodation.

Undulose (undulatory) extinction: A type of extinction of crystals under cross-polarized light that occurs successively in adjacent areas as the microscope stage is turned; also called wavy extinction or sweeping extinction.

Unit extinction: A type of extinction under cross-polarized light where large fragments of a fossil or cement reach extinction at the same time as microscopic stage is turned.

Upwelling: Dense, cooler, and typically nutrient-rich water is brought to the ocean surface and displaces the warmer, typically nutrient-poor water.

V-PDB, Vienna: The original standard for carbonate stable isotopes was the Cretaceous Pee Dee belemnite. This was all used up, however so an equivalent against which all carbonate isotopes are calibrated is now housed in Vienna.

V-SMOW, Vienna: Standard mean ocean water water that is a standard for carbon and oxygen stable isotopes and housed in Vienna.

Vadose zone: Zone above water table (air-filled or water-filled pores after rainfall).

Varves: Sediment couplets formed of summer and winter sediment layers.

Vaterite: Anhydrous form of $CaCO_3$, hexagonal system.

Vug: A small irregular hole produced by carbonate dissolution that cuts across original rock fabric.

Wavelength: Distance between crests of successive waves, ripples, or dunes.

Whitings: Clouds of milky white, microcrystalline carbonate-rich water in the open ocean.

Zooxanthellae: Symbiotic microorganisms (commonly dinoflagellates) that live in coral tissue.

INDEX

Origin of Carbonate Sedimentary Rocks, First Edition. Noel P. James and Brian Jones.
© 2016 Noel P. James and Brian Jones. Published 2016 by John Wiley & Sons, Ltd.
Companion website: www.wiley.com/go/james/carbonaterocks